2nd International Conference on Engineering and Applied Technology (ICEAT 2018)

IOP Conference Series: Materials Science and Engineering
Volume 674

Aceh, Indonesia
9 - 10 October 2018

Editors:

Yun Arifatul Fatimah
Agus Aktawan
Andri Pranolo
Leonel Hernandez
Lukas G. Swan
Pedro Hokama
Jazaul Ikhsan
Aris Widyo
Tri Widayanto
Ahmad Azhari
Adhi Prahara

ISBN: 978-1-7138-0926-5
ISSN: 1757-8981

Printed from e-media with permission by:

Curran Associates, Inc.
57 Morehouse Lane
Red Hook, NY 12571

Some format issues inherent in the e-media version may also appear in this print version.

Institute of Physics
Dirac House, Temple Back
Bristol BS1 6BE UK

Phone: 44 1 17 929 7481
Fax: 44 1 17 920 0979

techtracking@iop.org

Additional copies of this publication are available from:

Curran Associates, Inc.
57 Morehouse Lane
Red Hook, NY 12571 USA
Phone: 845-758-0400
Fax: 845-758-2633
Email: curran@proceedings.com
Web: www.proceedings.com

TABLE OF CONTENTS

CIVIL ENGINEERING

ELECTRICAL ENGINEERING

INDUSTRIAL ENGINEERING

INFORMATICS ENGINEERING

IOP Conf. Series: Materials Science and Engineering **674** (2019) 011001 doi:10.1088/1757-899X/674/1/011001

Prefeace

On behalf of the organizing committee, I would like to welcome all of you to the 2nd International Conference on Engineering and Applied Technology (ICEAT), to be held on Ocotber 9th – 10th October 2018 in Aceh, Aceh Province, Indonesia.

Catastrophe impacts to human, economic, social, nature and environment caused from natural disaster including earthquake, volcano eruptions, landslides, floods and storms have increased significantly in the last decades. Natural disaster Risk Management has received more attention from governments, NGOs, researcher, firms, national organization and international organization. This special issue is an interesting, latest and trend topic in which the Muhammadiyah Engineering Forum is also concerned on.

Thus, the 2nd 2018 of the International Conference on Engineering, Applied Science (ICEAT) organized by Muhammadiyah Engineering Forum held in Aceh Indonesia, has an overarching theme "Sustainable Engineering Science and Technology for Natural Disaster Risk Management" to be discussed in the conference. This theme highlights the role and contribution of Engineering Science and Technology in meeting this natural disaster challenges toward global sustainability.

The 2nd 2018 ICEAT topic is limited to Engineering scope which is including Architecture Engineering, Automotive Engineering, Chemical Engineering, Civil Engineering, Electrical Engineering, Industrial Engineering, Informatics Engineering, and Mechanical Engineering fields. The 72 papers published in this proceeding pick up sustainability diverts cases related to the risk management disaster through engineering science and technology approaches. It is our sincere hope that these special issues can provide fresh ideas to international readers on the natural disaster topics, fields, case, solution, innovation.

Lastly, I hope the conference will bring a new exciting experience, provide a good platform to make friends and learn from each other. I hope your stay in Aceh, Indonesia will be enjoyable and a memorable one.

With best wishes,

Organizing Committee Chairman
Yun Arifatul Fatimah, Ph.D

IOP Conf. Series: Materials Science and Engineering **674** (2019) 011002 doi:10.1088/1757-899X/674/1/011002

ICEAT 2018 Committee

Organizing Committee

General Chair

- Yun Arifatul Fatimah (Universitas Muhammadiyah Magelang, Indonesia)

General Co-Chair

- Hafnidar A. Rani (Universitas Muhammadiyah Aceh, Indonesia)

Technical Program Committee

- Andri Pranolo (Universitas Ahmad Dahlan, Indonesia)
- Rafał Dreżewski (AGH University of Science and Technology, Poland)
- Shi-Jinn Horng (National Taiwan University of Science and Technology, Taiwan)
- E.P. Nowicki (University of Calgary, Canada)
- Husni Thamrin (Universitas Muhammadiyah Surakarta, Indonesia)
- Dwi Anggraini (Universitas Muhammadiyah Malang, Indonesia)
- Gunawan Ariyanto (Universitas Muhammadiyah Surakarta, Indonesia)
- Zulfatman (Universitas Muhammadiyah Malang, Indonesia)
- Slamet Riyadi (Universitas Muhammadiyah Yogyakarta, Indonesia)
- Yun Arifatul Fatimah (Universitas Muhammadiyah Magelang, Indonesia)
- Pedro Hokama (University of Campinas, Brasil)
- Wolfgang Keller (University of Colorado, Colorado, US)
- Hari Prasetyo (Universitas Muhammadiyah Surakarta, Indonesia)
- Eko Setiawan (Universitas Muhammadiyah Surakarta, Indonesia)
- Ilyas Mas'udin (Universitas Muhammadiyah Malang, Indonesia)
- Siti Mahsanah (Universitas Ahmad Dahlan, Indonesia)
- Herry Purnama (Universitas Muhammadiyah Surakarta, Indonesia)
- Dan Mugisidi (UHAMKA Jakarta, Indonesia)
- Lukas G. Swan (Dalhousie University, Canada)
- Sudarisman (Universitas Muhammadiyah Yogyakarta, Indonesia)
- Aris Widyo Nugroho (Universitas Muhammadiyah Yogyakarta, Indonesia)
- Muji Setiyo (Universitas Muhammadiyah Magelang, Indonesia)
- Marwan Effendy (Universitas Muhammadiyah Surakarta, Indonesia)
- Tri Widayatno (Universitas Muhammadiyah Surakarta, Indonesia)
- Tri Yuni Hendrawati (Universitas Muhammadiyah Jakarta, Indonesia)
- Denny Vitasari (Universitas Muhammadiyah Surakarta, Indonesia)
- Nurul Hidayati Fithriyah(Universitas Muhammadiyah Jakarta, Indonesia)
- Haryanto (Universitas Muhammadiyah Purwokerto, Indonesia)
- Agus Setyo Munthohar (Universitas Muhammadiyah Yogyakarta, Indonesia)
- Sri Sunarjono (Universitas Muhammadiyah Surakarta, Indonesia)
- Nurul Hidayati (Universitas Muhammadiyah Surakarta, Indonesia)

IOP Conf. Series: Materials Science and Engineering **674** (2019) 011002 doi:10.1088/1757-899X/674/1/011002

- Samin (Universitas Muhammadiyah Malang, Indonesia)
- Jazaul Ikhsan (Universitas Muhammadiyah Yogyakarta, Indonesia)
- Ade FAH Alhashimy (Universias Muhammadiyah Sumatera Utara, Indonesia)
- Gunawan (Universitas Muhammadiyah Surabaya, Indonesia)
- Wisnu Setiawan (Universitas Muhammadiyah Surakarta, Indonesia)

IOP Conf. Series: Materials Science and Engineering **674** (2019) 011003 doi:10.1088/1757-899X/674/1/011003

ICEAT 2018 Editors

- Dan Mugisidi (UHAMKA Jakarta, Indonesia)
- Ade FAH Alhashimy (Universias Muhammadiyah Sumatera Utara, Indonesia)
- Agus Setyo Munthohar (Universitas Muhammadiyah Yogyakarta, Indonesia)
- Rafał Dreżewski (AGH University of Science and Technology, Poland)
- Yun Arifatul Fatimah (Universitas Muhammadiyah Magelang, Indonesia)
- Agus Aktawan (Universitas Ahmad Dahlan, Indonesia)
- Andri Pranolo (Universitas Ahmad Dahlan, Indonesia)
- Leonel Hernandez (ITSA University, Colombia)
- Lukas G. Swan (Dalhousie University, Canada)
- Pedro Hokama (University of Campinas, Brasil)
- Jazaul Ikhsan (Universitas Muhammadiyah Yogyakarta, Indonesia)
- Aris Widyo Nugroho (Universitas Muhammadiyah Yogyakarta, Indonesia)
- Tri Widayatno (Universitas Muhammadiyah Surakarta, Indonesia)
- Ahmad Azhari (Universitas Ahmad Dahlan, Indonesia)
- Adhi Prahara (Universitas Ahmad Dahlan, Indonesia)

IOP Conf. Series: Materials Science and Engineering **674** (2019) 011004 doi:10.1088/1757-899X/674/1/011004

The 2018 2nd International Conference on Engineering and Applied Technology (ICEAT)

9-10 October 2018, Aceh, Indonesia

THE PROCEEDINGS PUBLICATION LICENCE:

Licence terms and conditions

By submitting your paper to the conference organiser, you, as author/representative of all the authors, grant a royalty free licence to IOP Publishing Limited (IOP) to use the copyright in the paper for the full term of copyright in all ways otherwise restricted by copyright, including the right to reproduce, distribute and communicate the article to the public under the terms of the Creative Commons Attribution (CC BY) licence (http://creativecommons.org/licenses/by/3.0/) and to make any other use which IOP may choose world-wide, by all means, media and formats, whether known or unknown at the date of submission, to the conference organiser.

This licence does not transfer the copyright in the paper as submitted which therefore remains with the authors or their employer, as appropriate. IOP encourages authors to use the paper in any way provided that, where possible, he/she displays citation information and the IOP Proceedings Licence Notice, for electronic use, best efforts are made to include a link to the online abstract in the journal and no author offers the paper to another publisher (prior to withdrawal or rejection) or includes it in another publisher's website.

However, a re-written and extended version of the paper may be published in another journal provided such re-use is within generally accepted ethical scientific limits and provided further citation information and the IOP Proceedings Licence Notice is displayed if possible, and for electronic use best efforts are made to include a link to the online abstract in the journal.

By granting this licence, the author warrants that the paper he/she is submitting is his/her original work, has not been published previously (other than in a research thesis or dissertation which fact has been notified to the conference organiser in writing), all named authors participated sufficiently in the conception and writing of the paper, have received a final version of the paper, agree to its submission and take responsibility for it, and the submission has been approved as necessary by the authorities at the establishment where the research was carried out.

By granting this licence, the author also warrants that he/she acts on behalf of, and with the knowledge of, all authors of the paper, that the paper does not infringe any third party rights, it contains nothing libellous, all factual statements are, to the best of the authors' knowledge, true or based on valid research conducted according to accepted norms, and all required permissions have been obtained.

IOP Conf. Series: Materials Science and Engineering **674** (2019) 011004 doi:10.1088/1757-899X/674/1/011004

The 2018 2nd International Conference on Engineering and Applied Technology (ICEAT)

9-10 October 2018, Aceh, Indonesia

The IOP Proceedings Licence Notice

Editor : Dan Mugisidi, Ade FAH Alhashimy, Agus Setyo Munthohar, Rafał Dreżewski, Yun Arifatul Fatimah, Agus Aktawan, Andri Pranolo, Leonel Hernandez, Lukas G. Swan, Pedro Hokama, Jazaul Ikhsan, Aris Widyo Nugroho, Tri Widayatno, Ahmad Azhari, Adhi Prahara.

Publisher : IOP Publishing Ltd.

Secretariat : Fakultas Teknik
Universitas Muhammadiyah Aceh
Jalan Muhammadiyah No.91, Batoh, Lueng Bata, Batoh, Lueng Bata, Kota Banda Aceh, Indonesia
Email: iceat@fgdt-ptm.or.id

IOP Conf. Series: Materials Science and Engineering **674** (2019) 011004 doi:10.1088/1757-899X/674/1/011004

The 2018 2nd International Conference on Engineering and Applied Technology (ICEAT)

9-10 October 2018, Aceh, Indonesia

Statement of Peer Review

All papers published in this volume of IOP Conference Series: Materials Science and Engineering have been peer reviewed through processes administered by the Editors. Reviews were conducted by expert referees to the professional and scientific standards expected of a proceedings journal published by IOP Publishing.

Organizers and Sponsors

Organized by

FGDT PTM (Engineering Forum of Muhammadiyah Higher Education), Indonesia

Hosted by

Universitas Muhammadiyah Aceh, Indonesia

Technical Co-Sponsored by

IOPscience, IOP

Supported by

Majelis Pendidikan Tinggi Penelitian dan Pengembangan, PP Muhammadiyah, Indonesia
Universitas Ahmad Dahlan, Indonesia
Universitas Muhammadiyah Yogyakarta, Indonesia
Universitas Muhammadiyah Surakarta, Indonesia
Universitas Muhammadiyah Malang, Indonesia
Universitas Muhammadiyah Purwokerto, Indonesia
Universitas Muhammadiyah Magelang, Indonesia
Universitas Muhammadiyah Mataram, Indonesia
Universitas Muhammadiyah Makassar, Indonesia
Universitas Muhammadiyah Sumatera Utara, Indonesia
Universitas Muhammadiyah Jakarta, Indonesia
Universitas Muhammadiyah Prof. Dr. Hamka, Indonesia

The performance of ventilation in internal buildings affected by differences of building orientation

Q Aini[1] and N Nadia[2]

Department of Architecture, Universitas Muhammadiyah Aceh, Indonesia

Email: [1] iqur_ennie@yahoo.com, [2] nda02ars@yahoo.com

Abstract. Natural ventilation is a passive cooling strategy with maximize of airflow into the building. Profil of hills and building orientation are some factor that influences the air quality, and it can control the wind speed. By the author's previous research that building with configured in a grid pattern on the contoured land has enhanced the airflow quality than the grid pattern applied on the flat land. That condition affected building adaptation, such as good position of the opening in the building can optimize the utilization of airflow as the refreshment. This study purposed to evaluates performance of building ventilation of grid patterns on the hillside. This study using Jacky Chan Settlements that was located in Neuhen, Aceh Besar, Aceh Province, Indonesia. *Computational Fluids Dynamics (CFD) Fluent* used to analyze how far the performance of ventilation in the internal building. This study found that the pattern of airflow and the airflow quality in the room influenced by orientation. All orientation in the settlements created good cross ventilation with differences in wind speed. Contribute of this study is can be a reference for designing better settlements in a contoured land.

Keywords: Building Orientation, Computational Fluids Dynamics (CFD) Fluent, Ventilation

1. Introduction

Energy consumption is mostly absorbed in residential homes, where houses consume the largest energy of 28.8% of all energy consumption in the development sector [1]. Design adaptation to climate condition is can overcome excess energy consumption. Make use of airflow with a natural ventilation system is one solution that can minimize energy consumption. It can choose as passive control before use active system. Plan and orientation are other solutions that can considered as one of the adaptations to the physical environment [2].

The Ground surface condition effects to determine building configuration. It can change the behavior of air movement, which is it can accelerate the airflow even slow [3]. Radiation, wind, and topography are the main factors that determine building configuration. It can affect to consider the building orientation remain to protection of the sun, utilization of cross ventilation, and layout [4]. Minimize of sun radiation and maximize natural ventilation are the solution to building orientation in hot humid areas [5]. Differences in building orientation were influence to hit convection, wind speed and temperature [6]. Wind occurs because the heat of the sun that warms the air and the rotation of the earth, so that the air rise and move from one place to another [7]. The principle of the air movement is consists of [8] [9] [10]:

a. Reason the flowing air. This condition occurs because of differences in temperature. In this condition, the air moves from low pressure to high pressure.

b. Inertia. The air can be deflected because there is a barrier without influence a certain angle
c. High and low-pressure area. The air can be deflected because it prevented by vegetation and building.
d. Bernoulli effect. Acceleration flow needed to cover one form of building
e. Venury effect. Acceleration flow when through the opening with limited laminar flow
f. Stack effect. The exchange of air with natural convection caused indoor temperature higher than outdoor temperature.
g. Cross ventilation. This occurs when inlet and outlet openings in external walls have an internal flow path between them.

In addition, to distribute the fresh air into the room, natural ventilation also can dissipation of heat and give physiological cooling effect [11]. In choosing and designing of site to optimize the natural ventilation should be considered:

a. Take advantage airflow pattern with considered topography and building configuration
b. Conditioning of comfort in winter and summer
c. Avoid permanently for undesirable wind
d. Avoid highly wind velocity
e. Avoid the path of the wind that causes dust and pollution

The characteristics of the behavior of air movement are velocity, direction, and turbulence [10]. Air movement occur cause of surface condition that can divert the wind. That condition was forming freestream flow (laminar), shear layer flow (separated) and wake flow (turbulent) [3] [10]. Another factor that can divet the wind and change the wind speed is orientation, building dimension, overhangs and roof form (Figure 1).

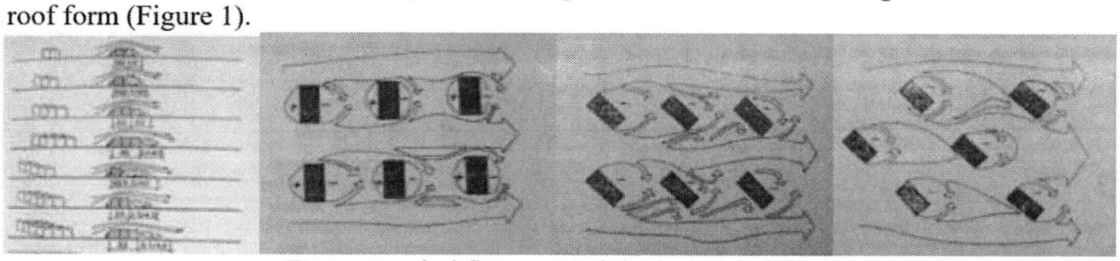

Figure 1. wind flow around the building [5][9]

Orienting buildings with certain angle to the wind direction can produce relatively homogeneous wind patterns. That orientation will affect airflow and wind velocity. In contour land, the grid pattern building that arrayed linear and angular to windward will create airflow better than grid pattern building in flat land. Based on that theories, optimization of the refreshment of the air highly building orientation, so that can reduce overheating in the building. This research-based on Author's previous research, that have done at the same location.

There are two types of natural ventilation that cross ventilation and single-sided ventilation. The flow of Cross ventilation is determined by the combined effect wind temperature difference. It depends on other openings on opposite side of the building (Figure 2). Single-sided ventilation occurs when large ventilation openings are situated on one external wall only. Exchange of air occurs by wind turbulence when outward openings interacting with the local external airflow and stuck.

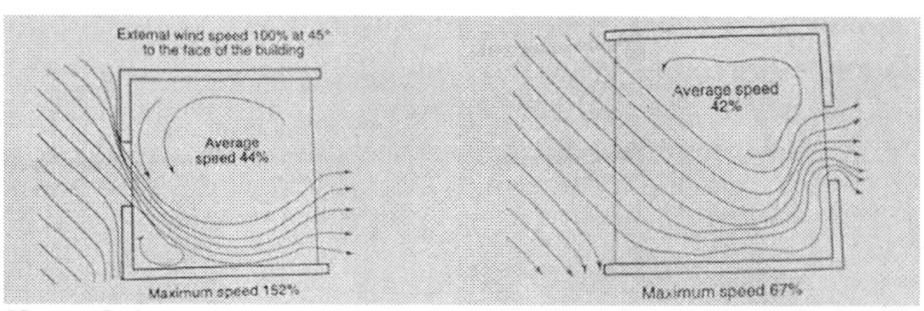

Figure 2. Effect of size of inlet and outlet on internal wind speed and distribution [9]

IOP Conf. Series: Materials Science and Engineering **674** (2019) 012001 doi:10.1088/1757-899X/674/1/012001

The location and orientation of obstacles can influence wind speed and airflow in the building (Figure 3). Parallel obstruction to windward will affect air movement [9]. Cross ventilation can be enhanced by obstacle, position of outlet in the sidewalls of building, and by the use of wind walls (Figure 4).

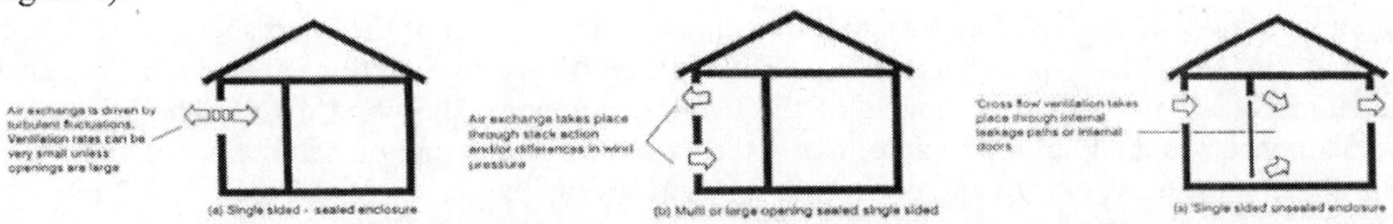

Figure 3. Sealed enclosure, multi-opening sealed single side, and single side unsealed enclosure [9]

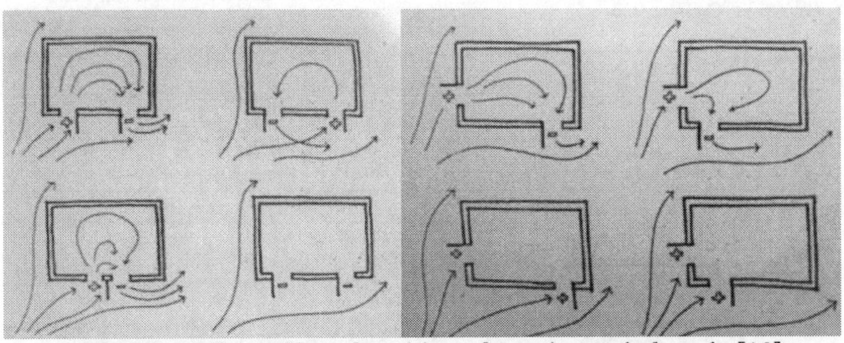

Figure 4. airflow affected position of opening and obstacle [10]

2. Methodology

The simulation method was utilised for this research. The CFD Fluent Simulation used to see the influence of contour and building orientation to airflow in the room. The location of this research is in Jacky Chan Settlement, Neuhen, Aceh Besar, Aceh Province, Indonesia. This settlement consists of three different building cluster orientations that arranged in grid pattern. The simulation model in this research is representation of house model in each orientation. The building area in each housing is 45 m^2.

Collecting data for this research is climate data such as wind speed and wind direction. Another date is the existing of Jacky Chan Housing Unit related layout, opening position and building orientation (Figure 5.). There are four openings in the model. This research purpose to evaluate the influence of opening position in building toward performance of ventilation. The analyze for this research are:

a. Wind speed that occurs in the building
b. Performance of ventilation in internal building

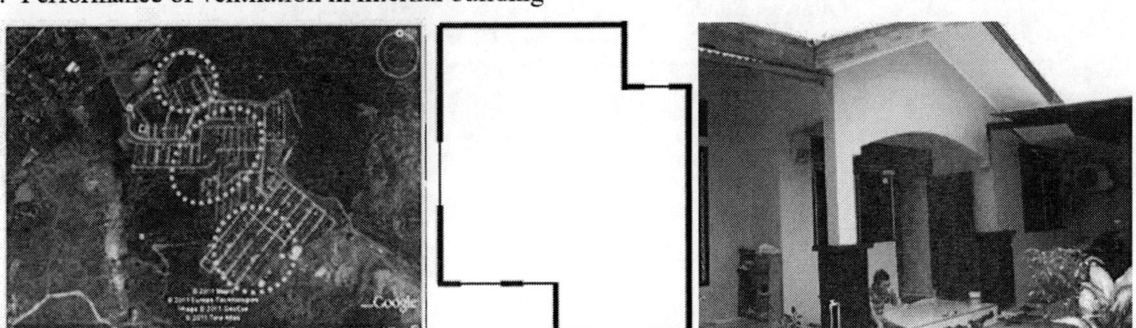

Figure 5. Existing of modeling for simulation (Authors survey, 2015)

IOP Conf. Series: Materials Science and Engineering **674** (2019) 012001 doi:10.1088/1757-899X/674/1/012001

3. Result and Discussion

3.1. Wind Velocity in The Building

The major of the wind speed is occur in Northwest-Southeast orientation (Table 1 and Table 2). In that orientation, wind speeds were created is 0,2 m/s-1,1 m/s. The lowest of the wind speed is 0.2-0.6 m/s that occur in O1. That speed occurs in Southwest-Northeast orientation. Based on the above explanation, it is known that the angels of the building influence the wind divert. The principle of movement occurs is the inertia principle, where the existence of obstruction can deflect of airflow and it will influence wind velocity [9]. An average of wind speeds that occur in the room is 0.2 m/s – 0.6 m/s.

Table 1. Wind speed in each orientation that occur in each opening

Building Orientation	**Wind Speed (m/s)**			
	O1	**O2**	**O3**	**O4**
Southwest-Northeast	0.2-0.6	0.4	0.2	0.4-1.4
West-East	0.2-0.4	0.2-0.8	0.2-1	0.2-1
Northwest-Southeast	0.2-0.8	0.2-1.1	0.2-0.4	0.2-0.8

.**Table 2**. The contour of wind in the building

No	**Orientation**	**Contour of wind**
1	Southwest-Northeast 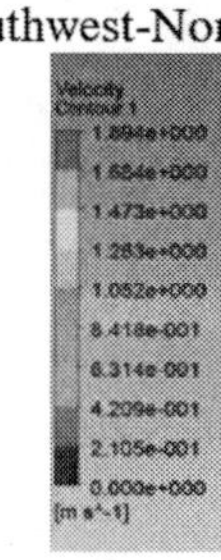	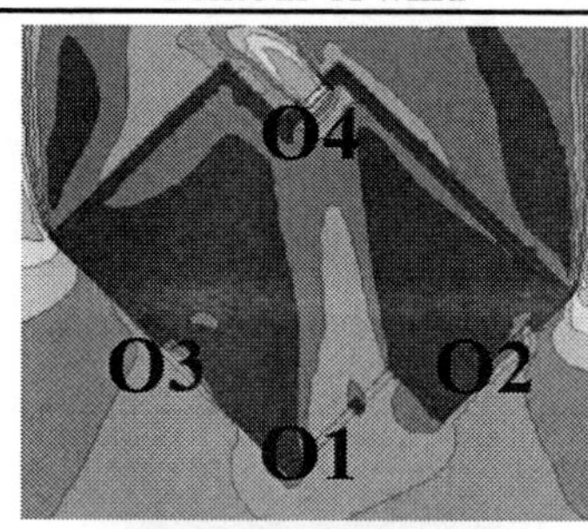
2	West-East 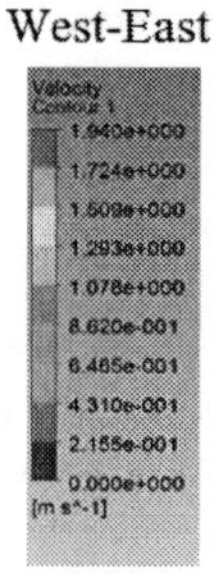	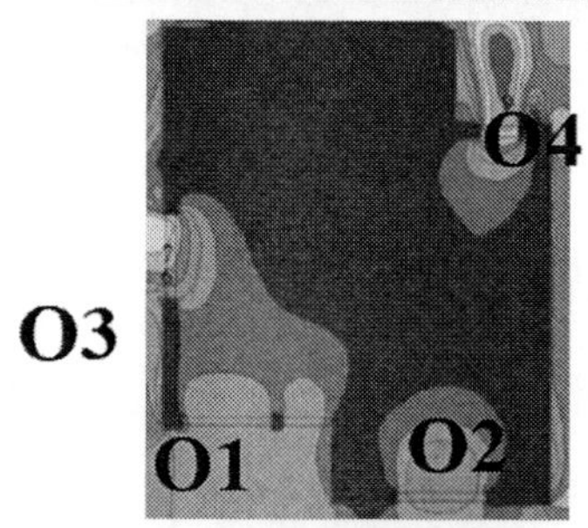
3	Northwest-Southeast 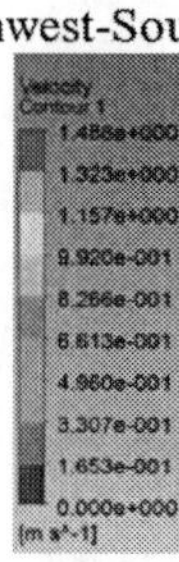	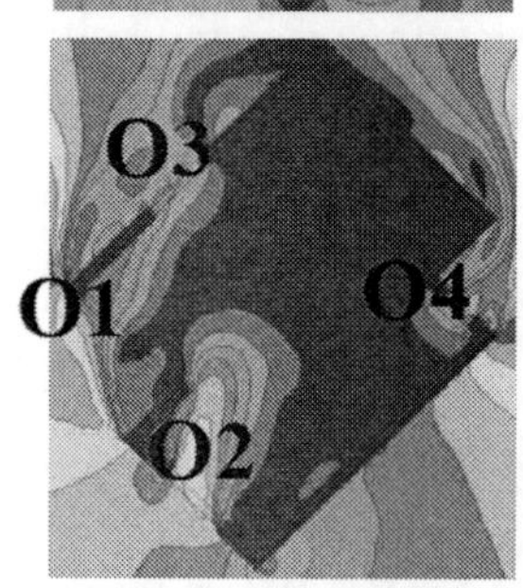

Information: O = Opening

3.2. Performance of Ventilation in Internal Building

IOP Conf. Series: Materials Science and Engineering **674** (2019) 012001 doi:10.1088/1757-899X/674/1/012001

The performance of ventilation has been reviewed by wind direction, wind speed, and airflow that create in the room. Table 3 shows about wind vector and wind direction that create in the room. Based on that result, it will be known positive area and area in the opening. The positive area is inlet and the negative area is outlet. The positive and negative area that happened depend on orientation and opening position to windward. The dimension of opening also influences airflow that occurs in the room. From four opening there are in the building O1 is the biggest opening than others.

Tabel 3. Wind vector in each building orientation

No	Orientation	Wind vector
1	Southwest-Northeast 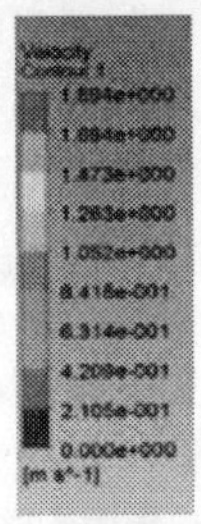	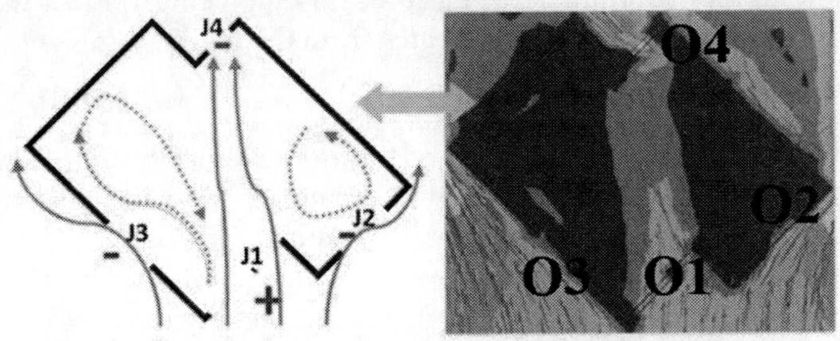
2	West-East 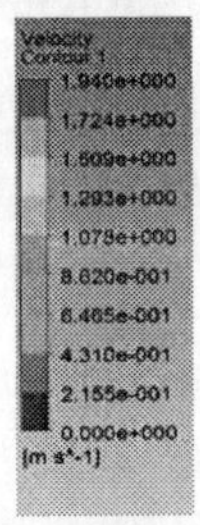	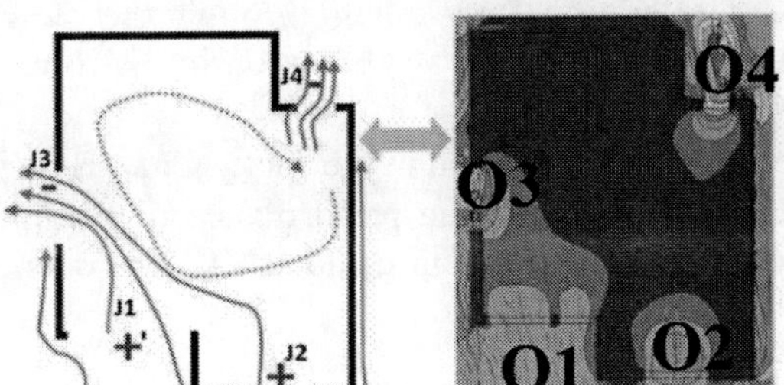
3	Northwest-Southeast 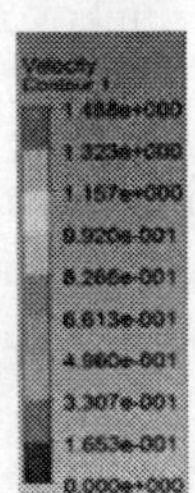	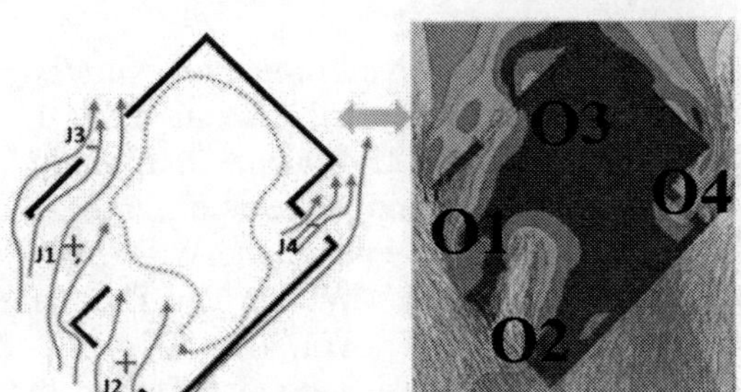

Information: O = Opening

3.2.1. Southwest-Northeast Orientation

In the southwest-northeast Orientation (Table 3), the positive flow is just created in the opening 1 (O1), while in another opening (O2, O3, O4) are created the negative flow. This is because opening 1 (O1) parallel to windward. In the opening 2 dan 3(O2 and O3) the wind just flowing in the wall surface. In this condition, natural ventilation is occurring from opening 1 (O1) to opening 4 (O4). Opening 4 is the outlet. In the room occur the air reduction and then the air flowing into O4. Wind speed is rise in the opening 4 (O4) with the wind speed is 1,4 m/s. That occurs because venturi effect, where occurs the acceleration flow when through the flow of freestream.

3.2.2. West-East Orientation

There is two inlets in building with west-east orientation (Tabel 3). The wind direction parallel to Opening 1 and Opening 2 (O1 dan O2). Natural ventilation directly creates of O1 to O3. In the middle room, the wind speed is declined, and airflow is slow. The wind speed returned increasing in O4 achieve 1 m/s. The venturi effect also occurred in this orientation.

3.2.3. Northwest-Southeast Orientation

In Northwest-southeast orientation, wind speed in inlet (O1 dan O2) is higher than other orientations. The flow of compelled because one corner of the building facing to windward. The inlet (O1 dan O2) is also parallel to windward, so that is creating airflow to the room. This condition is happening according to inertia principle. O4 also occurs the venturi effect with an increase of wind speed is 0,2-0,8 m/s. Natural ventilation also creates from O1 to O3.

4. Conclusion

The conclusion of the performance of ventilation analyze are:

a. The best Natural ventilation was created in opening that parallel to windward. That condition produces the airflow that optimum to cooling space.
b. The building that angled against to windward will deflect the flow and the airflow to the building surface. The deflect of airflow that influenced by the obstruction is occurred inertia principle.
c. In the outlet (O4) was create a venturi effect. That condition causes the wind speed is rise, where occur the acceleration flow when through the flow of freestream. O4 is smaller than other openings, so that will create air reduction in the room, however in O4, the wind speed is rise due to venturi effect.
d. All of the building orientation in the Jacky Chan Housing is to create the best cross ventilation.
e. The factor that influences the performance of ventilation is topography, building orientation to windward, dimension of opening and position of opening.

References

[1] M. Hegger, M. Fuchs, T. Stark, and M. Zeumer, *Energy Manual: Sustainable Architecture*. Walter de Gruyter, 2012.
[2] A. Krishan, *Climate Responsive Architecture: a Design Handbook for Energy Efficient Buildings*. Tata McGraw-Hill Education, 2001.
[3] T. S. Boutet, *Controlling Air Movement: A Manual for Architects and Builders*. 1987.
[4] G. Lippsmeier, K. Mukerji, and S. Nasution, *Bangunan Tropis*. Erlangga, 1997.
[5] N. A. M. Al-Tamimi, S. F. S. Fadzil, and W. M. W. Harun, "The Effects of Orientation, Ventilation, and Varied WWR on The Thermal Performance of Residential Rooms in The Tropics," *J. Sustain. Dev.*, vol. 4, no. 2, p. 142, 2011.
[6] H. Rahim, "Pengaruh Orientasi Bangunan Terhadap Temperatur Ruang pada Rumah Sederhana Type 21 Perumnas BTP Makassar." Tugas Akhir Program Pascasarjana Universitas Hasanuddin Makassar, 2002.
[7] T. H. Mulyani, *Seri Eko-Arsitektur 2 Arsitektur Ekologis*, vol. 2. Kanisius, 1998.
[8] R. M. Aynsley, W. Melbourne, and B. J. Vickery, "Architectural Aerodynamics." 1977.
[9] F. Moore, M.-H. Architecture, and urban planninig series, *Environmental Control Systems: Heating, Cooling, Lighting*. McGraw-Hill New York, 1993.
[10] N. Lechner, *Heating, Cooling, Lighting: Sustainable Design Methods for Architects*. John wiley & sons, 2014.
[11] S. Szokolay, "Introduction to Architectural Science the Basis of Sustainable Design, Linacre House, Jordan Hill." Oxford OX2 8DP, 2004.

Acknowledgments
The research was supported by the Engineering Faculty University of Muhammadiyah Aceh and Institute for Research, Publishing, Community Service and Development (LP4M) University of Muhammadiyah Aceh.

Design recommendation on accessibility for people with disability (PwDs) on public space in Banda Aceh (case study Taman Sari Kota Banda Aceh)

H Marlina

Department of Architecture, Engineering Faculty, University of Muhammadiyah Aceh, Banda Aceh, Indonesia.

Email: hmarlina_73@yahoo.co.id

Abstrak. Taman Sari located in the center of Banda Aceh city, Aceh Province, Sumatera, Indonesia. That is a public space. Most of the people spend their time to have some relax, family and friends gathering, it provided educational and recreation facilities. The existence of public space should be able to be accessible for all of the society including persons with disabilities. The accessibility for everyone including the elderly can be measured by safety, convenience, usefulness, and independence. This study is a continuation of previous research that provides design recommendations on accessibility to persons with disabilities in public spaces in Banda Aceh City, and the case study was focused on Taman sari. Kota Banda Aceh. The methodology used was identifying the results from the previous research on the Evaluation of Accessibility for People with Disabilities on the public space in Banda Aceh in the form of design recommendations according to the guidelines of technical facilities and disabilities. Measuring the comfort level on public space can be seen with shape, dimension, color, and texture. Design recommendations obtained might be useful for accessibility for disability in Taman sari Banda Aceh.

Keywords: Accessibility, Disability, Design Recommendations, Taman Sari Kota Banda Aceh.

1. Introduction

Taman Sari Kota Banda Aceh (TSKBA) is a public space where many activities of the society in Aceh held, either in the form of education or recreation. Located in the city center with a stretch of grass and shady trees make Taman Sari Kota Banda Aceh (TKSBA) as one of the options that often visited by the entire community in Aceh, including disability. The existence of public space is the area that could be used by various communities, both personally and groups [1]. The presence of the disability is often considered as a compliment due to their minority number, even though in the 1945 Constitution Articles 28 H paragraph 2 reads "Everyone shall have the right to get special facilities and privileges to obtain the equal opportunities and benefits in order to achieve the equality and justice" [2]. The public spaces should belong to all society and communities, both with normal physical as well as people with disabilities, and the facilities should also facilitate the disability. According to the mentioned problems and reinforced by the previous research entitled "Evaluation of Accessibility for People with Disability on Public Space in Banda Aceh especially in Taman Sari" the design recommendations on TKSBA formed so that afterward the accessibility needs for disability in public space TSKBA fulfilled.

The government has arranged the technical guidance for facilities and accessibilities about building and environment [3]. One of the guidelines related to accessibility, which should make it easy for everyone including the disability and elderly, in order to get equal opportunity for all aspects of life. The design control of the park can be considered feasible when it meets the principle of the facilities and disabilities, which are the safety principle, health, comfort, ease, security, beauty, and independent [4] [5].

The technical guidance of facilities and disabilities that must be considered in the public spaces [6] [4] [3] are:

1) A public toilet should be equipped with a handrail that has positions and altitudes adapted to the user of wheelchair or other disabled people. The recommended grip has a right-angled shape to help the wheelchair user to move from the toilet.
2) The basic size of the shape/free floor space in three-dimension refers to the size of the adult's human body equipment and functions of the space.
3) The pedestrian pathway should be paid attention to the road surface, slope, stopping point, lighting; drainage the minimum with is 120 cm, and safety handles along the pedestrian path.
4) The guided pathway should have steering tiles and warning tiles located in front of the vehicle traffic line, in front of the entrance and exit way, pedestrian connecting environment and buildings, and guides from public space to public transport.
5) The parking area for disability is equipped with an access point located between two parks and equipped with ramps leading to the circulation pathway/line that has differences in evaluations.
6) Ramp with slope requirement should not exceed 60; the maximum of ramp horizontal length should not exceed 900 cm, width /[=mum 95 cm without safety edge, and 120 cm with a safety edge. The borders on the prefix or suffix of the ramp should be free and flat, illuminated with sufficient lighting and equipped with handrails.
7) Signs and marks are needed on the directions and destinations of pedestrian-path public toilets, disability-specific parking, facilities and name of the places, public phone, and ATM.

Based on the problems mentioned above, the existence of the public space should provide accessibility by the technical guidelines for facilities and disabilities. This study will evaluate the appropriateness of the feasibly of the public space in Banda Aceh for people with disabilities.

2. Method

2.1. Research type
The research conducted to give design recommendations to accessibility on the public space in Banda Aceh based on the previous research.

2.2. Research Location
Public space reviewed was only in Taman Sari Kota Banda Aceh which is located in Kampong Baru Kecamatan Baiturrahman, Banda Aceh (Figure 1).

2.3. Research Variable and Operational Definitions
In this study, the dependent variable is accessibility, while the independent variable is the outer space element related to the design recommendation to improve the availability and conformity with the standardization and convenience implemented in the design plan for the improvement of public open space that is friendly to disability. Related to these variables, then the operational definition is:

a) Accessibility is the availability of a pathway for everyone, including the people with disabilities and the elderly, which can be measured by safety, convenience, usefulness, and independent to obtain equal opportunity in all life aspects and livelihood [5].
b) Availability is the presence of access to people with disabilities and the elderly as the study scope.
c) Compliance is the disability facilities available in the public space appropriate with the technical guidelines and standards.

IOP Conf. Series: Materials Science and Engineering **674** (2019) 012002 doi:10.1088/1757-899X/674/1/012002

d) Convenience is disability facilities available in the public space and does not interfere the user satisfaction by reviewing the form, dimension, color, and text.

Figure 1. Research Locations
(Source: Google Earth, 2018)

2.4. Data Collections
The collected data were accessibility measurement data through field observations to observe the environmental condition in TSKBA based on the previous research. The case studied related to the existing design to the availability and fulfillment of standardization according to the principle facility and accessibility. Surveys conducted with direct observation and measurement. Documentation is needed to determine the physical condition of the environment and fell the environment situation. Official data from government agencies in the case of Banda Aceh City Planning Agency and Plantation are also required.

2.5. *Data Analysis*
The stages of analysis performed are:
a) Study and analyze the results from the previous research by focusing on the old designs on TSKBA public space;
b) Planning for design recommendations on the old design by technical guidelines of facilities and disabilities considered in the public space.

3. Results and Discussion

3.1. The existing condition of TSKBA
The existing condition in figure 2 is illustrating the facilities and accessibilities in the research object.

3.2. Identify the availability access
The identification of the facilities was based on the survey results shown in Figure 2. That identification refers to Technical guidelines of facilities accessible to the building and environment [4]. The accessibility based on the technical guideline on TSKBA case studies is described in Table 1.

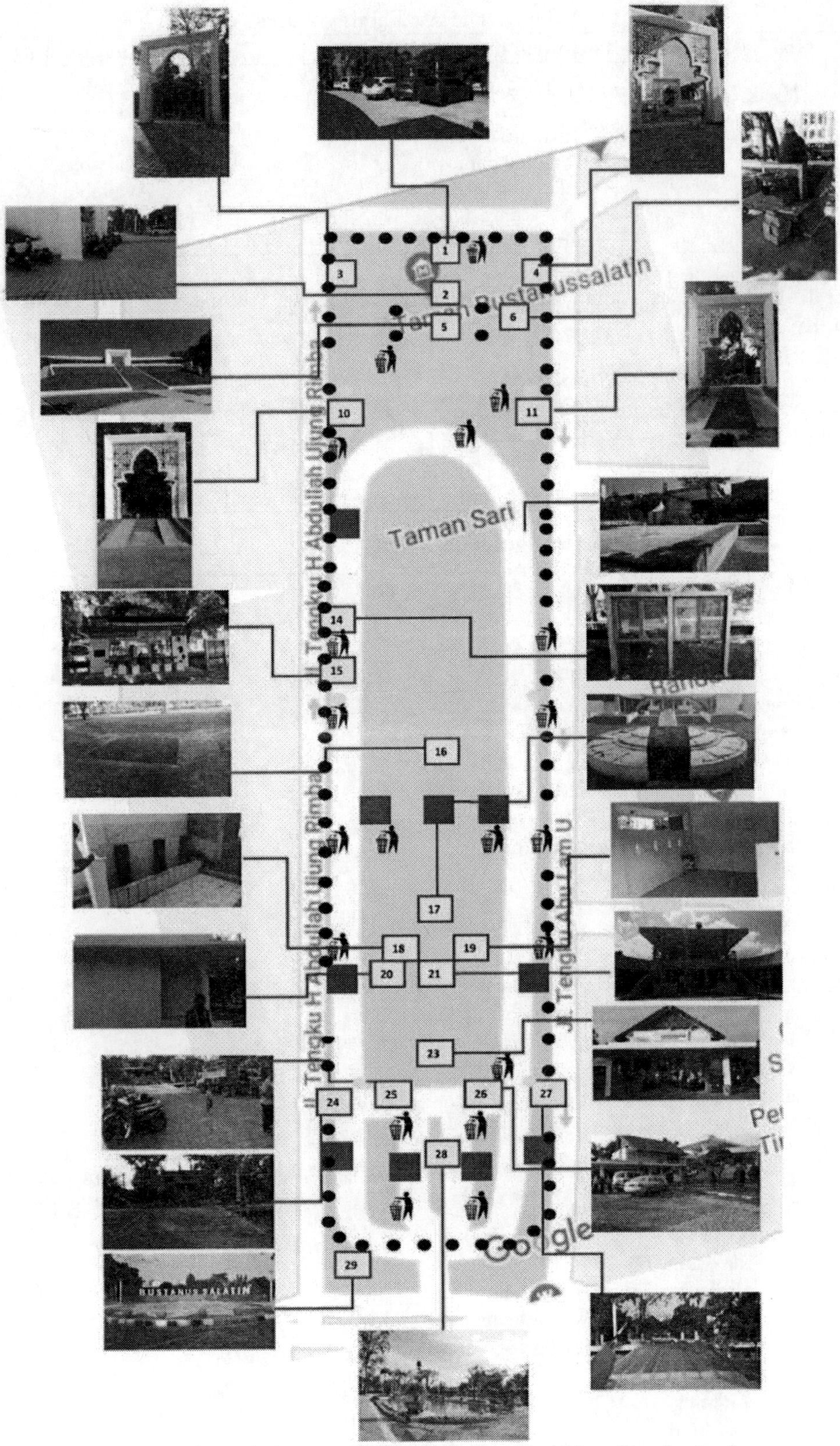

Figure 2. The existing condition of Taman Sari Kota Banda Aceh

IOP Conf. Series: Materials Science and Engineering **674** (2019) 012002 doi:10.1088/1757-899X/674/1/012002

Tabel 1. The Availability Access [1]

No.	Technical guidelines	Access availability
1	Public toilet	Available
2	Space basic size	Available
3	Pedestrian path	Available
4	Guide path	Available
5	Parking area	Available
6	Ramp	Available
7	Rambu dan Marka	Available

Tabel 1 shown that in Taman Sari there are available the technical guidelines for facilities and accessibility.

3.3. Evaluation of the availability access

Table 2 shows the evaluation of the available access which is containing technical guidelines and accessibility compliance.

Tabel 2. Accessibility compliance with technical guidelines [1]

No	Technical guidelines	Accessibility compliance
1	Public toilet	
	a. Embossed design	NS
	b. space	NS
	c. High closet 45-50 cm	NS
	d. Handrail	NS
2	Basic size	
	a. Math to the function	NS
	b. According to the needs	NS
3	Pedestrian path	
	a. Stable, strong, weatherproof	NS
	b. Mound <1,25 cm	NS
	c. Max slope 2°	S
	d. Seat	S
	e. Lighting	S
	f. Drainage far away from the pedestrian edge	S
	g. One way lane 120 cm, two-way lane 160 cm	S
	h. The safety edge is 10 cm high and 15 cm wide	NS
4	Guiding line	
	a. Directional block	S
	b. Warning is	S
5	Parking area	
	a. access point	S
	b. Ramp	NS
6	Ramp	
	a. Max slope 6°	NS
	b. Horizontal length < 900 cm	S
	c. Min width 95 cm without edge protection	NS
	d. Max width 120 cm with edge protection	S
	e. Handrail	NS
	f. Lighting	NS
7	Sign dan Marka	S

Note: S: Suitable, NS: Not Suitable

IOP Conf. Series: Materials Science and Engineering **674** (2019) 012002 doi:10.1088/1757-899X/674/1/012002

The accessibility recommendation design for TSKBA as follows:

a) Public toilets (Figure 3 and 4)

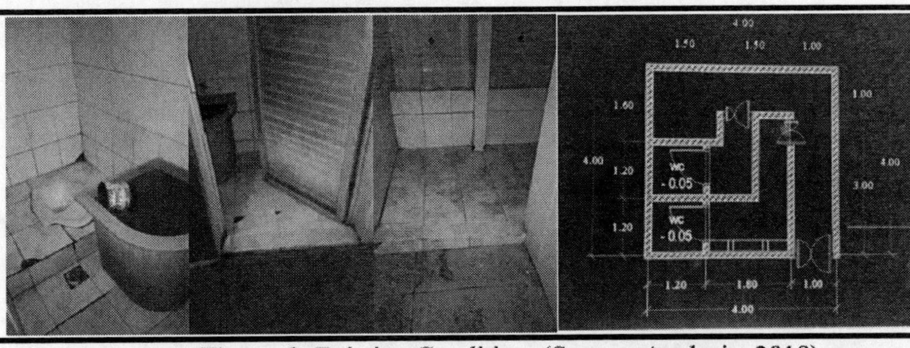

Figure 3. Existing Condition (Source: Analysis, 2018)

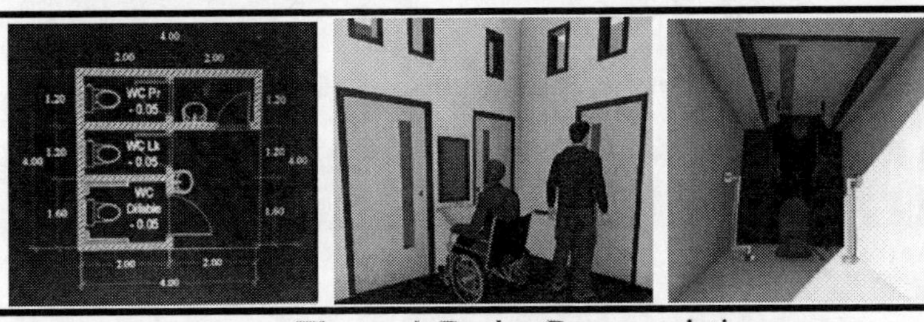

Figure 4. Design Recomendation

The planned design recommendations follow conformity with technical guidelines.

b) Basic Size
The basic size in the design that recommended adjusted with the Technical guidelines of facility accessibility.

c) Pedestriant path and Guiding line (Figure 5 and 6)

Figure 5. existing condition (Source: Analysis, 2018)

Figure 6. Design Recomendation

IOP Conf. Series: Materials Science and Engineering **674** (2019) 012002 doi:10.1088/1757-899X/674/1/012002

d) Parking Area (Figure 7 and 8)

Figure 7. Existing Condition
Source: Analysis, 2018

Figure 8. Design Recommendation

e) Ramp (Figure 9 and 10)

Figure 9. Existing Condition
Source: Analysis, 2018

Figure 10. Design Recommendation

f) Sign and *Marka*
This sign and *marka* to the evaluation of the available access in the TSKBA are suitable.

3.4. Evaluation of Access Convenience

Design recommendations arranged based on four factors that can interfere with comfort or satisfaction, namely form, dimensions, color and texture [4]. These factors are reviewed based on the availability of access reviews. The result of the design recommendation of TSKBA was planned able to give comfort to building users in particular for People with Disability (PwDs).

Tabel 3. Comfort Access [1]

No	Comfort Measurement	Comfort Access
1	Public toilet	
	a. Form	UC
	b. Dimension	UC
	c. Color	UC
	d. Texture	UC
2	Basic Size	
	a. Form	UC
	b. Dimension	UC
	c. Color	UC
	d. Texture	UC
3	Pedestrian Path	
	a. Form	UC
	b. Dimension	C
	c. Color	C
	d. Texture	UC
4	Guiding Path	
	a. Form	C
	b. Dimension	C
	c. Color	C
	d. Texture	C
5	Parking area	
	a. Form	UC
	b. Dimension	C
	c. Color	UC
	d. Texture	C
6	Ramp	
	a. form	C
	b. Dimension	UC
	c. Color	UC
	d. Texture	C
7	Sign and Marka	
	a. Form	C
	b. Dimension	C
	c. Color	C
	d. Texture	UC

Note: C: Comfort, UC: Uncomfort

4. Conclusion

Design recommendations are planned following the Facility Technical Guidelines standards and Accessibility in Building Buildings and the Environment. The concepts based on availability, conformity with standardization and comfort. Fulfillment of accessibility for persons with disabilities in TSKBA, which is a public space that is friendly to persons with disabilities.

Acknowledgments
The research supported by the Faculty of Engineering Muhammadyah Aceh University (UNMUHA) and Institute for research, publishing, community service, and development (LP4M) Muhammadiyah Aceh University.

References

[1] Q. Aini, H. Marlina, and A. Nikmatullah, "Evaluation of Accessibility for People with Disability in Public Open Space," in *IOP Conference Series: Materials Science and Engineering*, 2019, vol. 506, no. 1, p. 12018.
[2] *Undang-Undang Dasar 1945 Pasal 28H Ayat (2)*. Indonesia.
[3] E. M. Badley, "The ICIDH: Format, Application in Different Settings, and Distinction Between Disability and Handicap: a Critique of Papers on the Application of the International Classification of Impairments, Disabilities, and Handicaps," *Int. Disabil. Stud.*, vol. 9, no. 3, pp. 122–125, 1987.
[4] Ministry of Public Works and Housing Indonesia, *Ministerial Regulation 30/PRT/M/2006: Technical Guidance for Facilities and Accessibility in Buildings and Environments*. Indonesia, 2006.
[5] K. Widyawati and R. Laksmitasari, "Penilaian Ruang Bermain Anak di Kota Depok Sebagai Salah Satu Indikator Tercapainya Kota Layak Anak," *Fakt. Exacta*, vol. 8, no. 3, pp. 165–207, 2015.
[6] S. Thohari, "Pandangan Disabilitas dan Aksesibilitas Fasilitas Publik bagi Penyandang Disabilitas di Kota Malang," *Indones. J. Disabil. Stud.*, vol. 1, no. 1, 2017.

Total resistance analysis on bow form model ulstein X-bow with various angle of flare and stem angle

Kiryanto*, E S Hadi and A Firdhaus

Department of Naval Architecture, Faculty of Engineering, Diponegoro University, Semarang, Indonesia

* Email: kiryantodst@yahoo.com

Abstract. The current bow form named X-Bow can face waves coming towards the ship by breaking waves more subtly. Based on the Ulstein X-bow design patent, there are provisions such as angle of flare bow and stem angle. Continue the previous research that investigates comparison of total ship resistance between conventional bow and x-bow on ship; this study aims to show the impact from various angle of flare bow and stem angle on total ship resistance by Computational Fluid Dynamic. Results are summarized in tables and graphs, showing that angle variations on the bow with 6° stem and 10° flare has the most optimal ship performance at 14 knots with a total resistance value 383,13 kN compared to the original model of 385,62 kN, which decreased by 0.64%.

1. Introduction

A main problem for the practicing naval architect is the prediction of the resistance of new ship building an early stage in the project. When a new ship is ordered, a contract containing a specification of the ship is signed between the owner and the shipyard. The shipbuilding company continues to develop hull forms to reduce the global warming effect. IMO MEPC's latest regulation on Energy Efficiency Design Index (EEDI) has increased the interest of ship designers in estimating velocity losses based on actual sea conditions. Innovation in sea transportation especially ship with more optimal performance and maintain the existence of Indonesia as a maritime country in the eyes of the world. These efforts can be started from following the development of design innovation bow form. The bow of the ship is the most pressure and tension of the ocean waves because the bow serves to break the waves. Ships with a good bow will provide the efficiency of the resulting barriers so that ship operations and ship movements become better [1].

Therefore, the field of ship design should be open to innovation from other countries in order to compete. Ulstein group from Norway made an innovation in the ship's bow form called Ulstein X-Bow. With the X-bow form the waves coming towards the ship can be break more subtly and reduce the resistance than the conventional bow. The Ulstein X-bow concept needs to be studied more thoroughly to be implemented in Indonesia. One study has proven that the Ulstein X-bow is better at reducing the ship resistance than the bulbous bow [2]. Based on the Ulstein X-bow design patent there are several requirement such as the term flared bow and stem angle. These two angles will affect the diversity of shapes of the Ulstein X-bow.

In this study, the authors will investigate the effect of variations in the angle of the flared bow and stem on the X-bow to analyze the impact of the angular change on the resulting ship resistance.

IOP Conf. Series: Materials Science and Engineering **674** (2019) 012003 doi:10.1088/1757-899X/674/1/012003

2. Literature review

2.1 Ulstein X-bow definition
Overall the shape of the Ulstein X-bow is different from a conventional bow. The Ulstein X-bow is dominated by a high, rounded and expands slightly at the top of bow [3]. Ships with Ulstein X-bow have great buoyancy because of increasing volume of the bow shape. Ulstein X-bow was first introduced in 2005. Known as Inverted Bow because of the bow shape with the top flipped towards the rear. Ulstein X-bow was originally designed for offshore vessels. A vessel with a bow like this has better sea keeping than a conventional bow. In addition to sea keeping, this bow is also able to increase fuel efficiency and makes waves more subtly.

2.2 Ship Resistance
Ship resistance is the most important factor that determines the power of the ship needed [4]. Ship resistance is the study of fluid reactions due to the movement of the ship through the fluid. In terms of hydrodynamics, the ship is the amount of fluid force acting on the ship in such a way that it opposes the movement of the ship. The main and most significant factor is the hull geometry and the wet surface of the vessel [5]. The resistance is the same as the force component that works parallel to the axis of the ship's velocity. The total resistance on the vessel can be stated as follows.

$$R_T = 0.5\ C_t\ \rho\ V_s^2\ S \qquad (kN) \tag{1}$$

Where :

C_t = total resistance coefficient
S = wet surface area on hull(m^2)
ρ = density of sea water (kg/m^3)
Vs = service velocity (m/s)

2.3 Flare bow and stem angle
The definition of Flare Bow Angle is the slope angle in the bow at the ship's body plan measured from the intersection of the high line with the waterline. The change of the flare angle will affect the shape of the fore area of the vessel. The design that reduces Bow Flare is claimed to eliminate slamming opportunities and reduce resistance [5]. The stem of angle is defined as the tilt angle the requirement 9° - 45°. For stem angle that is set with range 0° - 55 ° [6].

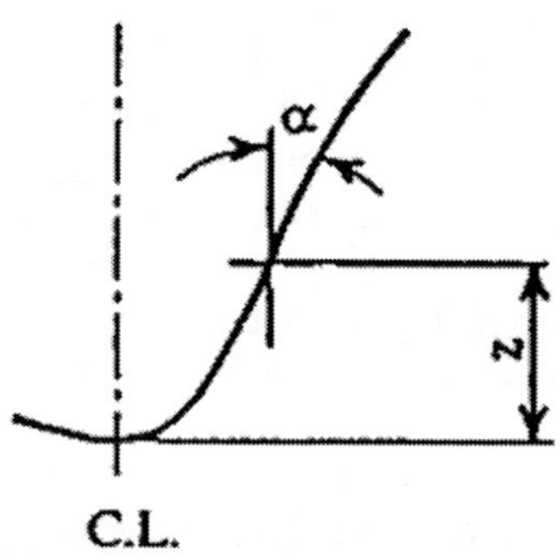

Figure 1. The angle of flare bow [7]

The flare bow angle is described in Figure 1 with the symbol α which states the magnitude of the flare angle in the cross-section of the vessel. Figure 2. Illustrates the angle of the stem at the Ulstein X-bow direction in the longitudinal section of the ship.

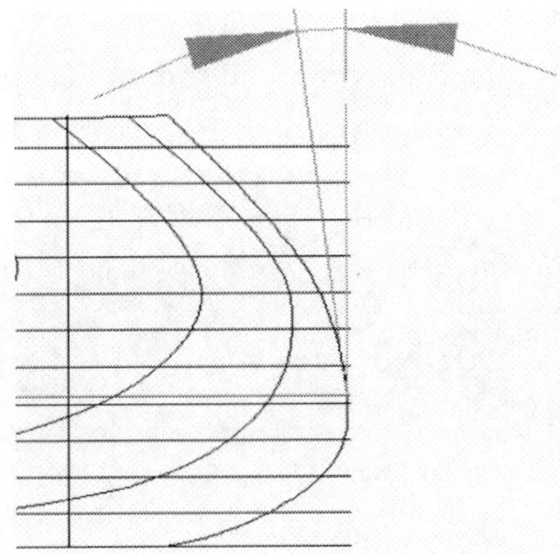

Flare variation

Figure 2. Angle of stem

3. Results and discussion

Table 1 shows the ship principal dimension which is used in the research.

Table 1. Principal dimension

Principal dimension	
Length of Perpendicular (LPP)	102,12 m
Breadth (B)	20,65 m
Draft (T)	4,00 m
Height (H)	11,52 m
Displacement	4500 ton

Furthermore, the ship's resistance analyzes with the variation of the angle of flares bow and stem in X-bow form.

3.1 Modeling with delftship

From the main dimension, ship modeling as in Figure 3 with the Delftship software and export to AutoCAD to get better vision of linesplan.

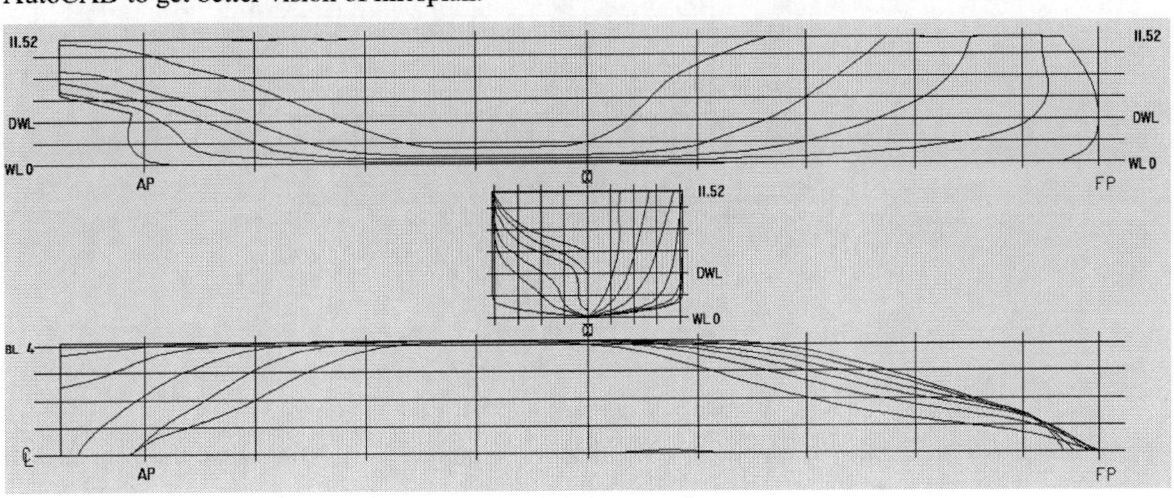

Figure 3. Linesplan of ship

3.2 Modeling with CAD software

Modeling is also done with the help of the Rhinoceros software in Figure 4 so that the model can be solid and run in CFD software.

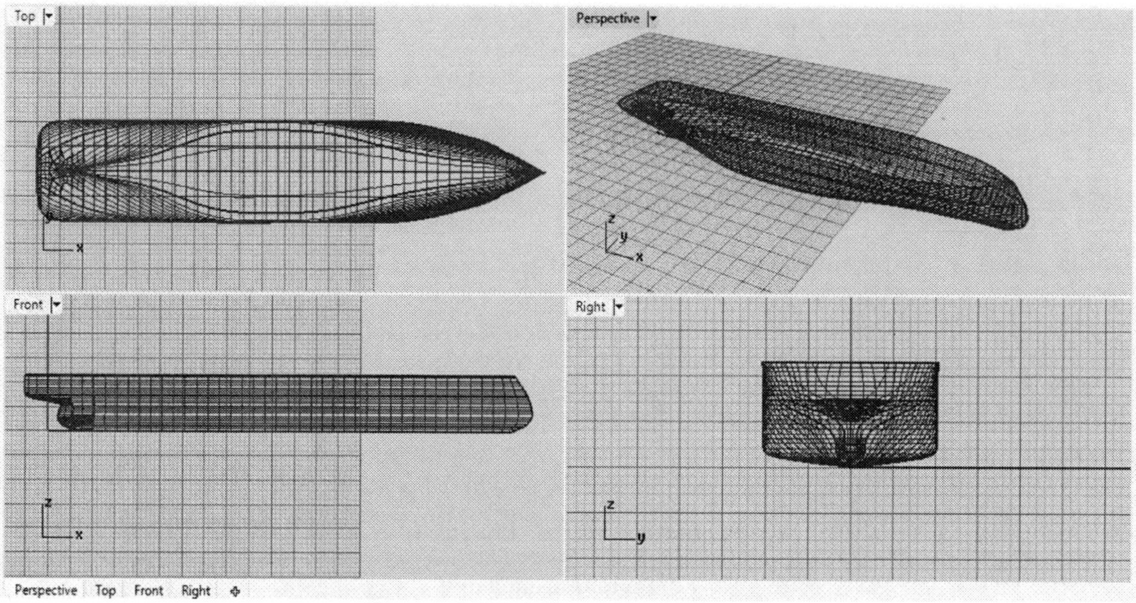

Figure 4. Modeling with CAD software

3.3 Making of model variation

The modeling was originally designed using Autocad software to perform angle measurements. Then it was transferred to the Delfship software to form a smoother model and to the Rhinoceros software so that it could run in the CFD software. This research examines as many as 10 models of variation and 1 original model with bow shape X-bow which changes the angle of flare and stem. Figure 5 describes the variation of the flare angle. Taken by 5 corners so there are 5 models of variation of flare angle.

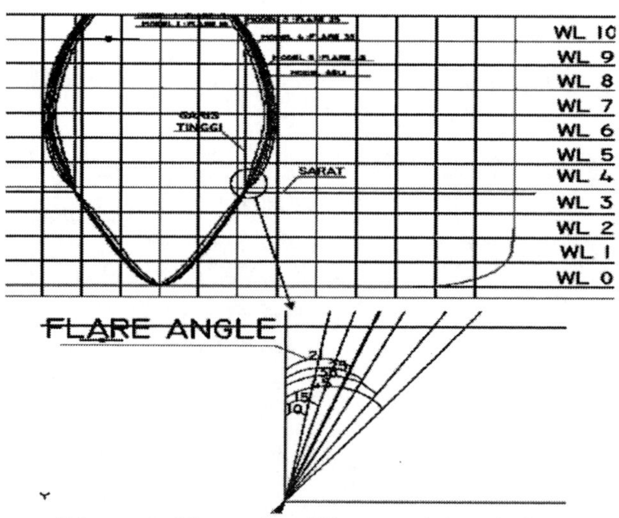

Figure 5. The angle of flare model variation

In table 2 is a breakdown of variations in the flare angle model. For this variation, the stem angle is not changed in order to see the effect of the flare angle clearly.

IOP Conf. Series: Materials Science and Engineering **674** (2019) 012003 doi:10.1088/1757-899X/674/1/012003

Table 2. Variation details of flare angle models

Model	Angle variation (degree)	
	Stem (°)	Flare (°)
Original	**6**	**21**
1	6	10
2	6	15
3	6	25
4	6	35
5	6	45

Figure 6 illustrates the variation of the stem angle. Taken as many as 5 corners so there are 5 models of stem corner variations.

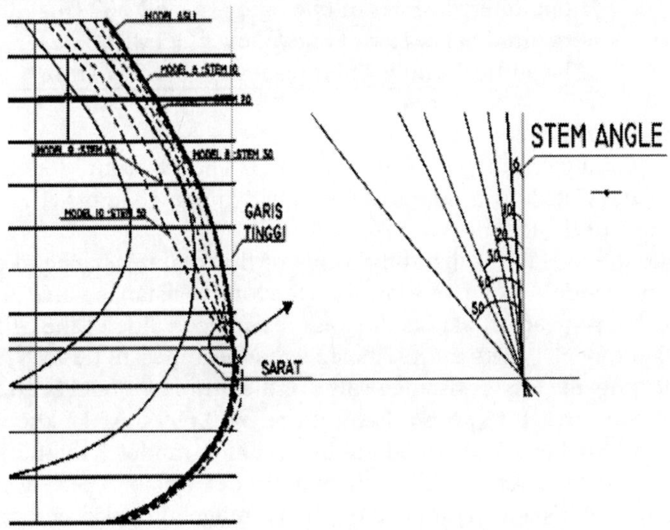

Stem variation

Figure 6. Stem angle model variations

In Table 3 is a detailed variation of the stem corner model. For this variation, the angle of the flares is not changed in order to see the effect of the stem angle clearly.

Table 3. Variation details of stem angle models

Model	Angle variation (degree)	
	Stem (°)	Flare (°)
Original	6	21
6	10	21
7	20	21
8	30	21
9	40	21
10	50	21

3.4 Model validation

Models obtained from modeling need to be validated first. Correction result on validation should not be more than 5%. If the analysis model is in accordance with the actual data it can be used in further analysis. In table 4 the ship displacement value after becoming a model is 4499.9 tons while the original data of the vessel is 4500 tons, the validation correction results are 0.002%.

Table 4. Hydrostatic data

Volume Properties	
Moulded volume	4390.1 (m^3)
Total displaced volume	4390.1 (m^3)
Displacement	4499.9 (tons)
Block coefficient	0.5205
Prismatic coefficient	0.6203
Vertical prismatic coefficient	0.7431
Wetted surfaced area	1886.3 (m^2)
Longitudinal center of buoyancy	52.607 (m)
Longitudinal center of buoyancy	1.461 %
The vertical center of buoyancy	2.367 (m)

3.5 Total resistance analysis

Resistance analysis running on software Autodesk CFD. The eleventh Model tested in software is on 1: 1 scale with each model tested 4 variations of speed ie 8 knots, 10 knots, 12 knots, and 14 knots. Below is presented the results of the test.

Table 5 describes the overall results of the value of the total resistance of each model at each speed. In the table shows that model 1 (stem 6°, flare 10°) the total resistance value always decreases compared to the original model at a speed of 8 knots the total resistance value of the original model is 131.50 kN, while in the variation model 1 the total resistance value becomes 131.29 kN, which is down by 0.16%. For the speed of 10 knots the total resistance value of the original model is 202.63 kN and has decreased in the variation 1 model to 201.35 kN so that it drops by 0.63%. At 12 knots the original model has a resistance value of 287.01 kN and dropped on the variation model 1 to 286.37 kN so that it decreased by 0.22% and at the maximum speed of 14 knots of model 1 also experienced a decrease in resistance from 385.62 kN to 383.14 kN which dropped by a percentage of 0.64%. From the results of the analysis on the variation model 1 to 5 with variables of the flare angle, then model 1 (stem 6°, flare 10°) the most constant at any speed always decreases in the value of resistance than the original model l (stem 6°, flare 21°). Thus flares with the smallest degree in the context of this research can produce a more optimal model in terms of the reduction in the value of total resistance.

Table 5. Model original total resistance result until the variation flare 1-5 model at 8 to 14 knots

Model	**Velocity (knot)**							
	8 (knot)		**10 (knot)**		**12 (knot)**		**14 (knot)**	
	R_T(kN)	**Percent**	**R_T(kN)**	**Percent**	**R_T(kN)**	**Percent**	**R_T(kN)**	**Percent**
Original	131.50		202.63		287.01		385.62	-
1	131.29	-0.16%	201.35	-0.63%	286.37	-0.22%	383.14	0.64%
2	132.98	1.12%	204.13	0.74%	289.75	0.95%	386.24	0.16%
3	131.83	0.24%	202.31	0.15%	287.92	0.31%	388.05	0.63%
4	132.36	0.65%	202.77	0.07%	290.59	1.24%	387.09	0.38%
5	132.74	0.94%	205.61	1.46%	290.51	1.21%	387.36	0.45%

Figure 7 illustrates the resistance graphic character for the original model to the variation of model 5 that the increasing speed increases the total resistance value, but model 1 is always at the bottom of the original model as well as other variation models. Furthermore, the analysis of the total resistance value was carried out in the second test, the ship model that experienced variations in the angle of the stem model 6 to model 10. Here are the results of the analysis of total resistance values.

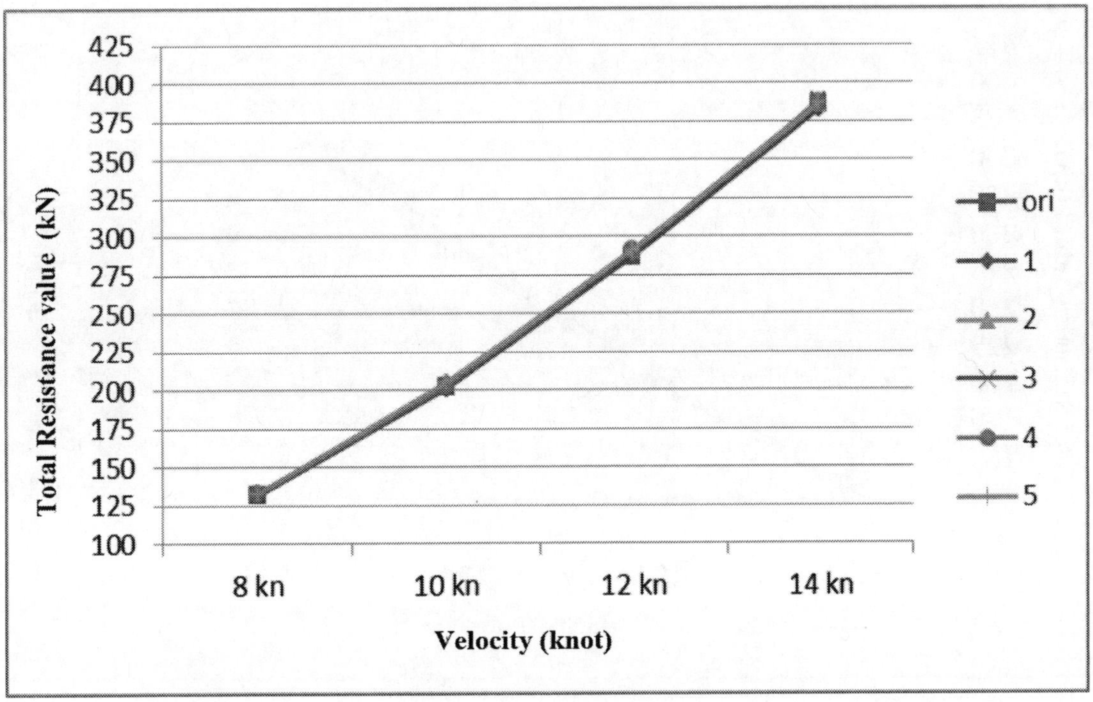

Figure 7. Resistance function graph for the speed of the original model up to 1-5

Table 6 describes the overall results of the value of the total resistance of each model at each speed. But in this analysis states that the variation models 6 to 10 have a higher total resistance value than the original model at each speed. The original model (stem 6°, flare 21°) has a smaller total resistance value than any other variation model for any speed variation. In this analysis variation of model 6 (stem 10°, flare 21°) has increased the total resistance value but the result does not vary much with the original model results. Thus, in the original model and variation model 6 to 10 for variable angular variables that have a small degree of value can affect the small value of the total resistance.

Table 6. Total resistance result original to 6-10 stem variation models at 8-14 knot

Model	Velocity (knot)							
	8 (knot)		10 (knot)		12 (knot)		14 (knot)	
	R_T(kN)	Percent	R_T(kN)	Percent	R_T(kN)	Percent	R_T(kN)	Percent
Original	131.50		202.63		287.01		385.62	
6	131.73	0.16%	203.64	0.49%	288.08	0.37%	386.8	0.30%
7	133.66	1.63%	205.76	1.54%	291.91	1.70%	390.97	1.38%
8	135.81	3.27%	208.00	2.64%	296.09	3.16%	398.6	3.36%
9	134.84	2.53%	207.65	2.47%	293.73	2.34%	395.06	2.59%
10	133.73	1.68%	206.66	1.97%	290.88	1.34%	394.31	2.25%

Figure 8 describes the character of the resistance graph for the original model up to the 6-10 models variation that the more speed increases the total resistance value increases. But with a small stem angle the resistance value also decreases in the context of this second test because the smallest angle of the stem is owned by the original model, the smallest resistance model is still owned by the original model.

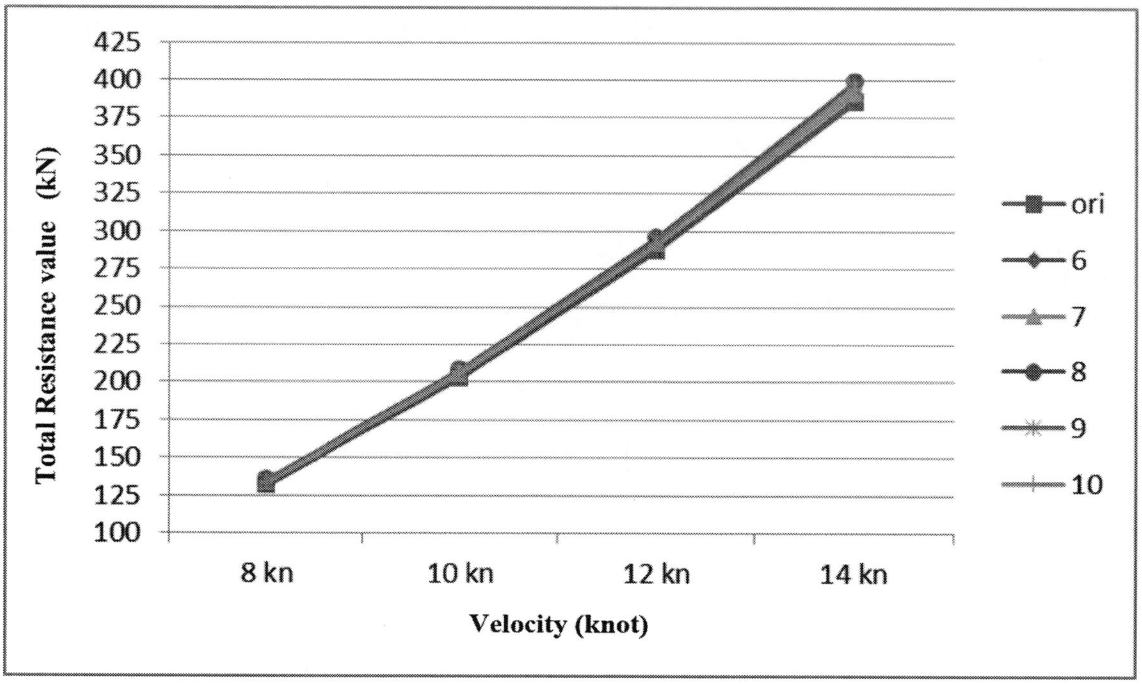

Figure 8. Resistance function graph for the speed of the original model up to 6-10

3.6. Holtrop validation

To validate the results of the test model of CFD software, the manual calculation of the total resistance value in one model is carried out, namely the variation 1 model with the angular variation (stem 6°, flare 10°) compared to the total resistance value obtained from the software. The formula for calculating the total resistance value used is the general resistance formula.

$$R_T = 0.5\ C_t\ \rho\ V_s^2\ S = 0.5 \times 0.007549208848 \times 1{,}025 \times 7.196^2 \times 1886.3 = 377.91 \text{ kN}$$

In CFD software the total resistance value of model 1 variation at 14 knots is 383.14 kN so based on the two values above there is an error difference of 1.36%

4. Conclusions

From the results of the analysis of ship resistance obtained results for variations in flare angles the highest resistance of the variation model compared to the original model is that it occurs at a speed of 14 knots in the variation model 1 (stem 6°, flare 10°) with a percentage decrease of 0.64% from 385.62 kN in the original model to 383.14 kN in the variation model 1. Whereas for the variation of the angle of the stem there is no decrease in the total resistance value in the variation model compared to the original model. The most optimal variation model performance in terms of reduced resistance that is equal to 0.64% compared to the original model occurs in model 1 with variation (stem 6°, flare 10°).

References

[1] D. Chrismianto, A. Trimulyono, and M. N. Hidayat 2014 *J. Teknik Perkapalan UNDIP*, vol. 11, no. 1, pp. 40–48,.
[2] A. P. Sidabalok, D. Chrismianto, and M. Iqbal 2016 *J. Teknik Perkapalan UNDIP*, vol. 4, no. 1, pp. 141–151.
[3] B. A. Perwira and Murdjianto 2014 *J. Teknik Perkapalan ITS*, pp. 1–4.
[4] C. B. Barrass 2004 *Ship Design and Performance for Masters and Mates.*
[5] Murdijanto, K. A. P. Utama, and A. Jamaluddin 2011 J. *Makara Seri Teknol.*, vol. 15, no. 1, pp. 25–30.

[6] Kamsyag 2006 *A Foreship Arrangement For A Vessel Of The Displacement Type* W02006/096066, *Ulstein Gr. Norwegia.*
[7] Croatia Register of Shipping 2012 *Rules For The Classification of Ships Part 2 Hull* Croatia.

IOP Conf. Series: Materials Science and Engineering **674** (2019) 012004 doi:10.1088/1757-899X/674/1/012004

Prototype development of smart brake controller (SBC) for light-duty vehicles (LDVs) with manual

S Munahar[1*], M Setiyo[1], B C Purnomo[1], B Waluyo[1] and W Himawan[2]

[1]Department of Automotive Engineering, Universitas Muhammadiyah Magelang, Magelang, Indonesia.
[2]Automotive Laboratory, Universitas Muhammadiyah Magelang, Magelang, Indonesia.

*E-mail: suroto@ummgl.ac.id

Abstract. This paper presents the development of a smart brake controller (SBC) that applied to passenger cars to reduce accidents while stopped on a hill road. Crash accidents often occur due to the lack of driver skill or negligence in holding the brake pedal. The smart brake controller works by electro-mechanically to lock the brake system even though the brake pedal is released. The results showed that SBC could work well on a variety of vehicle loads and variations in road slope.

1. Introduction

Recently, traffic accidents have increased very surprisingly both at regional and international levels. In 2016, traffic accidents that occurred in Greece on the highway have reached 59,316 cases [1]. Whereas in 2011, traffic accidents that caused deaths in the European Union reached 30,000 cases. Seeing the fact of traffic accidents that occur today cause technology safety and vehicle safety systems into a worldwide research topic. The technology was developed to improve driving safety and reduce the number of traffic accidents. Traffic accidents that occur on vehicles have several factors. The dominating factors include human error or driver failure and vehicle security system failure. While other factors include, environmental factors that are not good roads and weather/climate that does not support.

Efforts to develop vehicle safety technology in reducing traffic accidents have been widely practiced, including the development of brake system technology. Some brake system innovations have been carried out, one of them with Electronic Vacuum Booster (EVB) [2]. In high-speed vehicles, Support Vector Machine (SVM) has been applied to detect braking system failures that may occur[3]. The Anti Blocking System (ABS) has also been widely developed to prevent the wheels from locking during sudden braking[4–6]. Recently, brake technology with an autonomous system has also been developed [7]–[10].

The development of brake system technology is not limited to control systems but also to its mechanical systems, including research on hybrid vehicle disc brakes with an orientation to reduce braking loads [11]and brake systems with special materials to increase efficiency [12–15]. Another sophisticated brake system is electric parking technology to improve safety [16]and brake systems with regenerative control in electric cars[17]. Recently, fuel consumption control can also be done by the brake system so that the brake system has dual functions[18], [19].

To improve vehicle control on uphill roads in times of congestion or when stuck in red light, some modern cars have been equipped with Hill Start Assist - HSA technology [20]. This feature serves to prevent the vehicle from moving backward when the brake pedal is released. However, HSA is only used on cars with automatic transmission. Modeling studies Research on HSA technology has also been carried out, including modeling[21]. Currently, vehicles with pneumatic brake systems have also used HSA technology[22].

HSA is rarely applied to passenger cars with manual transmissions. Most HSA technologies experience multiple detection problems. Another problem is when the vehicle is in a flat position, HSA still functions so that fuel consumption is more wasteful. Another disadvantage is that when the car stops and the engine does not operate, the wheel will be locked so the car must be pulled. For this reason, the aim of this research is to develop a prototype of SBC technology that is applied to passenger cars with manual transmission. This research is important considering the number of passenger cars with manual transmissions dominating the population of commercially available vehicles today.

2. Method

2.1. Design of Smart Brake Controller (SBC)

In this study, an SBC was developed following the systematics presented in Figure 1. This system was developed based on the input signal from the controller. Brake sensors, clutch sensors, TPS sensors, and speed gear sensors are used as controller inputs. The brake sensor sends a digital signal to the controller to provide information about the position of the brake system. TPS sensor as a source of information to determine vehicle acceleration. Clutch sensors have a role in conveying information about the powertrain work system. The controller will process data coming from sensors. Then, data from sensors will be processed to solve problems in the form of braking system operation. Finally, the hydraulic system will be activated by an actuator consisting of DC motors and complementary mechanics.

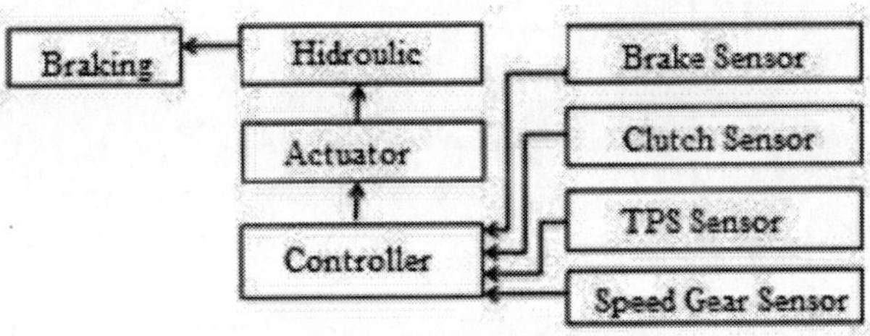

Figure 1. Block diagram of SBC.

The working principle of the SBC developed in this study is divided into three conditions. The first condition, when the driver slows down the car by stepping on the brake pedal. The brake system will work as usual, and the controller seems to not active. The second condition, when the car stops on the uphill road. The driver will release the accelerator pedal, step on the brake and clutch pedal with the speed gear still connected. TPS, speed sensor, brake sensor, and clutch sensor will send a signal to the controller. In this condition, the SBC starts actively. Next, the driver steps on the brake pedal, neutralize the speed gear and releases the clutch pedal. The brakes will remain locked until the car will run again. The third condition, the driver steps on the accelerator pedal, activates the speed gear and releases the brake and clutch pedals. This step is the most important; the SBC is still active for 3 seconds to prevent the car from moving backward. The logic of the controller is shown in Table 1.

Table 1. Logic control system.

No.	Selection switch	Ignition switch	Brake sensor	TPS sensor	Clutch sensor	Speed gear sensor	Controller
1.	On	Off	Off	Off	Off	Off	Off
2.	On	On	On	Off	Off	Off	On
3.	On	On	On	On	Off	Off	On
4.	On	On	Off	On	Off	Off	On
5.	On	On	Off	On	On	On	On
6.	On	On	Off	On	Off	On	On (3 seconds)

IOP Conf. Series: Materials Science and Engineering **674** (2019) 012004 doi:10.1088/1757-899X/674/1/012004

No.	Selection switch	Ignition switch	Brake sensor	TPS sensor	Clutch sensor	Speed gear sensor	Controller
7.	On	On	On	Off	Off	On	Off
8.	Off	Off	Off	Off	Off	Off	Off
9.	Off	On	On	Off	Off	Off	Off
10.	Off	On	On	On	Off	Off	Off
11.	Off	On	Off	On	Off	Off	Off
12.	Off	On	Off	On	On	On	Off
13.	Off	On	Off	On	Off	On	Off
14.	Off	On	On	Off	Off	On	Off

The brake controller developed in this study has a hydraulic motor as an actuator which is presented in Figure 2, while the wiring connecting the sensor, controller, and the actuator is presented in Figure 3.

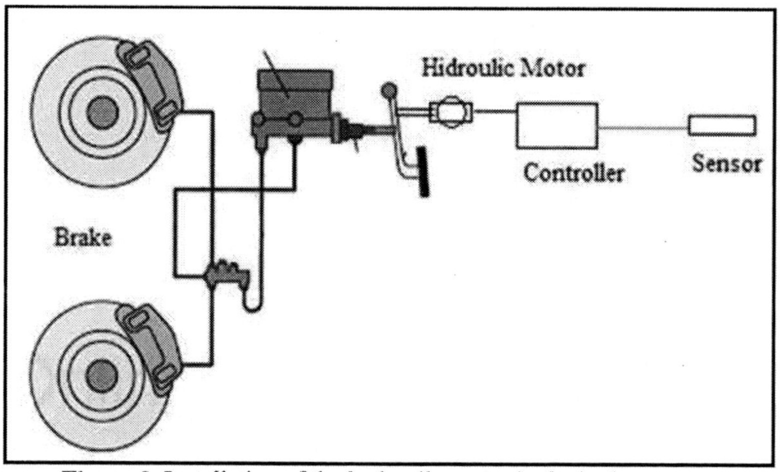

Figure 2. Installation of the hydraulic motor in the brake system.

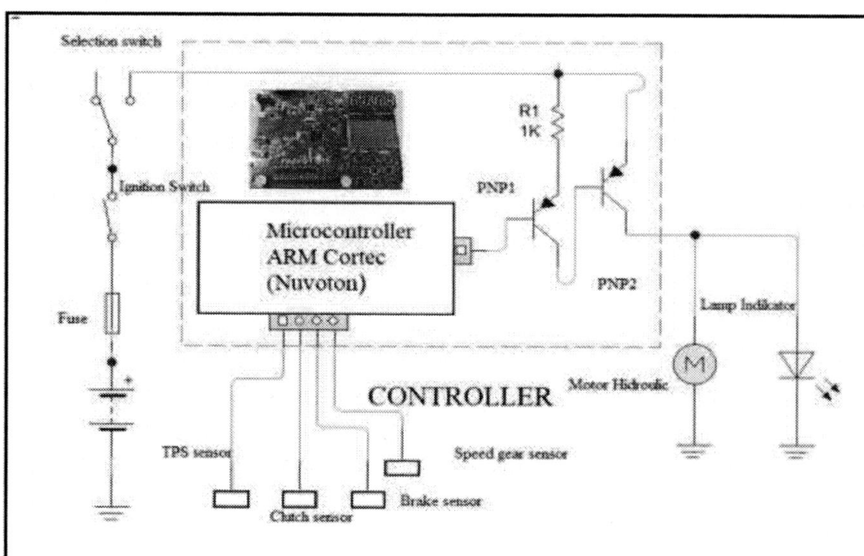

Figure 3. Wiring diagram of SBC.

IOP Conf. Series: Materials Science and Engineering **674** (2019) 012004 doi:10.1088/1757-899X/674/1/012004

2.2. Measurement of brake reliability and road slope

The developed SBC prototype was tested on the actual environment with a number of loading including the number of passengers, road angles, and the load of the vehicle itself. A brake pressure gauge is attached to the master cylinder to measure brake system reliability. Then, an inclinometer is used to measure road slope as shown in Figure 4.

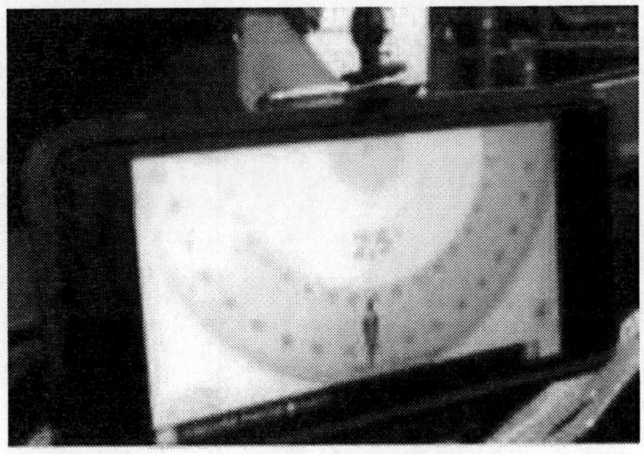

Figure 4. Inclinometer.

3. Result and Discussion

3.1. Initial validation

Vehicle testing was carried out at 7°, 9°, and 14° of road slopes. For measuring the slope of the road, a digital inclinometer was used on the vehicle body. Vehicle loads were varied on one, two, three and four passengers as shown in Figure 5a. Measurement of braking capability was carried out by observing the pressure hydraulic pressure mounted on the master cylinder. After braking, the brake pedal is released. Without SBC, the hydraulic pressure on the brake fluid drops rapidly while with SBC, the hydraulic pressure of the brake fluid is held for a few seconds. This is because this is when the brake pedal is released from the driver's leg; the brake pedal is still pressed automatically by the SBC. This condition is characterized by high brake pressure up to 10 seconds. Then the SBC will withstand the hydraulic pressure of the brake oil for 3 seconds when the driver steps on the accelerator pedal, releases the brake pedal, activates the speed gear, and releases the clutch pedal. Furthermore, the brake pressure test was carried out on various slope variations shown in the graph. Figure 5b. Brake fluid pressure has shown that the higher the vehicle load, the higher the hydraulic pressure of the brake fluid.

Figure 5a. Photographic view of initial validation by four passengers.

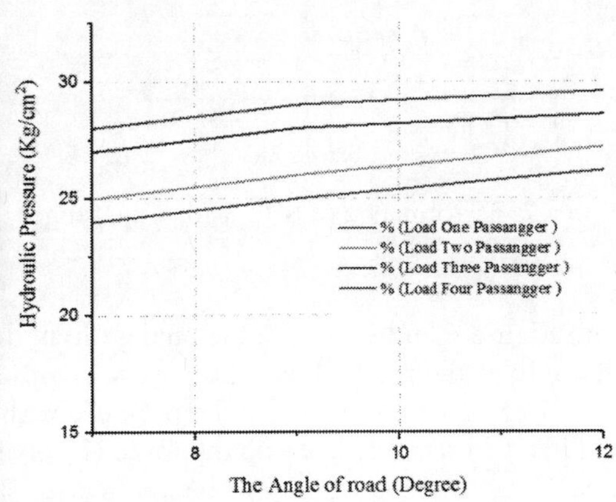

Figure 5b. Fluid pressure, vehicle load, and road slope.

IOP Conf. Series: Materials Science and Engineering **674** (2019) 012004 doi:10.1088/1757-899X/674/1/012004

3.2. Braking performance

The SBC test with a single passenger load (68 kg) shows that the hydraulic pressure remains stable when the brake pedal is released after 2 seconds, while without SBC, the hydraulic pressure drops rapidly when the brake pedal is released after 2 seconds. The same conditions are shown in variations in loads 2 passengers (137 kg), 3 passengers (199), and 4 passengers (265) as shown in Figure 6.

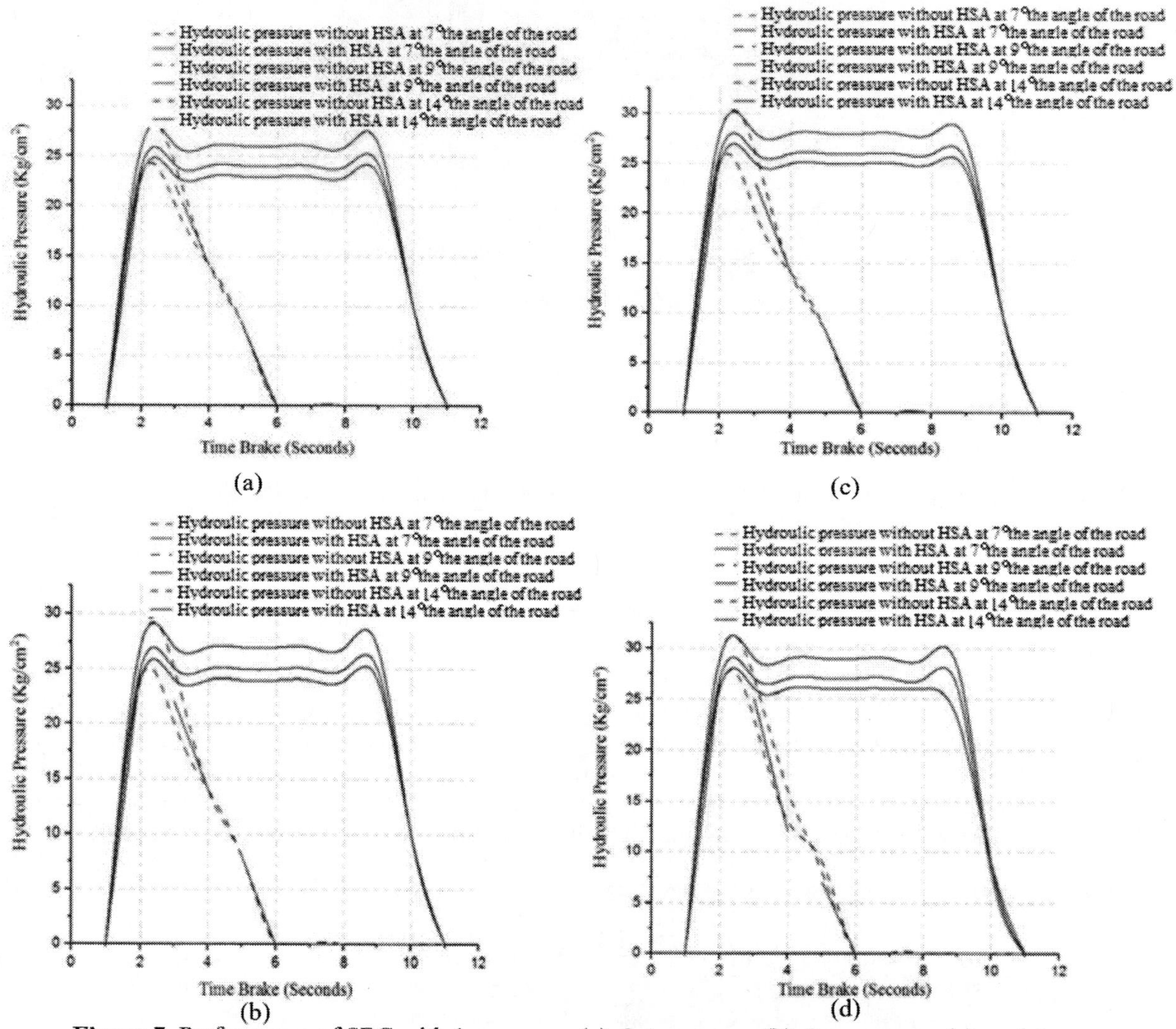

Figure 7. Performance of SBC with 1 passenger (a), 2 passengers (b), 3 passengers (c), and 4 passengers (d).

From Figure 6 and Figure 7, the change in hydraulic pressure is affected by the passenger load and the tilt angle of the road. Hydraulic pressure with a passenger load of 265 kg and a road slope of 14 ° reaches 25 kg / cm2, while hydraulic pressure with a passenger load of 69 kg at a slope of 14 ° is 21 kg / cm2. This difference is because the force (F) given to hold 265 kg is higher than the weight of 69 kg. Thus, SBC has the opportunity to be applied to LDVs with manual transmission. Implementation can be done semi-automatically to give the driver an option in sporty and eco mode.

IOP Conf. Series: Materials Science and Engineering **674** (2019) 012004 doi:10.1088/1757-899X/674/1/012004

4. Conclusion

Based on a series of tests that have been carried out, it is concluded that the Smart Brake Controller (SBC) can be applied to LDVs with the manual transmission. The increase in hydraulic pressure which is proportional to the increase in vehicle load indicates that the SBC is responsive to changes in vehicle load. With a load of more than 250 kg at 14° road slope, the hydraulic pressure can be maintained more than 10 kg/cm^2, and the vehicle is still able to stop without rolling backward when accelerating on the uphill road.

References

[1] Yannis G, Theofilatos A and Pispiringos G 2017 Investigation of road accident severity per vehicle type Yannis *Proc. of Transp. Research* vol 25 (Shanghai) pp 2081–88

[2] Bebei J, Guangquan L, Yunpeng W and Guizhen Y 2013 The Measurement for Performance Parameter of Automobile Braking System with Electronic Vacuum Booster *Proc. Soc. Behav* vol 96 (Cina / Elsevier) pp 2058–65

[3] Jie L, Yan F L and Enrico Z 2016 A SVM framework for fault detection of the braking system in a high speed train *J.Mech. Syst. Signal Process* **87** 401-9

[4] Bera T K, Bhattacharya K and Samantaray A K 2011 Simulation Modelling Practice and Theory Evaluation of antilock braking system with an integrated model of full vehicle system dynamics, *J.Stimul. Model. Pract. Theory* **19** 2131–50

[5] Giovanna F, Domenico T, and Giacomo R 2016 A fuzzy-genetic control system in the ABS for the control of semi-active vehicle suspensions *J. Mechatronics* **39** 89–102

[6] Mehdi M and Hossein M 2012 Optimal design of a non-linear controller for anti-lock braking system *J.Transp. Res. Part C* **24** 19–35

[7] Andrikov D, Andrikov DM and Droh M C P 2017 Design of flat wheel braking control system with three modes of motion : rolling , sliding , locking, *Proc. Comput* vol 103 (Moscow/Rusia Elsevier) pp 466–69

[8] Jessica B C 2017 Effectiveness of forward collision warning and autonomous emergency braking systems in reducing front-to-rear crash rates *J. Accid. Anal. Prev.* **99** 142–152.

[9] Monika D and Michal G 2017 If every car had autonomous emergency braking system for forward collisions avoidance *Proc. Computer Science* vol 110 (Berlin/London Elsevier) pp 386–93

[10] Nickolay P, Viktor D, Alexander P and Vladimir L 2017 Methods of Assessing the Influence of Operational Factors on Brake System Efficiency in Investigating Traffic Accidents *Proc. of Transp. Research* vol 20 (Petersburg/Rusia) pp 516–22

[11] Hiu L and Dejie Y 2016 Advances in Engineering Software Optimization design of a disc brake system with hybrid uncertainties J.Adv. Eng. Softw. **98** 112–22

[12] Wu L M, Ren C Z, Xu H Z and Zhou C L 2017 Surface wetting of the C/SiC brake lining with micro-scale heat dissipation fins to cool off the brake system: Influence of the fibre ending orientation and fin interval J. of the Franklin Institute **43** 10805-816

[13] Yevtushenko A A, Kuciej M and Och E 2014 Temperature in thermally nonlinear pad – disk brake system J. Int. Commun. Heat Mass Transf. **57** 274–81

[14] Nico L, Michael R, Krenkel, Jens R, Reinhard H, Walter K and Frank R 2016 Full-ceramic brake systems for high performance friction applications J. Eur. Ceram. Soc. **36** 3823-32

[15] Vytenis S and Edgar S 2016 Research of the Vehicle Brake Testing Efficiency, *Proc. Engineering* vol 134 (Transbaltica Elsevier) pp 452–58

[16] Bin W, Xiexun G, Chenchai Z, Zhe X and Jie Z 2014 Modeling and control of an integrated electric parking brake system *J. Franklin Inst.* **352** 626-44

[17] Jian C, Jiangze Y, Kaixiang Z and Yan M 2017 Control of regenerative braking systems for four-wheel-independently-actuated electric vehicles *J. Mechatronics* **1** 1–8

[18] Muji S and Suroto M 2017 Modeling of LPG-fueled engine based on engine , transmission , and brake system using fuzzy logic controller (FLC) *J. Mechatronics, Electr. Power, Veh. Technol* **8** 50–59

[19] Aris T, Enda W S, and Joga D S and Suroto M 2015 Smart Controller Design of Air to Fuel Ratio (AFR) and Brake Control System on Gasoline Engine,*Proc. Intern.Conf.on Information Technology, Computer, and Electrical Engineering (ICITACEE)*- Semarang Indonesia

[20] Xuefeng J, Xian X, Guodong J and Wei C 2012 Research on Hill Start Control for Heavy Truck with AMT *Proc. of the 2012 Intern. Conf. Computer Application and System Modeling* (Paris, France)

[21] Hu J J, CHen X, Du R and Zhan Y 2013 Control strategy HSA for electric car *J. Appl. Sci.* **13** 1429–35

[22] Zhao W, Changfu C, Hongyu Z, Huaji W and Shengnan Y 2012 Integrated HIL Test and Development System for Pneumatic ABS/EBS ECU of Commercial Vehicles *J.SAE Tech.* **01** 13

Acknowledgment

This paper is part of the automotive mechatronics development project at the Automotive Laboratory of the Universitas Muhammadiyah Magelang. The researchers would like to thank students laboratory assistants who helped in this work. May Allah bless the next research activity.

IOP Conf. Series: Materials Science and Engineering **674** (2019) 012005 doi:10.1088/1757-899X/674/1/012005

The role of ethanol as cosolvent on the separated gasoline methanol blend

B Waluyo[1,*], I N G Wardana[2], L Yuliati[2], M N Sasongko[2] and M Setiyo[1]

[1] Department of Automotive Engineering, Universitas Muhammadiyah Magelang, 56172, Indonesia

[2] Department of Mechanical Engineering, Brawijaya University, Malang, 65141, Indonesia

* Email: otobudy@ummgl.ac.id

Abstract. This article aims to reveal the role of ethanol as a cosolvent in maintaining the stability of a gasoline methanol blend. Polar methanol was mixed with polar gasoline with a fraction of 30, 50 and 70% v/v at room temperature and atmospheric pressure. Ethanol titration in gasoline methanol mixtures was conducted to determine the ethanol content needed to maintain the stability of the methanol-gasoline blend. Mixing stirring was done manually. A blend of gasoline-ethanol-methanol was waited for 120 minutes to ensure the mixture was not separated. The results of the study show that ethanol has a role as a cosolvent where the OH (hydroxyl) as the head (polar) is attached to the polar methanol molecule while the CH (hydrocarbon) group as the tail (non-polar) interacts with the non-polar gasoline molecule.

1. Introduction

To date, extensive biofuels utilization is still recognized as a way to reduce dependence on crude oil and as a source of renewable energy [1]. Although the use of ethanol as a fuel has only two-thirds of the energy content of gasoline, ethanol has the required positive character on IC engine burning. The positive characters are laminar flame propagation speed and higher octane rate compared to gasoline [2]. Another positive behavior of ethanol as a fuel is the ownership of the hydroxyl group of molecules ethanol so it is an oxygenate fuel and it's a polar substance. The chemical formula for alcohol is $C_n H_{2n+1} OH$. The addition of alcohol to gasoline allows perfect combustion, better combustion efficiency and friendlier emissions [3].

The disadvantage of using alcohol as a fuel, especially those with short carbon chains, is the low energy content that has an impact on increasing fuel consumption [4]. The increase in alcohol fraction in gasoline also affects the problem of mixture separation, metal corrosion and changes in fuel vapor pressure [5]. However, the use of alcohol as fuel is the most logical solution to overcome the crisis of fossil fuels without making any changes to the existing engine construction [6].

Methanol is a biodegradable fuel. The use of methanol has been suggested as fuel as a means of extending the supply of liquid fossil fuel consumption [7]. Most methanol is derived from natural gas, coal and some are obtained through biomass synthesis. Methanol has long been used as a fuel, both as a mixture of gasoline and as a dedicated fuel. Methanol shows a good market signal for it's easy to obtain, has a low price [8], easy to store, and distributed broadly [9]. Methanol also has a high-octane number allowing engine design to have a high compression ratio to improve thermal efficiency. Another advantage of methanol is a high evaporation enthalpy so that it positively impacts its volumetric efficiency [8]. The energy content of methanol is less than gasoline so that the higher fuel consumption would be theoretically predicted for methanol and gasoline blends than for regular gasoline. Lately, many researchers in the field of alternative fuels for automotive, are interested in studying methanol as a mixture of fuels both gasoline and diesel [10], so that methanol has the potential to become a favorite alternative fuel candidate in the future

One recent study by the authors is that polar ethanol is capable of inducing non-polar hydrocarbons ethanol and forming molecular clusters to change the evaporating character and other properties of fuel blends [6]. From the previous findings and the potential for ethanol as a cosolvent will provide a better mixing role between methanol polar and non-polar gasoline. Ethanol molecule as a cosolvent concept was presented in Figure 1.

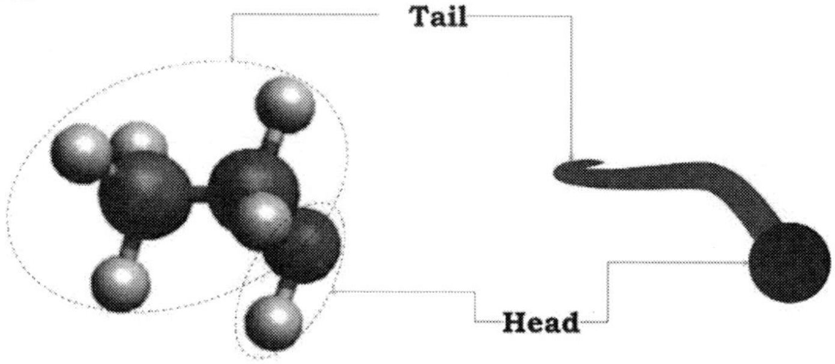

Figure 1. Ethanol molecule as a surfactant concept

The hydroxyl groups (head) in the ethanol molecule are polar and have the potential to form hydrogen bonds with other molecules that contain nitrogen, oxygen and flour elements. The concept of ethanol role as cosolvent on the separated gasoline methanol blend offered is presented in Figure 2.

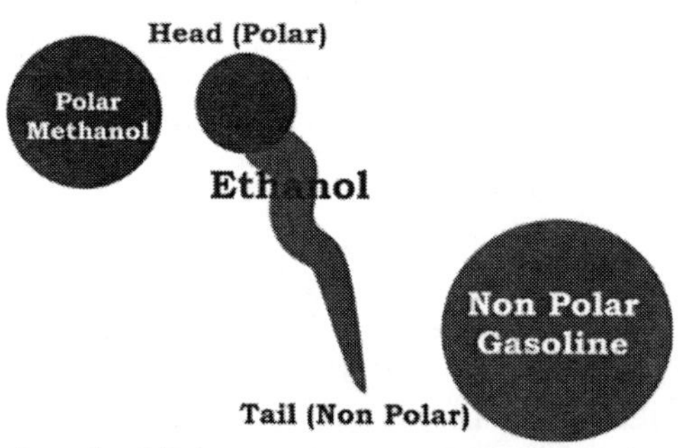

Figure 2. The role of ethanol addition on the separated methanol gasoline blend concept

Until now the implementation of methanol as a fuel blend is still limited to a light mixture of around 4-5% v / v, by using cosolvent to maintain the stability of the fuel blend [11]. Methanol has a high limit to mix blend with gasoline, so the mixture separation is easy [12]. Previous research also stated that an increase in the methanol fraction in gasoline would cause the problem of separating the fuel mixture [5]. This study presents essential findings from the addition of ethanol to the gasoline-methanol blends in several conditions of the fuel blend storage temperature. This article aims to reveal the role of ethanol as a cosolvent in maintaining the stability of a gasoline methanol blend.

2. Experiments

2.1. Fuel blends preparation

This study used commercial gasoline supplied by PT. Pertamina. Methanol and ethanol analytical reagent grade with a molar weight (MW) of 20.1 and 46.07-gram mole^{-1} and purity 99.5% and 99.7%

respectively. Smart-Lab, Indonesia supplied methanol and ethanol. The detailed properties of fuel which is used in this study were presented in Table 1.

Table 1 Properties of test fuels[12].

Properties	Gasoline	Ethanol	Methanol
Purity (%)	n/a	99.7	99.8
Chemical formula	$C_{5\text{–}10}$ $H_{12\text{–}22}$	C_2H_5OH	CH_3 OH
Boiling Temperature (°C)	27–225	78.2	65,2
Hydrogen (% mass)	12.5	34.7	12.57
Carbon (% mass)	87.5	52.2	37.5
Oxygen (% mass)	0	34.7	49.93
Low heating value (MJ kg^{-1})	44.0	26.9	20.1
Molecular Weight (gram $mole^{-1}$)	106.22	46.07	32.04
Density @ 20 °C (kg L^{-1})	0.737	0.785	0.792
Reid vapor pressure	53–60	17	32.4

2.2. Experimental apparatus
This research was carried out by mixing polar methanol and non-polar gasoline in a test tube as shown in Figure 3.

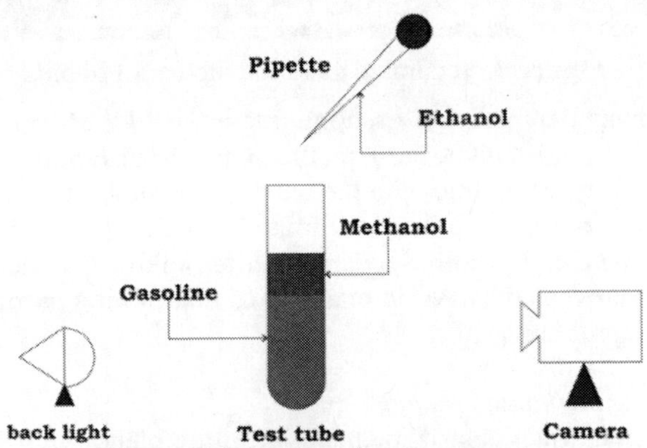

Figure 3. The experiment apparatus

The methanol fraction used in this study was 30, 50 and 70% v / v. Mixing methanol gasoline is done manually using a stirring spoon, then the fuel blend was allowed to stand for a while until the methanol-gasoline mixture was shown as shown Figure 4.

Figure 4. Separated fuel blends

IOP Conf. Series: Materials Science and Engineering **674** (2019) 012005 doi:10.1088/1757-899X/674/1/012005

A 5 ml pipette with the accuracy of 0.5 ml reading is used to drip ethanol into the fuel mixture until the separation mixture was removed. A 50 fps DSLR camera was used to take pictures of each mixture and treatment conditions. A rear light was used to sharpen the image from the camera.

3. Result and Discussion

3.1. Separated gasoline methanol blend

The first gasoline methanol blend in the reaction glass was set to 4 ml. After stirring, the gasoline methanol mixture was waited for up to 100 minutes, and images are taken using a Camera. Visualization of mixing methanol gasoline with various of fractions blend is presented in Fig. 5 below,

Figure 5. Separated gasoline methanol blends

Figure 5 shows that mixing the methanol-gasoline fraction 10, 15, 30, 50, and 70% v / v is separated. This phenomenon is easy to understand because methanol is a polar liquid while gasoline is a non-polar liquid. The results of this study are in line with the results of Bromberg & Cheng's study, 2010, which states that the use of gasoline methanol mixture is limited to 4-5% v / v methanol fraction [11]. Figure 5 also shows that mixing methanol gasoline separately after mixing the blend (energizing) is stopped. This phenomenon was also revealed by Awad et al., 2018 that mixing gasoline methanol would cause separating the fuel mixture problem [12].

3.2. Homogeneous gasoline methanol ethanol blend

The results of mixing various fractions of methanol gasoline blend with the addition of ethanol are presented in Figure 6.

Figure 6. Homogeneous gasoline methanol blends

Figure 6 clearly shows that the addition of ethanol to the gasoline methanol blend produces a homogeneous fuel mixture. The minimum ethanol requirement to produce a homogeneous fuel blend is presented in Table 2.

Table 2. The minimum ethanol fraction needed in various fractions of gasoline methanol blend

No	The volume of fuel blend (4 ml)				Methanol fraction v/v	Ethanol Min	
	Gasoline		Methanol				
	(ml)	(%)	(ml)	(%)	(%)	(ml)	(%)
1	3.6	90	0.4	10	10	0.425	11.0
2	3.4	85	0.6	15	15	0.308	8.0
3	2.8	70	1.2	30	30	0.300	7.5
4	2.0	50	2.0	50	50	0.208	5.0
5	1.2	30	2.8	70	70	0.100	2.5

Table 2 shows the minimum ethanol requirements to form a homogeneous mixture on various gasoline methanol fractions. The tendency of ethanol needs as a cosolvent in the gasoline methanol blend is presented in Figure 7.

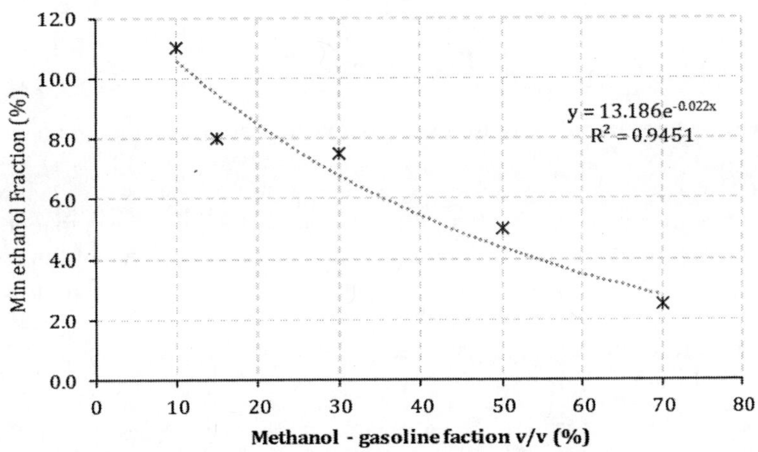

Figure 7. The tendency of ethanol needs as a cosolvent in gasoline methanol blend

Figure 7 shows that the increasing methanol fraction in gasoline the fewer ethanol needs as a cosolvent. The need for ethanol as a cosolvent of the methanol-gasoline mixture up to 12% v/v, is inversely proportional to the increase in the methanol fraction. This interesting phenomenon is caused by ethanol and methanol, each of which has a hydroxyl group so that both have the potential as cosolvent.

3.3. The role of ethanol as a cosolvent in maintaining the stability of a gasoline methanol blend

The process of mixing a fluid with another is a physical process. Molecular interactions between molecular substances that mix is physical activities that can explain a homogeneous and separate blend. Gasoline ($C_{5\text{-}10}$ $H_{12\text{-}22}$) is a non-polar hydrocarbon, while methanol (CH_3 OH) and ethanol (C_2H_5 OH) are short-group hydrocarbon chains that have hydroxyl groups at one end. However, ethanol has a hydrocarbon chain that is longer than methanol, so it has the potential to interact with gasoline hydrocarbons. This molecular interaction occurs between C_2H_4 as a tail (Figure 1) and the gasoline molecule. While the polar hydroxyl group at the end of ethanol becomes the head that interacts with the polar methanol molecule. The molecular interaction model between gasoline methanol and ethanol Figure 8.

Figure 8 shows that element H on the hydroxyl group of methanol has a negative partial charge interacting with element O on the hydroxyl group of ethanol which has a positive partial charge. This interaction is the strongest intermolecular interaction (hydrogen bond), while the interaction of non-polar groups of ethanol hydrocarbons with non-polar gasoline is a weak molecular surface force. In this molecular interaction housing, ethanol functions as cosolvent because, polar (hydroxyl) heads interact with polar methanol, while the non-polar tail side interacts with non-polar gasoline molecules (Figure 1). Figure 7 also gives an empirical explanation that using a high fraction methanol mixture is more beneficial because it only requires a small fraction of ethanol as a cosolvent. However, a good and satisfying scientific explanation needs to be carried out further study.

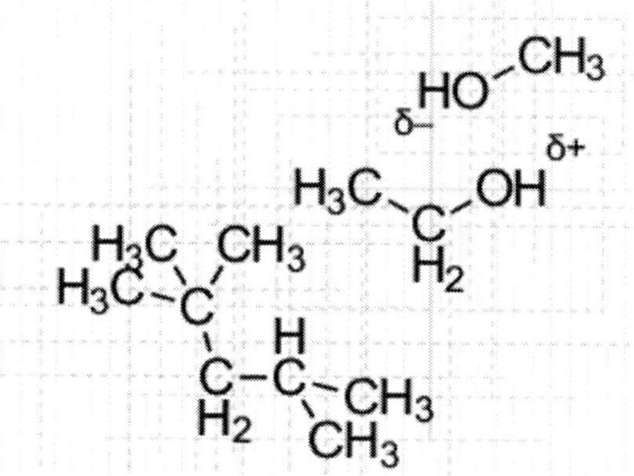

Figure 8. Molecular interaction model between gasoline methanol and ethanol

4. Conclusion

A series of simple experiments have been carried out to prove the hypothesis that ethanol can function as a cosolvent of the gasoline methanol blend to mix homogeneously. Ethanol functions as a cosolvent of methanol for perfectly blending with non-polar gasoline because it has a longer chain hydrocarbon. The role of ethanol as a cosolvent caused **by** the OH (hydroxyl) as the head (polar) is attached to the polar methanol molecule while the CH (hydrocarbon) group as the tail (non-polar) interacts with the non-polar gasoline molecule.

References

[1] Rao D C K, Syam S, Karmakar S and Joarder R 2017 Experimental investigations on nucleation, bubble growth, and micro-explosion characteristics during the combustion of ethanol/Jet A-1 fuel droplets *Exp. Therm. Fluid Sci.* **89** 284–94

[2] Cooney C P, Yeliana, Worm J J and Naber J D 2009 Combustion characterization in an internal combustion engine with ethanol - Gasoline blended fuels varying compression ratios and ignition timing *Energy and Fuels* **23** 2319–24

[3] Surisetty V R, Dalai A K and Kozinski J 2011 Alcohols as alternative fuels: An overview *Appl. Catal. A Gen.* **404** 1–11

[4] Elfasakhany A 2018 Exhaust emissions and performance of ternary iso-butanol e bio- methanol e gasoline and n-butanol e bio-ethanol e gasoline fuel blends in spark-ignition engines : Assessment and comparison *Energy* **158** 830–44

[5] Elfasakhany A 2015 Investigations on the effects of ethanol–methanol–gasoline blends in a spark-ignition engine: Performance and emissions analysis *Eng. Sci. Technol. an Int. J.* **18** 713–9

[6] Waluyo B, Wardana I N G, Yuliati L and Sasongko M N 2018 The role of molecule cluster on the azeotrope and boiling points of isooctane-ethanol blend *Fuel* **215** 178–86

[7] Iliev S 2015 A Comparison of Ethanol and Methanol Blending with Gasoline Using a 1-D Engine

Model *Procedia Eng.* **100** 1013–22

[8] Liu S, Cuty Clemente E R, Hu T and Wei Y 2007 Study of spark ignition engine fueled with methanol/gasoline fuel blends *Appl. Therm. Eng.* **27** 1904–10

[9] Li Y, Zhang C, Yu W and Wu H 2016 Effects of rapid burning characteristics on the vibration of a common-rail diesel engine fueled with diesel – methanol dual-fuel *FUEL* **170** 176–84

[10] Qu X, Gong C, Liu J, Cui F and Liu F 2015 Regulated and unregulated emissions from a DISI methanol engine under homogenous combustion and light load *Fuel* **158** 166–75

[11] Bromberg L and Cheng W K 2010 *Methanol as an alternative transportation fuel in the US:Options for sustainable and / or energy-secure transportation* (Massachusetts)

[12] Awad O I, Mamat R, Ibrahim T K, Hammid A T, Yusri I M, Hamidi M A, Humada A M and Yusop A F 2018 Overview of the oxygenated fuels in spark ignition engine: Environmental and performance *Renew. Sustain. Energy Rev.* **91** 394–408

Acknowledgments
Research equipment was supported by the Automotive Laboratory of Universitas Muhammadiyah Magelang and Mechanical Engineering laboratory of Brawijaya University. Therefore, the researcher would like to thank both institutions.

IOP Conf. Series: Materials Science and Engineering **674** (2019) 012006 doi:10.1088/1757-899X/674/1/012006

Characteristics of vapor compression refrigeration system with parallel expansion valves using refrigerant musicool 134

B C Purnomo, Muji Setiyo, Budi Waluyo, Saifudin and Noto Widodo

Department of Automotive Engineering, Universitas Muhammadiyah Magelang, Magelang, Indonesia.

E-mail: setiyo.muji@ummgl.ac.id

Abstract. This research was conducted to investigate the performance of vapor compression refrigeration systems with Musicool 134 refrigerant. The scope of this study was to compare the performance of double expansion valves installed parallel to the temperature expansion valve (TEV). The temperature and pressure of the refrigerant at the variable cooling load of 0.23, 0.33, and 0.39 Kg/s were observed during the system worked until steady conditions. The results of the study show that double expansion valves installed parallel produce better performance. The highest refrigeration effect produced was 257 kJ / kg and the COP was 5.84.

1. Introduction

In the last decade, the development of Air Conditioning technology, especially the use of refrigerants as working fluids, is strongly influenced by the issue of global warming potential (GWP) and ozone depletion potential (ODP) [1]–[4]. Now, the development of refrigerants is like the period 1830 to 1930, which is back to natural refrigerants such as hydrocarbon (HC), CO2, N2 and others [5]. The natural refrigerant is an alternative to CFCs and HFCs refrigerants that have higher environmental effects, even in 2020 refrigerants containing chlorine must not be reproduced. Table 1 presents some of the commercial refrigerants used in the last few decades and their effects on the environment [2], [6].

Table 1. Property of some commercial refrigerants

Refrigerants	R134a (HFC)	R290 (HC)	R600a (HC)	R290/R600a mixture (HC)	R152a (HFC)	R1234yf (HFO)
Global warming potential (GWP)	1400	11	8	7	140	4
Ozone Depletion Potential (ODP)	0	0	0	0	0	0
Atmospheric Life Time (years)	14	<1	<1	<0.04	2	<0.05
Acute Toxicity Exposure Level (ATEL) (ppm)	50,000	50,000	25,000	40,000	50,000	101,000
Lower Flammability Limit (LFL) (vol. %)	-	2.2	1.7	2.0	3.9	6.5
Minimum Ignition Energy (MIE) (mJ) Safety Group	-/A1	0.25/A3	0.35/A3	N/A/A3	0.38/A2	>1000/A2L
Molar Mass (kg/kmol)	102.03	44.096	58.122	N/A	66.051	N/A
Vapor Density at 25°C (kg/m^3)	32.35	20.65	9.12	N/A	18.47	N/A
Critical Temperature (°C)	101.1	96.8	135	114.8	113.3	N/A
Critical Pressure (kPa)	4061	4247	3647	4040	4522	N/A
Normal Boiling Point (NBP) (°C)	−26.11	−42.11	−11.78	−31.5	−24.0	N/A
Vapor Cp at 25 °C (kJ/KgK)	1.0316	2.0724	1.8189	1.77	1.2536	1.0

IOP Conf. Series: Materials Science and Engineering **674** (2019) 012006 doi:10.1088/1757-899X/674/1/012006

As an alternative refrigerant, hydrocarbons (HC) have long been tested and have been widely used in various fields of air conditioning systems. Propane (R-290) and butane (R-600) and mixtures are both hydrocarbons that have excellent properties as refrigerants [7], [8]. R-290/R-600 mixed refrigerants with various compositions have shown performance equivalent to R-134a [9], [10]. Other studies show a mixture of 40% / 60% of R-290/R-600 can replace R-12 with the same good performance [1].

HC refrigerant, besides having advantages over synthetic refrigerants, also has several disadvantages. A very prominent disadvantage is that it has high flammability compared to synthetic refrigerants [11], [12]. To reduce the flammability properties can be done by adding inert gases such as nitrogen (N2) and carbon dioxide (CO2) [2], [12]. Another way can be done by designing systems and equipment that can reduce the risk of burning due to the use of HC in the AC system [13].

The use of mixed refrigerants between propane and butane (LPG) has also been tested on hybrid car air conditioning systems using a half cycle system [14]–[17], even applied to expansion valves from PTFE material [18]. Meanwhile, in Indonesia, hydrocarbon (HC) refrigerants that have been commercialized are Pertamina products called Musicool 134 (M-134). Research using Musicool refrigerant has been carried out with variations in the mass flow of refrigerants entering the evaporator by adjusting the cross-sectional area of the expansion valve to improve the performance of the refrigeration system [19].

Meanwhile, the main problem in the refrigeration system is the irreversibility of the isenthalpic expansion process, thus reducing the performance of the AC system [20]. Therefore, this paper aims to reduce losses from the expansion process by installing double expansion valves in parallel so that it is expected to improve the performance of vapor compression refrigeration systems with Musicool refrigerants.

2. Method

This research was conducted to investigate the effect of using double expansion valves installed in parallel using musicool refrigerants that are compared to TEV performance. The study was conducted by measuring the temperature and refrigerant pressure on the variable cooling load of 0.23, 0.33, and 0.39 Kg / s taken from the amount of air mass that passes through the evaporator. The research apparatus is presented in Figure 1.

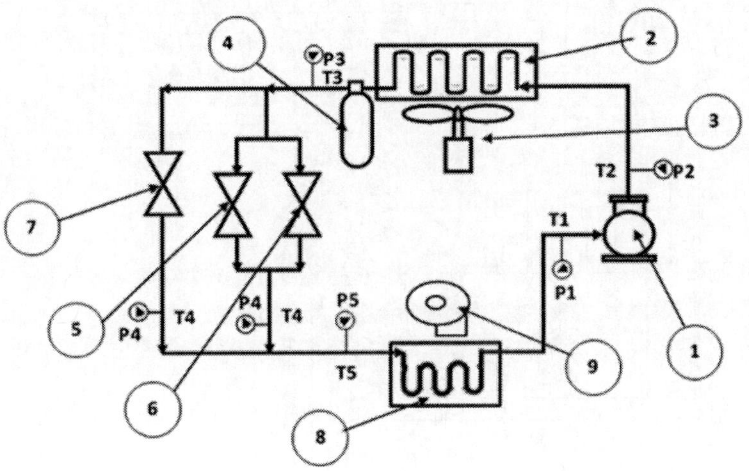

No	Part/ Component
1	Compressor
2	Condensor
3	Fan
4	Dryer
5	Expansion valve
6	Expansion valve
7	Expansion valve (TEV)
8	Evaporator
9	Blower

Figure 1. Research apparatus

3. Result and Discussion

3.1. Research Data

Table 2 and Table 3 show the results of data collection of temperature and pressure on variations in cooling loads (air flow rate across the evaporator) of 0.23, 0.33, and 0.39 kg/s.

Table 2. Temperature and pressure refrigerant with parallel expansion valves.

Cooling load	T1(°C)	T2(°C)	T3(°C)	T4(°C)	T5(°C)	P1(psi)	P2(psi)	P3(psi)	P4(psi)
0,23 kg/s	17	65	54	8	9	30	210	210	40
0,33 kg/s	24	69	54	7	17	30	210	205	40
0,39 kg/s	25	68	54	7	19	30	200	200	40

Table 3. Temperature and pressure refrigerant with TEV

Cooling load	T1(°C)	T2(°C)	T3(°C)	T4(°C)	T5(°C)	P1(psi)	P2(psi)	P3(psi)	P4(psi)
0,23 kg/s	14	55	55	5	3	30	200	200	40
0,33 kg/s	13	60	55	8	6	35	200	200	45
0,39 kg/s	12	60	58	9	7	35	200	200	50

Furthermore, Figure 2 shows plot data on the Musicool 134 P-h diagram for parallel expansion valves at variations cooling loads of 0.23 kg/s (left), 0.33 kg/s (right), and 0.39 kg/s (bottom).

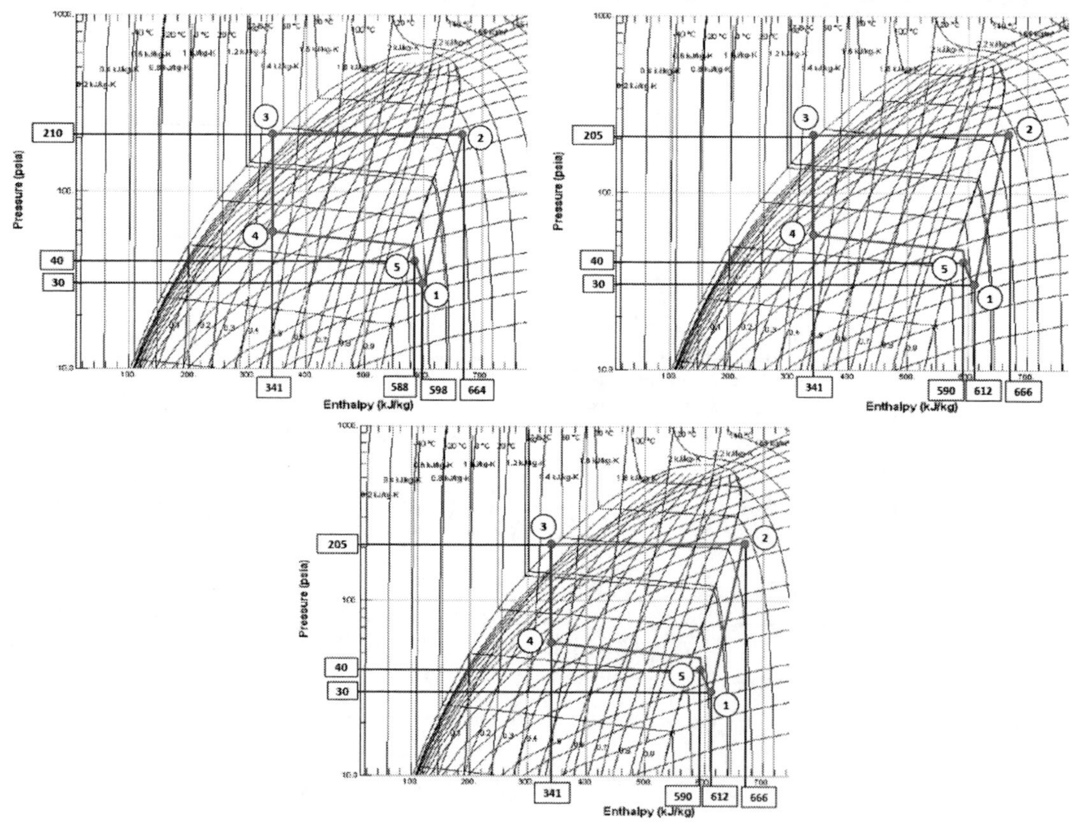

Figure 2. The plot of P-h diagram of Musicool 134 on the parallel expansion valve at variations of cooling load

Meanwhile, Figure 3 shows the plot data on the Musicool 134 P-h diagram for TEV at variations in cooling load 0.23 kg/s (left), 0.33 kg/s (right) and 0.39 kg/s (bottom).

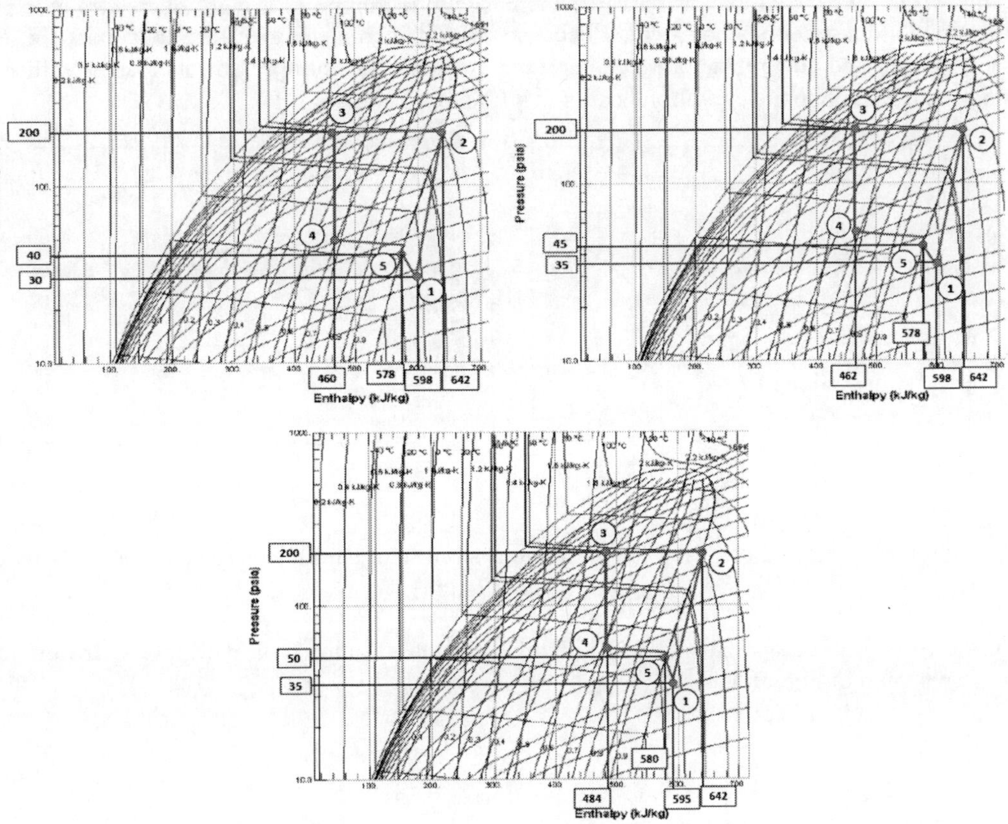

Figure 3. The plot of P-h diagram of Musicool 134 on TEV at variations of cooling load

3.2. Discussion

From the P-h diagram presented in Figure 2 and Figure 3, the specific enthalpy is presented in Table 4 and Table 5.

Table 4. Performance of the refrigeration system with parallel expansion valves

Cooling load	Compressor work (Wc)			Refrigeration effect (RE)			COP
	h_2	h_1	Δh	h_5	h_4	Δh	Wc/RE
0.23 kg/s	664	598	66	588	341	247	3.74
0.33 kg/s	666	612	54	590	341	249	4.61
0.39 kg/s	664	620	44	598	341	257	5.84

Table 5. Performance of an of refrigeration system with TEV

Cooling load	Compressor work (Wc)			Refrigeration effect (RE)			COP
	h_2	h_1	Δh	h_5	h_4	Δh	Wc/RE
0.23 kg/s	642	598	44	578	460	118	2.68
0.33 kg/s	642	598	44	578	462	116	2.64
0.39 kg/s	642	595	47	580	484	96	2.04

Figure 4 shows the performance of a vapor compression refrigeration system with Musicool 134 with a parallel expansion valve and TEV. From Figure 4, it can be seen that the refrigeration effects (RE) and COP by parallel expansion valve increase while compressor work decreases with increasing cooling load. This phenomenon occurs because the refrigerant temperature that exits the expansion valve is lower and the refrigerant temperature entering the compressor increases with increasing cooling load. The opposite condition occurs in TEV applications.

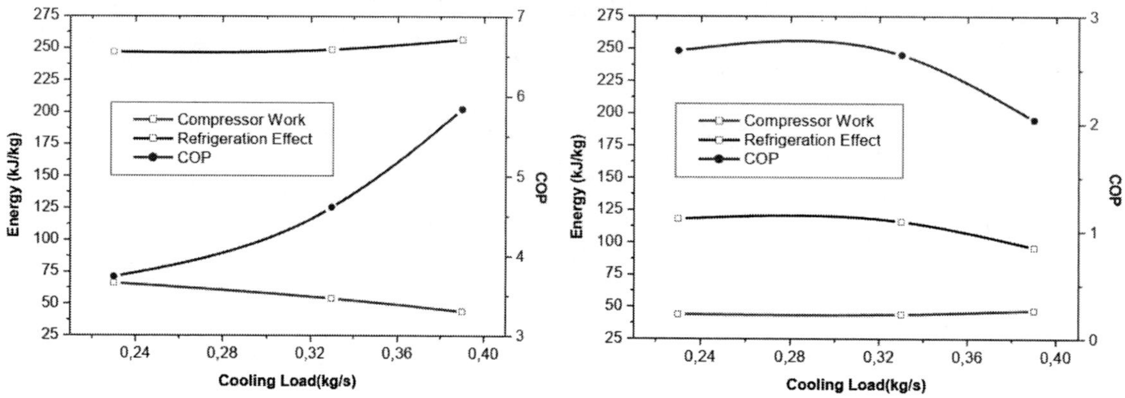

Figure 4. Performance of the refrigeration system with parallel expansion valves (left) and TEV (right)

Furthermore, the comparison of COP of refrigeration systems with parallel expansion valves and TEV is presented in Figure 5.

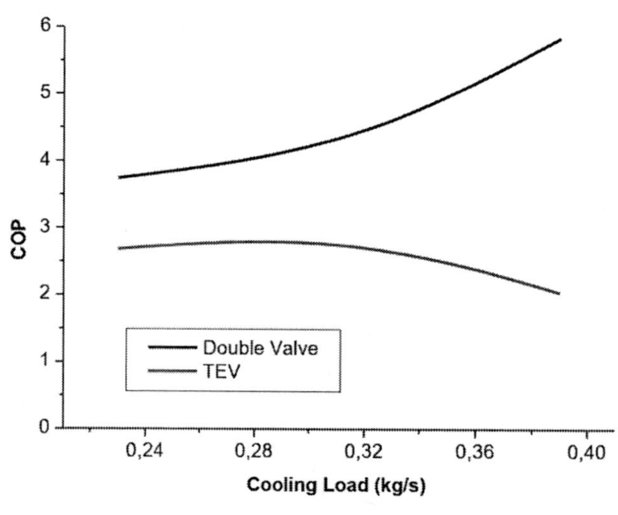

Figure 5. Comparison of COP

4. Conclusion

The results of this study indicate that double expansion valves installed parallel provide a positive effect on the performance of vapor compression refrigeration systems with musicool 134 refrigerants, especially for the effects of refrigeration and COP. The highest refrigeration effect produced was 257 kJ/kg and the COP value was 5.84, while the lowest compressor work was 44 kJ / kg for both systems, parallel expansion valves, and TEV. Looking at the results of this study, the use of double expansion valves is promising to conduct further research with various controlled variables.

References

[1] A. S. Dalkilic and S. Wongwises, "A performance comparison of vapour-compression refrigeration system using various alternative refrigerants," *International Communications in Heat and Mass Transfer*, vol. 37, no. 9, pp. 1340–1349, 2010.

[2] B. C. Purnomo and M. Setiyo, "Karakteristik sistem refrigerasi kompresi uap dengan refrigerant campuran musicool 134 - CO2," *Jurnal Teknologi*, vol. 9, no. 2, 2017.

[3] S. Daly, *Automotive air condiotining and climate control systems*, no. 1. 2006.

[4] X. H. Han, P. Li, Y. J. Xu, Y. J. Zhang, Q. Wang, and G. M. Chen, "Cycle performances of the mixture HFC-161 + HFC-134a as the substitution of HFC-134a in automotive air conditioning systems," *International Journal of Refrigeration*, vol. 36, no. 3, pp. 913–920, 2013.

[5] J. M. Calm, "The next generation of refrigerants – Historical review , considerations , and outlook 5 ′ ne ′ ration de frigorige ` nes – historique , analyse La prochaine ge et perspectives," vol. 31, pp. 1123–1133, 2008.

[6] K. Wang, M. Eisele, Y. Hwang, and R. Radermacher, "Review of secondary loop refrigeration systems," *International Journal of Refrigeration*, vol. 33, no. 2, pp. 212–234, 2010.

[7] Z. Yang and X. Wu, "Retrofits and options for the alternatives to HCFC-22," *Energy*, vol. 59, no. 2013, pp. 1–21, 2013.

[8] T. P. Teng, H. E. Mo, H. Lin, Y. H. Tseng, R. H. Liu, and Y. F. Long, "Retrofit assessment of window air conditioner," *Applied Thermal Engineering*, vol. 32, no. 1, pp. 100–107, 2012.

[9] S. Wongwises and N. Chimres, "Experimental study of hydrocarbon mixtures to replace HFC-134a in a domestic refrigerator," *Energy Conversion and Management*, vol. 46, pp. 85–100, 2005.

[10] M. K. Agrawal and A. G. Matani, "Evaluation of Vapour Compression Refrigeration System Using Different Refrigerants," *International Journal of Engineering and Innovative Technology*, vol. 2, no. 9, pp. 86–92, 2013.

[11] W. Zhang, Z. Yang, J. Li, C. X. Ren, D. Lv, J. Wang, X. Zhang, and W. Wu, "Research on the flammability hazards of an air conditioner using refrigerant R-290," *International Journal of Refrigeration*, vol. 36, no. 5, pp. 1483–1494, 2013.

[12] S. Y. Liao, Q. Cheng, D. M. Jiang, and J. Gao, "Experimental study of flammability limits of natural gas-air mixture," *Journal of Hazardous Materials*, vol. 119, no. 1–3, pp. 81–84, 2005.

[13] D. Colbourne and K. O. Suen, "Appraising the flammability hazards of hydrocarbon refrigerants using quantitative risk assessment model Part I: Modelling approach," *International Journal of Refrigeration*, vol. 27, no. 7 SPEC. ISS., pp. 774–783, 2004.

[14] M. Setiyo, S. Soeparman, N. Hamidi, and S. Wahyudi, "Cooling effect characteristics of a ½ cycle refrigeration system on an LPG fuel system," *International Journal of Refrigeration*, vol. 82, pp. 227–237, 2017.

[15] M. Setiyo, S. Soeparman, S. Wahyudi, and N. Hamidi, "A simulation for predicting potential cooling effect on LPG-fuelled vehicles," *AIP Conf. Proc. J. Chem. Phys. Rev. Sci. Instrum. J. Acoust. Soc. Am. J. Acoust. Soc. Am*, vol. 030002, no. 57, pp. 4704231–87, 2016.

[16] M. Setiyo, S. Soeparman, N. Hamidi, S. Wahyudi, and M. Hanafi, "Numerical study on cooling effect potential from vaporizer device of LPG vehicle," *Journal of Engineering Science and Technology*, vol. 12, no. 7, 2017.

[17] M. Setiyo, D. Syaka, B. Waluyo, N. Hamidi, B. Kiono, and J. Semarang, "Cooling effect potential from liquefied petroleum gas flow in the fuel line of vehicle," *International Journal of Automotive and Mechanical EngineeringOnline*, vol. 14, no. 4, pp. 2229–8649, 2017.

[18] M. Setiyo, Saifudin, B. C. Purnomo, B. Waluyo, and A. I. Ramadhan, "Temperature Distribution of R-134a Through Aluminum and PTFE Expansion Valve on Automotive Air Conditioning Applications," *ARPN Journal of Engineering and Applied Sciences*, vol. 12, no. 4, pp. 1046–1051, 2017.

[19] B. C. Purnomo and B. Waluyo, "Optimalisasi penggunaan refrigeran musicool untuk meningkatkan performa sistem refrigerasi kompresi uap dengan variabel katup ekspansi," *Semnastek Universitas Muhammadiyah Jakarta*, no. November, pp. 1–7, 2015.
[20] I. Journal, O. F. Heatmass, S. Wongwises, and K. Mongkut, "Performance of the two-phase ejector expansion refrigeration cycle refrigeration cycle," no. SEPTEMBER 2005, 2016.

Acknowledgment
The authors would like to thank the laboratory staff and students who have helped to collect data during this study.

Toward sustainable electroplating of decorative silver for SMEs in Indonesia

T Widayatno* and Hamid

Department of Chemical Engineering, Faculty of Engineering Universitas Muhammadiyah Surakarta, Surakarta, Indonesia

* Email: tri.widayatno@ums.ac.id

Abstract. Decorative silver electroplating has been a long time conducted by small and medium enterprises (SMEs) in Indonesia. Environmental pollution associated with the electroplating SMEs activities has created serious problems. For example SMEs of silver electroplating has contaminated groundwater which is no longer fit for domestic consumption due to its high level of cyanide contamination. Inefficient process and the use of high concentration of cyanide based electrolyte have been claimed to be the root cause. Therefore, the process of decorative silver electrodeposition in SMEs would be improved by developing an electrolyte with low cyanide concentration, determining onset potential and current densities, and modifying the process and the electrochemical cell so that the process would be more efficient and sustainable. A new electrolyte composed of $AgNO_3$ and KCN with a ratio of 5:39 were chosen for the experimentation. For this electrolyte, silver would possibly be electrodeposited at potentials ranging from -2.2 V to -2.8 V. Corresponding current densities for silver electrodeposition would be between -2 and 7 mA/cm^2. To improve the process, a new design of electrochemical reactor has been developed which ensure improvement of the observed electrochemical parameters i.e. the uniform current density distribution, the larger area of anode, and the stability of electrodes distance and position.

1. Introduction

The metallic coating industry has rapidly increased in Indonesia in recent decades. In general, the metallic coating application (i.e. silver, nickel, chrome, etc) is for decorative and functional purposes (e.g. corrosion resistance) [1]. The coating increases the quality of the material surface so that its appearance, wear and corrosion resistance are improved. This is crucial as metals commonly used for imitation jewelry, industrial equipment, and household appliances are easily oxidized and corroded due to exposure to the environment [1, 2].

One of the popular electrochemical coating industries in Indonesia is a silver coating. The silver layer plating has attracted great attention for decades particularly for handicrafts, accessories, and jewelry. This is the most popular decorative application of silver coating in some regions of Indonesia i.e. Kotagede Yogyakarta. Decorative silver electroplating has been long time conducted by small and medium enterprises (SMEs). SMEs have an important role in reducing poverty by creating job opportunities, however, increasing production of SMEs in manufacturing electrodeposited silver has resulted in significant environmental costs [3].

Although local newspapers regularly report these problems, no systematic monitoring has been conducted for tackling the environmental problems and improving health and safety of SME activities

[4]. The Ministry of Industry and Trade (MOIT) reported that pollution associated with SME activities, particularly in the electroplating sector, has created serious environmental problems. For instance, In Kotagede, Yogyakarta, where SME activities involving electroplating of silver are concentrated, groundwater is no longer fit for domestic consumption due to its high level of chemical contamination i.e. cyanide [5]. Workers in SMEs and the population near the SMEs' areas of activities are beginning to suffer from serious skin and respiratory problems caused by the pollution from electroplating industries. Hence it is crucial to find out the root cause and to tackle the issues. A traditional and conventional technology used by SMEs has been reported to contribute in process inefficiency and cause the environmental problem [2,6].

A technology for decorative silver plating used by SMEs is still relying on traditional and conventional method and equipment [c.f. Figure 1) which has some issues. A number of drawbacks have been identified such as 1) Inefficient process, 2) Low-quality product, 3) copious waste generation, 4) Frequent replace of the electrolyte, 5) High processing cost [2,6]. Figure 1 shows a simple electrochemical cell used in the process of silver plating in SMEs. The equipment setting implies to an unstable electrode distance and position, a smaller area of anode than cathode, and non-optimal cell geometry. These have been identified to be the causes of some issues that have been previously discussed [7].

Figure 1. Simple electrochemical cells and equipments for traditional silver electroplating method used by SMEs in Kotagede Yogyakarta [7]

Direct observation to the SMEs practices in the process of silver electroplating has also been carried out [7]. The observation revealed that electrochemical parameters and variables, as well as the operating condition, have not been optimally set. Table 1 shows the electrochemical parameters and operating conditions observed in the silver electroplating carried out by SMEs compared to those in literature.

A number of researches have been devoted to investigate the process of silver electrodeposition [8,9]. An electrolyte formulation, electrochemical parameters, operating condition and cell geometry are crucial to achieve more efficient and more environmentally friendly process [10]. There are several alternative options taken into account in this study: 1. the environmentally friendly process can be achieved by using low concentration of cyanide or by removing use of cyanides electrolyte. 2. The efficient process can be achieved by applying optimum electrochemical parameters and operating condition, 3. Electrochemical cell geometry has to be modified to ensure uniform current distribution.

In this paper, the process of silver electroplating improved by developing an electrolyte with low cyanide concentration, determining onset potential and current densities from the electrolyte, and modifying the electrochemical cell will be presented. Electroplating Process efficiency could be improved by thorough and comprehensive analysis and evaluation of SME practice based on the best practice of electrochemical method. The improvement of decorative electroplating process by SMEs have to take into account the principles and fundamental aspects of electrochemistry.

IOP Conf. Series: Materials Science and Engineering **674** (2019) 012007 doi:10.1088/1757-899X/674/1/012007

Table 1. Observed Parameters of Electroplating Process of decorative silver by SMEs [7] Compared to the literature [8]

No	Parameter	Observed SMEs Practices	Literature
1	Applied Current/current densities	Not monitored	0.5 – 1.5 A or 1.5 – 3 A/dm^2
2	Applied Potentials	30 V	4 – 6 V
3	Cell Geometry	Simple without considering current distribution	Well-designed consider uniform current distribution
4	Electrode area	Commonly cathode area> anode area	Anode area must be > cathode area
5	Electrodes distance	Always change, moved by operator	Relatively stable and steady
6	Electrodes position	Always change, moved by operator	Relatively stable and steady
7	Types of Electrodes used	Anode: stainless steel, cathode: work piece (Cu coated by thin nickel)	Anode: pure metallic silver, cathode: work piece (Cu coated by thin nickel)
8	Electrolyte composition and concentration	Initially according the standard, no systematic monitoring afterward	According to standard, and constantly monitored
9	Electrolyte make up/replacement	According to the operator feeling	Carefully estimated
10	Method	Potentiostatic	Galvanostatic

2. Theory

Electroplating or electrodeposition is a process of metal deposition through a nonspontaneous reduction reaction of dissolved metallic ions to form deposited metal on an electrode surface (cathode) in an electrolytic solution. It typically requires a set of electrochemical system composed of an external power supply, an electrolyte, cathode, anode, and reference electrode (optional). The equipment is set to be a system as shown in Figure 2. In order to achieve efficient process, the electroplating has to be carried out at the most appropriate electrochemical parameters and geometry of the cell [10]. The reduction reaction will take place as overpotential applied to the system so that the condition deviates from its equilibrium. The deviation leads to electrical current flow between the two electrodes and the reduction reaction begins to occur [11].

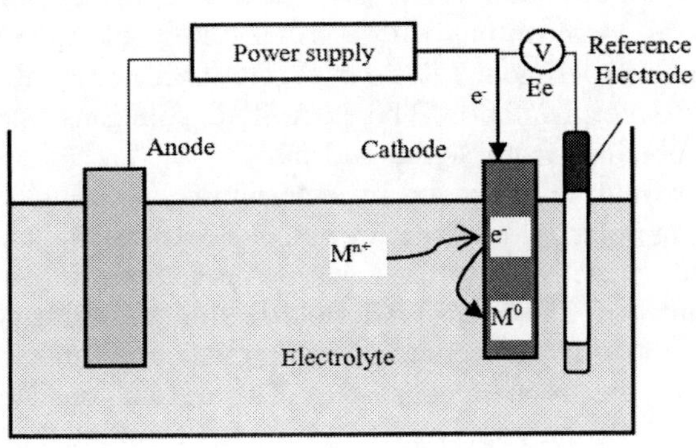

Figure 2. Scheme of typical electrochemical system for metal electroplating use [7]

Kinetics of the electrochemical reaction has been formulated by Nernst equation. According to Nernst's equation, the electrode surface reaction is determined by the value of overpotential. When negative overpotential (E<Ee) imposed to the working electrode, a reduction reaction occurs to achieve a new equilibrium condition. The cathodic current flows to supply the electron (ne^-) for the

reaction which is in the polarization curves represented by current densities. Based on Butler-Volmer equation, the current density depends on a number of parameters i.e. the temperature, the overpotential, the exchange current density, and the charge transfer coefficient. Therefore, the current density increases at higher overpotentials which represent the increase of the electrochemical reaction rate [11].

Decorative silver electrodeposition conducted by SMEs in Indonesia is commonly used to produce bright silver coating for imitation jewelry. In general, a number of silver electrolytic baths with various concentrations and compositions to obtain bright silver are used. Typical silver electrolytes used for decorative purposes are shown in Table 2.

Tabel 2. Typical silver electrolytes for decorative purposes [8]

No	Components	Concentration (g/l) /parameters
1	AgCN	31 – 55
2	KCN (free)	35 – 50
3	KCN (total)	50 – 80
4	Ag (metallic)	20 – 45
5	K_2CO3	15 – 90
6	Anode	Pure silver
7	Potentials (V)	4 – 6
8	Current densities (A/dm^2)	0,5 – 1,5/1,5 - 3
9	Temperature (°C)	20 - 28

It can be seen in Tabel 2 that typical silver electrolytes contain a high concentration of cyanide (50 – 80 g/l). Due to the present of AgCN, and KCN, a complex of $KAg(CN)_2$ is formed in the solution. Such cyanide baths have been used for more than 100 years application, as it is the most stable electrolyte to consistently produce high-quality silver deposit [9]. The process in cyanide has current efficiencies that are close to 100%. Thus, use of high purity silver anodes ($\geq$ 99.98%) is preferred [8]. However, the cyanide bath possesses extremely high risks due to its toxicity [9]. So it is vital to reduce the use of cyanide.

3. Experimental

In this study, more environmentally friendly process of silver electroplating would be achieved by using low concentration of cyanide. Therefore, a new silver electrolyte with low cyanide concentration would be formulated. Several compositions with a specific ratio of $AgNO_3$ and KCN were analyzed using steady-state polarization experiments. Ratio of $AgNO_3$: KCN that offers higher current densities would be chosen for further experimentation. The electrolytic solutions were prepared using analytical grade of AgNO3 and KCN obtained from Sigma-Aldrich.

The process efficiency would be improved by determining optimum electrochemical parameters and operating condition using polarization experiments. The electrochemical polarisation experiments for the chosen electrolyte carried out at various temperatures in a vertical cylindrical glass electrochemical cell (volume of 250 ml). All polarization experiments were conducted in the laboratory of chemical engineering Universitas Muhammadiyah Surakarta. The experiments were carried out in a two electrodes electrochemical system. The system consisted of a silver rod anode and a cathode of thin deposited nickel on copper plate. A range of potentials between 0 V and 3 V were supplied by a power supply of Sanfix SP3050 applied to the system with a potential interval of 0.1 V. Corresponding currents were recorded when steady state condition has been reached (for around 1 minute). The cell potentials and the corresponding current densities were then plotted into a graph to produce polarization curves. Using these curves, applied potentials and current densities at which the silver electrodeposition would be optimally achieved could be determined.

IOP Conf. Series: Materials Science and Engineering **674** (2019) 012007 doi:10.1088/1757-899X/674/1/012007

Since uniform current distribution is crucial to obtain a high coating quality, a new design of electrochemical cell for the process of silver deposition was developed. The new design considered electrode area (anode), distance and position of electrodes, as well as geometry of the cell that allows uniform current distribution.

4. Results and Discussion

A new formulation of silver electrolyte with low cyanide concentration has been developed. The electrolyte composed of $AgNO_3$ and KCN with a ratio of 5:39 were chosen for the experimentation. This has been discussed in the previous publication [12]. Polarization experiments for the chosen electrolyte have been carried out in the previously described electrochemical system. The polarization curves obtained from the experiments at different temperatures are shown in Figure 3.

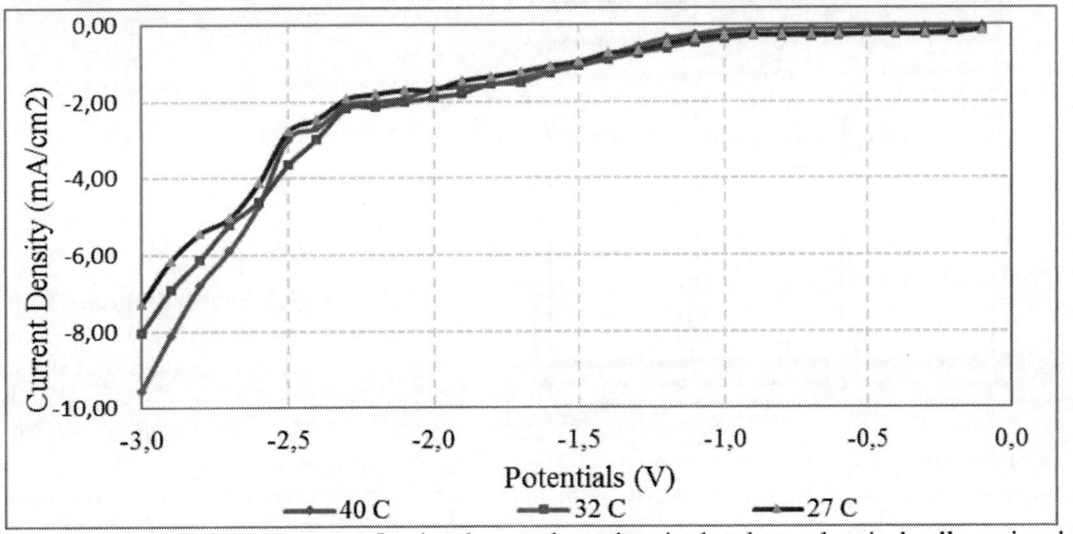

Figure 3. The polarization curves for the chosen electrolyte in the electrochemical cell previously described at various different temperatures

It can be seen in Figure 3 that cathodic current begins to flow at potential of around -1.0 V. This is an indication that there has been electrochemical reaction taking place. However at potential between -1.0 and -2.2 V, no silver deposit on the cathode surface was observed. This indicates that the current flow at potential between -1.0 and -2.2 V was consumed by competing reaction (s). The polarization curves show that silver would possibly be electrodeposited at potentials ranging from -2.2 V to -2.8 V indicated by a sharp increase of the current densities. Corresponding current densities for silver electrodeposition would be between -2 mA/cm^2 and 7 mA/cm^2. At cell potential greater than -2.8 V, gas evolution was clearly noticeable. Figure 1 also shows that temperatures do not significantly affect the current densities at lower applied cell potentials, however at higher potential cells the effect of temperature becomes more obvious [13].

To improve the electroplating process, a new design of electrochemical reactor has been developed. It consists of three compartments which allow parallel silver electroplating process (Figure 4). This design would improve the observed electrochemical parameters number 3, 4, 5, and 6 (*c.f.* Table 1) as it ensures the uniform current density distribution, the larger area of anode, and the stability of electrodes distance and position.

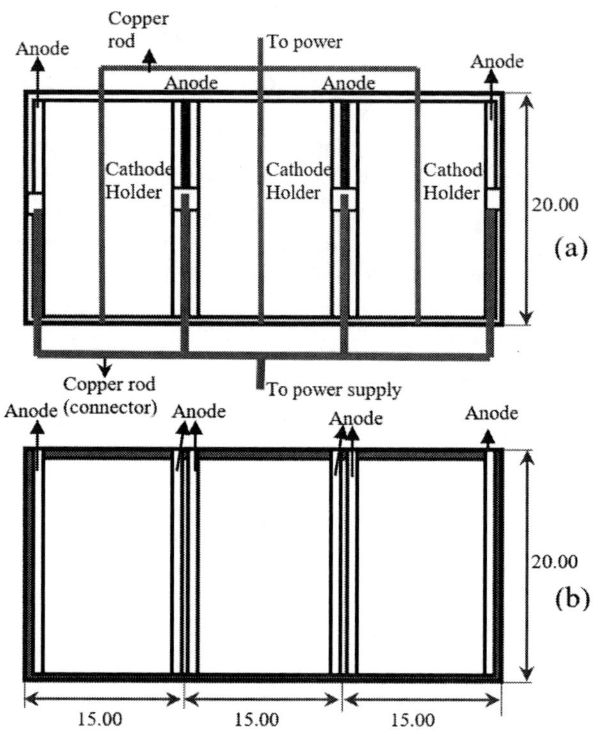

Figure 4. Schematic diagram of the new electrochemical cell (a) top view and (b) side view (cross sectional view)

5. Conclusion

A new formulation of silver electrolyte with low cyanide concentration has been developed. The electrolyte composed of $AgNO_3$ and KCN with a ratio of 5:39 were chosen for achieving more sustainable decorative silver electroplating. A range of applied potentials between -2.2 and -2.8 V was determined for silver electrodeposition using the chosen electrolyte. Applied current densities corresponding to the applied potential are between -2 mA/cm^2 and 7 mA/cm^2. A new geometry of an electrochemical cell for the silver electrodeposition has been designed which promises the uniform current density distribution, the larger area of an anode, and the stability of electrodes distance and position as well as parallel use of the three compartments.

References

[1] Mawadah, A. 2008. Thesis, Program Studi Fisika, Fakultas Sains dan Teknologi, Universitas Islam Negeri Syarif Hidayatullah, Jakarta. Online access on http://repository.uinjkt.ac.id/dspace/bitstream/123456789/15695/2/ADE%20MAWADAH-FST.pdf

[2] Istiyono, E., R Yosi Aprian Sari, and Banu Setyo Adi, 2008, Inoteks, Volume 12 No. 2. online access (27/04/2014): http://staff.uny.ac.id/sites/default/files/132048515/6_Inoteks_Elektroplating.pdf

[3] Ivone De Carlo, 2012. Dinamika Kerajinan dan Batik, Vol. 32, No. 2, Desember 2012

[4] Maksum, F, 2013, Manado Tribun News. accessed online 31 Maret 2018 on http://manado.tribunnews.com/2013/08/27/bahaya-sianida-bagi-manusia-dan-lingkungan

[5] BLH Propinsi DI Yogyakarta. 2014. Laporan tahunan 2014. Accessed online 31 Maret 2018 on https://blh.jogjaprov.go.id/po-content/uploads/Laporan-Tahunan-2014.pdf

[6] Lies Susilaning Sri, Hastuti .2012. *Seminar Nasional Teknik Kimia Soebardjo Brotohardjono IX,* 21 Juni 2012, Surabaya. accessed online (27/04/2014) on

http://eprints.upnjatim.ac.id/4147/1/C3.pdf
[7] Widayatno, T. and Hamid. 2016. Proceeding of *Simposium Nasional RAPI XV – 2016 FT UMS,* Surakarta, Indonesia.
[8] Schlesinger, M. and Paunovic, M., (2010), Modern Electroplating, 5th Edition, p. 131-138, Electrochemical Society Series, John Wiley and Sons, Inc. New York
[9] Liu, A., Ren, X., An, M., Zhang, J., Peixia Yang, Bo Wang, Yongming Zhu & Chong Wang. 2014. SCIENTIFIC REPORTS 4 : 3837 DOI: 10.1038/srep03837
[10] Widayatno, T. 2016. ARPN Journal of Engineering and Applied Sciences 11 (8), 5183 - 5189
[11] Bard, A. J. and Faulkner, L. R., (2001) Electrochemical Methods: Fundamentals and Applications, 2nd edition, 22 – 43, John Wiley and Sons, Inc.
[12] Widayatno, T., Hamid, Ibnu Ari Swasemba, dan Miftah Khairi Zufron, 2015. Simposium Nasional RAPI XIV - 2015 FT UMS
[13] Stern,M. And A.L.Geary. 1957. Electrochemical Polarization, J. Electrochem. Soc. 1957 volume 104, issue 1, 56-63. doi: 10.1149/1.2428496

Acknowledgments
Research funding from Ministry of Research technology and Higher Education through the scheme of Hibah Bersaing in the year of 2015-2016 is acknowledged by authors.

Electrochemical characterization of gold mining wastewater for recovery of mercury using electrochemical method

A I Rahmawati * and T Widayatno

Chemical Engineering Department, Universitas Muhammadiyah Surakarta
Jl. A. Yani Tromol Pos 1 Telp. 0271-717417 Pabelan Surakarta 57162, INDONESIA

* E-mail: d500140092@student.ums.ac.id

Abstract. An experimental study to determine optimum operating conditions for mercury recovery from gold mining wastewater has been conducted. Electrochemical characterization of gold mining wastewater was carried out by using steady-state polarization experiment in a batch system. The polarization curves obtained were then utilized to determine the optimum operating conditions. The polarization experiments were conducted by an electrochemical cell with two electrodes of copper (Cu), carbon (C), or stainless steel (SS) and an electrolyte solution of the wastewater. Variations of electrodes used were pairs of copper (Cu), carbon (C) and stainless steel (SS) which were arranged randomly. To obtain the polarization curves, the electrical potentials between 0 V and 3 V were applied and its corresponding currents were recorded at an interval potential of 0.1 V. The highest current density was achieved on electrode pairs of stainless steel-carbon (SS-C) and carbon-carbon (C-C). This is an early indication that the most optimum electrode pairs for mercury recovery possibly are SS-C and C-C. EDX analysis confirmed the presence of mercury in the deposited materials, this leads to a conclusion that mercury can be deposited on to carbon electrode from the wastewater using electrochemical method which then eventually it can be recovered.

1. Introduction

A large number of conventional gold mining activities in Indonesia cause severe environmental problems due to mercury (Hg) pollution [1]. One of the International media, Worldpolluted.org reported that mercury contamination occurred due to the gold mining waste in Kalimantan was ranked seventh in the 'Top Ten' Threats in 2013 or the 10thmost threatening thing in 2013. Estimated by The United Nations Industrial Development Organization (UNIDO) more than 1000 tonnes of mercury pollute the environment every year and 30% of it is anthropogenic mercury. In 2008, the Journal of Water and Environment Technology found a mercury concentration of 2,260 μg / L in the Kahayu river of Central Kalimantan [2].

Mercury is a heavy metal that has toxic effects on living things. Due to its high toxicity and ability to accumulate into the human body, continuous exposure to mercury may cause adverse effects to human health [3]. The element has been declared to be one of the most toxic pollutants [4]. The first identified case of mercury poisoning occurred in the Gulf of Manimata, Japan from 1930 to 1960. The case revealed that bioaccumulation of toxic materials (methyl mercury) naturally occurred in fishes could kill about a hundred people and caused genetic damage in large populations. The Minamata incident was followed by a more tragic report, Hg poisoning in Iraq in 1972 in which 450 villagers died after eating grain using mercury-containing pesticides [4,5].

Therefore, it is crucial to look for a technology that is capable of overcoming the mercury contamination. The technology has to be able to reduce the mercury in the wastewater without causing another environmental problem which so-called Green technology. The technology becomes a choice to gear up sustainable development. An electrochemical method was selected to be used in this study as it is green and offers a number of advantages [4]. Electroplating has attracted much attention because of its simple process, cheap equipment, ease of operation, shorter retention times, high

IOP Conf. Series: Materials Science and Engineering **674** (2019) 012008 doi:10.1088/1757-899X/674/1/012008

deposition speed and reduction of generated sludge [6]. There have been a number of studies dedicated to apply the electrochemical method, however the types of electrodes used are still in limited quantities, such as aluminum [7], magnetite [4] and graphite-titanium [8]. In this research various electrodes would be used which were graphite-stainless steel-copper. The electrodes were arranged by copper (Cu), carbon (C) and stainless steel (SS) which composed alternately as anode and cathode. In addition, operating conditions (current density and voltage) would also be determined systematically based on electrochemical fundamental theory through steady-state polarization experiments. So that an electrode pair that has the best efficiency and the most optimum operating conditions could be determined.

2. Experimental

All experiments were carried out at room temperature in the laboratory of chemical engineering, faculty of engineering, Universitas Muhammadiyah Surakarta. The gold mining wastewater used in this research was obtained from small and medium conventional gold mining industries from Wonogiri, Central Java. Initially, the waste was analyzed to determine its mercury content using Mercury analyzer. The analysis was carried out 3 times repetition to ensure the results were accurate.

The electrochemical cell used in the experimentation was made of vertical cylindrical glass with a volume of 250 ml. Stainless steel (SS), carbon (C) and copper (Cu) used for the electrodes were obtained from metal stores (Moro Dadi) in Surakarta with a minimum purity of 95%. At first, the metal was cut so that a surface area of 11.76 $cm^2$was obtained. The electrodes were then mechanically and chemically cleaned. Mechanical polishing was carried out by using sandpaper to obtain smooth surface. The chemical cleansing was conducted by dipping the electrodes into a solution of nitric acid (10%).Thereafter, the electrodes were washed by deionized water before installing it into the electrochemical system. The electrodes were attached to copper wire which was then connected to the power supply (Sanfix SP3050).

To achieve the aims and objectives of the research, first, polarization experiments were carried out in the previously described electrochemical cell. To see the effect of the electrodes used, the electroplating system was set with variations of electrodes pairs (anode/cathode) which were C-C, C-SS, C-Cu, SS-C, SS-SS, SS-Cu, Cu-C, Cu-SS, and Cu-C. The electrolyte solution used was 200 ml in a 250 ml beaker. The electrodes were connected to the power supply by using copper wire. An electrical potential was applied to the system started from 0.1 to 3 V with a potential interval of 0.1. Corresponding current responses were recorded when the system has reached steady-state conditions (usually at around 1 minute). The current responses were measured by using a digital multimeter of Fluke 17B+. The experiments were repeated several times so that three identical polarization curves were obtained for each applied potential to ensure its reproducibility. The polarization curves (potentials vs. current densities) obtained from the experiments were then utilized to determine the optimum operating conditions.

After each experiment, the mercury content in the waste quantitatively analyzed by using a mercury analyzer (LabAnalyzer/254). To see whether the mercury was deposited on to the electrode or not, the cathode was analyzed by using scanning electron microscopy (SEM) JSM-6510LA and EDX method. The EDX analysis was used to determine the composition of the deposited material.

3. Results and Discussion

The measurement of mercury content in the waste before treatment has been carried out. The initial concentration of mercury in the waste before the treatment was 0,052 ppm. It shows that the treatment of the waste was required as the mercury content exceeds the concentration that is allowed to be disposed to the environment, i.e. 0,000048 ppm under the threshold that's it 0,001 ppm. The next stage was to electrochemical characterization of the waste by using steady-state polarization experiments.

Polarization data obtained from the experiments were plotted to create polarization curves. The cathodic polarization curves for the wastewater containing mercury at various electrodes pairs are shown in Figure1. It can be seen in Figure 1 that at low potentials (0.1 – 0.4 V), small electrical

IOP Conf. Series: Materials Science and Engineering **674** (2019) 012008 doi:10.1088/1757-899X/674/1/012008

current flow was observed. This indicates that the reduction reaction of Hg has not occurred yet. The cathodic currents increase gradually with the increase of applied potentials from 0.4 V to 1.8 V. At potentials greater than 1.8 V, significant increases of the cathodic current densities are observed. However, at a potential of 2.2 -2.4 V gas (H_2) evolution was visible. It is generally seen that the current at the cathode flows since the low cell potential, however, at the potentials of the cell between 0-2 V do not show any mercury plating on the working electrode. The mercury reduction reaction most likely to occur at potentials between 1.8 V and 2.4 V which is indicated by a significant increase in cathodic current densities [4,9,10]. The corresponding current densities for mercury electroplating range from -0.075 to -0.175 mA/cm^2.

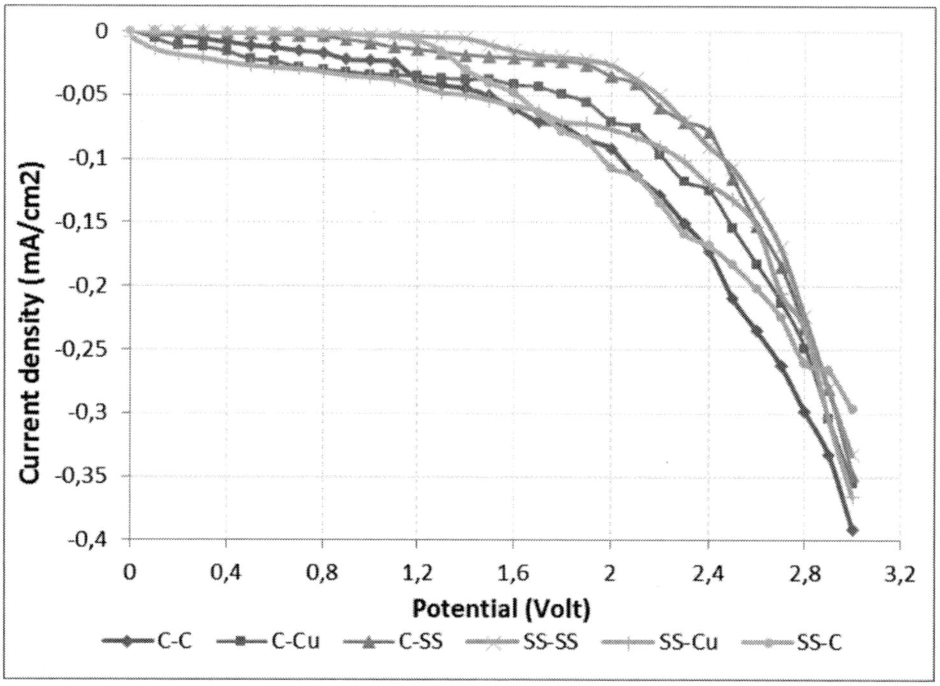

Figure 1. Steady-state polarization curves showing the relationship between the cell potentials with the current densities

The effect of electrodes pairs was also observed. Figure 1 also shows that the electrode pairs of SS-C and C-C are identified to be the most appropriate electrode pair characterized by an increase of the current flows significantly at a relatively low potential. The cathodic current flows indicate that the reaction of Hg electroplating began to take place and the more negative of the cathodic current results in higher rate of reduction reaction [11, 12, 13]. The characterization experiments showed that higher cathodic currents were obtained when the electrodes pairs of SS-C and C-C were used. This is an early indication that the SS-C and C-C pairs of electrodes are the most appropriate for use in the electrodeposition of mercury from the gold mining wastewater [9,10].

To see whether the mercury was deposited on to the electrode or not, the surface of the cathode was analyzed by using scanning electron microscopy (SEM) and EDX method. Figures 2 and 3 show the micrograph of SEM and EDX analysis results respectively. The SEM micrograph in Figure 2 indicates that due to the electroplating process, material from the wastewater was deposited on to the surface of the cathode. EDX analysis in Figure 3 confirmed the presence of mercury in the deposited materials [14]. This leads to the conclusion that mercury could be deposited on to carbon electrode from the wastewater using electrochemical method which then eventually can be recovered.

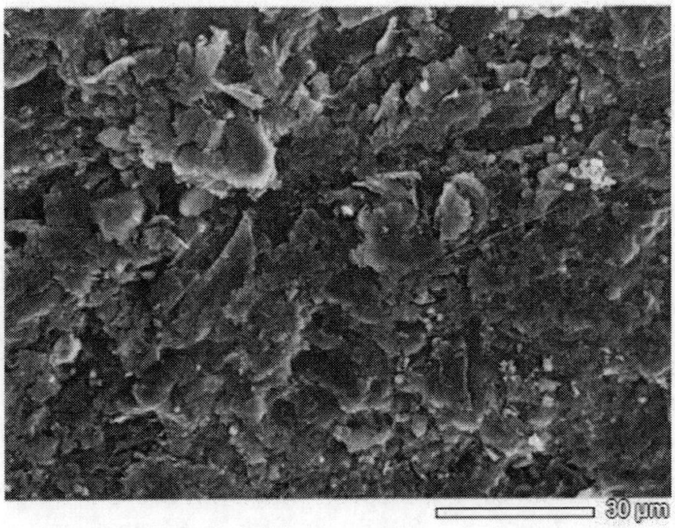

Figure 2.SEM Result of deposited material on the Carbon electrode

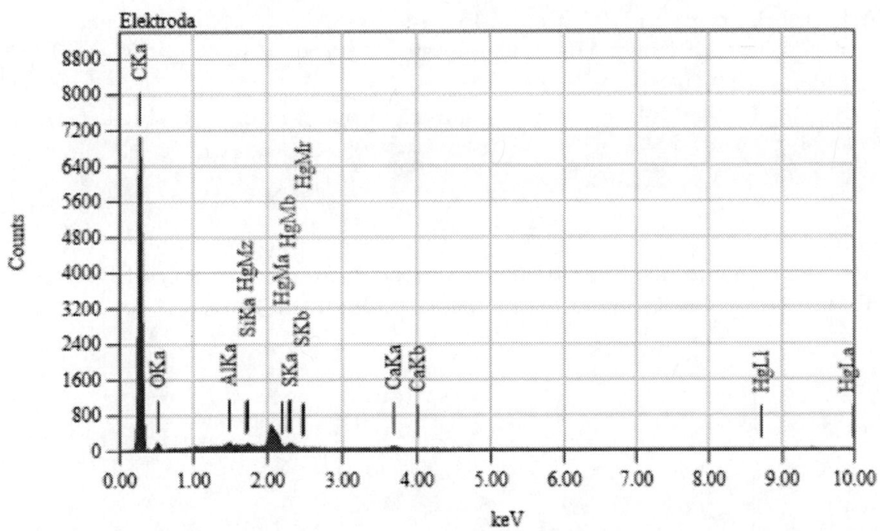

Figure 3. EDX Result of deposited material on the carbon electrode

4. Conclusion

Hg or mercury recovery from gold mining waste could be achieved by using the electroplating method. Stainless steel-Carbon and Carbon-Carbon couples electrode become the most potential cathode-anode pair to be used. The optimum electroplating potential for Hg ranges between 1.8 V-2.4 Vindicated by a significant cathodic current increase.The corresponding current densities for mercury electroplating range from -0.075 to -0.175 mA/cm^2. The experiment also showed that mercury could be deposited on to carbon electrode from the wastewater using electrochemical method which then eventually can be recovered.

References

[1] Blacksmith Institute. 2013. Top Ten Threatens. *WorstPolluted.org* - Projects Reports. Available online http://www.worstpolluted.org/reports/file/2007%20Report%20updated%202009.pdf.

[2] Cochrane, J. 2014. Small Scale Gold Mining pollutes Indonesian Land. The New York Times. 2

Jan 2014, available online on https://www.nytimes.com access 1 June 2016.

[3] Pressly, L. 2013. How mercury poisons gold miner and enter food Chain, BBC. 18 Sept 2013. Available online on http://www.bbc.com/news/magazine-24127661 accessed on 1 June 2016

[4] Hu, Z., Kurien, U., Murwira, K., Ghoshdastidar, A., Nepotchatykh, O., & Ariya, P. A. 2016.*ACS Sustainability Chemistry and Engineering, 0.* https://doi.org/10.1021/acssuschemeng.5b01612

[5] Rani, B., Singh, U., Chauhan, A. K., & Maheshwari, R. 2012. *Review Article Hazards Of Mercury Poisoning & Prevention Strategies*, *3*(1), 4–6.

[6] Robles, I., Lozano, M. J., Solís, S., Hernández, G., Paz, M. V, Olvera, M. G., & Bustos, E. 2015. *Electrochimica Acta*, 2–6.

[7] Kuokkanen, V., Kuokkanen, T., Rämö, J., & Lassi, U. 2013*A Review*, *2013*(May), 89–121.

[8] Ozgur, C., Coskun, S., Akcil, A., Beyhan, M., Serkan, I., & Civelekoglu, G. (2016). *Waste Management*, 1–5.

[9] Stern,M. And A.L.Geary 1957 *J.OF THE ELECTROCHEMICAL.* Electrochemical Polarization, 56–63.

[10] Lu, H., Wang, Y., & Wang, J. 2014. *Journal of Cleaner Production.* 1:1-10

[11] Widayatno, T. and Roy, S. 2014. J Appl Electrochem (2014) 44:807–820. https://doi.org/10.1007/s10800-014-0686-y

[12] Widayatno, T. and Roy, S. 2015. International Journal of Science and Engineering 8 (2), 135-140. https://doi.org/10.12777/ijse.8.2.135-140

[13] Widayatno, T. 2015. AIP Conference Proceedings 1699, 050002 (2016); https://doi.org/10.1063/1.4938338

[14] Huang, S., Gu, L., Zhu, N., Feng, K., Yuan, H., Lou, Z., Shan, A. 2014. *Green Chem*, 2696–2705.

Acknowledgment

The authors would like to thank the financial support from the Ministry of Research, Technology, and Higher Education of the Republic of Indonesia through the Student Creativity Program for the year 2017.

Mathematical modeling of drying behavior of coconut grate in a modified screw mixed dryer

I Santosa *, E Sulistiawati and A Aktawan

Department of chemical engineering, Universitas Ahmad Dahlan, Yogyakarta, Indonesia

* Email: imam.santosa@che.uad.ac.id

Abstract. The research aimed to examine the drying behavior of coconut grate in an oil bath pilot-scale modified screw mixed dryer using scientific mathematical models. Various drying temperatures of 80, 85, 90, 95 and 100 °C were used to investigate it using a recognized mathematical model of drying. Exploratory information of moisture ratio, temperature and relative humidity attained from various dryer conditions were fitted to the different experimental drying models. The drying model performances were contrasted based on their correlation coefficient (R^2) between the observed moisture ratios. The result of the research showed that Wang and Singh's model demonstrated the best fit model among the modified screw mixed dryer.

1. Introduction

The largest coconut plantation in the world is located in Indonesia with an area of 3.88 million hectares in which 97% is a people's plantation producing 3.2 million tons of copra. Nevertheless, the export of coconut oil to traditional markets of the USA and EU may be inhibited by the issue of high aflatoxin content in copra and coconut oil processed from copra raw materials [1]. A rapid coconut drying model ought to be constructed to attain a dry white coconut as a raw material for aflatoxin-free coconut oil.

Food properties are changed during the process of drying. The changes include discoloring, aroma loss, textural changes, nutritive value, and there are also changes in physical appearance and shape. Higher drying temperature lessens the drying time; however, it may result in higher energy consumption, heat damage to the surface, and poor product quality. Conversely, lower drying temperatures may enhance the product quality but decrease the drying rate and lengthen the drying period [2]. The absence of heating generates the clear color of pure coconut oil because the hydrolysis and oxidation undergone by the components of carbohydrates, proteins, and fats occurring during heating affect the color of oil [3]. The mathematical models used to describe the drying curve are presented in table 1.

Mathematical models of forced convection solar drying Cuminum cyminum applying the methods of mixed and indirect drying were presented by Zomorodian and Moradi [4]. A non-linear regression analysis technique was used to determine the pertinent coefficients for each model by studying eleven different mathematical models. The best model was selected by considering the high pertinent coefficients. The results showed that the diffusion model with $R2 = 0.995$, $\chi2 = 0.0023$ and RMSE = 0.0199 in mixed mode and the Midilli model with $R2 = 0.995$ $\chi2 = 0.023$ and RMSE = 0.0225 in indirect mode were the best.

Table 1. Mathematical models of drying curve.

No.	Model Name	Analytical expression	Reference
1	Page	$MR = \exp(-kt^y)$	[5]
2	Handerson and Pabis	$MR = a \exp(-kt)$	[6]
3	Wang and Singh	$MR = 1 + at + Bt^2$	[7]
4	Modified Page	$MR = \exp(-(kt)^y)$	[8]
5	Lewis	$MR = \exp(-kt)$	[9]
6	Logarithmic	$MR = a \exp(-kt) + c$	[10]

The research on drying model was conducted by Kaleta et al. [11], who investigated the drying behavior of apples in a fluidized bed dryer. There were three new types of drying models developed and compared with the accuracy of sixteen models derived from literature. The accuracy of the models was measured based on the correlation coefficient (R^2), root means square error (RMSE), and reduced chi-square (χ^2). In the last part of this study, the proposed model was considered the most suitable model with the value of $R^2 > 0.9977$, RMSE = 0.0094–0.0167, and $\chi^2 = 0.0001–0.0002$.

The research on mathematical modeling of drying was conducted by Kumar et al. [12] on thin layer hot-air drying carrot Pomace. The carrot pomace was dried at the average effective diffusivity values which range from 2.74×10^{-9} to 4.64×10^{-9} m2/sec and the activation energy value that was 23.05 KJ/mole. It was performed at the temperature of 60, 65, 70 and 75 degrees Celsius and carried out at 0.7 m/sec air velocity. The result of the experiment showed that the effective diffusivity increases in line with the increase of drying temperature, while the drying time of the carrot Pomace decrease.

The impacts of drying temperatures of 70, 80, 90 and 100°C by fluidized bed drying technique on dried coconut residue quality change regarding oil content and whiteness value was investigated by Assawarachan [13]. To determine the suitable drying empirical model, four empirical models including Newton, Henderson and Pabis, Page and Midilli et al. were studied. The study concluded that the *Page* model has the highest coefficient of determination (R^2), the least chi (X^2) and the lowest root mean square error. It means that shorter drying time requires higher drying temperature. The model has shown an excellent fit to predict the drying behavior of the coconut residue. The result of the research showed that based on the valuation of the quality, oil contents of dried coconut residue samples were higher in comparison with that of the fresh residue. Whereas dried coconut residue has the same range of whiteness values as that of fresh residue.

This study focused on the effects of temperature on the drying characteristics and drying time to undertake pilot-scale oil bath mixed drying by selecting a proper mathematical model of mixed drying which describes the drying kinetics of coconut grate.

2. Materials and methods

2.1. Materials

The coconut used in this study was obtained from a local market in Bantul, Yogyakarta, Indonesia. The coconut was broken into pieces and the meat was separated from the shell, washed and cleaned. It was then sliced to remove the tesla and grated by using a grating machine.

2.2. Experimental set-up and procedure

The sample of 3 kg coconut was dried in a pilot-scale mixing dryer. The experiments were performed at drying temperatures of 80, 85, 90, 95 and 100°C, and at a constant mixing velocity of 90 rpm. The coconut residue was dried to reach 0.002 from the initial 1,3 g_{water}/g_{dry} matter moisture contents. The samples were taken periodically in a 10 minutes interval during the drying process. They were subsequently measured and recorded to determine the moisture contents and fit drying models. The drying models adapted for the mixed drying grated coconut were Wang and Singh, Handerson and Pabis and Page models. They were used to obtain experimental drying data at different temperatures.

The non-linear regression was used to determine each constant of the tested models. The coefficient of determination (R^2) was used to evaluate the effectiveness of the model fit. The higher the R^2 the better the model fit [13], [14].

3. Results and Discussion

3.1. Drying process behavior

The initial moisture content (*Mi*) and equilibrium moisture content (*Me*) of coconut residue in the experiment were average 1.3 and 0.002 g_{water}/g_{dry} matter. The time required to dry coconut residue during fixed drying from the initial moisture contents of 1.3 to the final moisture content of 0.001 g_{water}/g_{dry} matter was 140, 120, 110, 110 and 100 minutes at the temperature of 80, 85, 90, 95 and 100°C respectively. As expected, the moisture ratio (*MR*) was rapidly decreasing from the beginning as illustrated in Figure 1. The figure shows the drying curves of coconut residue using mixed dryer at different temperatures under a constant mixing velocity of 90 rpm. The moisture ratio of coconut residue reduced exponentially as the drying time increased. Temperatures had a significant effect on drying. The drying time decreased as drying temperature increased because the increase of the pressure of water vapor within the coconut residue increased moisture removal.

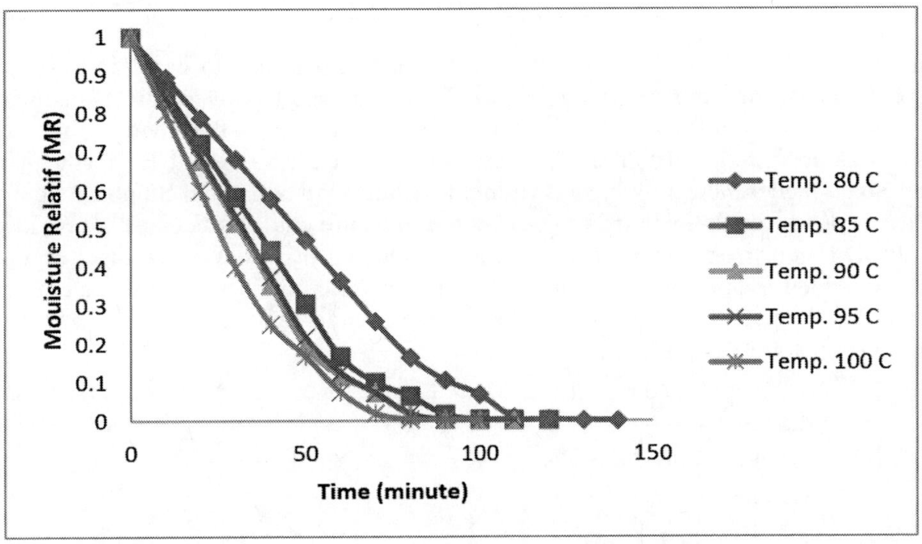

Figure 1. Effect of temperature on moisture ratio of coconut grate residue in the mixed dryer

There was no constant rate drying period in the drying curve of coconut residue during mixed drying. It was found only in the falling rate period. The moisture ratio was found diminished dramatically in the early stage of drying, and eventually changed a bit when it close to the desirable moisture content of around 0,039 g_{water}/g_{dry} matter. It showed that the higher drying temperature, the faster process to dry the sample to the final moisture content of around 0.039 g_{water}/g_{dry} matter. By using the highest level of drying temperature of 100°C, it resulted in the highest drying rate of coconut grate.

3.2. Modeling of Drying Kinetics.

To determine the suitable mathematical modeling that best describes the mixed drying of coconut grate, the non-linear regression analysis was used. Three empirical models were fitted to the drying curve to select the best model to describe it (Table 2). The average moisture content data were observed in g_{water}/g_{dry} matter during mixed drying at various different temperatures.

Table 2. Regression coefficients mixed drying models for coconut grate during mixed drying

Drying Models	Drying temperature	Empirical Drying Model Constants	R^2
Wang and Singh	80	b = -0.0145, c = 4.898×10^{-5}	0.9944
$MR = 1 + bt + ct^2$	85	b = -0.0191, c = 8.652×10^{-5}	0.9951
	90	b = -0.0215, c = 1.110×10^{-4}	0.9960
	95	b = -0.0211, c = 1.066×10^{-4}	0.9961
	100	b = -0.0240, c = 1.4122×10^{-4}	0.9984
Handerson and Pabis	80	k = 0.0528, a = 0.221	0.8874
$MR = a\,e^{-kt}$	85	k = 0.059, a = 0.288	0.9395
	90	k = 0.0698, a = 0.511	0.9435
	95	k = 0.0646, a= 0.298	0.9433
	100	k = 0.0732, a == 0.313	0.9534
Modified Page	80	a = 4.898E-05, b= -1.451E-02, c = 1.043	0.9798
$MR = \exp(at^2+bt+c)$	85	a = 3.844E-04, b = 9.068E-03, c = -0.0785	0.9767
	90	a = 5.124E-04, b = 1.858E-02, c = 9.898	0.9832
	95	a = 3.783E-04, b = 1.918E-02, c = - 2.266	0.9686
	100	a = 4.202E-04, b = 2.695E-02, c = - 2.388	0.9736

The model with the highest value of R^2 was the best model used to describe the mixed drying of agriculture product during mixed drying (Figure 2). The parameters of empirical mathematical model (*k, a, b* and *c*) and statistical analysis results were applied to the models by considering air temperatures as presented in Table 2. The table shows that the values of R greater than 0.99 are in Wang and Singh model indicating a good fitting. It means that Wang and Singh's model was the most appropriate model of the three models to predict the moisture contents because it provided the highest value of R^2. Different from the previous studies conducted by some researchers arguing that *Page* model was suitable for describing drying curve in many agricultural materials in laboratory scale, this study found that Wang and Singh's model was the most suitable model in describing the mixed drying behavior of coconut grate in pilot scale.

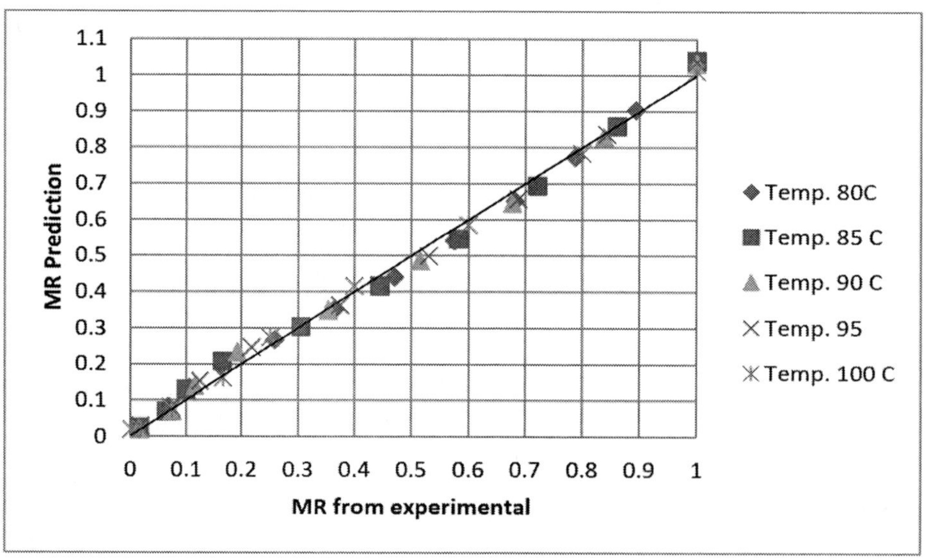

Figure 2. Experimentally determined and predicted moisture ratio of coconut grate during mixed drying

4. Conclusion.

The effects of the air temperatures on drying characteristics of coconut grate during mixed drying process at different air temperature levels were investigated in the pilot scale. Mathematical modeling for predicting the changes in the moisture contents of coconut residue was successfully developed. The time required to dry coconut grate from initial moisture content of 1,3 to 0.002 g_{water}/g_{dry} was 140, 120, 110, 110 and 100 minutes at the temperature of 80, 85, 90, 95 and 100 °C, respectively. According to these results, the Wang and Singh model was the most suitable model for the description of drying characteristics of coconut grate related to time.

Acknowledgment

The authors are pleased to acknowledge the Direktorat Riset dan Pengabdian Masyarakat, KEMRISTEKDIKTI, and Universitas Ahmad Dahlan for supporting this work.

References

[1] Directorate General of Agro-Industry 2010, Roadmap Industri Pengolahan Kelapa, 2010
[2] Kumar, C., Karim, M.A., Mohammad, U.H., Joardder, 2014. Intermittent Drying of Food Products: A Critical Review. Journal of Food Engineering 121, 48-57
[3] Budianto, A., Luwes, dan Rina, S., 2005, Pembuatan Minyak Goreng Dari Kelapa Dengan Proses Basah Menggunakan Enzim Papain, Jurnal IPTEK Vol. 8 No. 1
[4] Page, G.E., 1949. Factors Influencing the Maximum Rates of Air Drying Shelled Corn in Thin Layers M.S. Thesis. Department of Mechanical Engineering, Purdue University, Purdue, USA
[5] Zomorodian, A., Moradi, M., 2010, Mathematical modeling of forced convection thin layer solar drying for Cuminum cyminum, J. Agric. Sci. Technol. 12, pp. 401–408
[6] Henderson, S.M., Pabis, S., 1961. Grain drying theory. II. Temperature effects on drying coefficients. J. Agr. Eng. Res. 6, 169–174
[7] Wang, C.Y., Singh, R.P., 1978. Use of variable equilibrium moisture content in modeling rice drying. Trans. ASAE 11, 668–672
[8] White, G.M., Ross, I.J., Ponelert, R., 1981. Fully exposed drying of popcorn. Trans., ASAE 24, 466–468
[9] Bruce, D.M., 1985. Exposed-layer barley drying, three models fitted to new data up to 150 _C. J. Agr. Eng. Res. 32, 337–347
[10] Togrul, I.T., Pehlivan, D., 2002. Mathematical modelling of solar drying of apricots in thin layers. J. Food Eng. 55, 209–216
[11] Kaleta, A., Gornicki, K., Winiczenko, R., Chojnacka, A., 2013 Evaluation of drying models of apple dried in a fluidized bed dryer, Energy Convers. Manag. 67, 179–185
[12] Kumar, N., Sarkar, B.C., Sharma, H.K., 2012, Mathematical modeling of thin layer hot air drying of carrot pomace, J. Food Sci. Technol., doi:10.1007/s13197-011-0266-7.6.
[13] Assawarachan, R. 2013, Drying Kinetics of Coconut Residue in Fluidized Bed, International Journal of Agriculture Innovations and Resesarch, Volume 2, Issue 2, 262-266
[14] Jena, S., Das, H., 2007 Modelling for vacuum drying characteristics of coconut presscake., J. Food Eng, 2007, 79, 92–99

IOP Conf. Series: Materials Science and Engineering **674** (2019) 012010 doi:10.1088/1757-899X/674/1/012010

Biomass conversion of tamarind waste to syngas through gasification process on downdraft gasifier

A Aktawan *, **Maryudi and S Salamah**

Chemical Engineering Department, Faculty of Industrial Engineering, Universitas Ahmad Dahlan, Yogyakarta 55164, Indonesia

* Email: agus.aktawan@che.uad.ac.id

Abstract. National energy demand has been fulfilled by non-renewable energy sources, such as natural gas, petroleum, coal and so on. However, non-renewable energy reserves deplete increasingly which can cause an energy crisis. Conversion of biomass into energy becomes one of the solutions to overcome it. Indonesia has an enormous biomass potential, especially from tamarind. Tamarinds produce waste tamarind shells. These wastes can be treated with gasification technology to produce syngas. The gasification model used in this research is downdraft gasifier equipped with a cyclone to separate gas with solid or liquid gasification process. The more dry waste of Javanese tamarind, the more syngas produced during the gasification process. And the more dry waste of Javanese tamarind, then the longer the expenditure time of syngas. In syngas analysis known at 5 ml syngas contain 3,046% of CH_4, 14,337% CO and 9,031% H_2.

1. Introduction

The economic growth, population, development and national development causes the increase of energy demand from year to year in all sectors. During this time the national energy needs are met by non-renewable energy resources, such as natural gas, petroleum, coal and so on. However, fossil energy reserves deplete increasingly which can cause an energy crisis. The depletion of this energy reserve forces the Indonesian government and society to seek other alternatives as a source of energy. The hunt, development, and exploration of alternative energy sources must consider the main factors, such as energy, economy and ecology, in other words the developed system must be able to produce energy in large quantities, with low cost and have minimal environmental impact.

One of the alternative energy that developed intensively is biomass. Biomass is a material that can be obtained from plants either directly or indirectly and used as a large amount of energy or material. The resource base covers hundreds and thousands of species of land and oceans, various agricultural resources, forestry, and waste residues and industrial processes, waste and animal waste. Biomass is a renewable energy resource. Indonesia is one of the tropical countries, so it has a very large biomass potential. Agricultural and plantation industries such as oil palm plantations, coconut plantations, cane plantations, industrial plantation forests, and other species, produce a large amount of biomass waste. Tamarind (Tamarindus indica) is one of the sources of biomass waste which generates tamarind waste.

Sugarcane is one type of plant that can only be planted in tropical climate areas The total area of sugar cane in Indonesia is 344 thousand hectares with the main contribution is in East Java (43.29%), Central Java (10.07%), West Java (5.87%) and Lampung (25.71%). In the last five years, the whole area of sugarcane in Indonesia has stagnated in the range of about 340 thousand hectares. From all of sugarcane plantations in Indonesia, 50% of them are smallholder plantations, 30% private plantations, and only 20% of state plantations. In 2004, sugar production in Indonesia attained 2,051,000 tons of crystal [1].

Sugarcane obtained from plantations is generally processed into sugar in sugar factories. A bagasse or commonly called a bagasse, is a by-product of the sugar cane extraction process. Based on data from the Indonesian Sugar Plantation Research Center (P3GI), bagasse of 32% was generated from the weight of sugarcane. However, 60% of the bagasse is utilized as fuel, raw materials for paper, raw materials of brake canvas industry, mushroom industry, and others by sugar factories. Therefore, 45% of the bagasse is estimated not utilized. With the large potential of bagasse, the biomass of bagasse will be able to meet the current energy needs. One of the treatment methods that can be used to treat biomass is a gasification, which is a process to convert raw solid biomass into fuel gas (syngas).

Sugarcane is the residual waste juice, which generally contains 31 - 34% part of the sugarcane. The composition of 50% of bagasse is 47% of fibrous and 3% of residual sugars and other dissolved solids [2], which are presented in Table 1. Coronel and Feungchan et. al. [3] also described the composition of tamarind fruit which is shown in Table 2.

Table 1. Chemical composition of Bagasse

Composition	%
Ash	0,79
Lignin	12,70
Pentose	27,90
Core (alcohol, benzene)	2,00
Cellulose	44,80
Solubility in hot water	3,70

Tabel 2. Composition of tamarind fruit

Constiuents	Amount (per 100 gm)
Water	17.8-35.8 g
Protein	2-3 g
Fat	0.6 g
Carbohydrates	41.1-61.4 g
Fibre	2.9 g
Ash	2.6-3.9 g
Calcium	34-94 mg
Phosphorous	34-78 mg
Iron	0.2-0.9 mg
Thiamine	0.33 mg
Riboflavin	0.1 mg
Niacin	1.0 g
Vitamin C	44 mg

The utilization of sugarcane residues has been studied for various purposes. The cane cane has been made into charcoal and used for the collection of iron (Fe) and manganese (Mn) from water [4], as well as for the tri-chloropenol collection [5]. In addition, sugarcane residues are also used for chrome adsorption [6]. Tsai et al. [7] have conducted pyrolysis of sugarcane residues that produced bio-oil and charcoal.

Trends in the use of biomass as energy sources are increasing, as happened in India [8]. Bagasse is a potential biomass waste to be used as an energy source. Bagasse in Indonesia has also been used as an energy source for sugar mill by conventional combustion. The value calorie of bagasse in net caloric value is about 7600 kJ/kg at water content of 50%. The calorie value is lower than the calorific value of wood, which is 11.715 kJ / kg at water content of 30%. However, bagasse is a potential energy source because it is available in large and renewable sugar factories. In just 12 months, each hectare of land can produce not less than 30 tons of sugar cane. While at the same time, wood production is less than half and it must wait 8 - 9 years to cut it down [9]. Patel et al. [10] has studied the quality improvement of bagasse as a fuel through the torrefaction process.

Gasification is a thermo-chemical conversion of a carbon-based solid or liquid material (feedstock) into a combustible gas product with the supply of a gas gasification auxiliary material [11]. Thermo-chemical conversion changes the chemical structure of biomass at high temperatures. Gasification auxiliaries encourage carbon-based materials to be rapidly converted into gases through various heterogeneous reactions [11], [12].

The gasification unit consists of two types, i.e. updraft and downdraft. Updraft type that undergoes a counter-current flow between gas and solids, is more suitable for the conversion of less reactive char to gas. Nearly 90% of the world's gasified coal uses this configuration [12], [13]. While the downdraft type undergoes a concurrent flow between the feedstock and gas, which is less flexible for water content and size. However, the downdraft type is preferred for small-scale processes because it provides a cleaner gas yield and results an uncomplicated cleaning or purification process [12].

Gasification of biomass has been widely conducted. Adeyemi et al. [14] conducted a gasification comparison between coal and construction waste wood. Gasification of wood waste and efforts to reduce tar levels have also been done [15]. Anukam et al. [16] reviewed the potential utilization of waste bagasse as a energy source through gasification. However, researches on gasification of bagasse is still limited. This reports the gasification of bagasse using downdraft gasification system.

2. Method

2.1. Materials

Material needed for this research was tamarind waste which was collected from imogiri.

2.2. Conversion of Tamarind Waste to Syngas

This research was started with preparing raw materials. Tamarind waste was dried under the sunshine, and then followed by the reduction of tamarind waste size into 2 to 4 cm. The next step was to weigh biomass using a digital balance. Gasification was started by feeding tamarind waste feedstock into the gasifier. The blower was turned on for flowing the air into the gasifier and ignited the biomass through the burning hole, then temperature changes was recorded every minute during the gasification process. The final result of gasification was the resultant gas coming out of the blower, the tar which accommodated under the cyclone and ash mixed with the charcoal retained under the gasification reactor. The mixture of ash and charcoal was sieved using a 2 mm diameter sieve and the smooth ash could be separated from the charcoal which is then weighed and recorded. The sample of syngas was taken to check the gas content by using gas chromatographic analysis.

3. Results and Discussion

3.1. Effect of Tamarind Waste Weight on Syngas Weight

Cellulose content in bagasse is about 20-40%. Cellulose will be converted to syngas. Some data i.e. the weight of ash or residue and tar, were taken in the research to determine the amount of syngas production. From these two data, we will get data of effect of feed weight to syngas production. Data of effect of tamarind waste weight on syngas weight is presented in Figure 1.

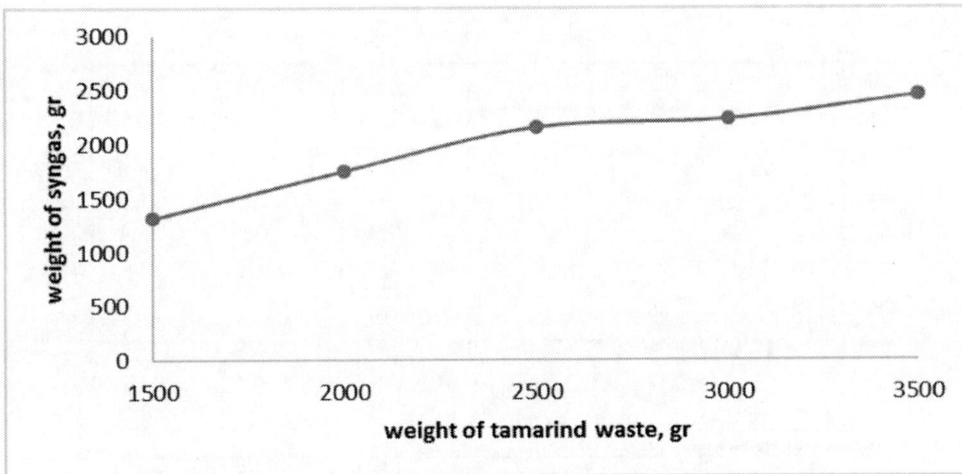

Figure 1. Weight of syngas for various weight of tamarind waste

Figure 1 shows the effect of weight of tamarind waste on syngas weight. The syngas weight increases with the increase of weight of tamarind waste that fed into the gasification unit. It can be seen that the more amount of tamarind waste produces more amount syngas. It is due to the higher cellulose content that results in a higher conversion of syngas.

3.2. Effect of Weight of Tamarind Waste on Duration of Syngas Combustion

The duration or time of syngas combustion/discharge were recorded. The results are presented in Figure 2. The syngas performs an easily burned characteristic when exposed to fire.

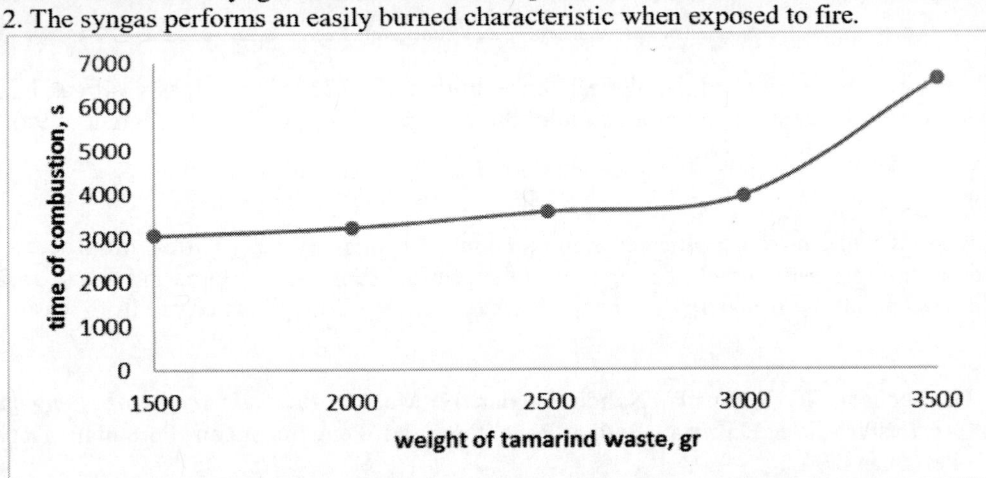

Figure 2. Syngas combustion time for various weight of tamarind waste

Figure 2 reveals that the time of syngas combustion increases proportionally with the increase of weight of tamarind waste. The syngas combustion time proportionally relates to amount of syngas. Data show that the weight of tamarind waste of 3.500 gram produced syngas with the longest time of combustion of 6.600 s or 110 minutes.

3.3. Syngas Composition

The concentration and content of compounds in syngas were determined by using gas chromatography. The analysis result of syngas is shown in Figure 3.

IOP Conf. Series: Materials Science and Engineering **674** (2019) 012010 doi:10.1088/1757-899X/674/1/012010

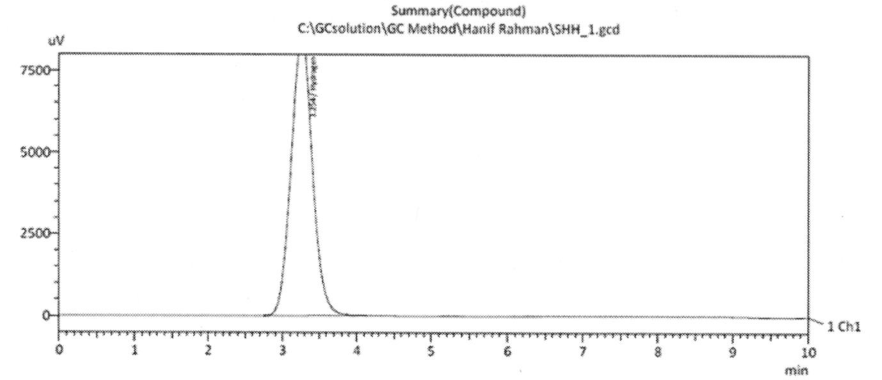

<< Channel 1 >>
ID#1 Compound Name: Hydrogen

Sample Name	Ret. Time	Area	Height	Mark	Dilution Factor	Conc.	Units
SHH_1	3.254	174735	8551	V	1.000000	9.031	%

Figure 3. Chromatogram of syngas of tamarind waste

The results of syngas analysis are summarized in Table 3.:

Table 3. Concentration of CO, CH_4, and H_2 in Syngas

No.	Composition	concentration (%volume *syngas*)
1.	CO	14,337
2.	CH_4	3,046
3.	H_2	9,031

As shown in Table 3, CO had the highest concentration of 14,337% in syngas gasification. In other words, CO had the biggest role on production of burning gas or syngas. The H_2 content of 9,031% and the CH_4 content of 3,046% in syngas.

4. Conclusion

The increase of tamarind waste, increases the amount of syngas by the gasification process. Carbon monoxide is the greatest content of syngas. In other words, syngas from tamarind waste gasification can be utilized as alternative energy or cheap alternative energy, since it is derived from waste.

References

[1] D. H. Goenadi, W. R. Susila, Nahdodin, dan H. Malian, *Prospek dan Arah Pengembangan Agribisnis Tebu*. Jakarta: Badan Penelitian dan Pengembangan Pertanian Departemen Pertanian, 2007.

[2] R. Moeksin, B. Rata, dan N. Kusuma, "Pengaruh Pemutihan Terhadap Warna Pulp Dari Ampas Tebu," *J. Tek. Kim.*, vol. 16, no. 3, hal. 31–34, 2009.

[3] G. Andaka, "Hidrolisis ampas tebu menjadi furfural dengan katalisator asam sulfat," *Teknologi*, vol. 4, hal. 180–188, 2011.

[4] C. Dalai, R. Jha, dan V. R. Desai, "Rice Husk and Sugarcane Baggase Based Activated Carbon for Iron and Manganese Removal," *Aquat. Procedia*, vol. 4, no. Icwrcoe, hal. 1126–1133, 2015.

[5] S. Mubarik, A. Saeed, M. M. Athar, dan M. Iqbal, "Characterization and mechanism of the adsorptive removal of 2,4,6-trichlorophenol by biochar prepared from sugarcane baggase," *J. Ind. Eng. Chem.*, 2014.

[6] I. Ullah, R. Nadeem, M. Iqbal, dan Q. Manzoor, "Biosorption of chromium onto native and immobilized sugarcane bagasse waste biomass," *Ecol. Eng.*, vol. 60, hal. 99–107, 2013.

[7] W. T. Tsai, M. K. Lee, dan Y. M. Chang, "Fast pyrolysis of rice straw, sugarcane bagasse and coconut shell in an induction-heating reactor," *J. Anal. Appl. Pyrolysis*, vol. 76, no. 1–2, hal. 230–237, 2006.
[8] A. Kumar, K. Kumar, N. Kaushik, S. Sharma, dan S. Mishra, "Renewable energy in India: Current status and future potentials," *Renew. Sustain. Energy Rev.*, vol. 14, no. 8, hal. 2434–2442, 2010.
[9] Y. Kurniawan dan H. Santoso, "Listrik sebagai ko-produk potensial pabrik gula," *J. Litbang Pertan.*, vol. 28, no. 1, hal. 23–28, 2009.
[10] B. Patel, B. Gami, dan H. Bhimani, "Improved fuel characteristics of cotton stalk, prosopis and sugarcane bagasse through torrefaction," *Energy Sustain. Dev.*, vol. 15, no. 4, hal. 372–375, 2011.
[11] V. Belgiorno, G. De Feo, C. Della Rocca, dan R. M. A. Napoli, "Energy from gasification of solid wastes," *Waste Manag.*, vol. 23, no. 1, hal. 1–15, 2003.
[12] C. Di Blasi, "Dynamic behaviour of stratified downdraft gasifiers," *Chem. Eng. Sci.*, vol. 55, no. 15, hal. 2931–2944, 2000.
[13] M. L. Hobbs, P. T. Radulovic, dan L. D. Smoot, "Combustion and gasification of coals in fixed-beds," *Prog. Energy Combust. Sci.*, vol. 19, no. 6, hal. 505–586, 1993.
[14] I. Adeyemi, I. Janajreh, T. Arink, dan C. Ghenai, "Gasification behavior of coal and woody biomass: Validation and parametrical study," *Appl. Energy*, vol. 185, hal. 1007–1018, 2017.
[15] J. Park, Y. Lee, dan C. Ryu, "Reduction of primary tar vapor from biomass by hot char particles in fixed bed gasification," *Biomass and Bioenergy*, vol. 90, hal. 114–121, 2016.
[16] A. Anukam, S. Mamphweli, P. Reddy, E. Meyer, dan O. Okoh, "Pre-processing of sugarcane bagasse for gasification in a downdraft biomass gasifier system: A comprehensive review," *Renew. Sustain. Energy Rev.*, vol. 66, hal. 775–801, 2016.

The effects of types and concentrations of adsorbents on aloe vera gel opacity

T Y Hendrawati[1,*], A D Utami[2], R A Nugrahani[3], U H Hasyim[4] and A I Ramadhan[5]

[1,2,3,4] Chemical Engineering Department, Faculty of Engineering, Universitas Muhammadiyah Jakarta, Indonesia
[5]Mechanical Engineering Department, Faculty of Engineering, Universitas Muhammadiyah Jakarta, Indonesia

*Email: yuni.hendrawati@ftumj.ac.id

Abstract. *Aloe vera* is one of the medicinal plants that are frequently used in the pharmaceutical industry, especially in cosmetic products. The type of Aloe vera that is commonly cultivated in Pontianak, Indonesia, is that of *Aloe Chinensis Baker*. The part of Aloe vera that is commonly used is its gel, which is an opaque mucilaginous gel. The objective of this research is to determine the effects of the addition of adsorbents on the Aloe vera gel opacity. Using Aloe vera as the main ingredient of aloe vera gel, gel is obtained by peeling off the green skin of the Aloe vera leaves that have been washed. A gel is then pureed and filtered to reduce dregs. The purified gel is added with adsorbents. This research employs two factorials, those are adsorbent types (Zeolite and Bentonite) and concentrations (0 %, 2.5 %, 5 % , 7.5 % and 10 %). The parameters used for analyses are refractive index, pH, Vitamin C content, density, and viscosity. Regression analyses are performed on the concentrations and yields. Results show that addition of Zeolite produces better clearance on Aloe vera gel than that of Bentonite, with the concentration of Zeolite at 2.5 %. The analyses further produce a refractive index of 1.332 for Zeolite treated gel. This value is close to the water refractive index of 1.3325.

1. Introduction

Aloe vera is one of the medicinal plants that are frequently used in the pharmaceutical industry, especially in the cosmetic industry. This is because Aloe vera's pharmacological properties make it one of the important raw materials for cosmetics products [1]. Aloe vera belongs to the family of *Liliaceae* [2]. The most commonly cultivated type of Aloe vera in Asia, including Indonesia is *Aloe chinensis* Baker, from China, although it is not native there. This type has been commercially cultivated in Indonesia in Kalimantan Barat (West Kalimantan) province and is commonly known as Aloe vera Pontianak, as described by Baker 1877 [3].

The parts of Aloe vera plants that are generally utilized are its leaves that can be directly used in the forms of extract, leaf sap, gel (the mucilaginous part inside the leaves after separating it from the skin), and consisting of 96 % water and 4 % solids containing 75 beneficial compounds. Aloe vera gel also contains 17 amino acids important for our bodies. Its contents make this plant multifunctional. These constituents include aloin, emodin, resin, lignin, saponin, anthraquinone, vitamins, minerals, and so on. Aloe vera is used in the industry in the form of gel, powder, and extract [3], [4], [5]. To produce Aloe vera Powder by microencapsulation, The optimal drying inlet temperature was found at

120^{0}C, which the active compounds of *Aloe vera* powder (aloenin (B), aloeresin A and chrysophanol) can be maintained [3]. Microencapsulation is a technology coating solids, liquids and gases by the capsule in the form of a small capsule that can release contents under specific conditions [6].

Adsorption is a process in which a substance adheres to the surface of a solid caused by atomic or molecular attraction forces without being integrated [7]. A couple of adsorbents that are generally used in the adsorption process are Zeolites and Bentonites. The types of Zeolites commonly found in Indonesia are clinoptilolite and mordenite [8]. Mordenite can be used in the adsorption of H_2O, CO, CO_2, and CH_4 gases, while clinoptilolite can be used for CO, CO_2, and NO gases [9].

The typical structure of Zeolite is marked by a large number canals and pores that result in a large surface area. This condition can be further explained that each pore and canal inside as well as between crystals is considered in cylindrical form, thus the total surface area of Zeolite is the accumulated sum of the wall area of the pores and canals that make up Zeolite. The larger number of pores and canals, the larger surface area of Zeolite [10]. Based on this characteristic, Zeolite can be used as an adsorbent, ion exchanger, and catalyst.

Bentonite is an aluminium silicate hydrate mineral that is classified as a phyllosilicate or sheet silicate consisting of $(SiO_4)^{2-}$ tetrahedral network built in an infinite space forming anionic $(SiO_3)^{2-}$ network with Si/O ratio of 2/5. The general chemical formula of Bentonite is $Al_2O_3.4SiO_2.H_2O$. The montmorillonite content in Bentonite is 85 % [11].

In a research conducted by Rajeswari (2012), on Aloe vera as a medicinal plant in India, the extract of Aloe vera is proven to heal many skin conditions. Aloe vera helps alleviate skin wounds caused by a burn, irritation, cut, and insect bite. Its antibacterial property can eliminate itch and reduce skin swelling [12]. Firdaus (2009) conducted an experiment on the applications of Bentonite and Zeolite to increase the quality of distilled vetiver oil in Garut regency, West Java. His result showed that the addition of Zeolites as adsorbent yielded a better result than that of Bentonite. The gel extracted from Aloe vera leaves usually has high opacity (cloudy), thus the objective of this research is to determine the effects of adsorbent additions on the opacity of Aloe vera gel.

2. Method

The research is conducted in the Applied Chemistry Laboratorium, Chemical Engineering Department, Faculty of Engineering, Universitas Muhammadiyah Jakarta (UMJ – University of Muhammdiyah Jakarta). The materials used in the research are Aloe vera (*Aloe Chinensis Baker*), distilled water, Bentonite, Zeolite, amylum/starch, concentrated HCl, Sodium Thiosulfate standard solution, and Iodine standard solution. The equipment used is beakers, graduated cylinders, volumetric flasks, measuring pipettes, drip pipettes, bulb, Erlenmeyer flasks, burette, funnels, filter papers, clamps and stands, analytical scale, pH meter, pycnometer, Brookfield viscometer, refractometer, blender, and centrifuge.

This research consists of two stages; making Aloe vera gel and the addition of adsorbents. It also contains two factorials; adsorbent types (Zeolites and Bentonites) and adsorbent concentrations (0 %, 2,5 %, 5 % , 7,5 % and 10 % m/v) .

The Aloe vera gel extraction starts with washing the harvested leaves. Peeling separates the green skin from the colorless mucilaginous gel, which the latter then placed in a blender for pureeing. The pureed gel is filtered to reduce dregs. The filtered gel is used in the research [3].

The amount of Aloe vera gel per sample is 200 ml. Zeolite and Bentonite at the concentrations of 0 %, 2,5 %, 5 %, 7,5 % and 10 % (m/v) are added in the gel and homogenized with a magnetic stirrer for one hour. After settling for two days, the Aloe vera gel is separated from the adsorbents by centrifugation. The resulting supernatant undergoes absorbancy and other analyses.
The following flow chart (Figure 1) describes the procedure.

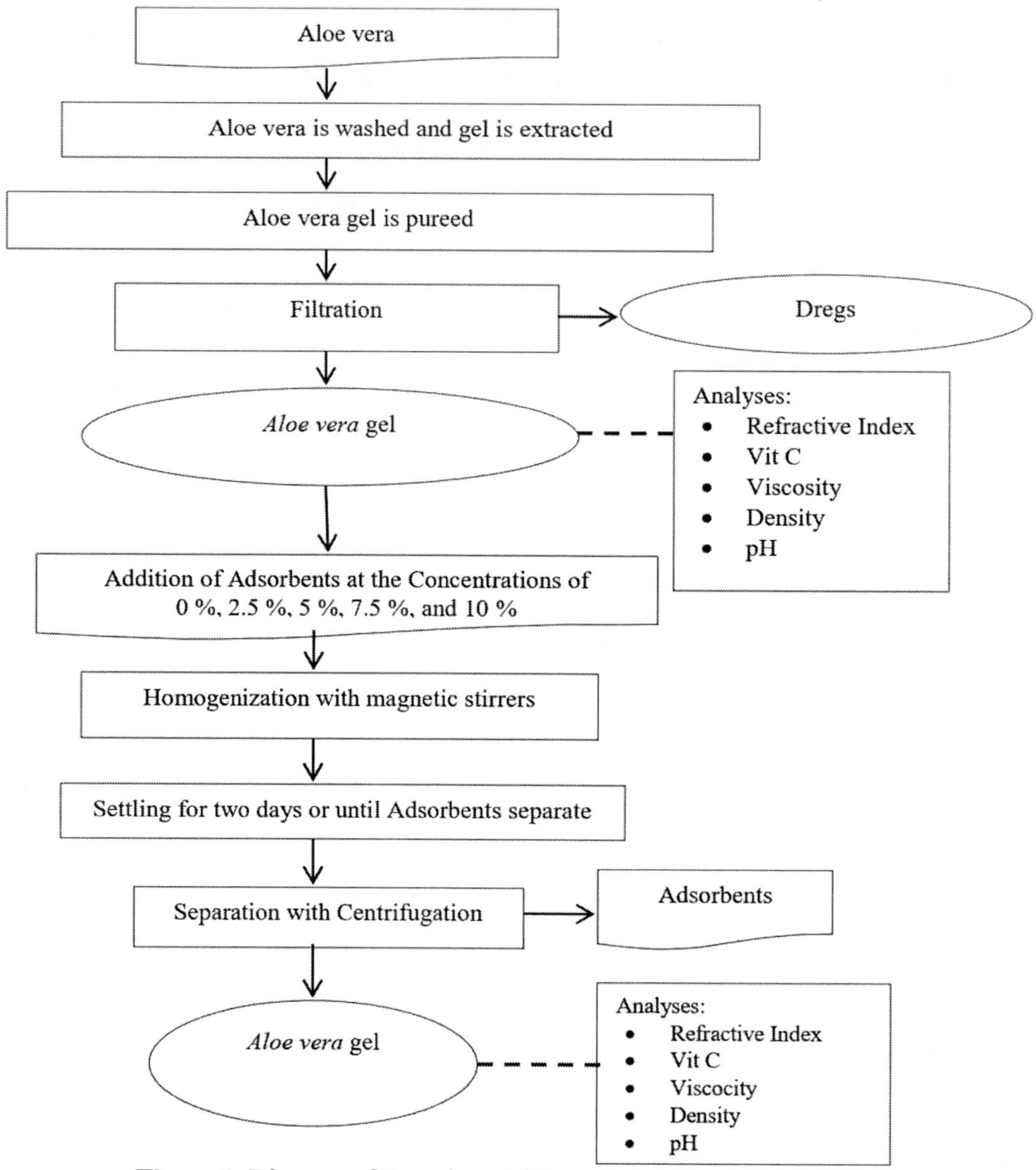

Figure 1. Diagram of Experimental Procedure

3. Results and Discussion

3.1. The effects of adsorbent types and concentrations on yields

This research is conducted with two types of variables; those are the types of adsorbents (Zeolite and Bentonite) and the concentrations of adsorbents (0 %, 2.5 %, 5 %, 7.5 %, and 10 %) with the sample volume of Aloe vera gel at 200 ml. The Aloe vera gel obtained from the adsorption process is differentiated between the adsorbent and Aloe vera gel. The yield is calculated from comparing the volume of gel from the adsorption to that of pure Aloe vera gel. Adsorption process takes place for 48 hours and at a maintained temperature of 14 ^{0}C.

The complete result of yields from different types and concentrations of adsorbents is presented in Table 1.

Table 1. The yield of Aloe vera gel at different adsorbent types and concentrations

No.	Sample	Final Yield (%)
1	Zeolite 0 %	100
2	Zeolite 2,5 %	80
3	Zeolite 5 %	78.5
4	Zeolite 7.5 %	77
5	Zeolite 10 %	75
6	Bentonite 0%	100
7	Bentonite 2.5 %	76
8	Bentonite 5 %	75
9	Bentonite 7.5 %	74.5
10	Bentonite 10 %	73.5

The data analyses of the effects of adsorbent types and concentrations on Aloe vera gel yield are presented in a graphic form shown in Figure 2.

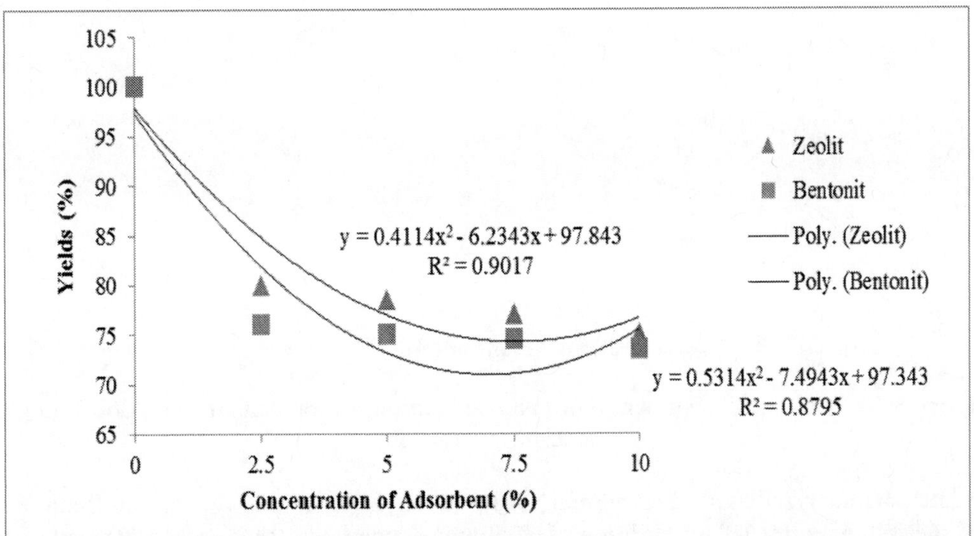

Figure 2. The effects of adsorbent types and concentrations on yields of Aloe vera gel from adsorption processes

The data are analyzed using Polynomial Order 2 to obtain the correlations between adsorbent concentrations and yields. The result is presented in Figure 2. Addition of Zeolite produces $y = 0.4114x^2 - 6.2343x + 97.843$ with $R^2 = 0.9017$, while addition of Bentonite produces $y = 0.5314x^2 - 7.4943x + 97.343$ with $R^2 = 0.8795$, with y representing yields and x representing concentrations. The above figure shows that the more adsorbent is added, the less yield is produced. These results are caused by the greater amount of adsorbent, the more molecules of Aloe vera gel are adsorbed thus reducing the volumes of Aloe vera gel yields.

3.2. The effects of adsorbent types and concentrations on densities

The density testings are performed by measuring the weight of an empty 25-ml pycnometer, then the empty pycnometer is filled with Aloe vera gel and weighted. The empty and filled pycnometer weights are used in the density formula to obtain results, which are presented in table 2.

Table 2. Results of Density Testing from different adsorbent types and concentrations

No.	Sample	Density
1	Zeolite 0 %	1.0024
2	Zeolite 2.5 %	0.9213
3	Zeolite 5 %	1.0028
4	Zeolite 7.5 %	1.0032
5	Zeolite 10 %	1.0036
6	Bentonite 0 %	1.0024
7	Bentonite 2.5 %	1.0024
8	Bentonite 5 %	1.0028
9	Bentonite 7.5 %	1.0028
10	Bentonite 10 %	1.0032

The effects of adsorbent types and concentrations on the densities of Aloe vera gel are presented in a graphic form shown in Figure 3.

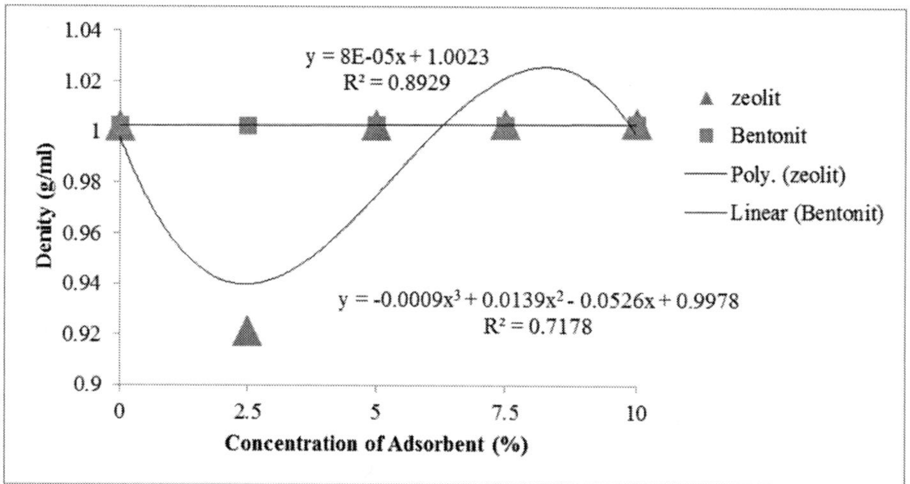

Figure 3. The effects of adsorbent types and concentrations on densities of Aloe vera gel from adsorption processes

The data are analyzed with Polynomial Order 3 on Zeolite and Linear on Bentonite, which produces the following formulas. Addition of Zeolite produces $y = -0.0009x^3 + 0.0139x^2 - 0.0526x + 0.9698$ with $R^2 = 0.7178$, while addition of Bentonite produces y = 8E-05x +1.0023 with R^2=0.8929, in which y represents densities and x represents adsorbent concentrations.

The measurement of Aloe vera gel densities is to determine the weight of Aloe vera gel per unit volume.

3.3. The effects of adsorbent types and concentrations on refractive indices

The measurement of the refractive index by utilizing light refraction. In the refractometer, measurements are read at the point where prism and solution meet. With a low concentration solution, the refractive index of the prism is much greater than that of the sample, creating a large refraction angle and a low reading. The reverse would happen with a high concentration solution. However, before reading takes place, the temperature of Aloe vera gel has to be the same as that of measurement process. Reading is taken when a stable temperature is achieved (Badan Standarisasi Nasional/National Standard Bureau, 2006). The refractive indices of Aloe vera gel from different adsorbent types and concentrations are presented in Table 3.

Table 3. Results of Refractive Index Testing from different adsorbent types and concentrations.

No.	Sample	Bias Index
1	Zeolite 0 %	1.3342
2	Zeolite 2.5 %	1.332
3	Zeolite 5 %	1.3335
4	Zeolite 7.5 %	1.3330
5	Zeolite 10 %	1.3325
6	Bentonite 0 %	1.3342
7	Bentonite 2.5 %	1.335
8	Bentonite 5 %	1.335
9	Bentonite 7.5 %	1.334
10	Bentonite 10 %	1.333

The refractive index of a material is the ratio between the speed of light in air and that in the tested material. Measurement of refractive index is used to evaluate the characters and purity of a medium, determine solute concentrations, compare the components between two solutions, and determine the yield of material extraction of a solution. The effects of adsorbent types and concentrations on the refractive index are presented in a graphic form shown in Figure 4.

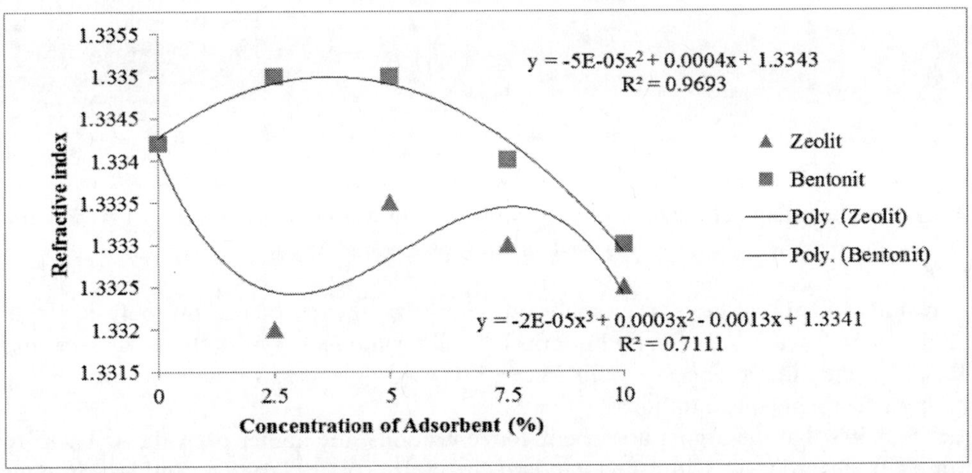

Figure 4. The effects of adsorbent types and concentrations on refractive indices of Aloe vera gel from adsorption processes

Polynomial Order 3 is used to analyse the addition of Zeolite, whereas Polynomial Order 2 is used to analyse the addition of Bentonite. Zeolite addition produces $y = -2E\text{-}05x^3 + 0.0003x^2 - 0.0013x + 1.3341$ with $R^2 = 0.7111$, while Bentonite addition produces $y = -5E\text{-}05x^2 + 0.0004x + 1.3343$ with $R^2 = 0.9693$, in which y represents refractive index and x represents adsorbent concentrations. The results show that the refractory indices that are close to that of water (at 1.3325) occur at 2.5 % Zeolite and 10 % Bentonite.

3.4. The effects of adsorbent types and concentrations on pH

The measurement of pH is performed with a pH-meter by placing the probe inside the Aloe vera gel. The results of pH testing are presented in Table 4.

Table 4. Results of pH Testing from different adsorbent types and concentrations

No.	Sample	pH
1	Zeolite 0 %	5
2	Zeolite 2.5 %	5.6
3	Zeolite 5 %	5.8
4	Zeolite 7.5 %	5.98
5	Zeolite 10 %	6
6	Bentonite 0%	5
7	Bentonite 2.5 %	5.4
8	Bentonite 5 %	5.5
9	Bentonite 7.5 %	5.7
10	Bentonite 10 %	5.9

The effects of adsorbent types and concentrations on the pH of Aloe vera gel is presented in a graphic form shown in Figure 5.

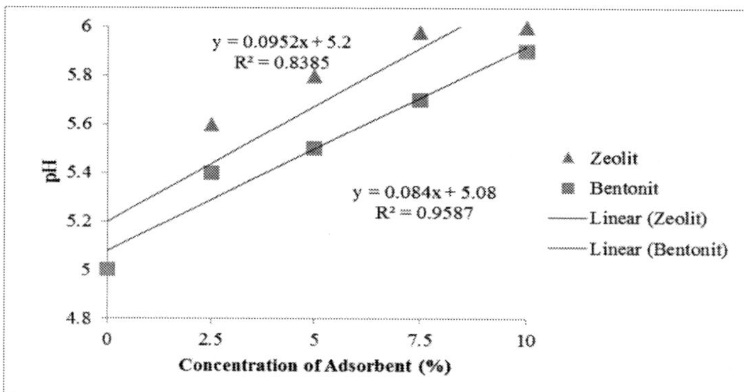

Figure 5. The effects of adsorbent types and concentrations on pH values of Aloe vera gel from adsorption processes

The results of pH testings are analyzed using a linear correlation between the adsorbent concentrations and pH values. The addition of Zeolite produces y = -0.092x + 5.2 with R^2= 0.837, while that of Bentonite produces y = 0.084x + 5.08 with R^2= 0.9587, with y representing pH and x representing adsorbent concentrations.

Figure 5 shows that the higher adsorbent concentrations, the higher pH values. These resulting pH values are in accordance with the standard in SNI 16-4399-1996 (SNI = *Standar Nasional Indonesia*/Indonesia National Standard). pH would influence adsorbent surface charge, degree of ionization, and the species that are adsorbed in the adsorption process. The values of pH also influence the chemical equilibrium of the adsorbates as well as adsorbents. In these pH variations, the possibility of chemical bonding between adsorbent and adsorbate may occur.

3.5. The effects of adsorbent types and concentrations on viscosities

The equipment used to measure the viscosity is the Brookfield Viscometer and takes place at the Testing Center of Jakarta Custom office. The working principle of a viscometer is to measure the torque needed to initiate the spinning motion of the spindle inside a solution. The spindle is activated by a motor synchronized with a calibrated springs; spring reflection is shown with a needle or digital gauge. Viscosity has a linear correlation with the spindle rotational speed, which is related to the size and geometrical form of the spindle [13]. The results of the viscosity testing on the adsorbent types and concentrations are presented in Table 5.

Table 5. Results of Viscosity Testing from different adsorbent types and concentrations

No.	Sample	Viscosity (cP)
1	Zeolite 0 %	117.6
2	Zeolite 2.5 %	18.9
3	Zeolite 5 %	11.4
4	Zeolite 7.5 %	10.8
5	Zeolite 10 %	11.1
6	Bentonite 0 %	117.6
7	Bentonite 2.5 %	24.6
8	Bentonite 5 %	12.6
9	Bentonite 7.5 %	13.5
10	Bentonite 10 %	20.4

The effects of adsorbent types and concentrations on viscosities are presented in the graphic form as shown in figure 6.

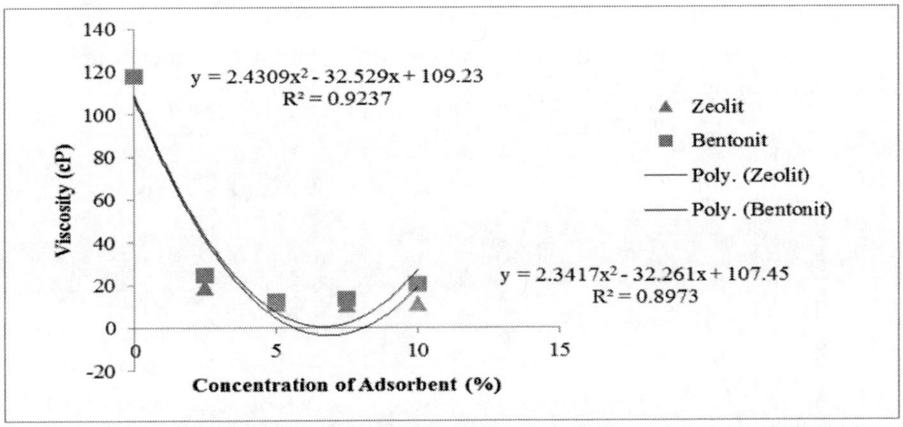

Figure 6. The effects of adsorbent types and concentrations on viscosities of Aloe vera gel from adsorption processes

The viscosity data are analyzed using Polynomial Order 2 for both adsorbents. The addition of Zeolite produces $y = 2.3417x^2 - 32.261x + 107.45$ with $R^2 = 0.8973$, whereas Bentonite addition produces $y = 2.4309x^2 - 32.529x + 109.23$ with $R^2 = 0.9237$, in which y is the viscosity and x is the adsorbent concentrations. Figure 6 shows that adsorbent additions reduce the viscosity values of Aloe vera gel. The viscosity values determine the rheological values of gel usage on skin.

3.6. The effects of adsorbent types and concentrations on Vitamin C contents

Vitamin C in Aloe vera gel is beneficial as a free radical inhibitor inside the body. Vitamin C contents are tested using Iodometric titration methods. Iodometric titration is a type of titration utilizing redox reactions by measuring the amount of Iodine left from redox reactions between Vitamin C and a reactant. The indicator used is amylum (starch) that is added at nearing the titration endpoint. This is to prevent the starch from enveloping the Iodine so that end reading is quickly achieved. These titration processes utilize standard Iodine (I2), which is used for compounds with a strong redactor characteristic like Vitamin C [14].

The testing of Vitamin C is performed by obtaining 20 grams of Aloe vera gel that has gone through the adsorption process. The gel is dissolved in a 100 ml graduated flask, and the resulting slurry is centrifuged to facilitate filtration. Twenty milliliters of filtrate is placed inside an Erlenmeyer flask, added with two milliliters of 1 % starch and 20 ml of distilled water. Titration is performed with

0.01N standard Iodine, with the calculation of one milliliter of 0,01N Iodine representing 0.88 mg of Ascorbic acid. The results of Vitamin C content testing from different adsorbent types and concentrations are presented in Table 6.

Table 6. Results of Vitamin C Content Testing from different adsorbent types and concentrations

No.	Sample	Vitamin C (mg/L)
1	Zeolite 0 %	0.2650
2	Zeolite 2.5 %	0.38345
3	Zeolite 5 %	0.3652
4	Zeolite 7.5 %	0.3287
5	Zeolite 10 %	0.3287
6	Bentonite 0 %	0.265
7	Bentonite 2.5 %	0.3652
8	Bentonite 5 %	0.3287
9	Bentonite 7.5 %	0.29215
10	Bentonite 10 %	0.3287

The effects of adsorbent types and concentrations on Vitamin C contents are presented in the graphic form as shown in figure 7.

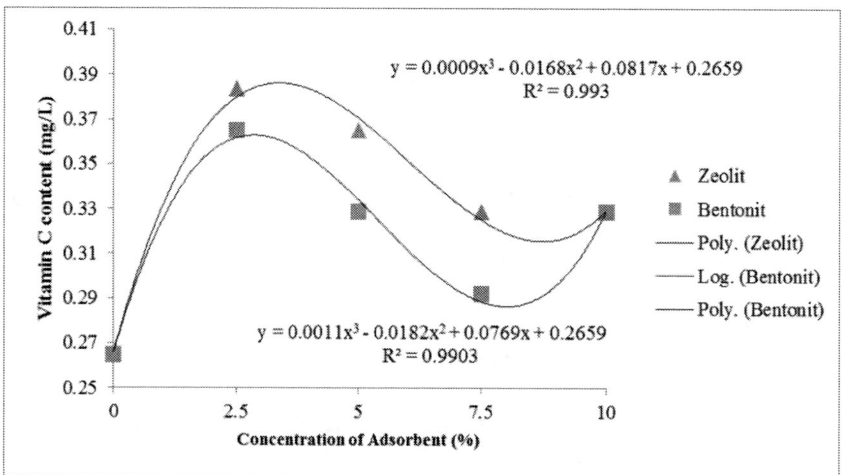

Figure 7. The effects of adsorbent types and concentrations on Vitamin C contents of Aloe vera gel from adsorption processes

The titration results are analyzed with Polynomial Order 2 to determine the correlation between adsorbent concentrations and Vitamin C contents. The addition of Zeolite produces $y = 0.0009x^3 - 0.0168x^2 + 0.0817x + 0.2659$ with $R^2= 0.993$, while the addition of Bentonite produces $y = 0.0011x^3 - 0.0182x^2 + 0.0769x + 0.2659$ with $R^2= 0.9903$, in which y represents the Vitamin C contents (mg/L) and x represents the adsorbent concentrations. Figure 7 shows that the highest Vitamin C contents are found at adsorbent concentrations of 2.5 %. Vitamin C content in Aloe vera gel functions as an antioxidant and inhibitor of free radicals.

4. Conclusions

The conclusions of this research are:

1. The absorbance processes for 48 hours at 14 ^{0}C produce the highest yield with Zeolite at the concentration of 2.5 % with 80 % yield and formula of $y = -2.12x + 92.7$ with $R^2= 0.6782$. This research shows that the higher the concentration of Zeolite, the lower the yield.

2. This research yields a refractive index close to that of water (1.3325) from addition of Zeolite at 2.5 % with refractive index value of 1.332. The formula for this refractory indeces for Zeolite is $y = -2E\text{-}05x^3 + 0.0003x^2 - 0.0013x + 1.3341$ with $R^2 = 0.7111$.
3. The optimal result from this research is from the addition of Zeolite at the concentration of 2.5 %; with the yield of 80 % ($y = -2.12x + 92.7$ with $R^2= 0.6782$), Vitamin C content of 0.38345 mg/L ($y = -0.0029x^2 + 0.321x + 0.2832$ with $R^2= 0.6287$), refractive index of 1.332 ($y = -2E\text{-}05x^3 + 0.0003x^2 - 0.0013x + 1.3341$ with $R^2 = 0.7111$), density of 0.9213 ($y = -0.0009x^3 + 0.0139x^2 - 0.0526x + 0.9698$ with $R^2 = 0.7178$), viscosity of 18.9 cP ($y = 2.3417x^2 - 32.261x + 107.45$ with $R^2= 0.8973$), and pH value of 5.6 ($y = -0.092x + 5.2$ with $R^2= 0.837$).

References

[1] T. Marwati, "Hermani. 2006. Pemanfaatan Bahan Aktif Lidah Buaya (Aloe vera) Sebagai Sediaan Kosmetik," in *Proceeding Seminar Nasional Tumbuhan Obat XXIX Indonesia*, pp. 24–25.

[2] I. Furnawanthi, "Khasiat dan Manfaat Lidah Buaya si Tanaman Ajaib," *BTPT Tangerang Agrobrmedia Pustaka*, 2002.

[3] T. Y. Hendrawati, "Aloe Vera Powder Properties Produced from Aloe Chinensis Baker, Pontianak, Indonesia," in *Journal of Engineering Science and Technology Special Issue on SOMCHE 2014 & RSCE 2014 Conference*, 2015, pp. 47–59.

[4] I. Ismiyati, T. Y. Hendrawati, and R. A. Nugraha, "Pelatihan Budidaya dan Pengolahan Aloe Vera Menjadi Bahan Tambahan Makanan dan Lotion di Aisyiah Kota Depok," *Pros. SNaPP Sains, Teknol.*, vol. 7, no. 1, pp. 163–170, 2017.

[5] S. D. Williams and W. H. Schmitt, *Chemistry and Technology of the Cosmetics and Toiletries Industry*. Springer Science & Business Media, 2012.

[6] T. Y. Hendrawati, A. Siswahyu, and A. I. Ramadhan, "Pre-Feasibility Study of Bioavtur Production with HEFA Process In Indonesia," *Int. J. Sci. Technol. Res.*, vol. 6, no. 04, 2017.

[7] P. W. Atkins, "Kimia Fisika Jilid 2," *Jakarta: Erlangga*, 1999.

[8] T. Las, "Mengenal Mineral Zeolit," *Saintika, J. Sains dan Teknol. UIN Syarif Hidayatullah Jakarta*, vol. 3, no. 1, pp. 64–75, 2006.

[9] W. Trisunaryanti, E. Triwahyuni, and S. Sudiono, "Preparasi, Modifikasi dan Karakterisasi Katalis Ni-Mo/Zeolit Alam dan Mo-Ni/Zeolit Alam," *J. Fak. Huk. UII*, vol. 10, no. 4, 2005.

[10] L. O. Sumarlin, L. Mukmillah, and R. Istianah, "Analisis Mutu Minyak Jelantah Hasil Peremajaan Menggunakan Tanah Diatomit Alami dan Tterkalsinasi," *J. UIN Syarif Hidayatullah Jakarta*, 2015.

[11] W. Endang, "Daya Adsorpsi Zeolit dan Bentonit Terhadap Alkil Benzena Sulfonat dalam Deterjen," *J. Penelit. Iptek dan Hum.*, vol. 1, no. 1, 1996.

[12] R. Rajeswari *et al.*, "Aloe Vera: the Miracle Plant its Medicinal and Traditional Uses in India," *J. Pharmacogn. Phytochem.*, vol. 1, no. 4, pp. 118–124, 2012.

[13] A. Martin, J. Swarbrick, and A. Commarata, "Farmasi Fisik, Edisi Ketiga," *Penerbit Univ. Indones. Jakarta, hal*, vol. 1019, pp. 1022–1077, 1993.

[14] A. Mursyidi and A. Rohman, "Pengantar Kimia Farmasi Analisis Volumetri dan Gravimetri." Gajah Mada University Press. Yogyakarta, 2008.

Acknowledgment

We would like to thank the LPPM Universitas Muhammadiyah Jakarta and Ministry of Research, Technology and High Education Republic of Indonesia for financial support of this work through research grant of Penelitian Terapan Unggulan Perguruan Tinggi (PTUPT) on the year 2017 - 2020.

Effect of tofu waste flour composition to hedonic test level and proximate test of Bread

A M Sari* **and Syamsudin AB**

Chemical Engineering Department, University of Muhammadiyah Jakarta, Jl. Cempaka Putih Tengah 27, Jakarta, 10510, Indonesia

* Email: alvika.metasari@ftumj.ac.id

Abstract. The tofu solid waste (TSW), known as *okara* in Japan, is a solid waste of the tofu industry, usually disposed or sold at a low price. Its shelf life is short due to the high moisture. In hence the protein, carbohydrate and fiber content of TSW is high; it is potential to be used as food products such as flour. To expand the shelf life of the TSW, it was utilized into flour. In this research, the TSW is processed into flour as raw material of bread. The purpose of this study was to determine the effect of TSW flour composition as raw material for bread, such as organoleptic tests (aroma, texture, and taste) and proximate test (protein, carbohydrate, fat, and dietary fiber) of bread produced. The experimental design was conducted by the composition of TSW flour variable, i.e. 0% (R1), 5% (R2), 10% (R3), and 15% (R4). The results showed that the greater the TSW flour composition, the lower the sensory value of panelists. The most favored sensory test was R2 sample, 5% tofu waste flour bread with the sensory score of favorite, texture, aroma, taste, and color of 2,71, 2,86, 3,14, 2,79 and 3. From the proximate test, the 5% TSW flour bread contains higher carbohydrate, protein, moisture content than 10% while the nutritional value was 44,39% carbohydrate, 6,45% fat, 9,16% protein, 39,3% moisture, 0,7% ash and 5,94% dietary fibre.

1. Introduction

Tofu is a popular food in Indonesia for thousands of years and it is produced in large quantities. Tofu consumption has been increased every year. The higher tofu production is corresponding to increasing the tofu solid waste. Tofu solid waste is also known as soybean curd residue or *okara* in Japan. Tofu industry is one of food industry which produced high solid waste, from 1 kg dried soy produced tofu will produce 1,2 kg wet solid waste [1]. It has become a problem for the environment while it is dumped into river or other environments. This problem should be solved due to its contamination such as smell and other organic materials. When it is dumped into a landfill, it will cause a potential environmental problem due to its easy putrefaction [2]. TSW has not been used optimally. TSW is usually disposed to environment without any treatment or sold as animal feed or processed again into food products such as *oncom* and *tempe* at very low price. TSW can be utilized as feed, fertilizer, and food in small quantities.

Although TSW is waste, it still has good nutrition as food, for example, the fiber and isoflavones [3]. TSW has several advantages as a food raw material, such as an industrial tofu waste containing 11.07% carbohydrate, 4.71% protein, 1.94% fat and 0.08% ash [4], which can be used as an alternative source of protein, carbohydrate, and dietary fiber. In hence it is delicious and safe as food. TSW is also abundant. Tofu industry is one of the industries that is growing and is usually on the scale of IKM (Small and Medium Industries). According to Faizal et al. [3] there are 84 thousand tofu industry units with a total production capacity of 2.56 million tons per year quantities tofu waste produced around 25-35% of tofu products. The other advantages are TSW is low price as it is a waste. TSW can be processed into varied food such as brownies [5], "cat`s tongue or *katte* tong" cookie [6] and nuggets [7]. Although the TSW can be processed as varied food, it is long-lasting. To increase its lifetime, it should be processed into TSW flour. TSW flour has many benefits. It is easy to be processed into other food products and durable.

In this study, the TSW was processed into TSW flour and then made into fiber-rich bread. The purpose of adding tofu flour to the bread-making process is to increase protein content. The other advantage of TSW flour is that the fiber content (dietary fiber) is greater than wheat flour. So that the flour is expected to be in the form of fiber-rich bread which has many health benefits compared to commercial bread.

Bread is food made from wheat flour, water, and yeast through a process of kneading, development (fermentation by yeast) and roasting [8]. The main raw material for bread is wheat flour. There are two types of wheat flour, low protein, and high protein. The high protein flour is used to make bread. Although it is high protein, it is low-fiber flour. Now, bread is consumed by many people as a constitute of rice. Therefore, bread is should be enriched with fiber-rich raw materials.

The fiber content in food has an important role in the process of human digestion. The fiber in the ingredients provides a positive effect on the human digestive system which is a laxative and prevents chronic diseases such as coronary heart disease and colon cancer. So there needs to be an alternative raw material for making bread that is rich in fiber and protein. The ingredients that can be used are koro flour which has high protein content [9], *keladi* flour [10], cassava flour and soy flour [11]. Although this flour is classified as protein-rich flour (PRF) which is potential as a raw material for bread, the raw material is not easy to find. Another alternative is to use TSW flour as a substitute for bread making.

2. Method

The main raw material in this study is tofu solid waste (TSW). TSW was taken from the tofu industry in Cakung, East Jakarta. TSW was made into flour and then processed into plain bread. Raw bread was TSW flour, high protein flour, instant yeast, softener ingredients, sugar, salt, margarine, and water.

The experimental design conducted in this study was a Randomized Block Design (RBD) with factorial 4 x 1 with 2 repetitions to obtain 8 treatments. The factorial design model of 4 x 2 with Randomized Block Design (RBD) can be seen in table 1.

Table 1. Factorial 4 x 1 Experiment Design Model with Randomized Block Design (RCBD) with 2 replications.

Composition	**Repetition**	
	1	**2**
0 %	R_{11}	R_{12}
5 %	R_{21}	R_{22}
10%	R_{31}	R_{32}
15%	R_{41}	R_{32}

2.1. Tofu Flour Making
The procedure for making TSW flour began with choosing good TSW for example good in terms of soybean aroma, texture, and color. Extortion was done so that TSW knows the water content decreases, then the steaming was carried out for 30 minutes so that the TSW can last longer and reduce the bad smell.

Steamed TSW then dried in a dryer with a temperature variation of 120 ° C and 120 minutes. The dried TSW was grinding using a dry blender. The flour produced is then sieved with a 40 mesh sieve. If there is still a rough one, it can be refined again. Tofu flour that has been finished is then stored.

2.2. Bread Making
Bread making process based on Skrbic et al process [13] with modification. The Bread was made from TSW flour, high protein wheat flour, instant yeast, bread improver, baking powder, sugar, and they were stirred using a mixer with low speed until well blended. The water was then put into the mixture, stirred at low speed until well blended. Speed was increased to medium speed, stirred again until smooth for 15 minutes. Salt and margarine were added, stirred with a mixer followed by a low speed mixed well for 10 minutes until the dough becomes smooth.

The dough has become big and initial fermentation was done by placing the mixture on the baking plate and then closed using a wet cloth for 45-60 minutes until it expands. The dough that has undergone initial fermentation was deflated and then kneaded again for 15 minutes. The oven was heated until 180 °C, the dough was put into oven and bake until 30 minutes.

2.3. Chemical Analyzes
The responsive design in the research that will be carried out includes the response of the organoleptic test and response to quality/chemical testing. The control was used an original bread without TSW flour. The TSW bread was evaluated for chemical composition and organoleptic response.

The chemical composition was carried out in TSW bread making is to do a proximate test, water content, ash content, protein content, carbohydrate content, and fat content, and food fiber content. The organoleptic response is carried out to analyze the specific quality characteristics and acceptance of panelists on tofu bread products based on color, aroma, taste, and texture. This organoleptic test uses a hedonic scale, a determination criterion based on panelist's assessment of organoleptic properties by analyzing the level hedonic scale.

2.4. Hedonic Test
The hedonic test was conducted to 25 random panelists. The hedonic test includes color, flavor, aroma, and texture. It was also asked for the favorite bread. They were given four sample R_1, R_2, R_3, R_4.

3. Result and Discussion

3.1. Hedonic Test Result
The result of the hedonic test to 25 random panelists can be seen in table 2. It can be seen that the bread without tofu flour is the most favorite of 25 panelists with score 2.92 but in the composition of 5% tofu flour on second favorite bread, its color, taste, aroma, texture can still be accepted by the panelists. Even on color sensory tests get a higher value than plain bread without tofu flour. So that on average the composition of adding TSW knows 5% is in the second position, it can be seen in Figure 1.

Tabel 2. The hedonic test result of tofu waste flour bread

Sensory test.	**R_1**	**R_2**	**R_3**	**R_3**
Waste tofu bread composition (% w)	0 %	5 %	15 %	15 %
Color	2.93	3.00	2.93	2.93
Flavor	2.93	2.79	2.43	2.43
Aroma	3.28	3.14	2.86	2.86
Texture	3.28	2.86	2.86	2.86
Favorite	2.92	2.71	2.5	2.5
Average	3.068	2.9	2.716	2.716
The panelist	10	9	5	5

Information : Assessment score for color, taste, aroma, and preference criteria are 1 = very dislike; 2 = dislike; 3 = rather like; 4 = like; 5 = really like it. Assessment score for texture criteria is 1 = very hard; 2 = hard; 3 = rather hard; 4 = soft; 5 = very soft).

Figure 1 shows that there are still many panelists who like white bread without dreg flour know as much as 40%. Following the second-order is 5% composition variation with 20% panelist.

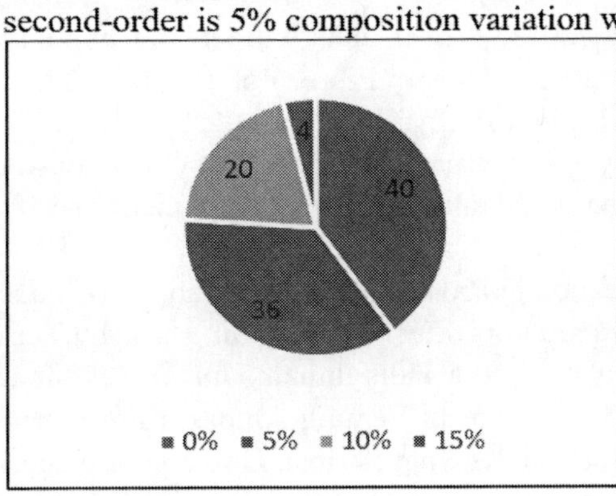

Figure 1. Percentage of the panelist of TSW bread

3.2. The Influence of Tofu Flour Composition on Proximate Tests and Food Fibre of TSW.

Of the two best compositions, besides without tofu flour, 5% and 10% are tested for proximate and dietary fiber, the results can be seen in Figure 1. The Proximate Test is carried out using the SNI 01-2891-1992 method, while the dietary fiber uses the testing method based on AOAC 985.29 19th e. 2012 (enzymatic).

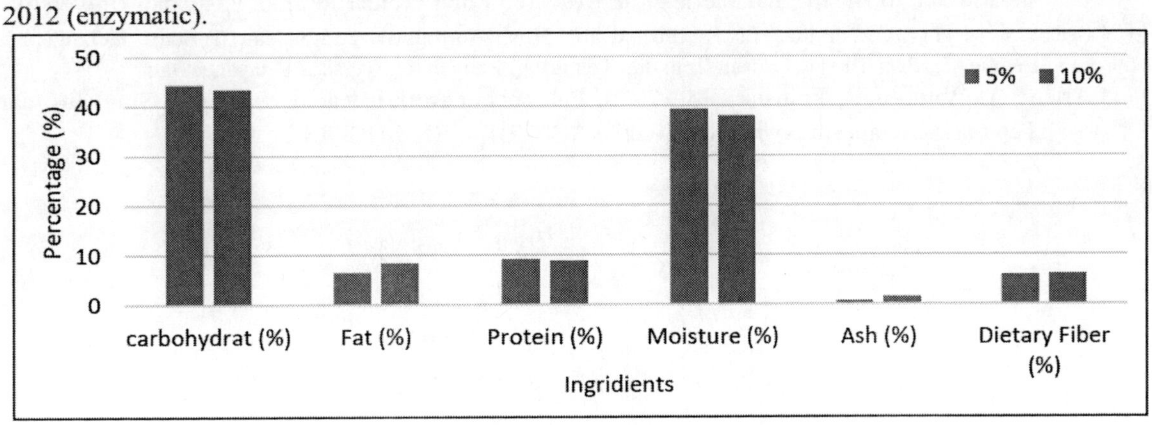

Figure 2. Proximate Test Results and bread fiber for composition of 5% and 10%

From Figure 2, it can be seen that the carbohydrate content of bread is still quite large, but the composition of 10% is lower than 5%; this is because the amount of tofu waste flour knows more. This is confirmed by the higher fiber content at 10% composition compared to 5%. Protein content in the composition is 5% higher because of the higher protein content of high-protein flour. The fat content of 10% is higher t 5% TSW flour than 10%.

4. Conclusion

The results showed that the greater the TSW flour composition, the lower the sensory value of panelists. The most favored sensory test was R2 sample, 5% tofu waste flour bread with the sensory score of favorite, texture, aroma, taste, and color of 2,71, 2,86, 3,14, 2,79, and 3. From the proximate test, the 5% TSW flour bread contains higher carbohydrate, protein, moisture content than 10% while the nutritional value was 44,39% carbohydrate, 6,45% fat, 9,16% protein, 39,3% moisture, 0,7% ash and 5,94% dietary fibre.

References

[1] Lu Fei, Yang Liu, BoLi. 2013. *Okara* dietary fiber and hypoglycemic effect of *okara* foods. Bioactive Carbohydrate and Dietary Fibre. Vol 2: 126 – 132.

[2] C.Song, Y. Kitamura, S.Li, and K.Ogasawara, 2012. Design of a cryogenic CO_2 capture system based on Stirling coolers. International Journal of Greenhouse Gas Control, Vol.7: 107 - 114

[3] Faisal, M., Gani, Maulana F., Maimon F., 2016. Treatment and Utilization of Indonesian. Asian Journal of Chemistry, vol 28 No. 3: 501 - 507

[4] Sutriswati, RE., 2012, Teknologi Proses Pembuatan Tahu, Yogyakarta, Kanisius.

[5] Ridhoresmi D, 2012, Pengaruh Subsitusi Tepung Ampas Tahu Terhadap Kadar Protein dan Daya Terima Brownies Kukus, Karya Tulis Ilmiah, Universitas Muhammadiyah Surakarta.

[6] Wati R., 2013, Pengaruh Penggunaan Tepung Ampas Tahu Sebagai Bahan Komposit Terhadap Kualitas Kue Kering Lidah Kucing, Skripsi, Universitas Negeri Semarang.

[7] Evawati and Roza,I. 2014. Utilization Of Solid Waste of Tofu and Out of Grade Carrot as a Source of Vegetable Protein and Fibre in Nugget Processing. International Journal on Advanced Science Engineering Information Technology. ISSN : 2088 – 5334, Vol. 4 : 71 - 74

[8] Mudjajanto ES dan Yulianti LN, 2004, Making Breads, Penebar Swadaya, Jakarta.

[9] Lestari CA, Widiantara T, dan Hasnelly, 2016, The Effect of *Canavalia ensiformis* flour substitution to Bread Characterization, Research Final Project Journal, Pasundan University

[10] Bramtarades, I.G.P.B., Putra, I.N.K, Puspawati, N.N, Formulasi Terigu dan Tepung Keladi pada Pembuatan Roti Tawar, Jurusan Ilmu dan Teknologi Pangan, Universitas Udayana, Bali.

[11] Arlene A, Witono JR, Fransisca M., 2009, Pembuatan Roti Tawar dari Tepung Singkong dan Tepung Kedelai, Simposium Nasional RAPI VIII, ISSN 1412-9612

IOP Conf. Series: Materials Science and Engineering **674** (2019) 012013 doi:10.1088/1757-899X/674/1/012013

Risk level analysis of work safety and health on national road routine maintenance project using self-management system

H A Rani*, M Isya and M Iqbal

Civil Engineering Department, Faculty of Engineering, Universitas Muhammadiyah Aceh, 23245 Banda Aceh, Indonesia

* Email: hafnidar.ar@unmuha.ac.id

Abstract. Routine maintenance refers to the repair of damages on roads' sections by means of steady service condition. It is always conducted on the Blangkejeren-Laweaunan Road Section on annual basis. It is often conducted using a self-management system by Executing Officer (PPK) 08, which, according to government statistics is 65 km long. While implementing the project, PPK worked with a competent subcontractor as well as a foreman from the community surrounding the project region. To reduce the risks and enhance safety, the contractor made use of the Management System of Safety and Health (SMK3), following all the regulations that apply in Indonesia. The study aims at identifying the K3 risk level in road routine maintenance work. This is done based on the PU Ministry Regulation No 05/PRT/M/2014 through interviewing the PPK. In the same context, determination of the compliance level is done by workers (subcontractor and foreman) is done based on the Construction and Building Guideline No. 004/BM/2006 by assessing the use of Personal Protector Equipment (APD) in the routine road maintenance activities, sidewalk routine maintenance, water channel routine maintenance of cut and fill, and road equipment routine maintenance. The low risk can be found in the routine maintenance the pavements, roadsides, and road infrastructure, while the moderate risk in water channels of the excavations and the heap's routine maintenance. Among the two, moderate risk has more portions and it must be conducted by workers who can fully comply with APD.

1. Introduction

The national road is the main artery of the economy. It must be well maintained to achieve the desired targets [1-4]. Routine maintenance is a core activity that can be done to improve the service period [5-10]. The road maintenance work is characterized by openwork location (which is affected by weather), limited time, untrained workers, high energy input, and the use of working equipment that may endanger workers' safety as well as health. There are many studies conducted around the world to improve [11-24].

To address the issue of workers' health and safety, PU Minister Regulation No 05/PRT/M/2014 on The Guideline of Work Safety and Health Management System (SMK3), provides that the Executing Officer (PPK) must apply SMK3 for the general construction work. The SMK3 fields entail the K3 policy, K3 planning, operational control, inspection, K3 performance evaluation, and K3 performance review [25].

The routine maintenance in Blangkejeren-Laweaunan Road Section has been conducted by the PPK 08, which is under the National Road Construction Working Unit of Regional I Aceh and the road section, with a length of 65 km. Notably, this segment is the only road connecting Blangkejeren to South

East Aceh District, an economic artery that must be protected and well reserved to boost the local economy of the area. In its implementation, the PPK undertook the project work with the help of a competent subcontractor who could effectively pay the foreman.

In the Work Accident Theory points out that one of the basic causes of accidents is the worker compliance level [26]. In this project, some activities do not uniformly comply with the requirements of the K3 standard. This is evidenced by the use of personal protective equipment that is inconsistent, thereby endangering workers along with other road users. This study is expected to apply the K3 implementation solutions for routine maintenance work on a self-managed basis that meets the provisions of the legislation. It will address the nature of K3 risk level in the national road routine maintenance project using self-management system in the Blangkejeren-Laweaunan Road Section. Furthermore, its main objective is to identify the K3 risk level in the routine national road maintenance work using self-management system on the Blangkejeren-Laweaunan Road Section.

2. National Road

According to the Regulation No. 38 of 2004, a national road is a land transportation infrastructure consisting of the road as well as the infrastructure and facility provided for the traffic located on the soil surface, under the ground and/or water, and above the water level except train railway, lane railway, and cable network [27].

3. Routine Maintenance

The PU Minister Regulation No. 13/PRT/M/2011 identifies routine maintenance as both preservation and repair activities following the damages occurring on the road sections by means of steady service condition road. A steady service condition road refers to the road sections with conditions that meet the design period calculated following the established standards [28].

4. Self-Management

According to the Republic of Indonesia President Regulation No. 54 of 2010 concerning The Procurement of Government Good/Service, self-management is the good/service procurement activity planned, carried out and/or supervised directly by K/L/D/I like the budget responsibility, other government institution and/or community group [29].

5. Work Accident

The Worker Minister Regulation No. Per. 03/MEN/1994 states that work accident is the mishap occurring in the work activities including the disease emerging from work activities, as well as the accidents occurring either on the way to the work location or on the way back home in the common street [30]. According to [31-33] acknowledge work accident as an unplanned, unpredictable, unexpected activity. In other words, it is something unforeseen and not intentional.

6. Work Safety and Health (K3)

According to RI Worker Minister No. Kep. 463/MEN/1993, Work Safety, and Health (K3) is a protection effort directed at the workers and other people located in the project location or company [34]. Health and safety measures should be maintained to ensure that every production resource can without harm be used efficiently. Acknowledge health and safety as two basic elements that all contractor companies must provide workers with [12, 20, 22, 23, 35].

6.1. Management System of Work Safety and Health 38

The Minister Regulation (SMK3) along with Minister Regulation No. 05/MEN/1996 perceives Management System of Work Safety and Health as a management system part. As a system, it includes activities such as organization, planning, implementation, taking responsibility, formulating the procedure

IOP Conf. Series: Materials Science and Engineering **674** (2019) 012013 doi:10.1088/1757-899X/674/1/012013

and resources required in development, application, achievement, review, maintenance, and policy of K3 [36].

The main purpose of these activities is to control the risks related to the work activity with a view to creating a safe and productive project location. The objectives and the target of SMK3 are basically to create a K3 system in the project location involving the elements of management, worker, and condition and project location integrated in order to avoid and reduce accidents and diseases arising at the project location. It also aims at creating a safe and efficient project location [37-38].

6.2. K3 Risk

According to the PU Minister Regulation No. 05/PRT/M/2014 on Management System of Work Safety and Health according to Minister Regulation of general sector construction explains, K3 risk of the construction is a measure of loss possibility to the public safety, property, human life, and the environment arising from the current dangerous resources within the project construction [25]. The measurement of K3 risk in the construction can be carried out by combining the value of the K3 risk occurrence and the damage or the resulting impact. The determination of the occurrence value ok K3 risk in the construction can be explained as the statistics in Table 1.

Table 1. The Occurrence Value of K3 Risk in the Construction [25]

Value	Occurrence
1 (One)	Seldom occurred in the construction activity
2 (Two)	Sometimes occurred in the construction activity
3 (Three)	Often occur in the construction activity

K3 risk level can be formulated as in (1).

$$TR = P \times A \quad (1)$$

where TR is K3 risk level, P for the occurrence of K3 risk in the construction, and A is severity raised the level. The calculation result of the K3 risk in the construction can be explained in Table 2.

Table 2. The Value of K3 Risk Level in the Construction [25]

K3 Risk Level in The Construction		Severity (Caused)		
		1	**2**	**3**
Frequency	1			3
	2		4	6
	3	3	6	9

Remark:

Low K3 Risk Level

Moderate K3 Risk Level

High K3 Risk Level

The calculation of severity level is done based on the average severity of the human, property, environmental and public safety. For the severity of the person resulting in death, the severity value level is 3 (high). If controlling efforts have been made and still high K3 risk level is high, it requires additional controlling efforts.

IOP Conf. Series: Materials Science and Engineering **674** (2019) 012013 doi:10.1088/1757-899X/674/1/012013

6.3. K3 Compliance
The compliance is a person's attitude to obey and participate in the specification, standard or regulation managed clearly in which regulations are issued by the related companies or other authorized institutions [39]. In addition, [11] mentioned that one component of safety behavior is safety compliance, that is, it is the activity that must be done by someone to protect his safety in the project location. The behavior follows the work standard procedure and the use of APD. A worker has two choices in creating the save and health project location. He or she can choose to obey K3 policy or avoid accidents and diseases due to work.

6.4. K3 Guideline for Road Routine Maintenance Project
The guideline of Construction and Building No. 004/BM/2006 states that the road routine maintenance has some references and provisions which have to be obeyed by the construction providers related to the K3 application [40].

7. K3 Risk Analysis
The K3 Risk Analysis is carried out to identify the K3 risk level in the national road routine maintenance project using a self-management system in the Blangkejeren-Laweaunan Road Section with a length of 65 Km. The routine maintenance is divided into 4 work items such as the routine maintenance of the pavement, the roadside, the water channel of excavations and the heaps, and the road infrastructure. The assessment of K3 risk level on each work item is conducted based on [25, 41, 42] using interview and observation methods. The risk level of each work item on national road routine maintenance project can be explained as follows:

7.1. K3 Risk Level on the Pavement Routine Maintenance
The pavement routine maintenance consists of 3 work items including assessment and stacking, stripping and closing, and returning the condition. The assessment of Risk Level (TR) in this item can be determined by multiplying the occurrence (K) and the effect (A) to the person, property, environment and public safety. This occurrence or K3 hazard frequency can be found out by interviewing the owner personnel. The occurrence value is marked by an area whose score is 1, 2, or 3. The impact of the occurrence can be determined by observing the road section of Blangkejeren-Laweaunan with a length of 65 Km. The score can be 1 for low, 2 for moderate or 3 for high. The result of the multiplication between the occurrence and the effect can then interpreted. The interpretation of K3 risk level can be divided into 3 levels starting from 1 and 2 for low, 3 and 4 for moderate, and 6 to 9 for high.

7.2. K3 Risk Level on the Roadside Routine Maintenance
The roadside routine maintenance consists of 3 work items, including assessment and stacking, stripping and closing, and returning the condition. It is explained that the total mean of K3 risk level achieved is 2. This implies that K3 risk level that occurred on the Roadside Routine Maintenance in the Blangkejeren-Laweaunan Road section has low risk.

7.3. K3 Risk Level on the Water Channel of the Excavations and the Heaps Routine Maintenance
The water channel of the excavations and the heap's routine maintenance consists of 3 work items including assessment and stacking, dredging the excavations and the heaps, and stone repair. The total mean of K3 risk level achieved is 3. That means that K3 risk level that occurred on the Roadside Routine Maintenance in the Blangkeujeren-Laweaunan Road section is moderate.

7.4. K3 Risk Level on the Road Infrastructure Routine Maintenance
The road infrastructure routine maintenance also consists of 3 work items, among them being assessment and stacking, stripping and closing as well as returning the condition. The total mean of K3 risk level

achieved is 2, meaning that the K3 risk level that occurred on the Road Infrastructure Routine Maintenance has a low risk.

7.5. K3 Risk Level on National Road Routine Maintenance.
This road routine maintenance is divided into 4 work items such as the routine maintenance of the pavement, the roadside, the water channel of the excavations and the heaps, and the road infrastructure. Based on the calculation, the statistics can be seen in Table 3.

Table 3. The Assessment of K3 Risk Level on National Road Routine Maintenance

No.	National Road Routine Maintenance	Total Mean TR K3	Risk Level
1	The Pavement Routine Maintenance	2	Low
2	The Roadside Routine Maintenance	2	Low
3	The Water Channel of the Excavations And The Heaps Routine Maintenance	2	Medium
4	The Road Infrastructure Routine Maintenance	2	Low

Table 3 shows that there are 2 risk levels on the National Road Routine Maintenance Project, low and moderate. The routine maintenance of the pavement, the roadside, and the road infrastructure have a low-risk level, that is, 2. Lastly, the water channel of the excavations and the heap's routine maintenance has a moderate risk level with a value is 3.

8. Conclusion
National road routine maintenance project in Blangkejeren-Laweaunan Road Section is done every year using self-management method. This road routine maintenance is divided into 4 work items such as the routine maintenance of the pavement, the roadside, the water channel of the excavations and the heaps, and the road infrastructure. The K3 risk is a condition of a national road routine maintenance project that occurred because of the uncertainty with the current frequency. It will create a negative impact on the person, property, environment and public safety.

Based on [25], K3 risk level is grouped into 3 levels: moderate and high. The assessment of the K3 Risk Level (TK) on the road routine maintenance project is determined by multiplying Occurrence (K) and Effect (A) to person, property, environment and public safety. The occurrence is the frequency of the activity and can be low, moderate or high. The occurrence can be found be established by interviewing the owner and the personnel. The effect can be determined by observation. This study establishes that there are 2 risk levels, low and moderate. The low risk can be found in the routine maintenance of the pavement, roadside, and road infrastructure. On the 0ther hand, the moderate risk can be found in the water channel of the excavations and the heaps routine maintenance.

References
[1] Rani H A 2017 Maintenance management of transportation effect to national road infrastucture cost performance in Aceh province *Impact Journals* **5 3** 1-8
[2] Rani H A 2017 Development priority of road infrastructure the Aceh post tsunami in Simeulue district *Int. J. Civ. Eng. (IJCE)*, **6 3** 1–12
[3] Tabti-Talamali A and Baouni T 2018 Public transportation in Algiers: towards a new governance approach *Case Stud. Transp. Policy* **9**
[4] Veeneman W and Mulley C 2018 Multi-level governance in public transport: governmental layering and its influence on public transport service solutions *Res. Transp. Econ.*

[5] Bryde D J and Brown D 2004 The influence of a project performance measurement system on the success of a contract for maintaining motorways and trunk roads *Proj. Manag. J.* ***35 4*** 57–65

[6] Issa A and Abu-Eisheh S 2017 Evaluation of implementation of municipal roads' maintenance plans in Palestine: a pilot case study *Int. J. Pavement Res. Technol.* **10 5** 454–463

[7] Gerami Matin A, Vatani Nezafat R and Golroo A 2017 A comparative study on using meta-heuristic algorithms for road maintenance planning: insights from field study in a developing country *J. Traffic Transp. Eng.* **4 5** 477–486

[8] Faith-Ell C, Balfors B Folkeson L 2006 The application of environmental requirements in Swedish road maintenance contracts *J. Clean. Prod.* **14 2** 163–171

[9] Nordin L and Arvidsson A K 2014 Are winter road maintenance practices energy efficient? A geographical analysis in terms of traffic energy use *J. Transp. Geogr.* **41** 163–174

[10] Fu L, Trudel M and Kim V 2009 Optimizing winter road maintenance operations under real-time information *Eur. J. Oper. Res.* **196 1** 332–341

[11] Griffin M A and Neal A 2004 *Safety Climate and Safety at Work,The Psychology of Workplace Safety* (Eds. Barling J & Michael R F) (Washington: American Psychology Association)

[12] Rani H A, Amin J and Gunawan R 2017 The factors identification of the work safety and health (K3) program application in the fly over project construction in Banda Aceh *Aceh Global Health International Conference*

[13] Rani H A 2015 Minimizing the cost for repairing road with pavement recycling *1st Jt. Conf. Indones.* 228 – 237

[14] Aziz R F and Abdel-Hakam A A 2016 Exploring delay causes of road construction projects in Egypt *Alexandria Eng. J.* **55 2** 1515–1539

[15] Victora L C, Cesar Gomes, Cardoso and Azevedo L J 2018 Portuguese mainland road network safety performance indicator *Elsevier* **6 3** 416–422

[16] Corazza M V, Musso A, Finikopoulos K and Sgarra V 2017 A methodology for the evaluation of road accidents health care costs for powered two-wheelers: The Rome case study *Case Stud. Transp. Policy* **5 2** 417–424

[17] McAndrews C and Deakin E 2018 Public health sector influence in transportation decision-making: the case of health impact assessment *Case Stud. Transp. Policy* **2**

[18] Mueller N 2015 Health impact assessment of active transportation: a systematic review *Prev. Med. (Baltim)* **76** 103–114

[19] Loprencipe G, Moretti L, Cantisani G and Minati P 2018 Prioritization methodology for roadside and guardrail improvement: Quantitative calculation of safety level and optimization of resources allocation *J. Traffic Transp. Eng.* **5 5** 348–360

[20] Gul M and Ak M F 2018 A comparative outline for quantifying risk ratings in occupational health and safety risk assessment *J. Clean. Prod.* **196** 653–664

[21] Moore L L, Wurzelbacher S J and Shockey T M 2018 Workers' compensation insurer risk control systems: Opportunities for public health collaborations *J. Safety Res.* **66** 141–150

[22] Botti L, Duraccio V, Gnoni M and Mora C 2018 An integrated holistic approach to health and safety in confined spaces *J. Loss Prevention in the Process Industries* **55** 25-35

[23] Parker A, Tones M and Ritchie G 2017 Development of a multilevel health and safety climate survey tool within a mining setting *J. Safety Res.* **62** 173-180

[24] Mohammadi A, Tavakolan M and Khosravi Y 2018 Developing safety archetypes of construction industry at project level using system dynamics *J. Safety Res.* **67** 17-26

[25] Permen PU No. 05/PRT/M/2014 *Tentang Pedoman SMK3 Konstruksi Bidang Pekerjaan Umum*

[26] Heinrich H W 1980 *Industrial Accident Prevention* (New York: Mc. Graw Hill Book Company)

[27] Undang-undang No. 38 Tahun 2004 *Tentang Jalan*

[28] Permen PU No. 13/PRT/M/2011 *Tentang Tata Cara Pemeliharaan dan Penilikan Jalan*

[29] Perpres R.I No. 54 Tahun 2010 *Tentang Pengadaan Barang/Jasa Pemerintah*

[30] Permenaker No. Per. 03/MEN/1994 *Tentang Kecelakaan Kerja*

[31] Endroyo B 2007 Analisis Faktor-faktor Penyebab Kecelakaan Kerja Konstruksi *Jurnal Unnes Semarang*

[32] Kaygisiz Ö, Senbil M and Yildiz A 2017 Influence of urban built environment on traffic accidents: The case of Eskisehir (Turkey) *Case Stud. Transp. Policy* **5 2** 306–313

[33] Vieira Gomes S 2013 The influence of the infrastructure characteristics in urban road accidents occurrence *Accid. Anal. Prev.* **60** 289–297

[34] Kepmenaker R.I. No. Kep. 463/MEN/1993 *Tentang Pola Gerakan Nasional*

[35] Andarini P 2017 Evaluasi Pelaksanaan Keselamatan dan Kesehatan Kerja (K3) pada Perusahaan Konstruksi Pemeliharaan Jalan di Dinas Kimpraswil Kota Yogyakarta *Jurnal Universitas Ahmad Dahlan Yogyakarta*

[36] Permenaker No. Per. 05/MEN/1996, *Tentang SMK3*

[37] Pangkey F 2012 Penerapan Sistem Manajemen Keselamatan dan Kesehatan Kerja (SMK3)pada Proyek Konstruksi Di Indonesia (Studi Kasus: Pembangunan Jembatan Dr. Ir. Soekarno-Manado) *Jurnal Ilmiah Media Engineering Universitas Sam Ratulangi Manado* **2 2**

[38] Xia N, Zou P, Griffin M, Wang X and Zhong R 2018 Towards integrating construction risk management and stakeholder management: A systematic literature review and future research agendas *International Journal of Project Management* **36 5** 701-715

[39] Wardani 2009 *Riset Sumber Daya Manusia* (Jakarta: Gramedia Pustaka Utama)

[40] *Pedoman Pelaksanaan K3 untuk Konstruksi Jalan dan Jembatan No: 004/BM/2006* (Jakarta)

[41] Sepang B A W 2013 Manajemen Risiko Keselamatan dan Kesehatan Kerja (K3) pada Proyek Pembangunan Ruko Orlens Fashion Manado *Jurnal Sipil Statik Universitas Sam Ratulangi Manado* **1 4**

[42] Rouhani O M, Geddes R R, Do W, Gao H O and Beheshtian A 2018 Revenue-risk-sharing approaches for public-private partnership provision of highway facilities *Case Stud. Transp. Policy* **4**

Acknowledgments
This research was supported by Universitas Muhammadiyah Aceh. We thank our colleagues from Universitas Syiah Kuala who provided insight and expertise that assisted this research, although they may not agree with all of the interpretation and conclusion produced of this paper.

Spectral matching and propagation ground motion earthquake from bedrock to surface

I T Saputro

Department of Civil Engineering, Faculty of Engineering, Universitas Muhammadiyah Sorong, Indonesia

Email: Imam.civil10@gmail.com

Abstract: Ground motion earthquake is the acceleration data of ground during the earthquake. The tool used to obtain this data is called the accelerogram. In Indonesia, this data is difficult to obtain due to the limitations of the tool. It needed to analyze high-rise building because analysis structure of high-rise building used the dynamic load from the earthquake. This data is important in analyzing earthquake-resistant buildings. Besides from the accelerograms tool, this data can be obtained through seismic hazard analysis process. This was obtained by taking ground motion from elsewhere and then uniforming the spectrum response results with the location being reviewed. Wave propagation from bedrock to the soil surface will see the amplification process so this needs to be reviewed as well. Based on seismic hazard analysis results obtained ground motion suitable for matching process is San Fernando 1971. Peak ground acceleration on bedrock obtained is about 0.2608 g. After wave propagation to the surface, the peak ground acceleration is about 0.3001 g. Based on that, Amplification factor earthquake waves is about 1.12.

Keywords: Seismic Hazard Analysis, shallow crustal, ground motion.

1. Introduction

High-rise Building Structure must be analyzed using a dynamic load. Ground motion acceleration earthquake is one of the dynamic loads that used for building structural analysis. Structure analysis using ground motion acceleration earthquake is rarely done because the data is difficult to obtain. Generally, ground motion acceleration data is adopted from earthquake recording elsewhere without the process of matching because it is difficult. This way should not be done because a character of ground motion acceleration is not the same. So in this study, the response of high-rise building structure was evaluated using the ground motion acceleration that comes from the analysis of earthquake hazard using PSHA method.

Ground motion acceleration that used is the result matching of earthquake data in bedrock based on hazard de-aggregation analysis. Then the ground motion data in the bedrock is vertically propagated to obtain ground motion acceleration data at the surface. Ground motion acceleration at ground level is used in the evaluation of the structural response of the building

Conducted seismic hazard analysis using the PSHA method (Probabilistic Seismic Hazard Analysis) with shallow crustal earthquake source for probability exceeded 10% in 500 years to know the acceleration of peak land that occurred in Apartments and hotels Mataram City Yogyakarta. Peak ground acceleration obtained for 0.2608 g. After the hazard de-aggregation process obtained the distance of dominant 19.49 km with a dominant magnitude (Mw) of 6.31 from the Opak fault. Ground motion suitable for this location is San Fernando 1971 [1].

This research will be analyzed spectral matching ground motion San Fernando 1971 toward response spectrum of research location to obtain artificial ground motion. Earthquake wave propagation is very influential on soil properties. Therefore, the analysis of earthquake wave propagation from the bedrock to the surface is needed to obtain amplification factor of earthquake waves.

2. Method

2.1. Research sites

This research is located at the Apartment and hotel of Mataram City in Yogyakarta area. This location was chosen because Apartment and hotel Mataram City is one of the highest buildings in Yogyakarta so it is important to know the ground motion of the area as an effort to evaluate the structure of the building. The location of the study is shown in Figure 1.

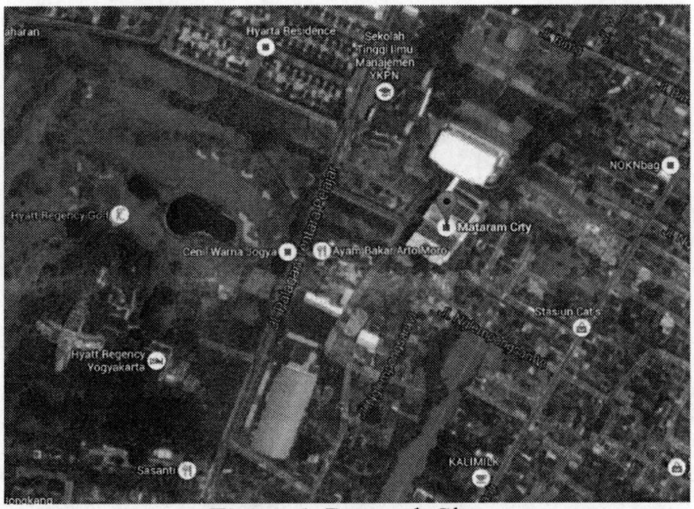

Figure 1. Research Sites
(Google Map, 2018)

2.2. Digitization Ground Motion

Digitization of ground motion is an acceleration recording data from the earthquake. There are 2 (two) methods to obtain ground motion data that is using ground motion data ever recorded around study location and using data of motion synthetic. The problem that arises is that almost all major cities in Indonesia do not have representative ground motion data for use in the seismic analysis [2]. Based on this, ground motion analysis can be done using ground motion synthetic derived based on hazard seismic analysis using PSHA method. After that done uniformity with the process of analysis of spectral matching.

2.3. Analysis of Dynamic Soil Response

Soil dynamics response is a soil behavior that occurs due to dynamic load/force. In geotechnical engineering of the earthquake, earthquake response is a parameter that must be calculated. The maximum acceleration and response spectrum occurring on the surface is strongly influenced by soil conditions such as the type of soil layer and the thickness of the soil layer [3]. This is evidenced by research conducted by Idriss and Kramer [4]. It is shown in Figure 2 and Figure 3.

IOP Conf. Series: Materials Science and Engineering **674** (2019) 012014 doi:10.1088/1757-899X/674/1/012014

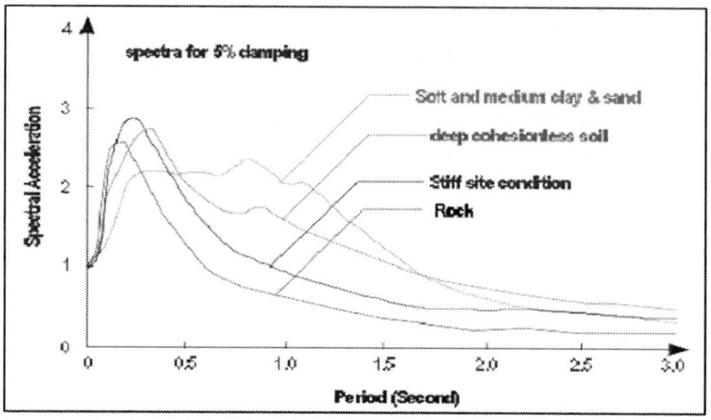

Figure 2. Effect of Soil Type on Spectral Acceleration Form (Kramer, 1996)

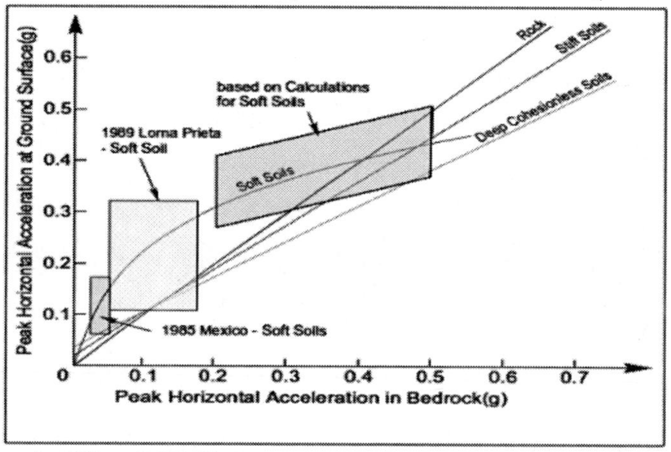

Figure 3. Effect Of Soil Type On Acceleration Change (Idriss, 1990)

The character and spread of earthquake damage are most influenced by the soil response to cyclic loads. The soil response is determined by the dynamic parameters of the soil. The dynamic parameters used in soil dynamics response analysis are the maximum shear modulus (G_{max}), shear wave velocity (V_S), and damping (ξ). Here is the correlation of N-SPT value with parameter acceleration of soil in Table 1.

Tabel 1. Correlation between G_{max} and V_S with SPT (Irsyam, 2000)

Reference	Gmax Correlation (kPa)	*Vs* Correlation (m/det)	R	Type of soil
Ohsaki, Iwasaki (1973)	$G_{max} = 11500N^{0.8}$		0.888	All (Japan)
Ohta, goto (1978)		$V_S = 85.3N^{0.341}$	0.72	All (Japan)
Imai, Tonouchi (1982)	$G_{max} = 14070N^{0.68}$	$V_S = 96.9N^{0.314}$	0.867	All (Japan)
Seed et al (1983)	$G_{max} = 6220N$			Sand (USA)
Sykora, Stokoe (1983)		$V_S = 101N^{0.29}$	0.84	Sand (USA)

3. Results and Discussion

3.1. Ground Motion Recording

After performing the de-aggregation process, the parameters obtained such as dominant distance (R) and dominant magnitude (Mw) are used as reference in determining ground motion soil/time history according to the condition of research location. The ground motion used was obtained from Ground Peer Motion Database that can be accessed and download at the site Ground Peer Motion Database.

From these results obtained that the recording ground motion/time history following the location of the location review is San Fernando earthquake data in 1971 recorded at Pasadena earthquake recording station - Old Seismo Lab. This quake has a magnitude 6.61 with a distance of 21.5 km. The following is the graph of the earthquake in Figure 4.

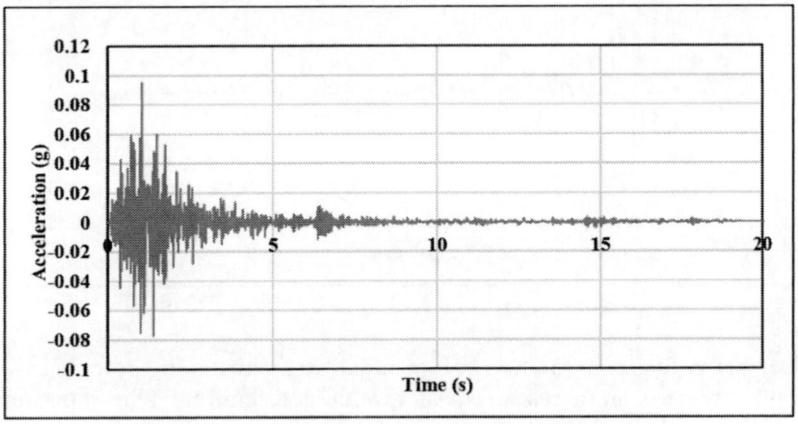

Figure 4. San Fernando 1971 Ground Motion

3.2. Spectral Matching

After obtaining ground motion data on the Peer Ground Motion Database site that matches the characteristics and source of the earthquake in the reviewed location, the ground motion can be used in the spectral matching process. The spectral matching process is done so that ground motion obtained according to the condition of the study site. This process does by matching the response of the target location spectral studied. In this study, the target spectral used is a spectral response in bedrock from the source of a shallow crustal earthquake. The matching process is presented in Figure 5.

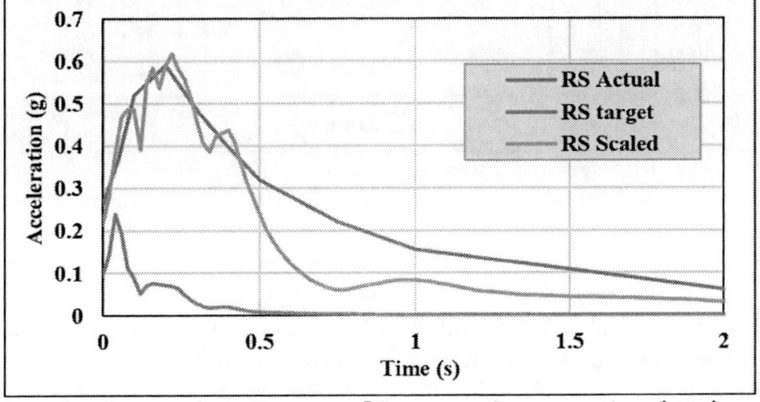

Figure 5. *Matching* Process of Response Spectrum Acceleration

IOP Conf. Series: Materials Science and Engineering **674** (2019) 012014 doi:10.1088/1757-899X/674/1/012014

The actual response spectrum is a response spectrum acceleration original of the 1971 San Fernando earthquake. Response spectrum acceleration of the target is obtained from the analysis of seismic hazard at the site which is reviewed in this study. Then response spectrum scaled is the spectral response of the matching result. The result of matching the response spectrum produces a motion synthetic ground in Figure 6.

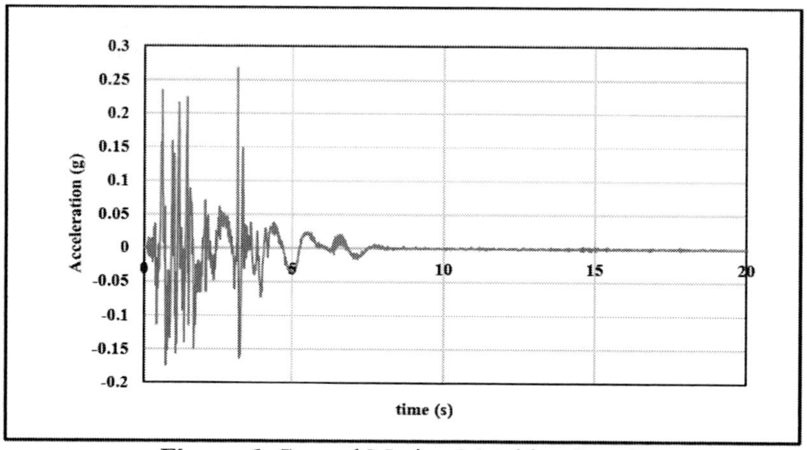

Figure 6. Ground Motion Matching Result

The comparison between ground motion before and after the matching process is shown in Figure 7. The graph shows that there is an increase in peak ground acceleration. This is because the location of the review has a higher peak ground acceleration value.

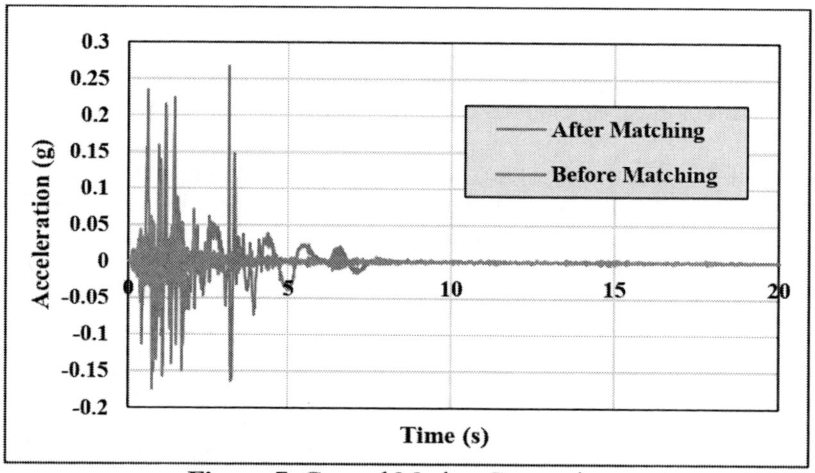

Figure 7. Ground Motion Comparison

3.3. *Dynamic Soil Parameters*

The soil has a layered texture. Each layer has different properties that have different patterns and behaviors in the wave propagation process. Therefore, the dynamic parameters of the soil need to be analyzed. The soil dynamic parameters required for dynamic soil response analysis are G_{max} and Vs. This value can be known through the results of land borlog at the research location which can then be correlated. The following results of N-SPT at the study sites are presented in Figure 8.

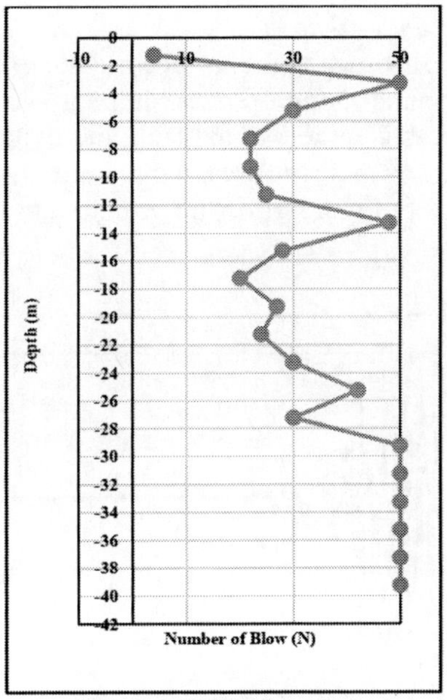

Figure 8. Borlog Data

To determine the correlation between the number of blow (N) with dynamic soil parameters in the form of shear wave velocity (*Vs*) using the equation correlation formula Ohta, Gotto (1978) [5] and Imai, Tonouchi (1982) [6]. Graph of propagation results in the form of correlation G_{max} and *Vs* depend on depth after analyzed in Figure 9.

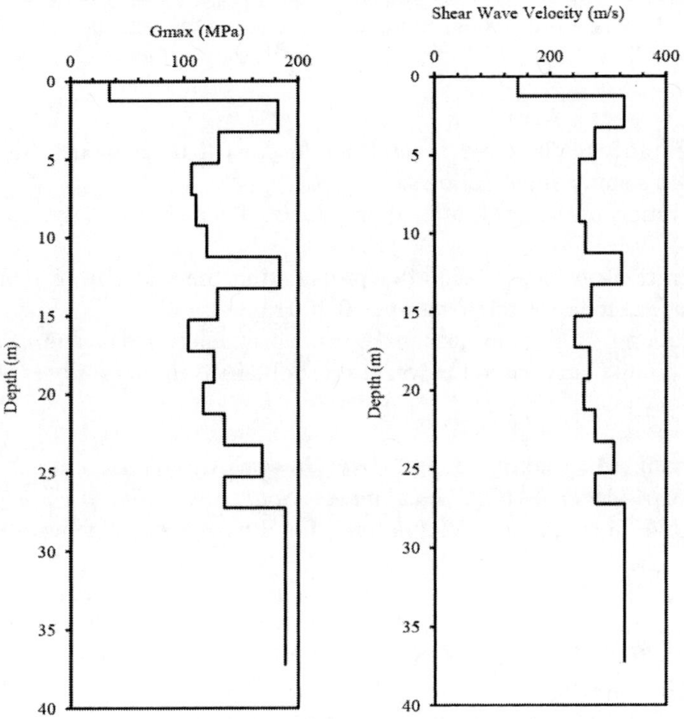

Figure 9. Correlation *Gmax* and *VS* to the depth at Apartment and Hotel Mataram City, Sleman, Yogyakarta

3.4. Ground Motion Acceleration on Surface

Analysis of wave propagation from bedrock to the surface is done to determine the effect of soil on wave propagation. After the analysis ground motion process in bedrock, then does wave propagation from *bedrock* to the surface. This process is done by analyzing the dynamic parameters of the soil in each layer. The ground dynamic parameters included are the maximum shear modulus value (G_{max}) and shear wave velocity (Vs). The following in Figure 10 is the ground surface acceleration motion after being propagated to the surface.

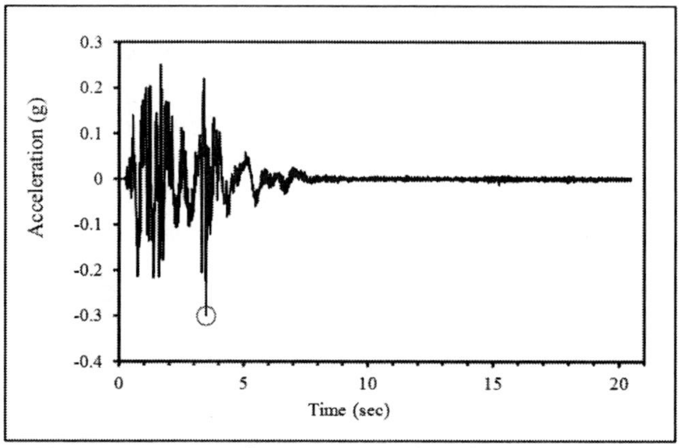

Figure 10 *Ground Motion* Acceleration On Surface

Based on the wave propagation results, the peak acceleration value on the surface has increased after being propagated from the bedrock. This increase is called the amplification factor. This factor may change depending on the soil condition of the study location. In this study, the peak ground acceleration at bedrock is about 0.2679 g while the peak ground acceleration on the surface increased is about 0.3001 g. Therefore, it can be concluded that the magnitude of amplification factor in the research area is 1.12.

4. Conclusion

After doing the analysis process, there are several conclusions that resulted from this research. The conclusions obtained in this study are as follows:

1) Peak ground acceleration on bedrock after doing spectral matching analysis process at building location is 0.2679 g.
2) Peak ground acceleration on the surface after propagating the wave from bedrock using NERA software to the surface at the research location is 0.3001 g.
3) There is an amplification factor in the earthquake wave propagation. The amplification factor of the wave propagation from the bedrock to the soil surface obtained in this study is about 1.12.

References

[1] Saputro, I.T., Teguh, M., Makrup, L., (2018). *Ground Motion Earthquake Recommendation Based On PSHA Method With Earthquake Source Of Shallow Crustal.* International Conference on Education and Technology for Environmental Sustainability (ICETES), Sorong.

[2] Irsyam, M., et al., (2008), *Usulan Ground Motion untuk Batuan Dasar Kota Jakarta dengan Periode Ulang Gempa 500 Tahun untuk Analisis Site Specific Response Spectra.* Seminar dan Pameran HAKI 2008 "Pengaruh Gempa dan Angin terhadap Struktur".

[3] Idriss, I.M., (1990), *Response of Soft Soil Sites During Earthquake*. J.M. Duncan, ed., Proceedings, H. Bolton Seed Memorial Symposium, BiTech Publishers, Vancouver, British Columbia, Vol. 2

[4] Kramer, S. L., (1996), *Geotechnical Earthquake Engineering*, Prentice-Hall, Inc., Upper Suddle River, New Jersey

[5] Ohta Y, Goto N (1978) Empirical shear wave velocity equations in terms of characteristic soil indexes. Earthq Eng Struct Dyn 6:167–187

[6] Imai T, Yoshimura Y (1970) Elastic wave velocity and soil properties in soft soil. Tsuchito-Kiso 18(1):17–22

Mechanical degradation of normal concrete due to seawater intrusion

F Hamdi and H A Imran

Civil Engineering Muhammadiyah University of Makassar, Sultan Alauddin Street No. 259, Makassar, Sulawesi Selatan, Indonesia.

E-mail: fauzanhamdi@unismuh.ac.id, Hamzah@unismuh.ac.id

Abstract. Problems that occur in concrete buildings in the coastal environment of concrete damage caused by degradation of concrete that interacted seawater. Real damage caused by seawater, consists of 3 (three) parts: submerged concrete part, concrete part affected by tidal seawater and part of the concrete affected by seawater splashing. The concrete damage discussed in this research is the mechanical degradation of concrete due to the intrusion of seawater in the submerged seawater. Damage can occur in concrete due to the reaction between aggressive seawater that is intruded into the concrete and compounds in the concrete that causes the concrete lose some of its mass, its strength, and stiffness and also accelerate the weathering process. The results of this study indicate that the reduction of compressive strength of concrete due to the intrusion of seawater tends to show a logarithmic graph. Concrete compressive strength will lose by 50% when the concrete is soaked seawater for 19,031 days. Normal quality concrete f'c = 25 MPa undergoes mechanical degradation due to seawater intrusion, in this case, the compressive strength of concrete is 12,063 % when soaked 28 days and 16,809 % when soaked 90 days.

Keywords: Normal Concrete, Mechanical Properties of Concrete, Sea Water

1. Introduction

The concrete for the Building structure is required to have strong properties in holding the load or working forces [1]-[3]. In addition, concrete must also have high durability so that the concrete is not easily damaged and can be used as a structure for a long time [4]-[7].

Very strong writing that is soaked with air that shows better results when compared to concrete soaked with sea air [8]-[13]. The decreasing difference for the compressive strength of sea air concrete reaches 7.187% of the strong value of fresh air concrete inspection [8]. Previous research reported that air cement factor 0.45; 0.50; and 0.55 gives some values against intense pressure and also between very strong pressure using strong water with concrete using clean water for pressures using clean water higher than drying using sea air [14]. As for the duration of curing with sea air for 1 day, 2 days, and 3 days there is a strong area. The writing is relatively small. There was a slight decrease from 1 to 3 days for the cement water factor of 0.45 and 0.5 and a slight increase from 1 to 3 days for the 0.55 cement water factor.

There is a compressive strength penetration, tensile strength, bending strength in submerged concrete sulfate [15]. The percentage decrease in seawater immersion [16] was 2.92%, 2.56%, -8.04%. The decrease in magnesium sulfate solution 5% was 3.86%, 7.51%, -2.68% and percentage decrease in magnesium sulfate solution 5% tidal system 9,66%, 18,9% and 24,11% . With the above statement, the authors need to research seawater to degradation of seawater intrusion normal concrete materials.

2. Method

2.1. Media & water immersion

This research uses 2 media pond, where one pond in the contents of seawater and one pond again filled with fresh water. Water used for soaking concrete is seawater from Barombong Beach, Makassar, South Sulawesi, and freshwater used from PDAM Gowa, Gowa Regency, South Sulawesi Province. The model of treatment of specimens can be seen in Figure 1 and Figure 2.

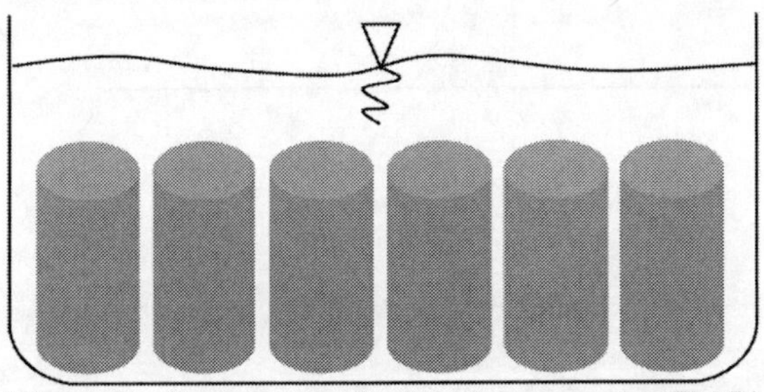

Figure 1. Concrete soaked in the seawater (BAL)

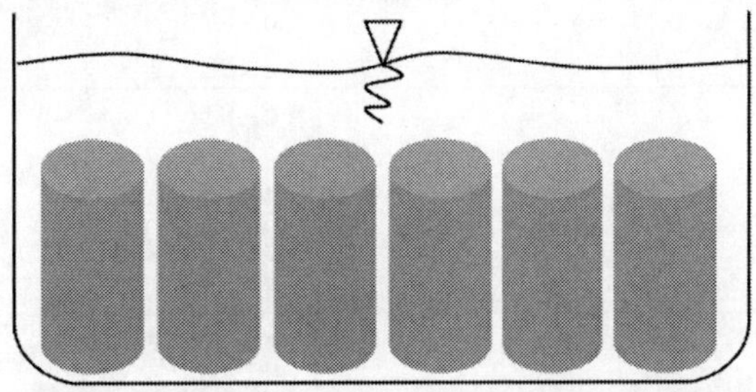

Figure 2. Concrete soaked in the freshwater (BAT)

The amount of concrete soaked in seawater as much as 20 specimens, and soaked freshwater (PDAM) as much as 20 specimens. Illustration of the treatment of test specimens can be seen in Table 1 and Tabel 2.

Tabel 1. Illustration of the test specimens being soaked in freshwater

Age of Concrete	Code	Compressive Strength Of Concrete Averages
14	FcAT14_1 FcAT14_2 FcAT14_3 FcAT14_4 FcAT14_5	$fcAT14$
28	FcAT28_1 FcAT28_2 FcAT28_3 FcAT28_4 FcAT28_5	$fcAT28$
56	FcAT56_1 FcAT56_2 FcAT56_3 FcAT56_4 FcAT56_5	$fcAT56$
90	FcAT90_1 FcAT90_2 FcAT90_3 FcAT90_4 FcAT90_5	$fcAT90$

Tabel 2. Illustration of the test specimens being soaked in seawater

Age of Concrete	Code	Compressive Strength of Concrete Averages
14	FcAL14_1 FcAL14_2 FcAL14_3 FcAL14_4 FcAL14_5	$f'cAL14$
28	FcAL28_1 FcAL28_2 FcAL28_3 FcAL28_4 FcAL28_5	$f'cAL28$
56	FcAL56_1 FcAL56_2 FcAL56_3 FcAL56_4 FcAL56_5	$f'cAL56$
90	FcAL90_1 FcAL90_2 FcAL90_3 FcAL90_4 FcAL90_5	$f'cAL90$

2.2. Material specification and test material
The concrete materials that will be used in this research are:
1) Cement to be used Composite Portland Cement in the market area of Makassar, South Sulawesi
2) Coarse aggregate used rubble stone originating from Bili-Bili area, Gowa regency, South Sulawesi
3) The fine aggregate used comes from the River Je'neberang, Gowa regency, South Sulawesi
4) Water mixture of concrete to be used fresh water from PDAM Makassar, South Sulawesi

Water used to soak the concrete :
1) Freshwater from PDAM Makassar, South Sulawesi
2) Seawater from Barombong, Makassar, South Sulawesi

Specification of specimen used :
1) The quality of normal concrete (f'c = 25 MPa)
2) Cylindrical concrete
3) Dimensions of concrete diameter 15 cm and height 30 cm

2.3. Variable testing
The variables studied in this research are compressive strength of concrete, duration of concrete immersion and concrete compressive strength reduction. The parameters used in this research are the normal concrete quality f'c = 25 MPa, the duration of soaking of concrete in freshwater and in the seawater that is 14 days, 28 days, 56 days and 90 days and the reduction of concrete compressive strength during the soak 14 day, 28 days, 56 days and 90 days

2.4. Research method
The methods to be used in this research are experimental research and literature study about concrete degradation due to seawater penetration. The stages of this research are preliminary survey and research preparation. This survey is conducted to determine the availability and feasibility of tools and materials that will be used in research so that it can support the implementation of research

2.5. Laboratory test of specification and characteristic of a constituent material
1) Fine aggregate examination includes specific gravity and absorption, filter analysis, organic content, mud content and weight of volume.
2) Coarse aggregate examination (rubble stone) including wear and tear, specific gravity and absorption, filter analysis, organic content, mud content, and weight volume.
3) Freshwater examination (PDAM) includes organic content.
4) Planning of concrete mix (make mix design).
5) Mixed design (mix design) using SNI Method (SNI 03-2834-2000) by National Standardization Agency [17].
6) Preparation of concrete test objects.

2.6. The test specimens will be made as follows
Analysis and test of mechanical degradation of concrete due to seawater intrusion. The concrete compressive strength test (f'c) uses a 15 cm X 30 cm (15 cm x 30 cm) normal concrete mixture soaked in 20 seawater and soaked in freshwater (PDAM) 20 pieces tested on the day to 14, 28, 56 and 90

2.7. Immersion of concrete test objects
Immersion of concrete test objects is done in the following manner:
1) Normal quality concrete (f'c = 25 MPa) cylindrical 15 cm X 30 cm submerged 20 seawater for concrete compressive strength test (f'c), concrete weight and weight of concrete volume.
2) Normal quality concrete (f'c = 25 MPa) cylindrical 15 cm X 30 cm water-soaked (PDAM) 20 samples for concrete compressive strength test (f'c), concrete weight and weight of concrete volume.

2.8. *Testing* the specimens
Testing of the concrete test object is done by concrete compressive test using Universal Testing Machine done on concrete soaked in freshwater (PDAM) [18]-[20] and seawater for 14, 28, 56 and 90 days. At each time tested the compressive strength of concrete as much as 5 pieces of test specimens soaked in seawater and 5 pieces of freshwater soaked specimens (PDAM water).

2.9. Analysis *and* data processing
The data analysis and processing were performed to determine the mechanical degradation of concrete due to seawater intrusion by testing the compressive strength of normal quality concrete (f'c = 25 MPa) soaked in seawater and freshwater (PDAM).

3. Result and discussion

3.1. Examination of material characteristics
The result of an examination of fine aggregate characteristic (sand) shows that it meets the requirement of SNI specification so that fine aggregate can be used as concrete mixture. The results of the examination can be seen in Table 3.

Tabel 3. Testing result of fine aggregate characteristics (sand)

No	Characteristic of Fine Aggregate	SNI Specification	Interval	Testing Result	Note
1	Clay Content	SNI 03-4141-1996	max 5%	1.25%	Qualify
2	Organic Content	SNI 03-2816-1992	< NO. 3	No. 1	Qualify
3	Water Content (Wp)	SNI 03-1971-1990	0.5% - 5%	1.60%	Qualify
4	Volume Wight	SNI 03-4804-1998			
	a. Unsolid Condition		1.4 - 1.9 kg/liter	1.42	Qualify
	b. Solid Condition		1.4 - 1.9 kg/liter	1.65	Qualify
5	Absorption (Rp)	SNI 03-1970-1990	0.2% - 2%	1.42%	Qualify
6	Specific dry weight of the surface	SNI 03-1970-1990	1.6 - 3.3	2.40	Qualify
7	Modulus of fineness	SNI 03-1968-1990	1.50 - 3.80	3.46	Qualify

The result of the examination of crude aggregate characteristic (rubble stone) indicates that it meets the requirement of SNI specification so that coarse aggregate can be used as a concrete mixture. The results of the examination can be seen in Table 4.

Tabel 4. Testing result of coarse aggregate characteristic (rubble stone)

No	Characteristic of Coarse Aggregate	SNI Specification	Interval	Testing Result	Note
1	Shrinkage	SNI 03-2417-1991	max 50%	20.38%	Qualify
2	Clay Content	SNI 03-4141-1996	max 1%	0.90%	Qualify
3	Water Content (Wk)	SNI 03-1971-1990	0.5% - 2%	1.34%	Qualify
4	Volum Weight	SNI 03-4804-1998			
	a. Unsolid Condition		1.6 - 1.9 kg/liter	1.61	Qualify
	b. Solid Condition		1.6 - 1.9 kg/liter	1.66	Qualify
5	Absorption(Rk)	SNI 03-1969-1990	max 4%	2.78%	Qualify
6	Specific dry weight of the surface	SNI 03-1969-1990	1.6 - 3.3	2.60	Qualify
7	Modulus of fineness	SNI 03-1968-1990	6.0 - 7.1	3.28	Qualify

3.2. Composition of concrete materials
The composition of the normal concrete mixture material f'c = 25 MPa can be seen in Table 5.

Tabel 5. The composition of the normal concrete mixture material f'c = 25

Concrete Material	Concrete Weight (Kg/m^3)	Ratio To Total Cement
Fresh Water	229	0,6
Cement	413	1,0
Sand	530	1,3
Rubble Stone	1058	2,6

3.3. Mechanical degradation of concrete due to seawater intrusion

The concrete mechanical degradation test is done by using normal concrete compressive strength test (f'c = 25 MPa) by two kinds of treatment variation ie freshwater and concrete soaked concrete seawater. The result of normal concrete compressive strength (f'c = 25 MPa) can be seen in Tables 6, Figure 3 and Figure 4.

Tabel 6. Testing result of compressive strength normal quality concrete

Age of Concrete (Days)	Compressive Strength Of Concrete Averages Fresh Water F'cAT (MPa)	Compressive Strength Of Concrete Averages Sea Water F'cAL (MPa)	Compressive Strength Of Concrete Reduction Averages (%)
14	14.66	13.86	5.41
28	17.83	15.67	12.06
56	26.03	22.07	15.22
90	26.60	22.13	16.81

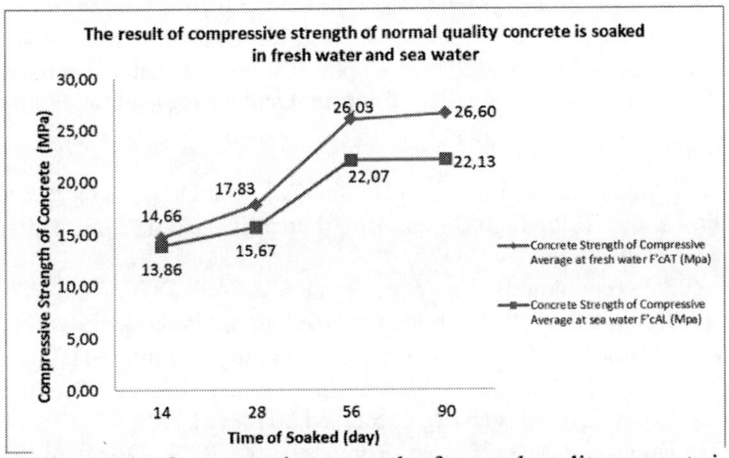

Figure 3. The result of compressive strength of normal quality concrete is soaked in freshwater and seawater

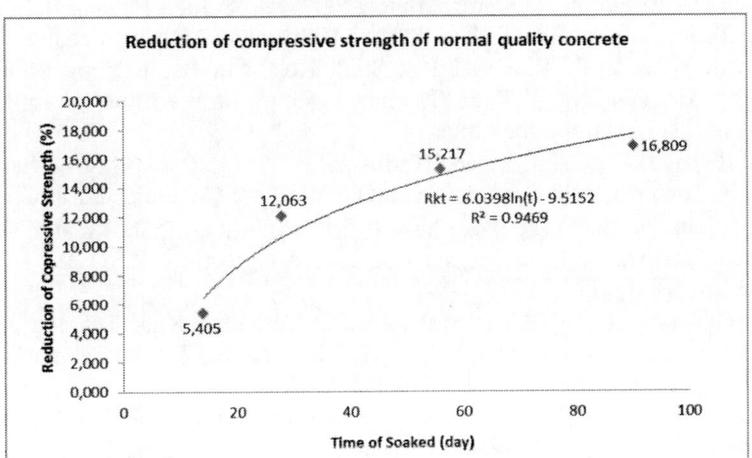

Figure 4. Reduction of compressive strength of normal quality concrete

IOP Conf. Series: Materials Science and Engineering **674** (2019) 012015 doi:10.1088/1757-899X/674/1/012015

Figure 3 and Figure 4 show that the reduction of compressive strength of concrete due to seawater intrusion tends to show the logarithmic equation pattern. The pattern of the equations is :

$$y = 6.0398 * \ln(x) - 9.5152 \tag{1}$$

Where y is the reduction of compressive strength of normal concrete post seawater intrusion (%), and x for duration of immersion (day). Equation (1) can predict that the normal compressive strength of the concrete will lose by 50% when the concrete is soaked in seawater for 19,031 days.

4. Conclusion

Normal quality concrete f'c = 25 MPa undergoes mechanical degradation due to intrusion of seawater, in this case, the compressive strength of concrete is 12.063% when soaked 28 days and 16.809% when soaked 90 days. Result of graph of mechanical degradation due to seawater intrusion, in this case graph relation between compressive strength of normal quality concrete f'c = 25 MPa with the length of immersion (days) yields logarithmic graph. Also, normal concrete strength of f'c = 25 MPa will experience mechanical degradation due to seawater intrusion by 50% when soaked in seawater during 19,031 days. In a further study of the degradation of physical and mechanical properties of other concrete need to be studied further on the various quality of concrete and an application of structural components

References

[1] Paul Nugraha & Antoni, 2004, Teknologi Beton, CV. Andi Offset. Yogyakarta

[2] Tjokrodimuljo, 2009, Teknologi Beton, Biro Penerbit Teknik Sipil Universitas Gajah Mada, Cetakan II, Yogyakarta.

[3] Tri Mulyono, 2003, Teknologi Beton, Yogyakarta, CV. Andi Offset. Yogyakarta

[4] Murdock, L.J., Brook, K.M., 1984, Concrete Materials and Practice, Edward Arnold.5thed.

[5] Neville, A.M. and Brooks, J.J., 2010, Concrete Technology, Prentice-Hall, 2nd Ed.

[6] Mehta, P. Kumar., 1991, Concrete in Marine Environment. Elsevier Science Publisher, Taylor & Francis e-Library, http://libgen.org/. accessed 26 Mei 2013

[7] Rompas R.M, Rumampuk, N.DC dan Rompas, J.R, 2009, Oseanografi Kimia, PT. Walaw Bengkulen, Jakarta Timur.

[8] Ade P.U., 2010, Pengaruh Rendaman Beton PC I PT. Semen Padang dalam Air Laut dan Air Tawar Terhadap Sifat Kuat Tekan, Skripsi Universitas Andalas, Padang.

[9] Annisa J., M. Wihardi T., Rita I., 2014, Studi Kekuatan Beton Yang Menggunakan Air Laut Sebagai Air Pencampur Pada Daerah Pasang Surut, Jurusan Sipil, Fakultas Teknik, Universitas Hasanuddin, Makassar.

[10] Otsuki, N., Furuya, D., Saito, T. and Tadokor, Y., 2011, Possibility of Sea Water As Mixing Water in Concrete, 36th Conference on Our World in Concrete and Structure, Singapore.

[11] Otsuki, N., Saito, T. and Tadokor, Y., 2012, Possibility of Sea Water as Mixing Water in Concrete, Journal of Civil Engineering and Architecture, ISSN 1934-7359, Vol 6, No.10 (Serial No. 59), pp.1273-1279, USA.

[12] Fahmi H., Gunawan T., 2013, Pengaruh Sulfat Terhadap Kuat Tekan Beton Dengan Variasi Bubuk Kaca Substitusi Semen Dengan W/C 0,4 Dan 0,5, Jurnal Teknik Sipil dan Lingkungan Vol. 1, No. 1, Desember 2013 ISSN: 2355-374X.

[13] Islam Md.S, Kaushik, S.K., and Islam, Md.M., 2005, Physical and Mechanical Behavior of Concrete In Sea Water Under Hydrostatic Pressure, Journal-The Institution of Engineers, Vol.66 No.2, Malaysia

[14] Ristinah S.,Agung W., Fauzan F.M., 1978, Pengaruh Air Laut Pada Perawatan (Curing) Beton Terhadap Kuat Tekan Dan Absorpsi Beton Dengan Variasi Faktor Air Semen Dan Durasi Perawatan, Jurnal Rekayasa Sipil / Volume 5, No.2-2011 ISSN 1978-5658

[15] David S., 2011, Pengaruh Pemakaian Portland Composite Cement (PCC) Terhadap Ketahanan Sulfat Pada Self Compacting Concrete (SCC), Skripsi Universitas Indaonesia 1049/FT.01/SKRIPS/07/2011. Depok.

[16] Prucker, F. danGjorv, O.E., 2003, Effect of CaCl2 and NaCl Additions on Concrete Corrosivity, Cement and Concrete Research, 34 (2004) 1209-1217

[17] Departemen Pekerjaan Umum, 2002, Tata Cara Perhitungan Struktur Beton Untuk Bangunan Gedung (Beta Version) SK SNI 03-2847-2002. Badan Standarisasi Nasional.

[18] Departemen Pekerjaan Umum, 2004, Semen Portland Komposit dengan Standar SK SNI 15-7064-2004. Badan Standarisasi Nasional.

[19] Tjaronge, M,W., 2012, Teknologi Bahan Lanjut Semen dan Beton Berongga, Telaga Zamzam, Makassar

[20] Dipohusodo, I., 1999, Struktur Beton Bertulang. PT Gramedia Pustaka Utama, Jakarta

IOP Conf. Series: Materials Science and Engineering **674** (2019) 012016 doi:10.1088/1757-899X/674/1/012016

Feasibility study and resilience disaster assessment toward shelter and settlement areas in Rinjani circumference, Sembalun district, East Lombok, West Nusa Tenggara Province

I R Sushanti*, **N H Imansyah, F P Hirsan and A Putranadi**

City And Regional Planning, Engineering Faculty University of Muhammadiyah Mataram, Mataram 83 114, Indonesia.

* Email: imarahmawati77@gmail.com

Abstract. Rinjani is one of the active volcanoes in the province of West Nusa Tenggara, Indonesia. In 2018, Rinjani designated as world geopark by UNESCO. Mount Rinjani has regional support Rinjani potential as a world geopark, which Rinjani circumference. This study aims to assess feasibility and disaster resilience towards shelter and settlement in Rinjani circumference. This study uses qualitative descriptive analysis by examining building physical elements and the environment. Based on the identification, it is necessary to organize shelter and settlement resilient among other things: 1) Land Use, is planning disaster vulnerability zones, emergency response settlements, and protected areas, and public facilities in safe zones , 2) Shape and Building Mass, is adjusting environment and local wisdom, climate and context, household and livelihood activities, occupancy problems, construction materials, construction activities, participatory design, venting and vector control, 3) Circulation and Parking, is integrating circulation paths with evacuation routes, assessing access, security and safety of local settlements, road infrastructure and artificial lighting requirements; 4) Open Space, is preserving sustainability, functions, and environmental value as the world's geopark region; 5) Supporting Activities and Signages, is making signage indicating route, tourist activities location, processing activities of agricultural and plantation as well as informative evacuation.

1. Introduction

The Rinjani circumference region in Lombok island is one of the destinations of the world which has been designated by UNESCO as a World Geopark area (UNESCO Global Geopark). The volcano located on the island of Lombok, West Nusa Tenggara province has Rinjani Lombok Potential Geopark of which have been identified, which consists of twenty-two sites Geology (GeoHerritage), eight sites Biology (BioHerritage), and seventeen cultural sites (CultureHeritage). The potential is dispersed and at some point the location of shelter, settlements and the environment in some areas.

In reference [1], is an event of the disaster, natural or humanmade, suddenly (progressive), which causes a devastating impact that the affected community or society or affected must respond with action outstanding. In the Spatial Plan of East Lombok years 2012 - 2032, Sembalun District defined as an area of geological disasters. In 2015, affected by the volcanic mountain Barujari exhausts. In 2017 Sembalun District flash floods. Based on data collected from Sembalun District of six villages affected by floods, there are six houses damaged, eleven moderately damaged, three hundred sixty-seven lightly damaged, two units house fire, house foundations undermined two and two units of houses were washed away.

The house is a refuge from the outside human influence, such as climate, enemies, and diseases. Every human being has the right to adequate housing [2]. In reference [3] there are five essential elements of settlement: Nature, is a land that can be used to build houses and other function, Human, both personal and group building or residence; Society, in which there are interactions between human and social relationships so as to form a specific bond as a community; Shell, which is building a human dwellings with their respective functions; Networks, in other words the infrastructure to support environmental functions both natural and man-made.

In terms of building or dwelling from residents of the area around the circumference Rinjani region, particularly in rural areas, functioned as a place of refuge, a place to move and socialize as well as businesses that support the potential geopark. Conditions settlements characterized by traditional architecture and tropical partially still in a stateless feasible, both from the aspect of architecture, structure and aesthetics, including the level of disaster resilience of the structure and the material used for the lack of public understanding of the home is worth and resilient, especially the area around hills, mountains or mountain. In reference [4] there are physical elements, namely land-use design, shape and mass of the building, circulation and parking, open space, pedestrian paths, support activities and bookmarks.

2. Method

In this research, the approach is through a qualitative approach, where the dominant data collection techniques derived from interviews, field notes, personal documentation, notes, memos, and other official documents. The purpose of this qualitative study describes the empirical reality behind the phenomena in-depth, detailed and complete. This study specifically examined the building and environments, using the overlay technique to determine the area included in disaster areas.

3. Results and Discussion

3.1. Identification Physical Elements

Under the existing conditions that existed at the shelter and settlements in the area of geopark corresponding physical design elements can be identified as follows:

3.1.1. Land Use

Sembalun land area in the district amounted to 216.61 hectares with the use of land for cultivation area, namely rice fields, building or yard by 57 percent and protected areas, namely legal or garden and others by forty-three percent. Land used for shelter areas and the other awoke still low. The land use is dominated by plantations, especially plants that support the potential of biodiversity, such as strawberries, tomatoes, potatoes, and so on competitive commodities from sub-Sembalun.

Sembalun district has a function designation as cultivation areas and protected areas. The cultivation area comprises shelter, agricultural, commercial and office buildings. While the protected area consists of plantations and forests. Rinjani circumference region as a world geopark has the potential distribution in the district geo product Sembalun include geo culinary, geo homestay and geo souvenir located in shelter areas. The buildings are not only used as a shelter function but also businesses. While geo-tourism is located at the distribution point natural tourist sites are located in protected areas, either in the form of hills, valleys, waterfalls, and more. The designated condition and the distribution area in the Sembalun District geo product potential can be seen in Figure 1.

In the Spatial Plan of East Lombok years 2012 - 2032, Sembalun District defined as an area of geological disasters. Hazard maps of the region and protected cultivation, including geo product distribution more clearly seen in Figure 2.

Based on Figure 2, it can be seen that the sub-Sembalun Districtexperiencing the potential hazard to the flow of material destruction, the movement of soil medium, high ground movement, tsunami, and river flood. Some of the potential disasters hit on cultivated areas and protected areas over which the settlement and distribution geo product.

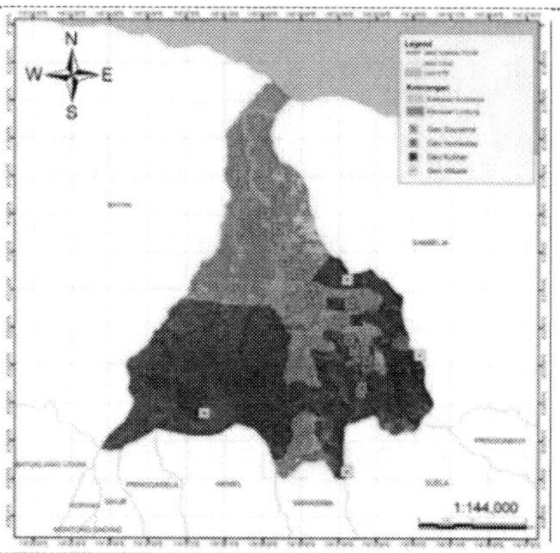

Figure 1. Map land use and distribution in geo product Sembalun. district

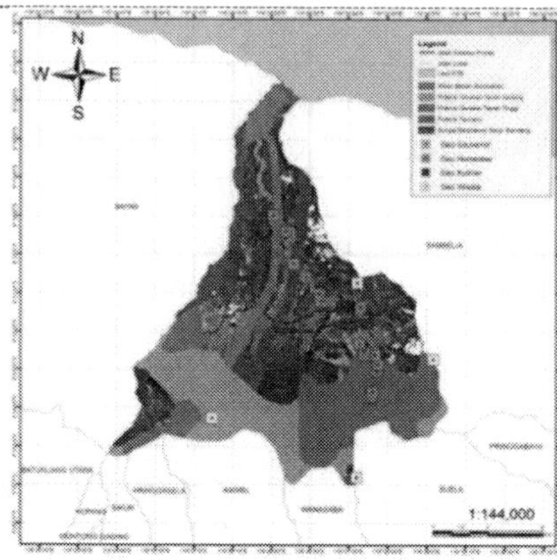

Figure 2. hazard map in Sembalun district

3.1.2. Shape and Mass Building

Building in Sembalun district consists of buildings that functioned as a shelter and business, such as homestay, commercial and household industries. The shape of the building uses traditional architectural features typical local tropical with most using the stage structure. The buildings using local materials and mostly located along the road and a small portion in the hills and along the river. These buildings using local materials and environmentally friendly housing is intended to adjust to the climatic conditions in Sembalun. A condition of the building with a shelter function and homestay in Sembalun can be seen in Figure 3 and Figure 4.

Figure 3. Homestay in Sembalun Bumbung.village

Figure 3 shows the homestay and the surrounding neighborhood that has integrated with the natural landscape and the use of environmentally friendly local materials with characteristics of tropical architecture building. Figure 4 shows the housing with a traditional tropical architectural feature with the use of local materials environmentally friendly structures and building houses on stilts low.

Figure 4. Shelter in Sembalun Lawang village.

The location of settlements located in a valley surrounded by hills with an altitude of 800 to 1200 meters above sea level. The settlement pattern is linear along roads and centralized in two locations administrative center but not organized regularly. Building density in Sembalun district is low five percent spread over administrative area with the distance between buildings between five to ten meters. The orientation of the building facing the road and leads to Mount Rinjani. Thus efforts should be made to the housing arrangement in Sembalun because most of the occupancy is in areas prone to natural disasters. The map layout and orientation of buildings in the Sembalun district can be seen in Figure 5.

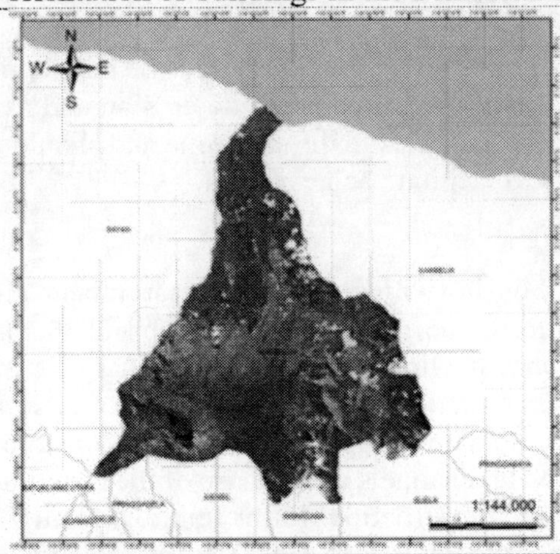

Figure 5. The map layout and orientation of buildings in the Sembalun district.

3.1.3. Circulation and Parking
Sembalun circulation patterns in the district itself consist of circulation that connects the Sembalun district with the surrounding district, where circulation is influenced by the presence of collector roads and local roads that serve as community liaison activities in Sembalun follow condition of contour and landscape. For the parking pattern in the district Sembalun district still utilize the community settlement to support tourism activities.

3.1.4. Open Space
With a low level of building density, namely, reach ten percent of the area, where the open space can optimally support the function room for shelter, settlement, and other designated functions. Most are protective vegetation and cover the open space to enhance the ability to carry capacity environment and provide a visual aesthetic value.

3.1.5. Supporting Activities and Signage
Supporting activities in Sembalun consists of tourism and agricultural activities, tourism activity itself consists of artificial nature and travel, while for the processing of agricultural activities of agricultural and plantation. Signage that describes identity and activities in Sembalun district has not found.

3.2. Structuring Shelter and Settlements
Structuring Efforts to manage shelter and settlements in disaster-prone areas are carried out as mitigation related to the following physical elements

3.2.1. Land use
Land use planning for shelter areas properly must be conducted with respect to each zone of disaster vulnerability, such as the zone of debris material flow and occupancy that is built to be within 50 meters of the river banks; zone potential ground movement being permitted to build settlements on the condition of houses on stilts and the selection of the type of low buildings and preserving the environment; zones of high potential ground movement is done with the relocation and engineering for slope; potential river flood zones, with occupancy, must awaken within fifteen meters of the river banks.

Land planning is also done for the settlement of emergency response, with planning of land for shelter communities while taking into account the social practices that exist and the shared use of resources, including water and sanitation with an assessment of the sewer line of surface water and the risk of inundation or flooding, the slope of the land should not be more than six percent; shelter areas and camps for planned or spontaneous, shall be made available extent feasible surface area at least 45 square meters per person; vector risk for temporary community settlements is a major control effort in reducing the impact of vector-borne diseases on affected populations and the safety of fires requiring a 30-meter area for firebreaks separating every 300-meter building space and at least two meters between occupancy

While land-use planning for protected areas must always be maintained continuity, even can also be added to the protective vegetation planted as a barrier to erosion and as a barrier to the cultivated area in the vicinity. Placement of public facilities in the safe zone and should be connected directly to shelter areas and other public buildings to facilitate achievement.

3.2.2. Shape and Building Mass

For the use of shelter space and housing with care for the environment and local wisdom as well as some aspects as disaster mitigation efforts, namely: 1) Climate to consider a range of extreme climate-related, the shape of occupancy must conform climate characteristics that exist at the site, maximize ventilation and minimize entry of direct light sun, the roof has a slope that is sufficient to provide channels for rainwater except in locations vulnerable to high winds, shelter construction in the form of lightweight construction, adequate surface water channels and the use of the raised floor height; 2) Troubleshooting occupancy, building materials and construction, for example, a family tent or building package ready for installation with the building materials must meet the specifications and national and international standards as well as technical and financial capabilities residents should be according to the specifications of materials and technology used and regular monitoring; 3) The design of a participatory, everyone must understand the advantages and disadvantages of forms of construction that are used in determining the type of shelter assistance will be provided; 4) Ventilation and vector control, providing adequate ventilation in the dwelling individual households and public buildings. In addition, consider the need for vector control efforts in building use, occupancy patterns of use by the displaced population and the selection of building materials.

While the arrangement of the layout and settlement patterns as well as the orientation of buildings adapted to each zone or building settlements, namely: 1) Building along the road, facing the road with regard to the building demarcation line; 2) The building along the banks of the river, overlooking the river with regard to the demarcation line of the river; 3) Building on the hills, with levels slopes minimum of fewer than thirty degrees and not in watersheds and away from cliffs to avoid the occurrence of landslides. It is also necessary to build retaining walls around the building; 4) Building on the waterfront, on a regular settlement patterns, parallel to the direction of propagation and tsunami tidal waves or perpendicular to the coastline, the wide side of the building is made parallel to the coastline, thereby reducing the water pressure and the tsunami tidal waves and monsoons hit the building which is the shortest side of the building perpendicular to the beach with the sea and the orientation of the circulation path. The location of settlements must further towards the most vulnerable zones of extreme waves, a minimum setback of hundred meters or with a radius of more than two hundred meters from the beach; 5) Side width should be perpendicular to the direction of the wind, to reduce the widest field of building. Regularity and parallel position of the building with the landscape and corridor reduces water pressure and winds that hit when a disaster occurs. The location of settlements must further towards the most vulnerable zones of extreme waves, a minimum setback of hundred meters or with a radius of more than two hundred meters from the beach.

3.2.3. Circulation and Parking

Some things should be done in circulation are: 1) Arrangement of the circulation path must adapt and integrate with the planned evacuation route when a disaster occurs, easy access to housing, infrastructure conditions, and sufficient conditions much nearby center - the transport hub for emergency response support; 2) For the settlement temporary community, the locations used and dots stock primary relief goods must be accessible by heavy trucks from the roads impassable in all types of seasons and to avoid the access point and an exit point which makes the emergence of regional isolated or limited to threaten the personal safety of its users; 3) Great trails and paths made in the settlements must provide secure access, ensure safety,

3.2.4. Open Space
In the presence of an open space consisting of field, garden, rice fields, plantations, and protected areas need to continue to be preserved by keeping function and environmental values of protected areas as part of a world geopark region.

3.2.5. Supporting Activities and Signages
Some things should be done on support activities, and signages are: 1) Making the signage consisting of hard signage and e-signage indicating the route and location of tourist activities and processing activities of agricultural and plantation as well as informative evacuation; 2) Preparation of signage to indicate the identity of the geopark area location is attractive and comfortable to see, and 3) Hard signage material should use natural materials.

4. Conclusion
Based on research that has been done it is necessary referrals towards shelter and settlements feasible and disaster-resilient as follows:
1) Land Use, with due regard to each zone of disaster vulnerability, planning for the settlement of emergency response, planning land use for the protected areas by maintaining sustainability, as well as the placement of public facilities that should be in the safe zone and is directly connected to shelter areas and other public buildings to facilitating the achievement.
2) Shape and Building Mass, with attention to the environment and local wisdom as well as some aspects of disaster mitigation efforts, taking into account the climate, household activities, and livelihood. While the layout and orientation of buildings is done by laying and settlement patterns as well as the orientation of buildings adapted to each zone settlements or buildings including attention to building orientation concerning the shape of the building and the direction toward the building
3) Circulation and Parking, with arranging the corresponding circulation path and integrated with the planned evacuation route when a disaster occurs, the assessment regarding access, security, and safety of the settlements and the condition of the local road infrastructure, including the need for artificial lighting.
4) Open Space, preserve the continues, functionality, and environmental values of protected areas as part of a world geopark region.
5) Support Activities and Signages with the provision of signage is using natural materials.

5. Suggestion
It takes several simultaneous and integrated efforts of all parties to continue to be active and participate in maintaining the existence of shelter and settlement resistant and worthy to disasters in the region as a destination region Rinjani circumference world geopark. Involve and provide insight to the public and all interested parties about the urgency of decent shelter and settlement and resilient as mitigation and development potential of world geopark that responds to disasters.

References
[1] Carter, W. N, 1991. Disaster Management. Asian Development Bank, Philipines.
[2] The Sphere Project, 2011. Humanitarian Charter and Minimum Standards in Humanitarian Response. Hobbs the printer, Southampton, United Kingdom.
[3] Doxiadis, Constantinos A. 1968. EKISTICS An Introduction To The Science Of Human Settlements. London: Hutchinson Of London
[4] Shirvani, Hamid.1985.The Urban Design Process, VanNostrand Compay Reinhold: New York.

Stiffness analysis comparison of masonry full infills frame and masonry open middle span frame using Lubuk Pakam Bricks with pushover analysis

F Sri * **and W D Rhini**

University of Muhammadiyah Sumatera Utara (UMSU)
Jl. Kapten Muchtar Basri No. 3, 20238, Medan, Indonesia

* Email: srifrapanti@umsu.ac.id, sri_prafanti@yahoo.co.id

Abstract. Local brick in Lubukpakam Indonesia has mechanical and physical characteristics that are different from the developed countries that have always been used as references, since a different behavior with other countries and the low power of local bricks, it widely needs research to get the stiffness of the structure factor of earthquake-resistant buildings. This research was conducted with the use of modeling in the Frame masonry full infills with Frame masonry opens on the Middle span of powerful data with the use of modulus of elasticity and press bricks with pushover analysis. Obtained values of elastic rigidity and stiffness of elastic post. The results obtained on the rigidity of walled portal bigger than an open portal on elastic Stiffness of Central and spans larger than the elastic stiffness of the post.

Keywords: Frame masonry full infills, Frame masonry Middle span open, pushover analysis, stiffness

1. Introduction

Reinforced concrete (RC) framed buildings with infill walls are usually analyzed and designed as bare frames, without considering the strength and stiffness contributions of the infills [1]-[2]. However, during earthquakes, these infill walls contribute to the response of the structure, and the behavior of infilled framed buildings is different from that predicted for bare frame structures. Therefore, based on the understanding of the actual response, design provisions need to be developed. Fortunately, a few countries already have codal provisions for seismic design of RC framed buildings with brick masonry infills. The present study evaluates these available provisions with a view to identifying design methodologies that exploit the benefits of infills in a rational manner, for improving the contribution of these infills and for reducing the detrimental effects. local bricks of Lubukpakam tested from Laboratory and by modeling the structures examined. On a wall full of middle walls open. The purpose of the experimental and analytical studies reported in the literature that attempts to understand the behavior of wall [2]-[4].

Studies show that the walls increase the stiffness and strength of the structure compared to a structure without walls. Elastic response during this phase, the presence of brick wall filler increases the lateral stiffness of the aircraft's structure and reduces fundamental, and as a result of the styles of the slide [1].

2. Review of Literature

2.1. Concept of the Planning Reinforced Concrete Frame

Reinforced Concrete Frame which is modeling a structure divided into two open frames and frame content. In the portal content, wall fillers are generally used as partitions or cover the outside of the structure of reinforced concrete frames. Installation will be done after the main structure is completed is done and is considered a non-component structure in essence, wall charger is a component of a non-structure that does not provide a meaningful contribution toward rigidity and the power structure, but in fact the wall charger provides a major influence on the collapse of the building so that it behaves differently with the frame open [2].

2.2. Brick Masonry Infills

Brick Masonry Infills used in this research is a red brick wall because it is very widely used in almost all the buildings on the territory of the country of Indonesia. This is because red brick has an economical price, easily obtainable, and impervious to weather. The behavior of a portal with a brick wall against lateral loading has been investigated, due to lateral loads that occur resulting in the onset of interaction between a wall charger with portal [5]-[7]. Whereas in the event of an earthquake the wall charger can affect the stiffness and strength of the structure that the effect is not profitable in the structure so that it can cause damage [8]-[9].

2.3. The Basic Concept Of Pushover Analysis Method

The method of static analysis is not linear (pushover analysis) is not a linear method is very popular used in planning or assessment of buildings located in areas prone to earthquakes[9]. As described, the underlying idea of this method is to explain the State of earthquake load acting on the order of the structure. The response order structure of the dynamic load is a combination of the dynamic range of vibrating system that vibrates[10]. So this method is also based on the fundamental concepts of analysis of spectrum of vibration on the structure. Explanation of the underlying theory of static analysis is not linear here is based on [10]–[11]

2.4. Concept planning of earthquake-resistant

Structure earthquake-resistant Building was designed based on the earthquake zone, the character of location, soil type, occupation building, the building uses factors, the natural period of the structure, and others. The entire structure of the elements designed with a prisoner who according to withstand lateral displacement occurs due to ground motion having regard to the response of inelastic structure, redundant factors, more powerful and dactylitis structures [7]

The response of a building due to earthquake loads is happening is very complex, so new methods continue to evolve in order to know the behavior of the structure due to the earthquake which occurred [9]. Structure with soft-storey will enlarge the lateral deformation and sliding on a column style The collapse of the building due to the earthquake, one of which can also be caused by the selection of the structure of the soft-storey [10].

3. Method

In this research, the analysis used to obtain good results is a Pushover Analysis. The method used is the method of a numeric approach. The purpose of the analysis was a pushover for estimating maximum style and deformations that occur as well as to obtain information, which is the critical part. The next, identifiable parts that require special attention to acid etching or stability.

3.1. Modeling Structure

On the research on the structure of the buildings are modeled as 2 dimensional frame and consists of 6 grid model but analyzed only 2 grid model in the model and model masonry infills frame the masonry opens on the Middle span, all the models structure consisting of 3 floors and 3 landscape including the structure of the portal, Figure 1.

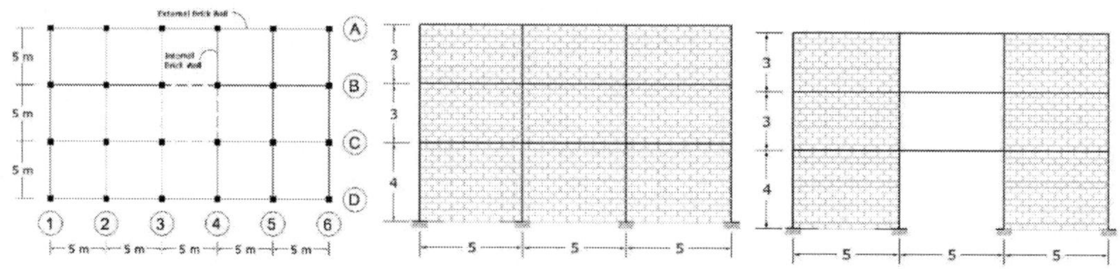

Figure 1. Floor plans and Look at Modeling and outdoor Walled Portal Diftesh Das and CVR CONTENT Murty (2004)[2]

3.2. Analysis of Testing Properties of Bricks

In this study used red bricks as wall material filler. Characteristics of the brick walls that will be used in this study are based on the characteristics of the brick wall on how to test the elasticity modulus of bricks[3] with single-axis pressure test objects, the bricks using 10 samples of the bricks from the region of LubukPakam, Deli Serdang, North Sumatera, Indonesia. From the results of Experiments Testing Powerful press Bricks obtained as in Table 1.

Table 1. Test results of Strong Brick Press

TEST RESULT DATA STRONG BRICK PRESS				
NO. TEST OBJECTS	SIZE OF THE BRICKS			ACCEPTED LOAD (KN)
	P (cm)	L (cm)	A (cm²)	
TYPE 1				
1	19.6	9.7	190.12	580
2	19.7	9.7	191.09	590
3	19.7	9.5	187.15	575
TYPE 2				
4	19.6	9.5	186.2	460
5	19.4	9.6	186.24	378
6	19.5	9.6	187.2	377
7	19.3	9.5	183.35	368
TYPE 3				
8	19.4	9.5	184.3	355
9	19.6	9.6	188.16	350
10	19.6	9.7	190.12	350

Table 2 shows the physical properties of brick were used.

Table 2. Physical Properties average brick red

Properties	Brick Red	Unit
Length	195.23	Mm
High	52.28	Mm
Thick	95.29	Mm
Heavy Type	1373.55	kg/m³

3.3. *Research Stages*

Stages are implemented so that the achieved the goals and purpose of the research. To achieve these goals then done stages, namely:

1) The First stage: This stage is done is to model a structure with its own modeling and determining technical data used in the analysis.
2) The Second stage: This stage is done the calculation of loading structures consisting of gravity load (dead load and live load) and earthquake load and done to determine the horizontal earthquake load takes shakes the building.
3) The Third stage: This stage is carried out to determine the horizontal load due to the earthquake needed to vibrate the building.
4) The Fourth stage: This stage is carried out to determine the occurrence of plastic joints on beams, columns and brick walls.
5) The five-stage: At this stage, do the procedure for calculating a pushover analysis
6) The sixth stage: At this stage, the pushover analysis of the output obtained i.e. the capacity curve shows the relationship of the basic styles of sliding towards the junction, showing the change in behavior of linear structure into a nonlinear, a decreased stiffness.

4. Results and Discussion

A study on modeling the structure in question is modeling the structure of 3-dimensional reinforced concrete building in comparison with the existing building (actual) and with the model design based on SNI 2012. After the run was completed analysis, the next step to control the results of the analysis, i.e. the natural period of vibration, slide style (base shear), and displacement.

4.1. *Frame masonry full infills*

Figure 2 shows and compares that to the third model Frame masonry full infills on a Grid of 1 to several types of brick fully infilled wall frames. Image shows value style slide base and the junction of the largest rooftop capacity in receiving the brunt of the earthquake or the styles there are lateral on a brick type 1 (M1-G1-B1).

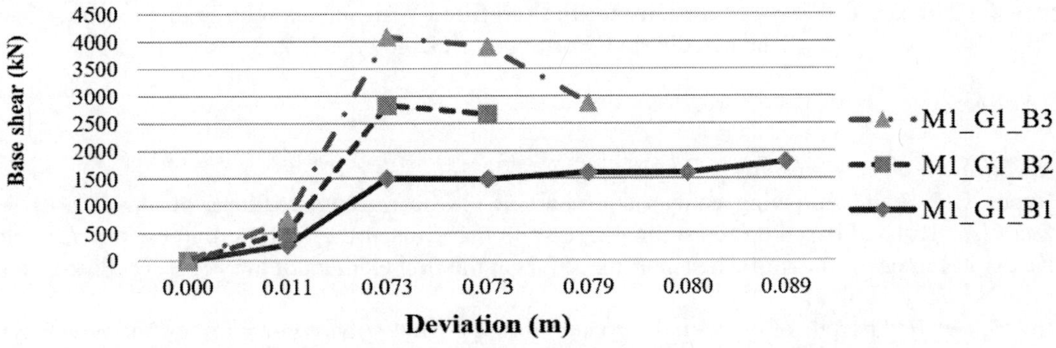

Figure 2. Comparison of the curves on the Grid Capacity Combined with 1 Some types of Bricks (type 1, 2, 3)

4.2. *Frame masonry opens on the Middle span*

The following described some of the capacity curves is the relationship between Frame masonry full infills structure of some brick that is type 1, type 2 and type 3 as shown in Figure 3.

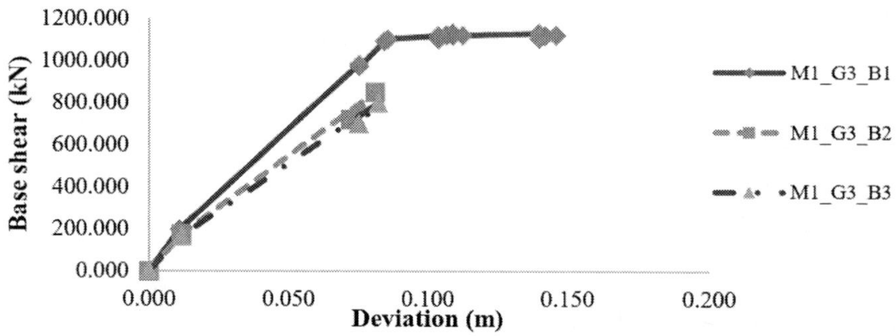

Figure 3. Capacity Curves on a Grid with combined several types of Bricks (type 1, 2, 3)

Figure 3 shows that the third model of Frame masonry opens on the Middle span on the Grid 3 (G3) for some types of bricks that have the largest capacity in receiving the brunt of the earthquake or the lateral style present on the portal open brick type 1 (M1-G1-B1). Comparison of the capacity curve for compared to the open portal as shown in Figure 4.

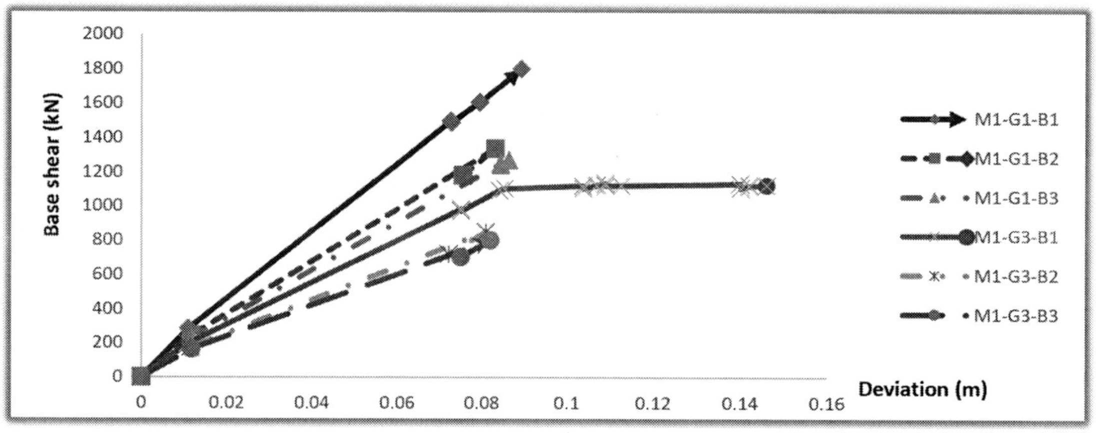

Figure 4. Comparison Curves of capacity for Fully Infilled Wall frames with Frame masonry opens on the Middle span Portal on bricks to type 1, 2, 3

4.3. Stiffness

4.3.1. Stiffness Frame masonry full infills.

On types 1, 2, 3 the following comparison chart of elastic rigidity, stiffness of elastic post. Elastic stiffness is calculated based on the sliding base style that caused the first melting on structure element with a displacement of the roof when melting occurs at the first element of the structure. Based on Figure 5, it can be analyzed that:

1) In brick-walled portals of type 1 have greater elastic stiffness than type 2 bricks and type 3 bricks.
2) The biggest percentage difference in elastic stiffness in the portal with brick walls in type 3 is reaching 24.43% less than the portal with a brick wall in type 1.
3) In brick-walled portals of type 1, the post-elastic stiffness is greater than that of type 2 bricks and type 3 bricks.
4) The biggest percentage difference in elastic stiffness in the portal with brick wall in type 3 is 28.57% smaller than the Port with a brick wall in type 1.

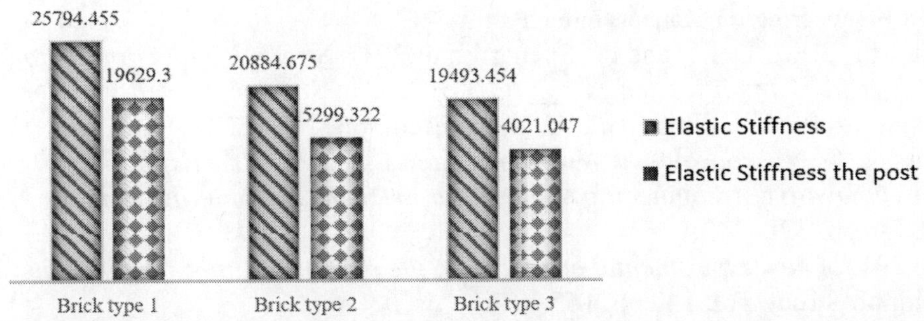

Figure 5. Elastic Rigidity and stiffness of elastic-wall Post on the Portal (grid 1)

4.3.2. Stiffness Frame masonry opens on the Middle span in type 1, 2, 3

Figure 6 shows the comparison of the elastic stiffness and rigidity of post for brick elastic type 1, type 2 and type 3. The following comparison chart of elastic rigidity, stiffness of elastic post. Elastic stiffness is calculated based on the sliding base style that caused the first melting on structure element with a displacement of the roof when melting occurs at the first element of the structure in Figure 6.

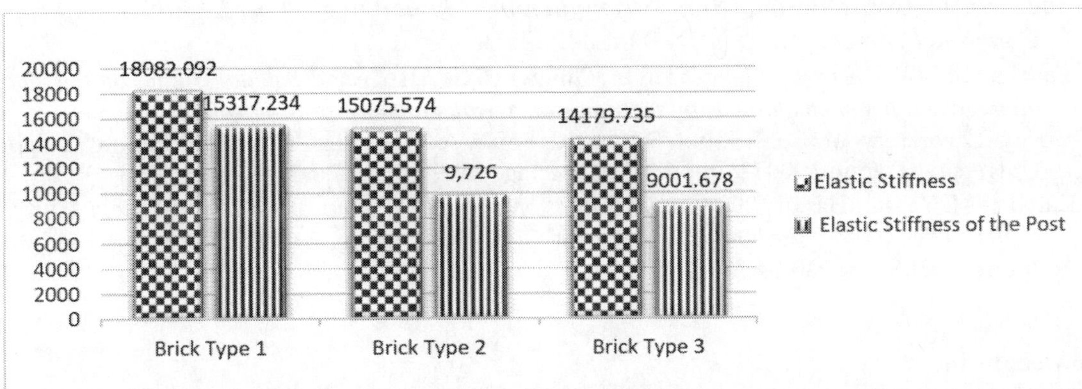

Figure 6. Elastic stiffness and rigidity of post on the Frame masonry open on the Middle span Elastic (Grid 3).

5. Conclusion

Based on the results of the analysis and discussion in accordance with the problems that have been raised against the walls of the Portal model is Frame masonry full infills and Frame masonry opens on the Middle span. The results of the study showed that the stiffness of brick walls,

a. Stiffness of elastic. From the results of the analysis of stiffness of elastic then it can be inferred that the elastic stiffness values larger Frame masonry full infills on the model is compared to Frame masonry open on the Middle span and brick larger than type 1 type 2 and type 3.
b. Post Stiffness elastic. From the results of the analysis of the elastic stiffness of the post then elastic stiffness values can be summed up the larger Frame masonry full infills on the model is compared to the Frame masonry opens on the Middle span and brick larger than type 1 type 2 and type 3.

References

[1] Muhammad H.Jinya, V.R. Patel 2014 *Analysis of RC frame with and without masonry infill wall with different stiffness with outer central open*, International Journal of Research in Engineering and Technology. eISSN: 2319-1163 | pISSN: 2321-7308

[2] Diptesh Das, C.V.R Murty 2000 *Brick masonry infills in seismic design of RC frame buildings* Civil Engineering IIT Kanpur, India, Part 2

[3] Wisnumurti, S. M. Dewi, and A. Soehardjono 2014 *Strength reduction factor (R) and displacement amplification factor (Cd) of confined masonry wall with local brick in Indonesia* Procedia Eng., **vol. 95** no. Scescm, pp. 172–177

[4] A. Martins, G. Vasconcelos, R. Fangueiro, and F. Cunha 2015*Experimental assessment of an innovative strengthening material for brick masonry infills Compos.* Part B Eng., **vol. 80** pp. 328–342

[5] D. Foti 2015 *A new experimental approach to the pushover analysis of masonry buildings* Comput. Struct. **vol. 147** pp. 165–171

[6] T. Sipil 2002 *Analisa Inelastis Portal - Dinding Pengisi* J. Tek. Sipil ITB, **vol. 12** no. 4, pp. 229–240

[7] E. Bernat-Maso, L. Gil, and P. Roca 2014 *Analytical method for the assessment of unreinforced brick masonry walls subjected to eccentric compressive loads* Constr. Build. Mater., **vol. 73** pp. 180–186

[8] B. Belletti, C. Damoni, and A. Gasperi 2013 *Modeling approaches suitable for pushoveranalyses of RC structural wall buildings* Eng. Struct., **vol. 57** pp. 327–338

[9] Elev. World Conf. Earthq 2003 *Strengthening Methodology For Lightly Reinforced Concrete Frames* . Eng. ISBN 0 08 042822 3

[10] Yanchao Shi ,Wei Xiong , Zhong-Xian Li, Qingfeng Xu 2016 *Experimental studies on the local damage and fragments of unreinforced masonry walls under close-in explosions* Key Laboratory of Coast Civil Structural Safety of the Ministry of Education, Tianjin University, Tianjin 300072, Impact Engineering, **vol.90** pp. 122-131

[11] Kaltakei M. Yassar, Hasan . 2007 *An Investigation of the behaviour of stell Frames with Masonr Infills under lateral Building* Departement Civil Engineering, Selcuk University, Medwell Journal, **vol.5** pp. 930-943

Acknowledgment

Submission of this paper is supported by the University of Muhammadiyah Sumatera Utara.

The flexural buckling comparison between open and close cross sections in high column structure

W D Rhini * and F Sri

Department of Civil Engineering, Faculty of Engineering, University of Muhammadiyah Sumatera Utara
Jl. Kapt. Muchtar Basri #03, Medan 20238, North Sumatera, Indonesia

* Email: rhiniwulan@umsu.ac.id

Abstract. Selection a cross-section of column on building construction needs to keep on a constant review to flexural buckling. The flexural buckling of column affects construction behavior overall. The aim of this study is comparing open and close of a cross-section column profile against flexural buckling. The hollow and pipe are representing close cross-sectional column profiles while I and T profiles are other represent. All profiles used the same material and properties, 400 Mpa of tensile strength and 740 mm^2 area of cross-sectional. The high of column was 24 m which, both ends be fixed. Load Resistance Factor Design (LRFD)- SNI 1729:2015 method was applied to analyze the flexural buckling. The T profile was a control to determine percentage of column slender other profiles. The result showed that the slender was 53.43%, 75.57%, and 86.67% for hollow, pipe, and I respectively to T profile. The study indicates that the high column and inertia of cross-sectional were important role in flexural buckling failure.

Keywords: Flexural buckling, open and close cross-section, SNI 1729:2015, high column

1. Introduction

Steel is one of the most widely used construction materials in Indonesia. Bridges, warehouses, factories, airport terminals, office buildings, and apartments typically use structural steel. In Indonesia, steel is widely used since the 1970s [1].

The stell structures, either columns, and beams must be a strength, stiff, and durable for life service of structures. In designing stell columns must be safe and able to accept service load, the columns should have the ability to the predictable overload or understrength. Overload can occur due to changes in building construction, too low for the assessment of the effects of excessive simplification in the analysis of his structure, and cause of the variations in the construction procedure [2].

Tables enlisting the section properties of commercially available structural steel sections are frequently used to help the designer obtain the required section properties such as weight, area, inertia, and radius of gyration. The design code of structural steel in Indonesia, SNI 1729:2015[3], uses the Load and Resistance Factor Design (LRFD) concept. In LRFD, the factored load is compared with the nominal strength of the selected steel section [1].

2. Profile cross-section performance

The hollow and pipe are representing close cross-sectional column profiles while I and T profiles are other represent. Each section has the same area properties. Ratio b/t must be in control to classify whether section properties are with or without elements slender elements [4]-[5]. Control value r_{min} against r_t. If $r_t > r_{min}$ going flexural buckling. Table 1 presents data on properties of each cross-section. Calculation of analysis done on a numeric based on SNI 1729-2015 [3].

Tabel 1. Section Properties

Section Properties	**Dimension**						
	B (cm)	**H (cm)**	**t (cm)**	**r_x (cm)**	**r_y(cm)**	**b/t without slender elements**	**A_g (cm^2)**
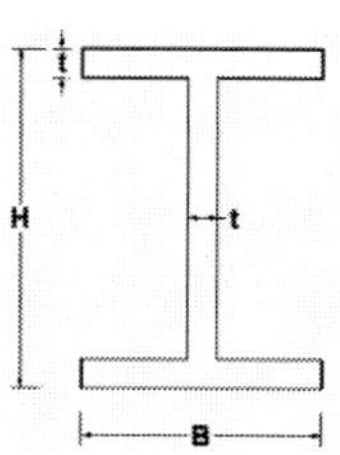	62	60	4	24.9	15	Flens $\frac{31}{4} < 0.56\sqrt{\frac{200000}{400}}$ Web $\frac{52}{4} < 1.49\sqrt{\frac{200000}{400}}$	704
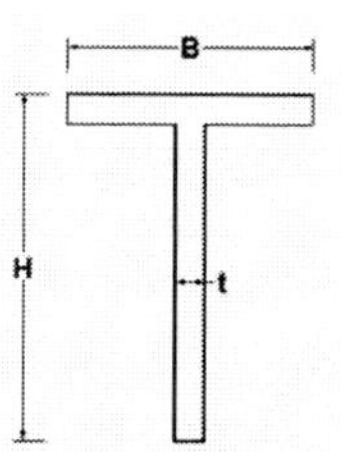	100	80	4	24.5	21.85	$\frac{50}{4} < 0.56\sqrt{\frac{200000}{400}}$	704
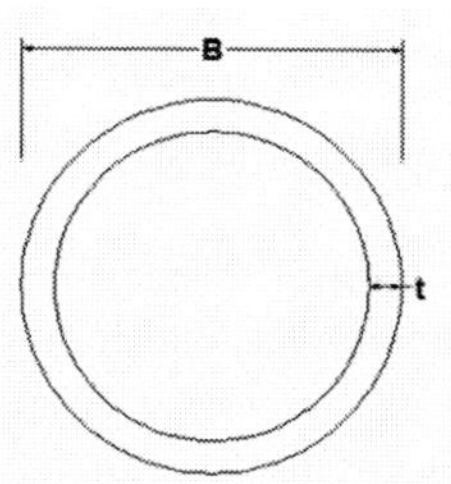	60	-	4	19.85	-	$\frac{60}{4} < 0.11\frac{200000}{400}$	704
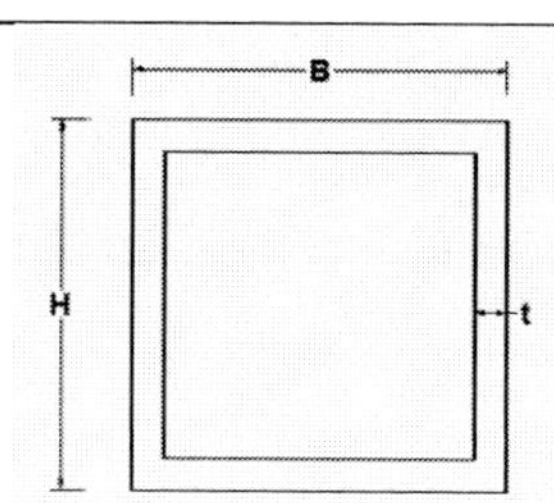	61.5	61.5	4	24.33	24.33	$\frac{(61.5-8)}{4} < 1.40\sqrt{\frac{200000}{400}}$	704

3. Assumptions and analysis

The hollow and pipe are representing close cross-sectional column profiles while I and T profiles are other represent. All profiles used the same material and properties, 400 Mpa of tensile strength and 740 mm^2 area of cross-sectional. The high of column was 24 m which, both ends be fixed. Load Resistance Factor Design (LRFD) method was applied to analyze the flexural buckling failure.

3.1. Nominal Compressive Strength

The design compressive strength $\Phi_c P_n$. The nominal compressive strength P_n which shows at Equation (1), shall be the lowest value obtained based on the applicable limit states of flexural buckling [6].

$$P_n = F_{cr} A_g \tag{1}$$

3.2. Critical Stress

The critical stress, F_{cr} , is determined as Equation (2) and (3).

$$\text{If } \frac{KL}{r} \leq 4.71\sqrt{\frac{E}{F_y}} \; ; \; F_{cr} = \left[0.658^{\frac{Fy}{Fe}}\right] F_y \tag{2}$$

$$\text{If } \frac{KL}{r} > 4.71\sqrt{\frac{E}{F_y}} \; ; \; F_{cr} = 0.887\, F_e \tag{3}$$

3.3. The radius of gyration equivalents against torsional buckling (r_t)

In determining the radius of gyration equivalents against torsion is influenced by the constant warping (Cw), the torsion constant (J), the length of the rod (L) and polar moment of inertia (Ips). Equation (4) shows that the radius of gyration of the flange components in flexural compression plus one-third of the web area in compression due to the application of major axis bending moment alone [6].

$$r_t = \sqrt{\frac{C_w + 0.04J(KL)^2}{I_{ps}}} \tag{4}$$

4. Result and Discussion

Table 2 shows the results of the parameters used to determine the cross-section are flexural buckling, torsional or flexural-torsional buckling. By comparing the value of r_t with r_x or r_y buckling behavior can be known where that will occur in the column. Views to the I profile at the time of KL value 0 up to 200 cm, the value of $r_t < r_{min}$, torsional buckling occurs. For KL 400 up to 1200 cm flexural buckling behavior going on. So is the T profile, flexural-torsional buckling is very dominated the start of the KL 0 up to 1200 cm. For the close cross-section $J = I_p = I_x + I_y$; then it is clear that I_x (flexural stiffness) only 0.5 from I_p (torsional stiffness) then that determine flexural buckling [7].

Table 2. Flexural and torsional buckling parameter

KL	The radius of gyration equivalents against torsional buckling (r_t)			
	I Profile	T Profile	Pipe	Hollow
0	14.47	1.30	For the close cross-section $J = I_p = I_x + I_y$; then it is clear that I_x (flexural stiffness) only 0.5 from I_p (torsional stiffness) then that determine flexural buckling is[7].	
200	14.82	3.10		
400	15.81	5.78		
600	17..33	8.55		
800	19.26	11.34		
1000	21.49	14.15		
1200	23.93	16.95		

Figure 1 shows the relationship curve nominal compressive strength with slender. Of the four section properties, profile T has high slenderness, which affects the nominal compressive strength to be low, so flexural-torsional buckling occurs. Whereas the hollow profile has low slenderness, at KL 200 cm the hollow profile reaches the maximum nominal compressive strength value. The T profile was a control to determine percentage of column slender other profiles. The result showed that the slender was 53.43%, 75.57%, and 86.67% for hollow, pipe, and I respectively to T profile.

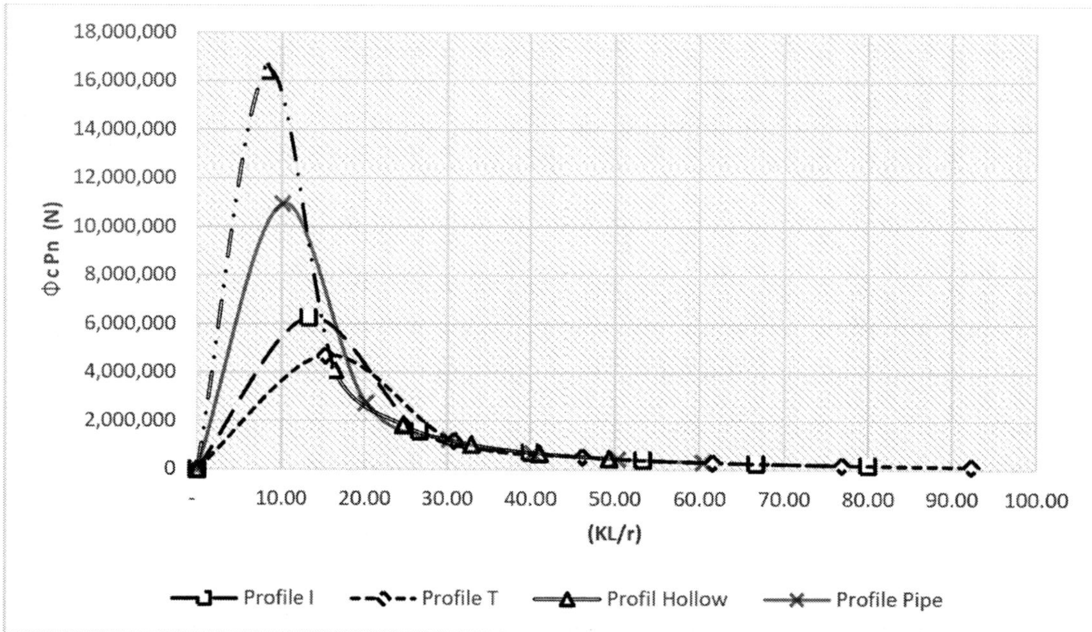

Figure 1. Relationship curve nominal compressive strength with slender

Close cross-sections have smaller slenderness compared to open sections. From the two close sections, the hollow profile has the smallest slenderness and has a nominal compressive strength greater than the other cross-section because of the influence of column height and cross-section inertia. The greater the slenderness ratio, the greater the buckling effect experienced by a column either flexural buckling, torsional or flexural-torsional buckling and the smaller the nominal compressive strength

References

[1] E. Tanojo, P. Pudjisuryadi, B. P. Candra, and W. William 2015 *Design capacity tables for structural steel based on SNI 03-1729-2002: built-up sections,* Procedia Eng., vol. **125**, pp. 1142–1148.

[2] R. Suryanita, and A. Kamaldi 2003 *Analisis Kekuatan Nominal Balok Lentur Baja dengan Metode Desain Faktor Analisis Kekuatan Nominal Balok Lentur Baja dengan Metode Desain Faktor Beban dan Tahanan (LRFD) dan Metode Desain Tegangan Ijin (ASD)*, Jurnal Natur Indonesia 5(2) 157-161 (2003), ISSN 1410-9379.

[3] S. N. Indonesia and B. S. Nasional 2015 *Spesifikasi untuk bangunan gedung baja struktural.*

[4] F. Alhussainy, M. N. Sheikh, and M. N. S. Hadi 2017 *Behaviour of Small Diameter Steel Tubes Under Axial Compression*, Structures, vol. **11**, pp. 155–163.

[5] K. J. R. Rasmussen and G. J. Hancock 1992 *Plate slenderness limits for high strength steel sections*, J. Constr. Steel Res., vol. **23**, no. 1–3, pp. 73–96,.

[6] A. American and N. Standard 2010 *Specification for Structural Steel Buildings.*

[7] W. Dewobroto 2016 *Struktur Baja Perilaku, Analisis & Desain - AISC 2010*, Edisi ke-2.

Acknowledgment

I would like to thank the University of Muhammadiyah Sumatera Utara and acknowledgment of staffing Departement of Civil Engineering, Faculty of Engineering.

The landslide risk analysis based on human activity using Arc-GIS method

B H Widayanti[1], A Yuniarman[2] and S A P Lestari[3]

Urban And Regional Planning, University of Muhammadiyah of Mataram, Mataram 83114, Indonesia

Email: [1] baiqharlywidayanti@gmail.com, [2]ardhi.yuniard@gmail.com, [3] reni.pujilestari@gmail.com

Abstract. This research aims to map the risk level of landslide based on human activity in the Guntur Macan Village, West Lombok. The method used in this study was descriptive quantitative research by mapping and assessing a human activity to classify the risk level of the landslides. The research variables were cropping pattern, excavation, and slope cutting, pond making, drainage, construction, population density, and mitigation. The result showed that the risk level of landslides caused by human activity was divided into three zones. Zone A is a place with a 2.1 risk level of landslides (moderate level), while Zone B is 2,0 (moderate level), and Zone C is 1,3 (low level).

1. Introduction

Guntur Macan is a village in West Lombok that has a risk of landslides. Based on the local regulation (PERDA) No.11/2011 about RTRW of West Lombok, Guntur Macan has been assigned as a protected area because most of the land is a hill with a 40% slope [1]. The rainfall of this village is relatively high (2000-3000 mm). Moreover, the soil is clay and a sandy type that makes it highly vulnerable to the landslides. The landslides in Guntur Macan Village in 2016 had destroyed many properties of the villagers. Therefore, in 2015, many residents were also killed and destroyed their properties. The recent landslides were not the first case in the Guntur Macan Village because in 1974 and 2000 there were landslides occurred that hit the village [2].

The Minister of Public Works Regulation No. 22/PRT/M on Guideline for Spatial Planning of Landslide Area said that the spatial planning need to consider the aspect of space usage by considering the ecosystem balance and the social welfare assurance through the assessment of spatial structure and space pattern of a landslide area based on its typology and the level of its vulnerability as well as maintaining the suitability between the implementation activities and the designated area functions [3]. The risk level classification is determined by 2 criteria, namely natural physical aspect and human aspect [4]. Physical, social, economic and environmental factors play a role in susceptibility to the landslide risk [5] [6].

The landslides that happened in the Guntur Macan village was actively influenced by the community activities, where 270 families (33%) live in the hills. The hills not only used as a residential area but also as farming and stockbreeding. Furthermore, a logging activity by the villagers in the hilly area will deplete the tree vegetation that supposes to conserve the soil and water. A large number of human activities without consideration of environmental sustainability is one of the reasons for the increasing frequency of landslide intensity occurring in the area [7].

Thus, the research aim is to analyze the risk level of the landslides based on the human activities in Guntur Macan Village.

2. Method

2.1 Types of Research

Based on the introduction, the purpose of this research is "to map the risk level of the landslide based on the human activity in the Guntur Macan Village". The method that used in this research was descriptive quantitative by mapping and assessing a human activity to classify the risk level of the landslides. The variables of the human activity were mapped based on The Minister of Public Works Regulation No. 22 / PRT / M on Guideline for Spatial Planning of Landslide Area, namely:

a. Cropping pattern (10%)
b. Excavation and slope cutting (20%)
c. Pond making (10%)
d. Drainage (10%)
e. Construction (10%)
f. Population density (20%)
g. Mitigation (10%)

2.2 Research Focus

The research was located in the Guntur Macan Village, Gunung Sari Subdistrict, West Lombok, with a total area of 2,749 ha, consisting of 7 hamlets: Guntur Macan, Barat Kokoq, Ladungan, North Poan, Pancor, Apit Aik, and Southern Poan.

2.3 Analysis

The analysis steps in this research were divided into several parts, which were:

a. A collection of primary data for the human activity indicator to determine the risk level classification. The data were displayed in the human activities map for each indicator.
b. A typology determination of disaster-prone areas based on the zonation determination by considering the character and the physical condition of nature, such are:
   - Zone type A, a place with a mountain slope, hillside, and riverbank, altitude above 2,000 meters above mean sea level and slope including more than 40%.
   - Zone B type, a place in the mountain foot, foothill, and riverbank, the height of 500 – 2,000 meters above mean sea with the slope between 21 - 40%.
   - Type C zone, a place of the highland, lowland, river banks or river valleys, elevations 0 – 500 meters above mean sea and incline slope between 0 - 20%.
c. An assessment of each zone by using the weight of the predetermined criteria indicators. Each indicator is rated as below:
   - 3 (three) if it is considered to have a significant impact on the landslide
   - 2 (two) if it is considered to have a moderate impact on the landslide
   - 1 (one) if it is considered to have less impact on the landslide
d. An assessment of the risk level of the landslide based on a human activity aspect which was done by calculating the sum of the 7 indicators with a total between 1-3, meanwhile to indicate the risk level of each zona was conducted using these criteria below:
   - Risk level zone with high mudslide potential between 2.4 - 3.0
   - Risk level zone with medium mudslide potential between 1.7 - 2.39
   - Risk level zone with low mudslide potential between 1.0 - 1.69

3. Results and Discussion

3.1 Typology of Guntur Macan Landslide Disaster-prone areas

Geographically, the Guntur Macan Village is a hilly area that forms the "U" letter between the hills (Figure 1). The altitude of the area varies between 57 - 513 meters above sea level (mdpl). The Guntur Macan village is flown by two rivers that flow between the hills and form a basin. 18.11% of the area is used as a residential area, while 67,94% is used as a farming area that spread equally in the hills and the lowland.

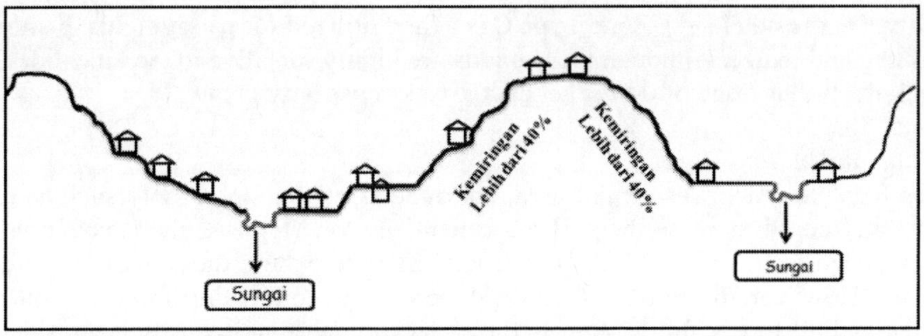

Figure 1. The Illustration of the landform of the Guntur Macan village.

The Guntur Macan Village has hillside slope variation where most of the slope has a 40% tilt with the area around 105.68% ha (30.19%). The area with a slope of 0-8% is 43.63 hectares, 8-15% is 44.75 hectares, 15-25% is 89.19 hectares, 25 - 40% is 89.19 hectares.

According to the guidelines for the spatial planning of landslide areas, the Guntur Macan Village is classified into 3 zones (Table 1), namely:

a. A type zones, it is an area that is located on the slopes, hillside, and river banks with a slope of more than 40% tilt. However, based on the area characteristic, which is hilly, the zone has variation contours which are foothill, hillside, and hilltop. The hillside slope with more than 40% tilt is designated as a type A landslide-prone area, while the area above the hill that has a slope of 0-8% is included conditionally as a type A with the following considerations:
   - The uphills area is adjacent to the hillside that has an extreme slope.
   - The uphills area has a similar characteristic to the hillside in terms of the potential of a landslide disaster.
   - The hillside and the uphills are an area that intrigued a landslide to the below area.

   Based on the landscape characteristic, all of the hamlets in Guntur Macan Village are part of type A zone, especially the North Puan Hamlet and the Southern Puan Hamlet.

b. Zone Type B is a foothill that dominated by slope tilt 21% to 30%, although in some residential areas the contour is relatively flat. In addition, the area of the type B zone is the area that is located in the basin between two hills which has a relatively flat slope between 0-15 % which directly adjacent to a slope of above 40%. Thus, the Pancor, Apit Aik, Guntur Macan, Barat Kokoq are included in the type B zone.

c. Zone Type C is a lowland area with a slope of 0- 15%, which are the characteristic of the Barat Kokoq and Apit Aik Hamlet.

Table 1. Area Threatened zones of the landslide zones of Guntur Macan Village.

No	Zone Typology	Land surface form	Broad (hectare)
1	Type A	Hillside	229.76
2	Type B	Foothills	63.00
3	Type C	Lowland	55.90

Source: analysis result, 2018

IOP Conf. Series: Materials Science and Engineering **674** (2019) 012019 doi:10.1088/1757-899X/674/1/012019

3.2 The Landslide Risk Analysis Based on Human Activity Aspects in Guntur Macan Village

3.2.1 Cropping Pattern
The cropping patterns are designed based on the existing land types. The Type A Zones and Type B Zones are mostly utilized for plantation. The hillside is planted with taproot plants such are, sengon, and teak where these types of plants are very effective in preventing landslides. Hence, the area will have a low sensitivity to the landslide risk. However, the cropping pattern is not complemented by the community activities in utilizing the logs for sale, mainly sengon and jati. The legal logging in Guntur Macan Village that supported by the government has been depleting the trees which disturbing the hillside stability. On the other side, Zone Type C is a land utilized for paddy fields, planted with fibrous plants such as rice and maize. Although these plants are highly sensitive to the landslides, the plants are planted in a relatively flat slope of 0-8%; hence, the risk sensitivity is low.

3.2.2 Excavation and cutting slopes
Cutting slopes for cultivation, residential area, and road construction activity such as excavation and mining, technically could increase the soil movement. In zone type A, the excavation and the slope cutting are conducted in the natural hillside/artificial hillside, where those activities are used to build houses and roads. However, the construction was done without concerning about the soil/rock layer, and without calculating the slope stability. Hence, the excavation/cutting intensity is considerably high.

At Zone Type B, the intensity of the excavation/slope cutting is also high in which the excavation and cutting slope is conducted for the construction of houses and roads, as well as brick production in the hillside which will increase the vulnerability sensitivity. On the other side, zone type C sensitivity is low due to the absence of excavation and slope cutting activity in the area. Furthermore, the area has a slope characteristic that relatively flat.

3.2.3 manufacturing and placement of ponds
Pond printing on steep slopes affects the potential for landslides because the water will affect the physical and generical properties of the soil, transforming the soil into soft and loose; hence, the soil strength is decreasing as the soil move quickly. Based on the field identification, the pond printing only found on the zone type C in the flat slope 0-8%. On the other side, zone type A and B has no activity of pond printing; thus, the risk level is low.

3.3.4 Drainage
In the Zone type A, the drainage system is inadequate, covered by dirt soil. There is no effort to repair it either from the government or the community. Hence, based on the drainage condition aspect, this region has high-risk sensitivity. On the other side, Zone Type B and Zone Type C drainages are quite adequate. Moreover, there are efforts to improve drainage; thus, the risk level in both zones is medium.

3.3.5 Construction Build
At Zone Type A, there is construction ongoing with considerably low loads but not exceeded the soil bearing capacity. The construction is conducted to build roads and houses, elementary school buildings, prayer facilities, and health care facilities like *Posyandu* (Integrated Healthcare Center). Although the construction activity is relatively infrequent, as the growth of the human needs of shelters, the construction will be increasing. Thus, the hillside loads increased as well as the landslide risk level. Even though the villagers are constructing in the flat area, as the hill slope more than 40% tilt, it is considered a steep hill. Hence, the risk level is still high.

At Zone Type B, the construction and the load capacity are considerably small. The construction is focused on the road, houses, elementary schools, health care facilities such as *Posyandu* (Integrated Healthcare Center) and village health center, and governmental facilities like village offices. Based on the field observation, the construction has been exceeded the load capacity of the land, because most of the residents are building their houses in the slope area (more than 40% tilt) without calculating the hillside stability. This area is highly vulnerable to landslides compared to other zones. Hence, the risk level is high.

In zone type C, the construction has a relatively small load capacity and has not exceeded the standard. The constructions are focused on houses, roads, elementary school facilities, prayer facilities, and health care facilities such as *Posyandu* (Integrated Healthcare Center). The landslide sensitivity in this area is medium based on the construction aspects because the area is considerably flat.

3.3.6 Population Density

Based on the area calculation, the population density in this area is considerably low, less than 20 residents/hectare. However, the population density was calculated based on the living area. Based on the number of inhabitants in each zone, the number of houses in Guntur Macan Village is corresponding to the zone typology, which interpreted by satellite imagery and primary survey with the help of ArcGIS, resulting population density in zone A is 34 resident/ha, zone type B is 37 resident/ha, and zone type C is 49 resident/ha. Consequently, the risk level of each zone is considerably medium due to the range of the population in the three zones (20-50 resident/ha).

3.3.7 Mitigation

After the landslide occurred in the Guntur Macan Village that killed the resident and destroyed their properties, the West Lombok government has been trying to keep the disaster away by coordinating the mitigation activities. The government has been forming a community of DESATANA (the resilient village disaster) in Guntur Macan Village. This forum is expected to be a forerunner of disaster management in Guntur Macan Village that vulnerable to the landslides (Figure 2 and Table 2).

The National Disaster Management Agency has been installing a natural disaster detection system (EWS) in the local hilly area. EWS installation is collaborative work between the National Disaster Management Agency and a team from Gadjah Mada University. There are four EWS tool mounting spots. Two spots for ground cracking detection, one spot for the land slope analysis, and the last spot for the rainfall. Disaster detection is expected to give an early warning to the community when the disaster comes. Consequently, all the zones will have a low-risk level of disaster.

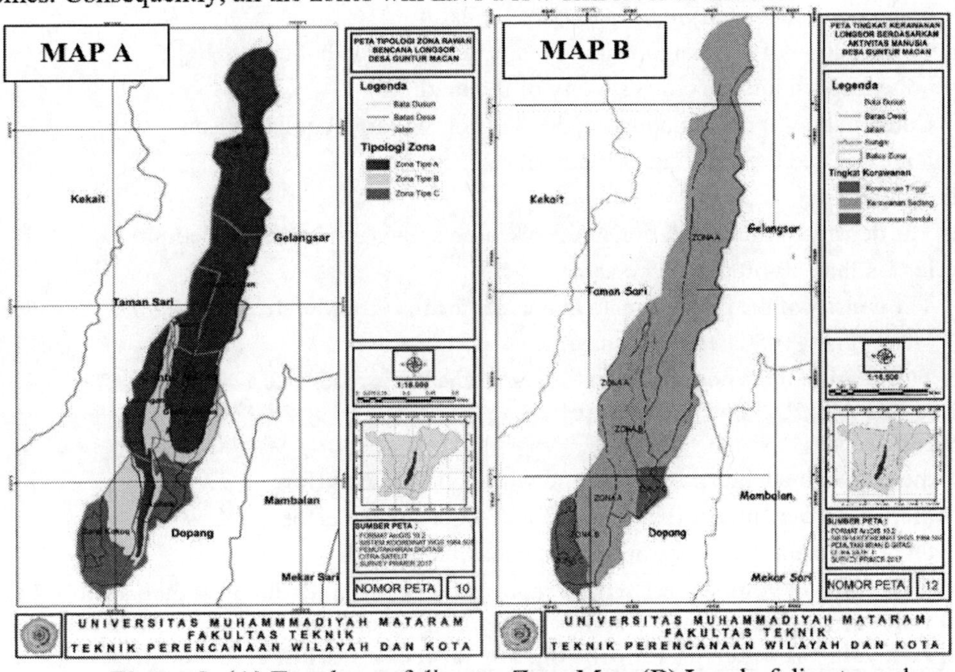

Figure 2. (A) Typology of disaster Zone Map, (B) Level of disaster vulnerability based on activity level map.

IOP Conf. Series: Materials Science and Engineering **674** (2019) 012019 doi:10.1088/1757-899X/674/1/012019

Tabel 2. The level of vulnerability of landslide disaster in the Guntur Macan village.

Zona	Condition	Sensitivity	Velue	Point	Total
Planting Pattern					
A	The slopes planted by the Tree are rooted, but there is a branching tree	Hight	3	0.1	0.3
B	The slopes planted by the Tree are rooted, but there is a branching tree	Hight	3	0.1	0.3
C	The slopes are planted Vegetable rooted, but on a flat area	Low	1	0.1	0.1
Excavation and Cutting Slope					
A	Excavation / Cutting Slope High	Hight	3	0.2	0.6
B	Excavation / Cutting Slope High	Hight	3	0.2	0.6
C	No Excavation / Cutting Slope	Low	1	0.2	0.2
Building and placement of pools					
A	Nothing	Low	1	0.1	0.1
B	Nothing	Low	1	0.1	0.1
C	There is a Pond printing but in a flat area	Medium	2	0.1	0.2
Drainage					
A	Inadequate and no improvement effort	Hight	3	0.1	0.3
B	Quite adequate and there is an improvement effort	Medium	2	0.1	0.2
C	Quite adequate and there is an improvement effort	Medium	2	0.1	0.2
Construction					
A	Construction of construction and load is not too large but has not exceeded the carrying capacity of the land	Hight	3	0.1	0.3
B	Construction of construction and load is not too large but has not exceeded the carrying capacity of the land	Hight	3	0.1	0.3
C	Construction of construction and load is not too large but has not exceeded the carrying capacity of the land	Low	1	0.1	0.1
Population Density					
A	The density of the population in the research area is moderate, ie less than 20-50 people/hectare	Medium	2	0.2	0.4
B	The density of the population in the research area is moderate, ie less than 20-50 people/hectare	Medium	2	0.2	0.4
C	The density of the population in the research area is moderate, ie less than 20-50 people/hectare	Medium	2	0.2	0.4
Mitigation					
A	there are a landslide mitigation and well-coordinated effort	Low	1	0.1	0.1
B	there are a landslide mitigation and well-coordinated effort	Low	1	0.1	0.1
C	there is a landslide mitigation effort coordinated with	Low	1	0.1	0.1
A total risk value of zone A= 2.1 with a medium level of landslide risk (the area is 229.76 ha)					
A total risk value of zone B = 2.0 with a medium level of landslide risk (the area is 63.00 ha)					
A total risk value of zone C = 1.3 with the low level of landslide risk (the area is 55.90 ha)					

Source: analysis result, 2018

4. Conclusion

The landslide risk level in Guntur Macan village based on the human activity assessment is divided into three zones as a result of seven indicator analysis, namely cropping pattern, excavation, and slope cutting, pond making, drainage, construction, population density, and mitigation. Zone A has a risk level around 2.1, while zone B is 2.0 which classified as medium level. On the other side, zone C has 1.3 risk level that categorized as low-risk level of landslide.

References

[1] Anonim, "Perda Nomor 11 Tahun 2011 tentang Rencana Tata Ruang Kabupaten Lombok Barat Tahun 2011 - 2031," Bappeda Kabupaten Lombok Barat, Gerung, 2011.

[2] B. H. Widayanti, A. Yuniarman dan F. Susanti, "Faktor Pemilihan Lokasi Bermukim Pada Kawasan Rawan Bencana Longsor di Desa Guntur Macan Kabupaten Lombok Barat," Journal of Regional and Rural Development Planning, vol. 2 No 1, pp. 34-44, 2018.

[3] Anonim, "Peraturan Menteri Pekerjaan Umum Nomor 22/PRT/M/2007 tentang Pedoman Penataan Ruang Kawasan Rawan Bencana Longsor," Departemen Pekerjaan Umum Republik Indonesia, Indonesia, 2007.

[4] D. Karnawati, Bencana Alam Gerakan Massa Tanah di Indonesia dan Upaya Penaggulangannya, Yogyakarta: Teknik Geologi UGM, 2005.

[5] P. D. Raharjo and S. Winduhutomo, "Social-Population Condition on The Physical Enviroment Characteristics in Influence The Risk of Landslide in Karangsambung," Jurnal Manusia dan Lingkungan, vol. Vol 23 No 1, pp. 1 - 11, 2016.

[6] S. D. Santha and R. K. Sreedharan, "Population Vulnerability and DIsaster Risk Reduction : A Situation Analysis Among The Landslide Affected Communities in Kerala, India," Journal of Disaster Risk Studies, vol. Vol 3 No 1, pp. 367-380, 2010.

[7] B. H. Widayanti, A. Yuniarman and F. Susanti, "Faktor Pemilih Lokasi Bermukim Pada Kawasan Rawan Bencana Longsor di Desa Guntur Macan Kabupaten Lombok Barat," Journal of Regional and Rural Development Planning, vol. Vol 2 No 1, pp. 34 - 44, 2018.

IOP Conf. Series: Materials Science and Engineering **674** (2019) 012020 doi:10.1088/1757-899X/674/1/012020

Global ductility demands of RC frames with various post-yield stiffness ratio and ductility capacity ratio under near-field earthquakes

A Faisal[1,*], F V Riza[1] and B Hadibroto[2]

[1] Program Studi Teknik Sipil, Fakultas Teknik, Universitas Muhammadiyah Sumatera Utara, Medan, Indonesia

[2] Civil Engineering Programme, Faculty of Engineering, State University of Medan, Indonesia

[*] Email: adef@umsu.ac.id

Abstract. The directivity pulse motion in near-field earthquake affects the responses of structures. The response includes the roof drifts and interstory drifts, which is so-called the ductility demand. Although many studies have done in this topic, the effect of various post-yield stiffness ratio and ductility capacity ratio to the global ductility demands of reinforced concrete (RC) frames under near-field ground motion is not specifically investigated yet. The trend of global ductility demands of RC frames under near-field earthquake is the objective of this study. The frames are modeled with various fundamental period, behavior factor, plastic rotation capacity, post-yield stiffness ratio, and ductility capacity ratio. The study reveals that the effect of ductility capacity ratio on the global ductility demands is apparent as as the plastic rotation capacity changes. The global ductility demand is also found influenced by global post-yield stiffness ratio and behavior factor.

1. Introduction

Many studies show that inelastic demands of structures depend on the type of induced ground motion (e.g. far-field and near-field ground motions). The parameters governed the near-field ground motions (NFGM) and its effect on the inelastic demand of structures have been extensively studied. In the 1970s, a pioneering work in investigating NFGM affecting a building was conducted by Bertero et al. [1]. The effect of NFGM on the structural response is further explored by Hall et al. [2]. The study found that the effect of NFGM can be very effective in causing damage in structure when the fundamental period of vibration of structure is comparable to the duration of the NFGM's pulse. Maholtra [3] found the directivity pulse motion affected the responses of structures in increasing the base shear and interstory drifts in high-rise buildings, increasing the ductility demand. Alavi and Krawinkler [4], Krawinkler et al. [5], and Krawinkler et al. [6] revealed that large elastic storey shear forces in the upper storey of the structure were found to be generated by FDE with a fundamental period of the structure higher than the pulse period.

Using the real records of forward directivity earthquake (FDE), similar studies were conducted by Akkar et al. [7] and Kalkan and Kunnath [8] in exploring the response of generic multi-storey structures due to FDE and their results support previous findings [4]. Approximately, similar findings were also reported in some studies which were using generic shear frame models induced by real and synthetic FDEs [9]. Majid et al.[10] employed the real NFGM and far-field ground motion in evaluating the response of irregular RC frames. The response showed that NFGM produced higher drift compared with far-field ground motion. The response of the 4- to 18-storey RC framed structures under single and sequential near-field earthquakes conducted by Zahid et al. [11] and Faisal et al. [12] in order to identify its displacement ductility and storey ductility demand. Beiraghi et al. [13] investigated the seismic behavior of the 20-story, 30-story and 40-story of core-wall buildings with fixed bases subjected to NFGM with respect to the energy concepts.

Although many studies have done in the topic of effect of NFGM on the response of buildings, the relation of post-yield stiffness ratio and ductility capacity ratio to the global ductility demands of building under NFGM is not specifically investigated yet. Therefore this study aims to indentify the trend of global ductility demands of RC frames under FDE. The demands is explored in the perspective of global post-yield stiffness ratio and global ductility capacity ratio. The study employs various fundamental periods of structures, behavior factors, and plastic rotation capacities in order to achieve a general insight.

2. Method

2.1. Inelastic Structural Models

2.1.1. Properties of the Model.

The generic frame model used in this study is proposed by Faisal et al. [12]. It focuses on the regular geometric system and horizontal dimension of the system in any story is <130% of an adjacent story or no setback. Considered fundamental periods of the 3D generic MDOF model are T_1 = 0.45, 0.75, 1.26, and 1.71 seconds. The elevation and plan of 3D generic MDOF structure is as shown in Figure 1. The model is assumed to be built in Zone III in Greece with spectra design (Figure 2). Columns and beams at each story have the same stiffness in order to reduce the uncertainty in modelling. The plan shape of floor and roof is squared plan size, H_i x H_i. By assuming ratio of the span to story height is equal to 2.0 for all stories of the generic model hence parameter H_i is equal to the twice of column height.

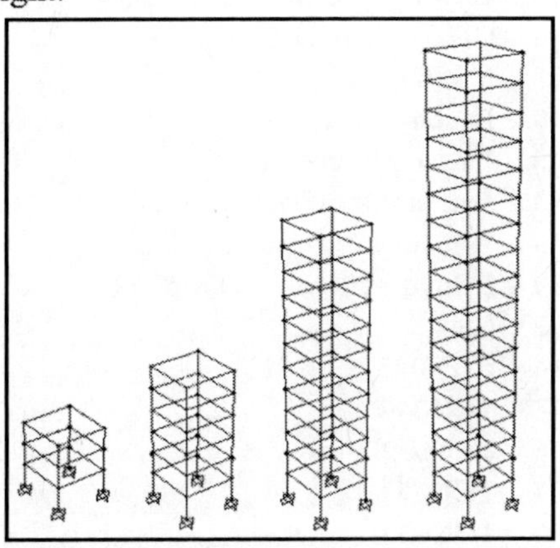

Figure 1. Models of 3D generic 3, 6, 12, and 18-story single-bay frames [12]

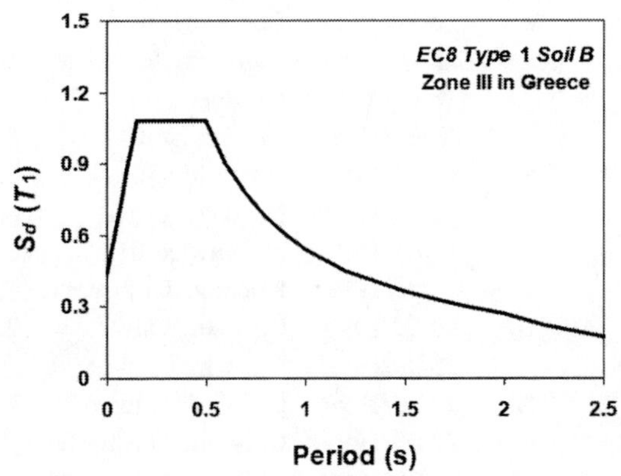

Figure 2. Design response spectrum of Eurocode 8 for Type 1 and Soil B at Zone III of Greece

2.1.2. Weight and Stiffness.

The story weight of RC structure is assumed to be identical (1240 kN) at all stories and all models, whereas stiffness is assumed to be the same for every 3 levels (stepwise distribution). The models are set up to be as the fundamental period of model increases the beam-to-column stiffness ratio decreases. The model sets to behave dominantly in flexure as the fundamental period (or number of story) increase.

2.1.3. Seismic Design and Strength.
The seismic base shear is defined from ordinate design spectrum at period T_1 of Type 1 spectrum of Eurocode 8 for condition of Soil B with peak ground acceleration (PGA), a_g=0.36 g, as shown in Figure 2. The a_g is based on 475-years return period of earthquake that reflecting the condition of Seismic Zone III at Greece. Greece represents the highest seismic region in Europe, along with Turkey and Italy, whereas Zone III is the highest seismic zone in Greece. The moment and rotation of this linear elastic static analysis at q = 1 means that both result from the same elastic and yield forces, $F_e = F_y$. The bending moment of member at this stage is considered as the yield strength of member.

2.1.4. Plastic Hinge and Rotation Capacity.
The plastic hinge is represented by moment-rotation relationship, whereas the shear deformation as well as moment-axial interaction are disregarded. The plastic hinge is modelled using lumped plasticity model. To simulate the cyclic behavior of members in plastic hinge under load reversals, Modified-Takeda hysteresis rule is employed. The unloading and reloading parameters (α and β) in hysteresis rule for beam member are equal to 0.3 and 0.6 respectively. This study employs the rotation capacities represent the capacity of general RC structures.

2.1.5. Ground Motion.
Selected ground motions for FDE are presented in Table 1 sourced from seismic station built in stiff soil. FDE motions are dominated by California and Taiwan earthquakes and mainly from source-to-site distance less than 10 km and magnitude larger than M6.0. The scaling factor of selected ground motions used in the study is corresponding to the fundamental period of models.

Table 1. Selected FDE motions on the stiff soil from PEER-NGA

No.	Date	Earthquake Name	Mag. (Mw)	Closest Dist. (km)	Station
1	24/04/1984	Morgan Hill	6.2	0.53	Coyote Lake Dam (SW Abut)
2	18/10/1989	Loma Prieta	6.9	9.96	Gilroy - Gavilan Coll.
3	18/10/1989	Loma Prieta	6.9	3.88	LGPC
4	28/06/1992	Landers	7.3	2.19	Lucerne
5	17/01/1994	Northridge-01	6.7	5.43	Jensen Filter Plant
6	17/01/1994	Northridge-01	6.7	5.43	Jensen Filter Plant Generator
7	17/01/1994	Northridge-01	6.7	5.19	Sylmar - Converter Sta East
8	17/01/1994	Northridge-01	6.7	5.30	Sylmar - Olive View Med FF
9	17/08/1999	Kocaeli, Turkey	7.5	10.92	Gebze
10	20/09/1999	Chi-Chi, Taiwan	7.6	3.14	CHY028
11	20/09/1999	Chi-Chi, Taiwan	7.6	3.78	TCU049
12	20/09/1999	Chi-Chi, Taiwan	7.6	0.66	TCU052
13	20/09/1999	Chi-Chi, Taiwan	7.6	5.97	TCU053
14	20/09/1999	Chi-Chi, Taiwan	7.6	5.30	TCU054
15	20/09/1999	Chi-Chi, Taiwan	7.6	0.32	TCU068
16	20/09/1999	Chi-Chi, Taiwan	7.6	0.91	TCU075
17	20/09/1999	Chi-Chi, Taiwan	7.6	2.76	TCU076
18	20/09/1999	Chi-Chi, Taiwan	7.6	5.18	TCU082
19	20/09/1999	Chi-Chi, Taiwan	7.6	1.51	TCU102
20	20/09/1999	Chi-Chi, Taiwan	7.6	6.10	TCU103

2.1.6. Global Ductility and Post-Yield Stiffness Ratio Capacities.
The ductility and post-yield stiffness ratio are resulted from nonlinear static analysis or pushover analysis [12]. To relate the global ductility capacity with the story ductility capacity, the ratio of story

ductility and global ductility capacities is employed in the study and denoted as ϑ_c. The ratio is defined based on the regression analysis in polynomial fourth-order form. The relationship of ϑ_c with number of story (N) for $0.45 \leq T_1 \leq 1.71$ s is in the following formula ($R^2 = 0.977$):

$$\vartheta_c = 6.177T_1^3 - 17.543T_1^2 + 16.450T_1 - 3.210 \quad (1)$$

The parameter of ratio of global post-yield stiffness to elastic stiffness, r_K, to represents global stiffness of the structures [12] is used in this study. The relationship of r_K as a function of the fundamental period of vibration, T_1, behavior factor, q, and plastic rotation capacity, θ_p, of the system ($R^2 = 0.925$ and Sig.F-ratio ≈ 0.00) is as follow:

$$Log(r_K) = -0.508T_1 - 0.135q - 11.252\theta_p - 0.952 \quad (2)$$

3. Result and Discussion

3.1. *Effect of Global Ductility Capacity Ratio*

The ratio of maximum story ductility to roof ductility denotes global ductility capacity ratio, ϑ_c. Figure 3 shows how global ductility capacity ratio under two types of plastic rotation capacity (Rot. = 0.02 and 0.06) affects the roof ductility demand in logarithmic form. This figure also reflects the fundamental period of models under consideration in its each point of post-yield stiffness ratio. Although ϑ_c is apparent in influencing μ_Δ, the trend of how ϑ_c governed μ_Δ is not clear as the plastic rotation capacity changes. It seems other factor also plays an important role to this trend, e.g. fundamental period of vibration, T_1. In statistical analysis point of view, the effect of ϑ_c on the roof ductility demand was found about similar with the effect of T_1. Therefore, the effect of ϑ_c on the roof ductility demand might be has dependency to T_1.

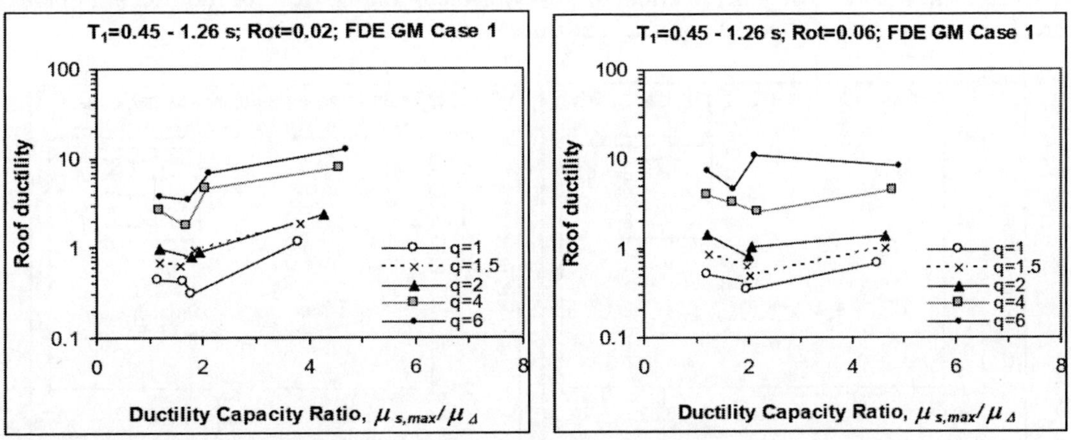

Figure 3. Mean roof ductility demand as a function of ductility capacity ratio and behavior factor. Each point represents fundamental period of models T_1 = 0.45, 0.75, 1.26 and 1.71 s.

3.2. *Effect of Global Post-Yield Stiffness Ratio*

Effect of global post-yield stiffness ratio, r_K, on the roof ductility demand, μ_Δ, can be seen in Figure 4. This figure explains the roof ductility demand as a function of global post-yield stiffness ratio and plastic rotation capacity (denotes Rot.=0.02, 0.04, and 0.06). It demonstrates that the post-yield stiffness ratio decreases as the roof ductility demand increases. This mechanism is correct because when the fundamental period of system under the same strength increases, the global stiffness decreases (more flexible), hence roof and story drifts increases. In Figure 4, each point of post-yield stiffness ratio also represents a behavior factor of the system, which explains the change in ductility class (e.g. DCM to DCH in Eurocode 8) affects the post-yield stiffness and thus it influences the roof ductility demand due to its post-yield stiffness ratio.

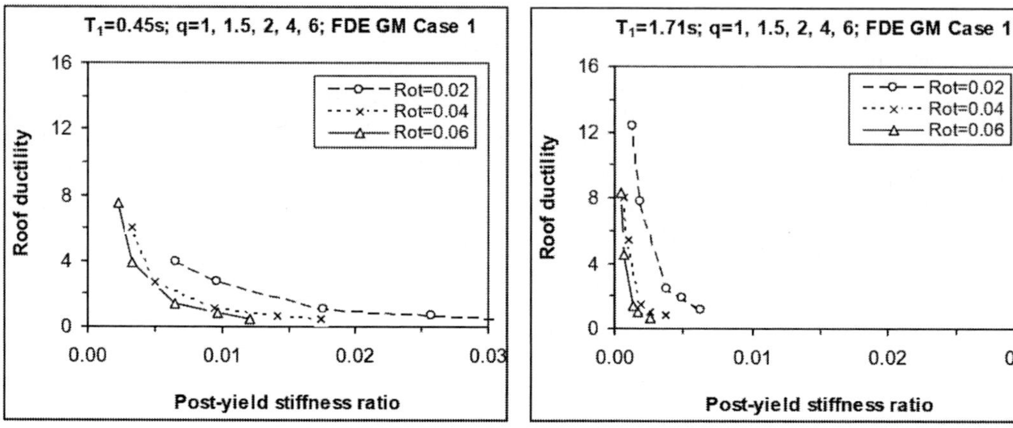

Figure 4. Mean roof ductility demand of models with $T_1 = 0.45$ and 1.71 s as a function of global post-yield stiffness ratio and plastic rotation capacity. Each point represents a behavior factor.

Figure 5 depicts the of ductility demand as a function of global post-yield stiffness ratio and behavior factor for two types of plastic rotation capacity (Rot.=0.02 and 0.06). Each point in this figure represents fundamental period of models under consideration and hence the trend follows the same way as previously discussed. It is evident that no regular trend is demonstrated as the post-yield stiffness ratio is changed. In this case, the irregularity of effect post-yield stiffness ratio on the roof ductility demand explains its dependency to fundamental period since its shape of line indicates the same trend as in Figure 5. In general, effect of behavior factor on the trend of roof ductility demand in Figure 5 is superior to the effect of post-yield stiffness ratio.

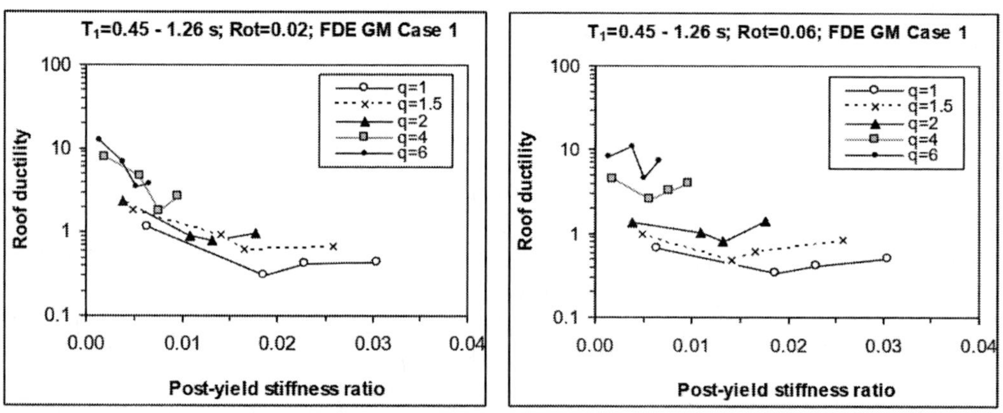

Figure 5. Mean roof ductility demand as a function of global post-yield stiffness ratio and behavior factor. Each point represents fundamental period of models $T_1 = 0.45$, 0.75, 1.26 and 1.71 s.

It should be noted that the variation of post-yield stiffness ratio of the system under the same behavior factor or ductility class is reflecting the variation of strength (i.e. flexural strength) of the elements represented by the ratio of maximum bending moment to the yield moment (M_c/M_y). It is because the M_c/M_y of elements has involved in the formation of the backbone curve of the hysteresis system, which in turn has participated in the global backbone curve. Unfortunately, the effect of M_c/M_y on the roof ductility ratio could not be presented since this parameter was not varied in designing the model. However, the variation of post-yield stiffness ratio in this study could reflect the variation of

M_c/M_y by assuming a single value of plastic rotation capacity is used. Therefore, it can be said that as the M_c/M_y of element in the system under the same plastic rotation capacity increases, the post-yield stiffness ratio increases and hence the roof ductility demand decreases.

4. Conclusions

This study investigates the global ductility demands of RC frames with various post-yield stiffness ratio and ductility capacity ratio under forward directivity motion of near-field earthquake. The models include the various fundamental period, behavior factor, and plastic rotation capacity. The following conclusions can be drawn from the study: 1) As the plastic rotation capacity changes, the effect of ductility capacity ratio on the global ductility demands is clearly apparent; 2) The change in ductility class (e.g. DCM to DCH in Eurocode 8) affects the post-yield stiffness and thus it influences the roof ductility demand due to its post-yield stiffness ratio.

References

[1] Bertero, V.V., Mahin, S.A. and Herrera, R.A., 1978. Aseismic design implications of near-fault San Fernando earthquake records. *Earthquake engineering & structural dynamics*, *6*(1), pp.31-42.

[2] Hall, J.F., Heaton, T.H., Halling, M.W. and Wald, D.J., 1995. Near-source ground motion and its effects on flexible buildings. *Earthquake spectra*, *11*(4), pp.569-605.

[3] Malhotra, P.K., 1999. Response of buildings to near-field pulse-like ground motions. *Earthquake Engineering and Structural Dynamics*, *28*(11), pp.1309-1326.

[4] Alavi, B. and Krawinkler, H., 2001. *Effects of near-fault ground motions on frame structures*. Stanford: John A. Blume Earthquake Engineering Center.

[5] Krawinkler, H., Medina, R. and Alavi, B., 2003. Seismic drift and ductility demands and their dependence on ground motions. *Engineering structures*, *25*(5), pp.637-653.

[6] Krawinkler, H., Alavi, B. and Zareian, F., 2005. Impact of near-fault pulses on engineering design. *Direction in Strong Motion Instrumentation*, *58*(1), pp.83-106.

[7] Akkar, S., Yazgan, U. and Gülkan, P., 2005. Drift estimates in frame buildings subjected to near-fault ground motions. *Journal of Structural Engineering*, *131*(7), pp.1014-1024.

[8] Kalkan, E. and Kunnath, S.K., 2006. Effects of fling step and forward directivity on seismic response of buildings. *Earthquake spectra*, *22*(2), pp.367-390.

[9] Sehhati, R., Rodriguez-Marek, A., ElGawady, M. and Cofer, W.F., 2011. Effects of near-fault ground motions and equivalent pulses on multi-story structures. *Engineering Structures*, *33*(3), pp.767-779.

[10] Majid, T.A., Wan, H.W., Zaini, S.S., Faisal, A. and Wong, Z.M., 2010. The effect of ground motion on non-linear performance of asymmetrical reinforced concrete frames. *Disaster Advances*, *3*(4), pp.35-39.

[11] Zahid, M.Z.A.M., Majid, T.A. and Faisal, A., 2012. Effect Of Repeated Near Field Earthquake To The High-Rise Rc Building. *Australian Journal of Basic and Applied Sciences*, *6*(10), pp.129-138.

[12] Faisal, A., Majid, T.A. and Hatzigeorgiou, G.D., 2013. Investigation of story ductility demands of inelastic concrete frames subjected to repeated earthquakes. *Soil Dynamics and Earthquake Engineering*, *44*, pp.42-53.

[13] Beiraghi, H., Kheyroddin, A. and Kafi, M.A., 2016. Forward directivity near-fault and far-fault ground motion effects on the behavior of reinforced concrete wall tall buildings with one and more plastic hinges. *The Structural Design of Tall and Special Buildings*, *25*(11), pp.519-539.

[14] ASCE. (2005). *Minimum design loads for buildings and other structures, ASCE/SEI 7-05*. Reston, VA.: ASCE.

[15] CEN. (2004). *Eurocode 8: Design of structures for earthquake resistance. Part 1: General rules, seismic actions and rules for buildings* (Vol. 1). Brussels.: European Committee for Standardization.

Acknowledgement

Support for this project was provided by the Grant of Penelitian dan Pengabdian Kepada Masyarakat Internal UMSU tahun 2016.

Study of alternative building for tsunami evacuation in Kuta Alam sub-district Banda Aceh

W Soviana * and H A Rani

Civil Engineering Department, Faculty of Engineering, Universitas Muhammadiyah Aceh, 23245 Banda Aceh, Indonesia

* Email: hafnidar.ar@unmuha.ac.id

Abstract. Kuta Alam Sub District has many potential hazards, especially from the sea. The location of the area that is directly adjacent to the sea causes it to be very vulnerable to tsunami. Its topography is lowland less than 1 meter above the sea-level. Tsunami in 2004 had taken a great number of people lives in the area about 14.95 % of the total number of inhabitants which is 55,030 people that time. However, there is still no any evacuation building available in Kuta Alam to escape from Tsunami. We know tsunami which might be occurred anytime unpredictable. In order to save people's lives, it is needed to conduct a study to find alternatives buildings to be used for evacuation and rescue in the area. The aim of this study to identify the locations and number of alternative buildings that are potential as the tsunami evacuation building in Kuta Alam Sub-district. The study used the survey system toward the buildings with qualitative and quantitative approaches. Data analyzing is conducted spatially through geographic information systems. Based on observation results toward evacuation building in Kuta Alam Sub-district, there are some buildings renovated in order to be functioned as tsunami evacuation buildings. The renovation should be done on roof, school, and ladder evacuation of the mosque that are selected as evacuation buildings. Until today, those buildings have not been renovated to be eligible as evacuation places. The result found that there are 45 public buildings that can be used for alternative evacuation buildings in Kuta Alam Sub-district. That can accommodate 94.06 % of the total number of people this time which is 49,011 people.

1. Introduction

A region will potentially risk whether from natural or unnatural disasters. The condition of topography and geography of the area potentially impacted by the natural disaster [1]-[3]. For the area that is bordered on the ocean like Kecamatan Kuta Alam is very vulnerable toward the disaster come from the sea such as a hurricane, sea abrasion, and tsunami.

Tsunami in 2004 caused massive destruction and killed so many people in almost every area of Banda Aceh. One of the impacted areas was Kuta Alam Sub District. There were 8,227 people killed or 15% of the inhabitants [4]. There are some factors why there were so many people killed at that time, such as lack of information or knowledge of the people about the disaster potential risk and how to reduce the risk [5]-[7]. Moreover, there was no clear evacuation direction. For the overcrowd area, was conducted by running, however for the target location, more than 4 km would be hard to reach on foot if the golden time less than 30 minutes [8]-[10].

From the ancient history (paleo tsunami), a tsunami occurred in 1000 AD, between 1350 -1550 AD and after 1800 AD in Aceh that washed away Aceh coastal area [11][12].While this information was never known by Acehnese people that caused how unprepared people when the tsunami strike. In order to avoid that, National Disaster Management Coordinating Board (BNPB) has stated that Aceh Province is disaster-prone area [13]. Aceh, located near to the subduction zone in the Nothern of Sumatera island, the convergence of two very active tectonic plates,Indo-Australia and Eurasia that potentially cause earthquakes and tsunami.

As the vulnerable tsunami impacted, it is very important to have mitigation base site planning and urban design program that is integrated with disaster mitigation principles [14][15]. Actually, in Qanun Banda Aceh District Government No. 4 in 2009 about Urban Planning in Banda Aceh city has included the problem of tsunami threat and mitigation. According to the Qanun, city planning design in the future must give safety and less vulnerable to the communities. It can be done by providing evacuation space and escape buildings to save more lives [16][17]. In fact, there is not all of this implemented in reality for example in Kuta Alam Sub District which isn't any escape building yet to anticipate the inevitable tsunami attack.

In order to apply the standard requirements of the buildings in the tsunami vulnerable area, it needs a very high cost. The government as the funder should decide to prioritize providing public building as the alternative for escape building in order to reduce the risk of life loss [18][19]. So that public buildings with two floors such as school and government offices expectedly can be alternative as evacuation place however with the calculated and measured standard capacity resistance of the buildings [20][21].

Based on [22] the research's result shows that the dominant criteria of the potential escape building could be applied in Banda Aceh are it is located in the position > 500 from the coastline with distance range < 1 km and adapted to the building capacity; road access to the building; evacuation signs clearly seen; it can be accessed by the community 24 hours without any obstacles; it has the emergency stairs outside the building and can be used as the access toward the save place located above the tsunami water average (run-up) predicted.

This study aims to identify the locations and number of alternative buildings that are potential as the tsunami evacuation building in Kuta Alam Sub-district. Before that, the study conducted with observation toward the number of public buildings in Kuta Alam District area. Literature research conducted in order to know how the exposure of hazards to the buildings in vulnerable tsunami areas. The vulnerability of the buildings can be identified from the study. Basic assumptions of determination of building exposure depend on buffer zone in the seashore. For the building, vulnerability is based on its height and numbers of floor and material construction.

2. Design of Experiment

This study used qualitative and quantitative approachment. This method measures the level of building resistance toward the hazard threat of tsunami with primary reference toward the tsunami on December 26th, 2004. The survey activity conducted directly toward every public building in Kecamatan Kuta Alam. The endurance level decided based on an initial survey which is determined by building structure and the height of the building. Based on initial survey of structure, there five types of building available in Kuta Alam. Table 1 explains in detail endurance assessment from building type as the reference to conduct survey.

Table 1. Type of buildings based on structure endurance level and its height [23]

Type of Building	Structure	Level of building endurance	Classification (Fcb)
Type A	- Reinforced concrete ≥ 3 floors - Reinforced concrete column diameter ≥ 30 cm - Brick wall - Reinforced concrete foundation	- Tsunami-resistant buildings - Good column structure condition and connected with the reinforcement on the foundation - The height of the building> 9 m	Very good (1)
Type B	- Reinforced concrete≥ 2 floors - Reinforced concrete column diameter ≥ 30 cm - Brick wall - Reinforced concrete foundation	- Tsunami-resistant buildings - Good column structure condition and connected with the reinforcement on the foundation - The height of the building 3 - 9 m	Good (2)
Type C	- Concrete floor 1 - Reinforced concrete column diameter ≤ 20 cm - Brick wall - Rock foundation	- Buildings are less tsunami-resistant - Good column structure condition and connected with the reinforcement on the foundation - The height of the building< 3 m	Not good (3)

3. Experimental Procedure

There are three parameters used as a benchmark for determining level of building vulnerability toward tsunami threat; they are (i) building conditions, (ii) inundation zones, and (iii) sea defense [24]. Each has different values toward the effect of building vulnerability as displayed in Table 2.

Table 2. Parameters of building tsunami vulnerability [24]

Parameter	Score
Building condition	$F_{wb} = 3$
Inundation zones	$F_{wi} = 2$
Sea defence	$F_{ws} = 1$

The selection of alternative evacuation building is based on tsunami vulnerability analysis. The proper tsunami evacuation buildings are not the ones classified into high vulnerability. For inundation zones based on the condition Aceh tsunami in 2004. The height of inundation reference taken from tsunami height memorial poles [25]. The damaging impact of the building caused by the height of tsunami struck referred to the Master Plan of tsunami impact risk reduction, which was published by National Board for Disaster Management [26], as presented in Table 3.

Table 3. Indicator of building vulnerability for inundation zones [26]

Type of inundation	The risk level of inundation	Inundation height	Power failure	Vulnerability (Fci)
Type A	Low	≤ 1 m	Little	1
Type B	Medium	1 – 3 m	Medium	2
Type C	High	≥ 3 m	Big	3

Buffer zones can be functioned as a natural brigade that would reduce tsunami wave pressure [27]. There two types of buffer zones conditions in Kuta Alam, first, the areas with mangroves and the other without mangroves. So that the assessment indicator for tsunami vulnerability buildings involved by buffer zone condition as displayed in table 4.

Table 4 Indicator of building vulnerability for sea defense [24]

Sea defense	Characteristic	Level of defense ability	Vulnerability factors (Fcs)
Mangrove	The presence of Mangrove vegetation	Good	1
Cultivated land	Dryland and aquaculture pond	Bad	2

The data obtained were analyzed using the equation building tsunami vulnerability (BTV) against tsunami hazard by using building condition inundation zones parameters and sea defense [24]. Based on the parameter above, the equation for calculating the level of vulnerability of the public buildings assessed used equation 1 as followed.

$$BTV\,\% = \frac{(Fwb\ x\ Fcb)+(Fwi\ x\ Fci)+(Fws\ x\ Fcs)}{\sum_{k=1}^{3}(Fcmaks\ x\ Fw)k} \text{x } 100 \qquad (1)$$

Where Fwb = building weigh factor; Fcb = building condition factor; Fwi = inundation weigh factor; Fci = inundation factor; Fws = sea defence weigh factor; Fcs = sea defence factor; k = constant; and $Fcmaks$ = indicator maximum.

4. Results and Discussion

A map of the spread of tsunami alternative evacuation buildings in Kuta Alam Sub-district areas can be seen in Figure 1.

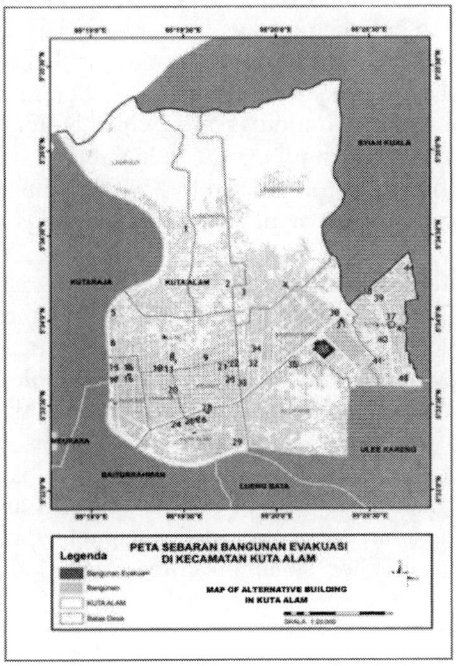

Figure 1. Map distribution of evacuation buildings in Kuta Alam Subdistrict

Results obtained were analyzed further in spatial using Geographic Information System (GIS) for Kuta Alam Sub-district with scale 1: 250,000 following the requirements stated by BNPB for describing the areas in detail in small scope of the area. Furthermore, followed creating map to spot tsunami alternative evacuation building in Kuta Alam Sub-district.

Evacuation buildings alternative was able to accommodate 46,100 people, while the total number of Kuta Alam population is 49,011. Based on the result above, evacuation buildings in Kuta Alam can accommodate 94.06% of the total number of population. The calculation of evacuation building capacity in each area in Kuta Alam in detail can be seen from Table 5.

Table 5. The capacity of alternative evacuation buildings

No	Name of Village	Building Capacity	Population Number	Difference	Description
1	Lampulo	500	5,071	-4,571	Not enough
2	Lamdingin	200	2,738	-2,538	Not enough
3	Lambaro Skep	1,700	4,855	-3,155	Not enough
4	Mulia	7,300	3,295	4,005	more
5	Peunayong	3,200	4,658	-1,458	Not enough
6	Laksana	1,200	5,096	-3,896	Not enough
7	Keuramat	3,600	4,816	-1,216	Not enough
8	Kuta Alam	8,500	6,617	1,883	More than enough
9	Bandar Baru	8,400	4,348	4,052	More than enough
10	Beurawe	3,000	5,864	-2,864	Not enough
11	Kota Baru	8,500	1,653	6,847	More than enough
Total	**Kuta Alam Subdistrict**	**46,100**	**49,011**	**-2,911**	**Not enough**

Based on observation results toward evacuation building in Kuta Alam Sub-district, there are some buildings renovated in order to be functioned as tsunami evacuation buildings. The renovation should be done on roof, school, and ladder evacuation of the mosque that are selected as evacuation buildings. Until today, those buildings have not been renovated to be eligible as evacuation places.

Alternative tsunami evacuation buildings are public buildings which are appropriate as safe building. They are determined according to vulnerability building assessment results. The vulnerability which is classified to be middle and low would be used for evacuation.There 45 units of building in Kuta Alam that can be as tsunami evacuation buildings, such as school, government office, hospital, etc.

The capacity of evacuation buildings available in Kuta Alam is still less in some areas, especially in high-risk tsunami areas. The area was tsunami puddles > 3 meters, such as Lamdingin, Lambaro Skep, and Lampulo. So that the evacuation process should be appropriately designed to prevent more casualties, vulnerable communities such as senior citizens, disable, and children should be priority to be evacuated to public buildings in the area while the ones who are physically good would be evacuated to other safe areas.

Kuta Alam Sub-district is included as prone to tsunami disaster because it borders directly with the sea; besides it has the greatest number of populations in Banda Aceh. Despite not having any particular evacuation building, but there are adequate public buildings available to accommodate 94.06% of population as alternative for tsunami disaster evacuation.

References

[1] Rusdi M, Roosli R and Ahamad M 2015 Land evaluation suitability for settlement based on soil permeability, topography and geology ten years after tsunami in Banda Aceh, Indonesia *The Egyptian Journal of Remote Sensing and Space Science* **18** **2** 207-215

[2] Marto R, Papageorgiou C, Klyuev V 2018 Building resilience to natural disasters: An application to small developing states *Journal of Development Economics* **135** 574-586

[3] Zhou L, Wu X, Xu Z and Fujita H 2018 Emergency decision making for natural disasters: An overview *International Journal of Disaster Risk Reduction* **27** 567-576

[4] GTZ-SLGSR. 2007. *Rencana Detail Tata Ruang Wilayah Kecamatan Kuta Alam Kota Banda Aceh Tahun 2007-2016*. Banda Aceh: Pemerintah Kota Banda Aceh

[5] Arce R, Onuki M, Esteban M and Shibayama T 2017 Risk awareness and intended tsunami evacuation behaviour of international tourists in Kamakura City, Japan *International Journal of Disaster Risk Reduction* **23** 178-192

[6] Wood N, Jones J, Peters J and Richards K 2018 Pedestrian evacuation modeling to reduce vehicle use for distant tsunami evacuations in Hawai'i *International Journal of Disaster Risk Reduction* **28** 271-283

[7] Løvholt F, Setiadi N, Birkmann J, Harbitz C and Bach C 2014 Tsunami risk reduction – are we better prepared today than in 2004? *International Journal of Disaster Risk Reduction* **10** 127-142

[8] Di Maero, M., Megawati, K., Cedillos, V., Tucker, B. 2013. *Tsunami Risk Reduction for Densely Populated Southeast Asian Cities: Analysis of Vehicular and Pedestrian Evacuation for The City of Padang, Indonesia, and Assessment of Interventions*. Journal Nat Hazard. Vol. 68, pp 373-404 : SpringerLink

[9] Syamsidik, Rasyif T and Kato S 2015 Development of accurate tsunami estimated times of arrival for tsunami-prone cities in Aceh, Indonesia *International Journal of Disaster Risk Reduction* **14** 403-410

[10] Yasufuku K, Akizuki Y, Hokugo A, Takeuchi Y and Takashima A 2017 Noticeability of illuminated route signs for tsunami evacuation *Fire Safety Journal* **91** 926-936

[11] Monecke, K.; Finger, W.; Kongko, W.; McAdoo, B.; Moore, A. L.; Sudrajat, S. U. 2007. *A Paleotsunami Record from Marshlands in West Aceh Province, Indonesia*. American Geophysical Union

[12] Syamsidik, Fahmi M, Fatimah A and Fitriyansyah A 2018 Coastal land use changes around the Ulee Lheue Bay of Aceh during the 10-year 2004 Indian Ocean tsunami recovery process *International Journal of Disaster Risk Reduction* **29** 24-36

[13] BNPB. 2011. *Panduan Perencanaan Kontinjensi Menghadapi Bencana*. Jakarta

[14] Sagala, S. dan Bisri, M. 2011. *Perencanaan Tata Ruang Berbasis Kebencanaan di Indonesia dalam Anwar, H. dan Harjono, H. (ed), Perspektif terhadap Kebencanaan dan Lingkungan di Indonesia*. Jakarta: Lembaga Ilmu Pengetahuan Indonesia

[15] Wood N, Jones J, Schelling J and Schmidtlein M 2014 Tsunami vertical-evacuation planning in the U.S. Pacific Northwest as a geospatial, multi-criteria decision problem *International Journal of Disaster Risk Reduction* **9** 68-83

[16] McCaughey J, Daly P, Mahdi S and Patt A 2017 Trust and distrust of tsunami vertical evacuation buildings: Extending protection motivation theory to examine choices under social influence *International Journal of Disaster Risk Reduction* **24** 462-473

[17] Leelawat N, Latcharote P, Abe Y, Sugiyasu K and Imamura F 2018 Tsunami evacuation experiment using a mobile application: A design science approach *International Journal of Disaster Risk Reduction* **29** 63-72

[18] Zulkaidi, D. dan P. Natalivan. 2005. *Zoning Regulation dan Building Code dalam Pembangunan Kembali Pasca Gempa dan Tsunami di Provinsi Nanggroe Aceh Darussalam*. Jurnal Perencanaan Wilayah dan Kota, Vol.16/No. 1, hlm 1-20 : Departemen Teknik Planologi ITB

[19] Ismail N, Okazaki K, Ochiai C and Fernandez G 2018 Livelihood changes in Banda Aceh, Indonesia after the 2004 Indian Ocean Tsunami *International Journal of Disaster Risk Reduction* **28** 439-449

[20] Oktari R, Shiwaku K, Munadi K, Syamsidik and Shaw R 2018 Enhancing community resilience towards disaster: The contributing factors of school-community collaborative network in the tsunami affected area in Aceh *International Journal of Disaster Risk Reduction* **29** 3-12

[21] Takabatake T, Shibayama T, Esteban M and Ishii H 2017 Simulated tsunami evacuation behavior of local residents and visitors in Kamakura, Japan *International Journal of Disaster Risk Reduction* **23** 1-14
[22] Rani, HA and Meillyta 2016 Identification of Potential Criteria and Assessment of Escape Building in Banda Aceh *Conference of The 3rd International Conference on Earthquake Engineering and Disaster Mitigation* **3**, Nusa Dua, Bali
[23] Amri, S. 2006. *Teknologi Audit Forensik, Repair dan Retrofit untuk Rumah dan Bangunan Gedung*. Buku 1. Jakarta: Yayasan John Hi-Tech Idetama
[24] Omira, R., Baptista, M.A., Miranda, J.M., Toto, E., Catita, C., Catala, J. 2010. Tsunami vulnerability assessment of Casablanca-Morocco using numerical modelling and GIS tools. Jurnal Nat Hazards. Vol.54, pp 75-95 : SpringerLink
[25] Iemura, H., Takahashi, Y., Pradono, M.H., Sukamdo, P., Kurniawan, R. 2006. *Earthquake and Tsunami Quessionaires in Banda Aceh and Surrounding Areas*. Disaster Prevention and Management Vol. 15 No. 1, 2006 pp. 21-30. Emerald Group Publishing Limited
[26] BNPB. 2012.*Menuju Indonesia Tangguh Menghadapi Bencana Tsunami.* Master Plan Pengurangan Risiko Bencana Tsunami. Jakarta
[27] Tanaka, N., Sasaki, Y., Mowjood, M.I.M., Jinadasa, K.B.S.N., Homchuen, S. 2007. Coastal *Vegetation Structures and Their Functions In Tsunami Protection: Experience of The Recent Indian Ocean Tsunami.* Article of Landscape and Ecological Engineering and Springer

Acknowledgments
This research was supported by Aceh government. The authors would like to be obliged to Muhammadiyah Aceh University for providing laboratory facilities and financial assistance under project.

Evaluation of asphalt porous mixture properties due to addition of Arenga pinnata and coconut fibers

F Rachman* and T Syammaun

Department of Civil Engineering, Engineering Faculty, University Muhammadiyah Aceh, Banda Aceh, Aceh Province, Indonesia

* Email: firmansyah@unmuha.ac.id

Abstract. Traffic condition, as well as tropical climate condition in Indonesia, demands high care to produce water resistance based surface layer in the flexible pavement industry. Asphalt porous is one of various technologies that could become promising prospects in this regard. However porous asphalt features might result in lower ability of pavement to endure traffic load. Thus, this study tries to find solution to previously mentioned issue by adding natural fibers to porous asphalt mixture by measuring the permeability, binder drains down, and void in mixture. The natural fiber is known to have in own strength to endure tensile force, and among many natural fibers available, we choose two of the most abundant material in Indonesia, which are Arenga Pinnata fibers and Coconut fibers. The result in this study is the addition of Arenga Pinnata fibers, or Coconut fibers results in decrease of both permeability and binder drain down test value for 8% to 35% and 32% to 73% respectively while it increases the void in mixture

1. Introduction

Road infrastructure is critical to support social activities and economic improvement. Damage to the surface layer will have an impact on the construction of pavement structures and the use of materials used [1]. In Indonesia, excessive traffic loads, high air temperature, and high intensity of rain are often that result to flooding in several areas, while it is well-known that flexible pavement is susceptible to water, thus consideration should be given in planning the asphalt mixture. One type developed flexible pavement technology on the surface layer is porous asphalt.

Porous asphalt is closely related to the properties of an asphalt mixture using aggregate gradation with a coarse fraction amount of 70-85% and fine aggregate ranges of 15 - 30% to the total weight of the mixture so that the resulting structure is hollow. The existing cavity provides the ability to flow water properly for both vertical and horizontal directions so that it will keep the surface layer dry.

From several literature studies and research, it was found that natural fibers can be used as modifier for asphalt mixtures either as additive, stabilizer or reinforcement [2]-[6]. In his research, natural fiber in the form Arenga Pinnata fiber and Coconut fiber is used as the material of reinforcement. Reinforcement herein meant as a process in which behavior is reinforced by the consequences that immediately follow such behavior Arenga Pinnata fibers and Coconut fibers both have the comportment to bind each other and made the fibers strengthen.

The purpose of this study is to find out how much influence the use of Arenga Pinnata as well as Coconut fibers on porous asphalt mixtures that utilize natural fibers available as an added ingredient.

IOP Conf. Series: Materials Science and Engineering **674** (2019) 012022 doi:10.1088/1757-899X/674/1/012022

2. Materials and Method

2.1. Material Selection

There are four primary materials used in this study which is, Aggregate, Asphalt Binder, Arenga Pinnata fiber and Coconut fiber. Aggregate is obtained from a stone quarry located at Blang Bintang, Aceh Besar District. While Asphalt used in this study is penetration grade asphalt 60/70 (Pen 60/70), which is commonly used asphalt material for road construction in Indonesia, this binder was provided by Syiah Kuala University.

As for Arenga Pinnata fiber and Coconut fiber, we obtain those items from the community's own land field at Darul Imarah Sub-District, Aceh Besar District, and Gampong Tiba Mesjid Beureunun, Pidie District respectively. The physical and chemical properties of all material used in this study are shown in Table 1, Table 2, and Table 3.

Table 1. Average Chemical Composition of Selected Fibers [7]

Constituent	Arenga Pinnata Fiber (%)	Coconut Fiber (%)
Alpha Cellulose	-	37.36
Hemi Cellulose	15.88	27.68
Holo Cellulose	-	65.04
Cellulose	51.54	-
Lignin	43.09	48.21
Water	8.9	18.2
Ash	2.54	2.87

Table 2. Properties of Aggregate

Test	Result
Aggregate Specific Gravity (ASTM C 127)	2,741 gr/cm^3
Aggregate Absorbtion (ASTM C 127)	1.0708 %
Aggregate Unit Weight (ASTM C 29-71)	1,568 (kg/ dm^3)
Resistance to Degradation (AASHTO T 96-74)	16.908 %
Flat Particle in Coarse Aggregate (D 4791)	21.98 %
Elongated Particle in Coarse Aggregate (D 4791)	23.86 %

Table 3. Properties of Asphalt Binder

Test	Result
Asphalt Specific Gravity (ASTM D 2041)	1,020 gr/cm^3
Penetration Test (ASTM D 5)	64 (0.1 mm)
Asphalt Softening Point (ASTM D 36)	48°C
Ductility (ASTM D 113)	130 cm
Flat Particle in Coarse Aggregate (D 4791)	21.98 %
Elongated Particle in Coarse Aggregate (D 4791)	23.86 %

In order to maintain the effectiveness of this paper, this section aims to give information about the sample combination and how the sample name structure used in this test. Table 4 shows the combination of samples uses for this study.

Table 4 Samples Combination and Samples Name

Fibre Type	Fibre Content	Sample Name
-	0%	0%
Arenga Pinnata	1%	1% AF
Arenga Pinnata	3%	3% AF
Arenga Pinnata	5%	5% AF
Coconut	1%	1% CF
Coconut	3%	3% CF
Coconut	5%	5% CF

2.2. Experimental Plan
In this study the permeability test, as well as Binder Drain down tests, were performed in order to investigate the effect of adding Arenga Pinnata fiber or Coconut fiber to the porous asphalt mixture aside from that standard Marshall testing also performed, however only the result of Void in Mixture testing was discussed in this paper.

2.2.1. Fibers Preparation
Before adding into the porous asphalt mixture, the Arenga Pinnata fiber and Coconut fiber were retrieved from the field and cleaned in order to free it from any unwanted materials; then, the fibers were dried in the oven until it completely free from moisture. After that, Arenga Pinnata fiber, as well as Coconut fiber, was cut until its only around 1 cm to 1,5 cm length. Then the fibers were added into the mixture with a quantity of 1%, 3%, and 5% of asphalt binder weight.

2.2.2. Permeability Test
This testing procedure aims to investigate the capability of porous asphalt mixture is draining underground water; the proposed steps for this testing are ASTM C1781/C1781M [8][9].

2.2.3. Binder Drain down Test
This test aims to determine the amount of drain-down occurring on un-compacted asphalt mixtures, during the production, transport, and mixing placement with the test probe following the standard of AASHTO T 305 [10].

2.2.4. Marshall Test
This examination is intended to determine the resistance (stability) to the plastic melt (flow plastic) of the asphalt mixture. This test is based on AASTO- 245-74. Marshall tests include stability tests, plastic melt (flow), density, VIM, and Marshall quotient. However, in this study, only VIM value was discussed.

3. Result and Discussion
Analysis of Variant (ANOVA) was done in order to identify the significant factors that affect the testing result to be further analyzed. The analysis was performed using Microsoft® Excel® software.

The ANOVA test assumed the null hypothesis (H_0) that all the mean of samples is equal. The confidence interval of 95% ($\alpha = 0.05$) was chosen in this study. In order to reject the null hypothesis, the p-value obtains should be smaller than the level of confidence ($\alpha = 0.05$). In this study, it is mean that the difference in result in regards to discussed variable is significant.

3.1. Permeability Test
The Permeability testing result shows in Figure 1 indicates that the addition of Arenga Pinnata fiber or Coconut fiber results in decrease of permeability value in porous asphalt mixture, moreover the augmentation of fibers concentration in mixture adding to more decrease of Permeability value. This result was expected since the fibers were filling the void in mixture thus result in less air void happen in

asphalt mixture. Even though asphalt porous mixture feature expected higher permeability in order to fasten the water drainage process. In this study, almost combinations with asphalt content 4.5% to 5.5% still give desired permeability value of 0.0575 cm/sec – 0.2493 cm/sec [11]. In the study, AF group show the lowest decrease of Permeability value with the average range of 9.37% to 35.65% compared to control mixture while CF group results from 8.54% to 33.2%.

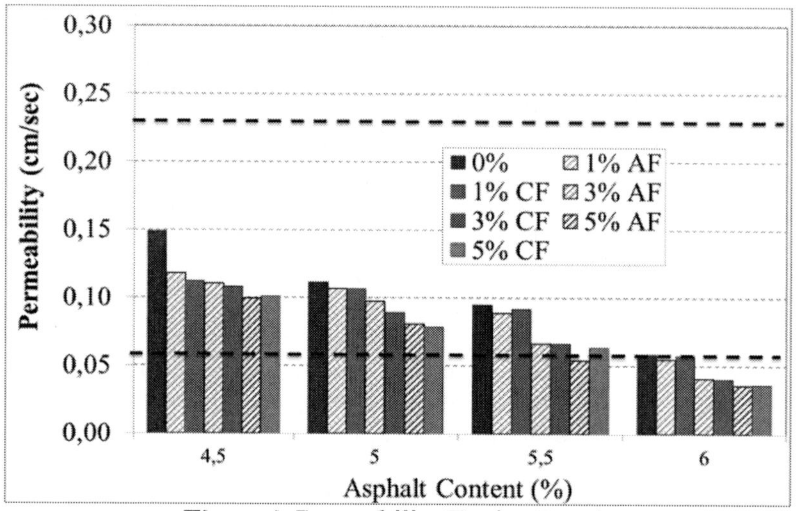

Figure 1. Permeability Testing Result

3.2. Binder Drain down Test

Figure 2 visibly shows that the addition of Arenga Pinneta fiber or Coconut fiber will result in less Binder Drain down value. The Binder Drain down testing results in this study evident that AF group perform the best since it has the lowest result in average 0.053 gr, 0.049 gr, and 0.045 gr for 1%AF, 3%AF, and 5%AF respectively, while the result of CF group is 0.125 gr, 0.067 gr, and 0.05 gr for 1%CF, 3%CF, and 5%CF respectively. Overall, only control mixture at 6% binder content that did not meet the maximum requirement of 0.3 gr. We believe that this result indicates that Arenga Pinnata fiber and Coconut fiber can bind a portion of asphalt binder, thus result in less Binder Drain down testing results compare to control mixture.

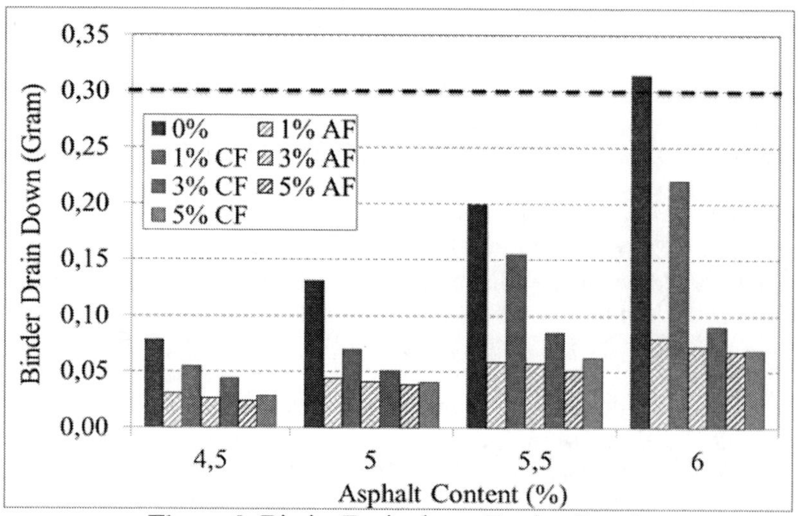

Figure 2. Binder Drain down Testing Result

3.3. *Void in Mixture (VIM)*

Figure 3 shows that addition of fibers regardless whether Arenga Pinnata nor Coconut tend to increase the VIM value compare to control mixture, however the increase was varied in each combination of asphalt content and different type of fibers and this study, all sample result is successfully meeting the desired VIM value of 10% - 25%.

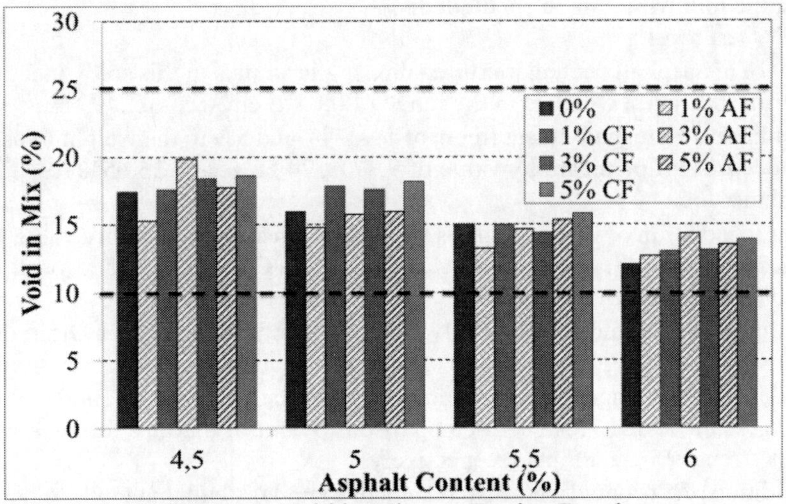

Figure 3. Void in Mixture Testing Result

In AF group, almost all sample shows the decrease VIM value in 1% and 3% addition of Arenga Pinnata Fibre with the average of 10% compare to control mixture outside the mixture with 6% binder content that gives the result of higher VIM in all samples with addition of Arenga Pinnata fiber. This trend might be due to the larger surface area occurs when more fibers were added into the mixture, and it must be covered by asphalt binder, thus result in higher air void in the mixture. These findings were similar to previous research [12] that demonstrates a similar trend when adding oil palm fibers into porous asphalt mixture.

On the other hand, the CF group shows more obvious trends in the escalation of VIM value with an average of 5% to 10% increase compared to a control mixture from 1%CF to 5%CF, respectively. This result strongly suggests that in this study the addition of Coconut fiber in mixture up to 5% can increase VIM value that results in more void in mixture.

Table 5 and Table 6 show that statistically, the addition of Arenga Pinnata fiber or Coconut fiber into asphalt porous mixture result in significant differences of VIM value, which is shown in F value that higher than F-Crit and P-value that lower than 0.05 (95% confidence interval) at both tables.

Table 5. ANOVA: Two-Factor Without Replication for AF Group

Source of Variation	SS	df	MS	F	P-value	F crit
Asphalt Content	41.76	3	13.92	19.04	0.000309	3.86
AF Group	10.03	3	3.34	4.57	0.032	3.86
Error	6.57	9	0.73			
Total	58.37	15				

Table 6. ANOVA: Two-Factor Without Replication for CF Group

Source of Variation	SS	df	MS	F	P-value	F crit
Asphalt Content	62.73	3	20.91	92.77	4.35E-07	3.86
AF Group	4.51	3	1.50	6.67	0.01	3.86
Error	2.02	9	0.22			
Total	69.27	15				

4. Conclusion and Suggestion

4.1. Conclusion
Permeability, Binder Drain down, and Void in Mixture value are some of the most important measurements in analyzing Asphalt Porous Mixture properties and also closely other. From those three measurements, the following can be concluded;

a) Permeability test result:
- Accept of 6% asphalt content mixture, almost all samples in this study meet the requirement for the Permeability test which is in the range of 0.0575 cm/sec – 0.2493 cm/sec.
- The addition of Arenga Pinnata fibers of 1%, 3%, and 5% to the weight of asphalt binder results in the decrease of permeability value of 9,37%, 24,58%, and 35,65% respectively compared to control mixture.
- Similar to addition of Coconut fibers that also decrease permeability value for 8,54%, 27,11%, and 33,20% due to the addition of 1%, 3%, and 5% Coconut fibers respectively

b) Binder Drain down result:
- All samples in this study successfully meet the standard for Binder Drain down test result that is a maximum of 0,3 grams with the exception of control mixture at 6% Asphalt content.
- The addition of Arenga Pinnata fibers results in the decrease of Binder Drain down value to 68,27%, 70,88%, and 73,43% in comparison to control mixture for each addition of 1%, 3%, and 5% Arenga Pinnata Fibers respectively.
- The addition of Coconut fibers also results in a similar trend that tends to decrease Binder Drain down testing value to 32,23%, 58,39%, and 69,87%, respectively, to the addition of Coconut fibers of 1%, 3%, and 5%.

c) Void in Mixture (VIM) measurement result:
- In the Arenga Pinnata group, void in mixture measurement shows a decreasing trend when the fibers were added at the rate of 1% to 3%, however VIM value increases when Arenga Pinnata fiber added at 5% of asphalt binder weight.
- In the Coconut fibers group, the trend is entirely consistent that is the VIM value keeps increasing when higher percentage of Coconut fiber was added.

4.2. Recommendation
For future recommendation, it is necessary to conduct time base measurement to investigate the medium or long-term effect of adding natural fibers to Asphalt Porous Mixture.

References

[1] A. T. Papagiannakis and E. A. Masad, *Pavement design and materials*: John Wiley & Sons, 2017.
[2] R. M. Xavier, B. Martin, L. A. Babu, L. E. Jose, and L. Roy, "A Review on Fiber Modified Stone Matrix Asphalt," 2018.
[3] T. Ting, P. Ramadhansyah, A. Norhidayah, H. Yaacob, M. Hainin, M. W. Ibrahim, *et al.*, "Effect of Treated Coconut Shell and Fiber on the Resilient Modulus of Double-layer Porous Asphalt at Different Aging," in *IOP Conference Series: Earth and Environmental Science*, 2018, p. 012065.
[4] M. Aliha, A. Razmi, and A. Mansourian, "The influence of natural and synthetic fibers on low temperature mixed mode I+ II fracture behavior of warm mix asphalt (WMA) materials," *Engineering Fracture Mechanics,* vol. 182, pp. 322-336, 2017.
[5] A. Mansourian, A. Razmi, and M. Razavi, "Evaluation of fracture resistance of warm mix asphalt containing jute fibers," *Construction and Building Materials,* vol. 117, pp. 37-46, 2016.
[6] M. Lavasani, M. L. Namin, and H. Fartash, "Experimental investigation on mineral and organic fibers effect on resilient modulus and dynamic creep of stone matrix asphalt and continuous graded mixtures in three temperature levels," *Construction and Building Materials,* vol. 95, pp. 232-242, 2015.

[7] L. Suhardiyono, *Tanaman kelapa: Budidaya dan pemanfaatannya*: Kanisius, 1991.
[8] I. W. Diana, "Studi rongga menerus dan kinerja permeabilitas perkerasan aspal porus lapis ganda," *Jurnal Transportasi,* vol. 4, 2004.
[9] L. A. Sañudo-Fontaneda, V. C. Andres-Valeri, C. Costales-Campa, I. Cabezon-Jimenez, and F. Cadenas-Fernandez, "The Long-Term Hydrological Performance of Permeable Pavement Systems in Northern Spain: An Approach to the "End-of-Life" Concept," *Water,* vol. 10, p. 497, 2018.
[10] J. Tanzadeh and R. Shahrezagamasaei, "Laboratory Assessment of Hybrid Fiber and Nano-silica on Reinforced Porous Asphalt Mixtures," *Construction and Building Materials,* vol. 144, pp. 260-270, 2017.
[11] A. A. P. Association, *National asphalt specification*: The Association, 2004.
[12] T. Syammaun, "Effect of Oil Palm Fiber on Some Physical Properties of Porous Asphalt," Jabatan Kejuruteraan Awam, Fakulti Kejuruteraan, Universiti Malaya, 2013.

IOP Conf. Series: Materials Science and Engineering **674** (2019) 012023 doi:10.1088/1757-899X/674/1/012023

Thermal conductivity of aerated concrete containing rice husk ash as partially sand replacement

J Hadipramana[1,a], F V Riza[1,b], N Ali[2,3], A A A Somad[2,3] and N H Basri[2]

[1] Department of Civil Engineering, Faculty of Engineering, University of Muhammadiyah Sumatera Utara, Jalan Kapten Muchtar Basri #03, Medan 20238, North Sumatera, Indonesia

[2] Faculty of Civil and Environmental Engineering, Universiti Tun Hussein Onn Malaysia 86400 Parit Raja- Batu Pahat, Johor, Malaysia

[3] Jamilus Research Centre, Universiti Tun Hussein Onn Malaysia, 86400, Parit Raja-Batu Pahat, Johor, Malaysia.

E-mail: [a] josef@umsu.ac.id, [b]fetra@umsu.ac.id

Abstract. Limited natural energy sources give impact on the selection of the material that can subtract energy consumption. Regarding to conservation energy, in the field of building and construction, the selection of concrete with good heat capacity is a requirement, besides light in weight. The objective of this study was to investigate the thermal conductivity of Foamed Concrete containing Rice Husk Ash (FCRHA). Rice husk ash (RHA) is the type fly ash that can absorb heat due to moisture. RHA The measurement of thermal conductivity on FC and FCRHA was conducted using box insulation and recorded using thermal data logger. The result showed that FCRHA absorb heat from the beginning when the heat is distributed, while FC slow absorp the heat. Modification FC by replacing sand partially with RHA provides an evolution to profitable use of construction materials in the heat absorption.

1. Introduction

In the last several years, some countries give special attention to energy consumption on the building that related to enhance the comfort. For that reason, building industry focus on the improvement and efficiency of building heating and insulation [1][2]. Furthermore, the use of materials in accordance with maximum heat transfer will be used to reduce loss of heat. Therefore, knowledge of physical property of materials is essential to the engineer, because conservation energy is very important policy strategy to all countries [3]-[5].

The application of concrete as thermal mass materials in buildings is regarded as a strategy to reduce energy consumption. Thus, thermal conductivity analysis on material concrete is a way to get information about heat transfer Essentially, thermal conductivity is a calculation about the profile of the heat and the flow of heat through a material [6]. Analyzed thermal conductivity on the concrete for the purpose of the green technology building is very important, however, heat transfer analysis on concrete is very complex. Analysis of thermal conductivity depends on the variables of the material, density, pressure and temperature, the structure of the material, and moisture [6]-[8]. While concrete is a material that is not homogeneous and is influenced by all the thermal conductivity variables. Therefore, to identify the phenomena of heat transfer on concrete have to consider material in the form of the concrete itself [6],[7],[9],[10].

Related to the type of concrete used in green building and consideration large sum of money then required concrete that can reduce energy consumption and light in weight. Therefore, the present study

IOP Conf. Series: Materials Science and Engineering **674** (2019) 012023 doi:10.1088/1757-899X/674/1/012023

is using aerated concrete namely foamed concrete (FC) that fulfills those needs. The objective this study to find out the thermal conductivity of FC and FC containing rice husk ash as a partial sand replacement (FCRHA) [11]. The previous researcher resulted that the materials constituent of FC influences the thermal behavior of FC [12]-[14]. FC as a lightweight concrete very potential in thermal resistance [13]. However, the use of FCRHA for green materials for low-rise buildings gives hope in the evolution of the construction materials and building a reliable.

2. Mix Proportion and Experimental method

2.1. Mix Proportion and Material Preparation

Aerated concrete that is intended in this experiment is the type of foamed concrete, as used in previous experiments include the use of RHA [15] to replacement sand partially [16]. Specimens produced by two densities specifically 1600 kg/m^3 and 1800 kg /m^3 for curing 28 days. The blending of ingredients slurry used the pre foamed method. A mixture of mortar stirred first was then added RHA. After a mixture of mortar and RHA mixed with evenly, next added foam. Foam made separate formerly of a mixture of foam agent and water. Comparison material that is used for mixing foamed concrete with rha as in Table 1.

Table 1. Mix Admixture FC and FCRHA

No	Property	FC	FCRHA
1	Cement-sand ratio	0.25	0.25
2	Water-cement ratio	0.6	0.6
3	Cement-sand-RHA	NA	1:3:1
4	RHA-water ratio	NA	1.25
5	Foam agent ratio	0.05	0.05
6	Foam density (Kg/m^3)	50	50
7	Dimension (mm^3)	150x150x150	150x150x150
8	Curing days	28	28
9	Curing Temperatures °C	23 ± 2	23 ± 2

2.2. Thermal Conductivity Test

Test of Thermal conductivity used heat transfer apparatus as shown in Figure 1. Each sample of FC and FCRA was put into the box heat transfer apparatus, which its interior incrust by insulation. The thermal conductivity test illustrated as shown in Figure 2. Quantity of heat transfer was produced by power supply channeled into the sample for 300 minutes. Furthermore, the temperature difference was recorded by temperature data logger. A difference in temperature was obtained from to differences of temperature between channel enter with the channel out.

Figure 1. Box insulation and temperature data logger.

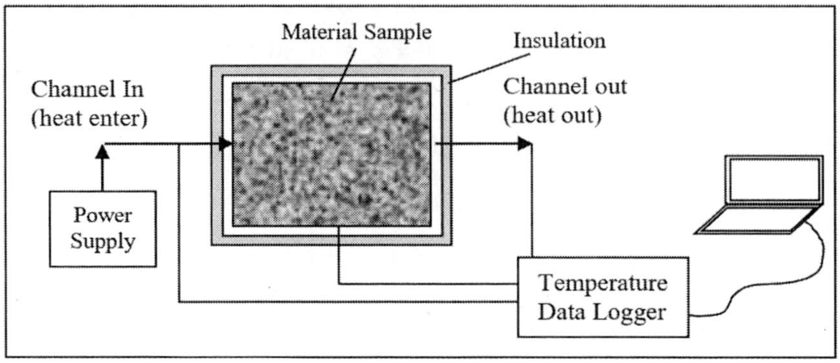

Figure 2. Schematic thermal conductivity test

3. Result and Discussion

3.1. Thermal Conductivity

The difference in materials and their density on the test gives results of thermal conductivity varying as shown in Figure 3. The result shows that higher material density improves heat transfers. This agrees with the previous investigation [17]. The thermal conductivity of 1800 Kg/m^3 material density is higher than1600 Kg/m^3. FCRHA transfers the heat better than FC due to fly ash in FCRHA is good thermal conductivity and affect the concrete overall [14]. The RHA as the fly ash in FCRHA keeps moisture and increased thermal conductivity [18]. Heat transfer behavior of FC is different from FCRHA, Figure 3 shows, that heat absorption of FC was delayed. At first, FC 1600Kg/m^3 very slowly absorb heat, this is due to the low density and more porous [17]. In 240 until 300 minutes reach the stable absorption.

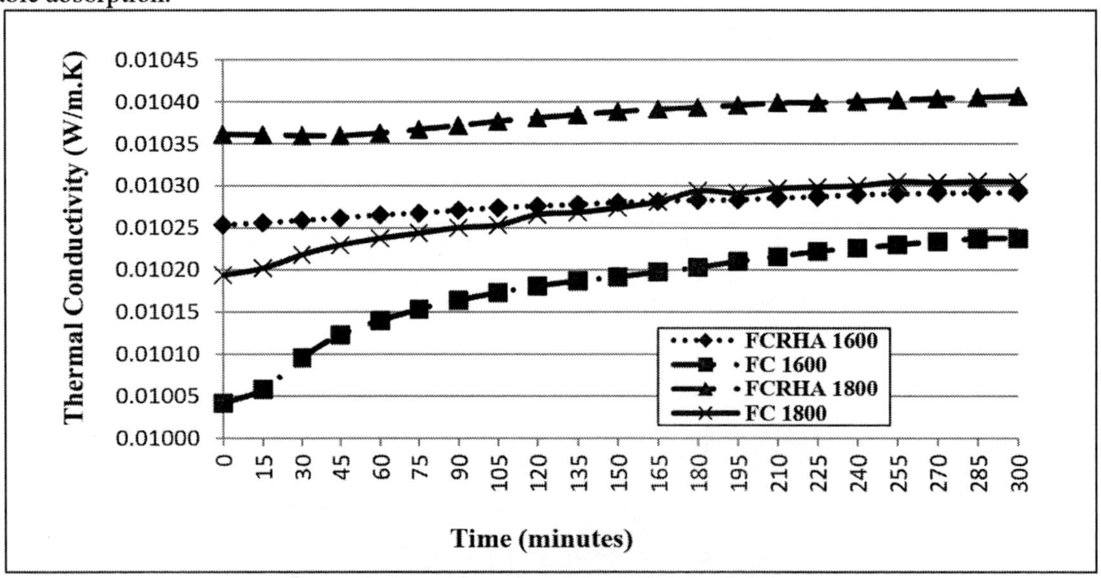

Figure 3. Thermal conductivity of FC and FCRHA.

3.2. Thermal Resistance

Thermal resistance is the inverse of thermal conductivity, which heat transfer occurs at a lower rate in materials. Figure 4 shows thermal resistance of FC is higher than FCRHA. This indicates, that density and humidity gives an important role in the insulation of material aerated concrete.

Obviously, FC with 1600Kg/m^3 of density is the higher thermal resistance than other materials. At the start, FC has a high thermal resistance, then decreases with increasing heat capacity. FCRHA 1800 Kg/m3 has a low heat capacity. When the heat was distributed all heat be accommodated and streamed.

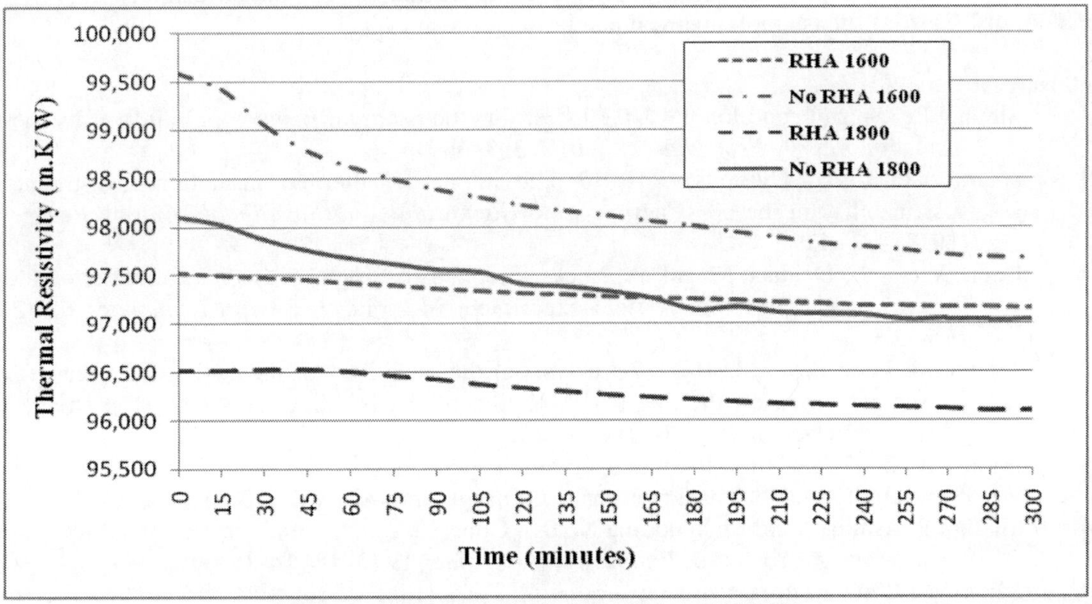

Figure 4. Thermal Resistivity FC and FCRHA

3.3. *Temperature*

Figure 5 shows that the temperature of material influences the thermal conductivity. The thermal conductivity of all materials tends to increase when increasing the elevation of temperature. FCRHA with 1800 Kg/m^3 accommodates the heat when the temperature reaches 39.33°C and channeled the energy heat. It has low heat capacity [17] due to the density and presence of RHA as fly ash.

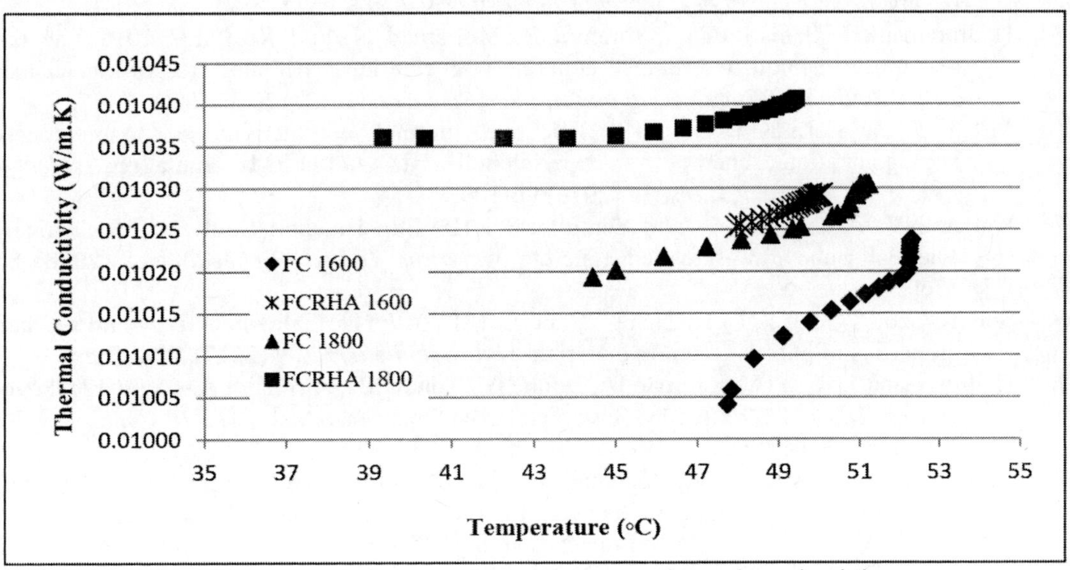

Figure 5. The influence of temperature on thermal conductivity.

4. Conclusion

Increasing the difference temperature, increase the value of thermal conductivity, then decreased the heat resistant material. FCRHA as foamed concrete modification is lighter weight than FC. FCRHA is good heat transfer properties and high thermal conductivity compared to FC. The higher density FCRHA improves thermal conductivity. RHA in FCRHA content is very influential to the thermal conductivity. RHA as fly ash took overheat and influence heat capacity FCRHA overalls.

References

[1] Simona PL, Spiru P, and Ion, IV 2017 Increasing the energy efficiency of buildings by thermal insulation *Energy Procedia* 128 (2017) 393–399.

[2] Adamczyka J and Dylewski R 2017 The impact of thermal insulation investments on sustainability in the construction sector *Renewable and Sustainable Energy Reviews* 80 (2017) 421–429.

[3] Chen, Victor L, Delmas, Magali A, Locke, Stephen L, Singh, A 2017 Information strategies for energy conservation: A field experiment in India, *J. Energy Economics* 68 (2017) 215-227.

[4] Adamczyk J and Dylewski R 2017 Analysis of the sensitivity of the ecological effects for the investment based on the thermal insulation of the building: A Polish case study, *J. of Cleaner Production* 162 (2017) 856-864.

[5] Berardi U 2017 The impact of temperature dependency of the building insulation thermal conductivity in the Canadian climate *J. Energy Procedia* 132 (2017) 237–242.

[6] Shafigh P, Asadia I, and Mahyuddina N 2018 Concrete as a thermal mass material for building applications - A review *J. Building and Engineer* 19 (2018) 14-25.

[7] Khan M I 2002 Factors affecting the thermal properties of concrete and applicability of its prediction models *J. Building and Environment* 37 (2002) 607–614.

[8] Asadi I, Shafigh P, Hassan ZFA, and Mahyuddin NB, Thermal conductivity of concrete - A review, *Journal of Building Engineering* 20 (2018) 81-93.

[9] Zhu L, Dai J, Bai G, and Zang F 2015 Study on thermal properties of recycled aggregate concrete and recycle concrete block *J. Construction and Building Materials* 94 (2015) 620 -628.

[10] Shin A H C, and Kodide U 2012 Thermal conductivity of ternary mixtures for concrete pavements *J. Cement & Concrete Composites* 34 (2012) 575–582.

[11] Hadipramana J, Samad, AAA, Ibrahim R, Mohamad N, and Riza, FV 2016 The energy absorption of modified foamed concrete with rice husk ash subjected to impact loading *ARPN Journal of Engineering and Applied Sciences* 11(12) (2016) 7437- 7442.

[12] Miled K, and Limam O, 2016 Effective thermal conductivity of foam concretes: homogenization schemes vs experimental data and FEM simulations. *Mechanics Research Communications* 76 (2016) 96-100.

[13] Kumar NV, Arunkumar C, and Senthil SS 2018 Experimental Study on Mechanical and Thermal Behavior of foaned concrete *Materials Today: Proceedings* 5 (2018) 8753–8760.

[14] Sun Y, Gao P, Geng F, Li H, Zhang L, and Liu H 2017 Thermal conductivity and mechanical properties of porous concrete materials *Materials Letters* 209 (2017) 349-352.

[15] Hadipramana J, Riza FV, Rahman IA, Loon LY, Adnan SH, and Zaidi AMA (2017) Pozzolanic Characterization Of Waste Rice Husk Ash (RHA) From Muar, Malaysia *IOP Conf. Ser.: Mater. Sci. Eng.* 160 012066

[16] Hadipramana J, Samad AAA, Mokhatar SN, Riza FV, Mohamad N, and Wahab MMY (2017). An investigation of Crater Diameter on Plain Slab Foamed Concrete Rice Husk Ash (FCRHA) Exposed to Low Impact Loading. *MATEC Web of Conferences* 103, 02025 (2017) DOI: 10.1051/matecconf/20171030.

[17] Shafigha P, Iman Asadi I, and Mahyuddina NB 2018 Concrete as a thermal mass material for building applications - A review *Journal of Building Engineering* 20 (2018) 14-25.

[18] Kim KH, Jeon SE, Kim JK, and Yang S 2003 An experimental study on thermal conductivity of concrete *J. Cement and Concrete Research* 33 (2003) 363–371.

Acknowledgment

The author would like to thank and acknowledgment to the research and staffing resources provided by the Department of Civil Engineering, Faculty of Engineering University of Muhammadiyah Sumatera Utara, Medan, Indonesia and also to Faculty of Civil and Environmental Engineering, Universiti Tun Hussein Onn Malaysia, Johor, Malaysia.

Land suitability based on land function using geographic information system (GIS) in landslide potential area

F Susanti*, R Ridha and A Kurniawan

Department of City and Regional Planning, Muhammadiyah University of Mataram, No 1 K.H.Ahmad Dahlan Street, West Nusa Tenggara, Indonesia 83117

* E-mail : febrita_s@yahoo.com

Abstract. Located in West Lombok, Gunung Sari is an area that frequently experiences landslides. Diverse land characteristics and the incompatibility of land allotment allegedly can be some factors causing the landslides. Therefore, it is necessary to conduct research to analyze the land suitability in some areas with a high frequency of landslides. The research approach used in this research is descriptive quantitative with overlay analysis techniques using Arc Gis application program. Based on the data analysis results, it was discovered that Gunung Sari has an area of 2.376.36 Ha or 29% of the total land area which is not in accordance with its designation, an area of 4.032.49 Ha or 49% of the total land area can be categorized into a conditional land and an area of 1.786 Ha or 22% of the total land might be classified into a suitable area land. According to the land function of this area, it was found that the largest area that is unsuitable land and conditional land are located in the conservation zone covering an area of 2249.48 Ha for the unsuitable land and 2433.24 Ha for the conditional land.

1. Introduction

Since 2000, the landslides in West Lombok have occurred for more than 59 times. West Lombok consists of 10 sub-districts, including Gunung Sari, which is a disaster-prone area. In 2015, the landslides occurred in Gunung Sari causing four deaths and two people injured. It was also reported that five houses were damaged [1].

Differences in the characteristics of the land in Gunung Sari cause some of these areas to be prone to landslides. Base on level susceptibility determination matrix, areas with a high-level susceptibility are Guntur Macan, Bukit Tinggi, Kekeri, Jeringo, Dopang, Taman Sari, Kekait, Gelangsar, Membalan and Mekar Sari [2]. Various factors causing landslides are frequent rainfall, land cutting, and steep slopes as in the case of Gunung Sari sub-district [1]. The landslides that can trigger deaths can also be caused by unsuitable designation of land use [3]. Therefore, it is important to have accurate information about the suitability of land use in disaster-prone areas so that the number of victims due to landslides can be minimized. The purpose of this study is to describe the level of potential susceptibility of landslide disaster and analyze the suitability of land use in areas that are susceptible to landslide based on the area function.

2. Methodology

Geographic Information System (GIS) is used for mapping disaster susceptible areas. These areas consist of five categories, including Very Low susceptible, Low susceptible, Moderate susceptible,

High susceptible and very high susceptible (Table 1) [4]. This categorization is obtained from multiplication of value and scores divided by one hundred [5].

Table 1. Category of Susceptibility Based on Scoring

No	Score	Category of Susceptibility
1	>4,3	Very High
2	3,5-4,3	High
3	2,6-3,4	Moderate
4	1,7-2,5	Low
5	<1,7	Very Low

The landslide parameter used in this study is in accordance with the landslide susceptibility factors shown in Table 2 [5].

Table 2. Landslide Susceptibility Factors

Classification	Score	Value	Classification	Score	Value
Rainfall			**Geology**		
<50	1	25%	Alluvial	1	10%
50-99	2		Limestone Hills	2	
100-199	3		Granit Hills	3	
200-300	4		Sedimentary Hills	4	
>300	5		Basal Hills	5	
Slope			**Depth of Soil**		
<25	1	15%	<1	1	5%
25-44	2		1-2	2	
45-64	3		2-3	3	
65-85	4		3-5	4	
>85	5		>5	5	
Land use			**Density of settlements**		
Forest	1	5%	<2000	1	5%
Bush	2		2000-5000	2	
Plantation	3		5000-10000	3	
Moor	4		10000-15000	4	
Settlement	5		>15000	5	
Fault			**Infrastructure**		
With fault	5	5%	No access	1	15%
Without fault	1		Cut ting slope	5	

The map of land suitability in disaster susceptible areas will be attained from the overlay results based on susceptibility criteria and land function (cultivation zone, buffer zone, and conservatory zone) [6].

3. Analysis

3.1. Analysis of Land Functions

The result of area function analysis in Gunung Sari, it is found that the cultivation area covers 2073,41 Ha or 25 percent, buffer area covers 1339,12 Ha or 18 percent, and conservatory area covers 4682,73 Ha or 57% (Figure 1).

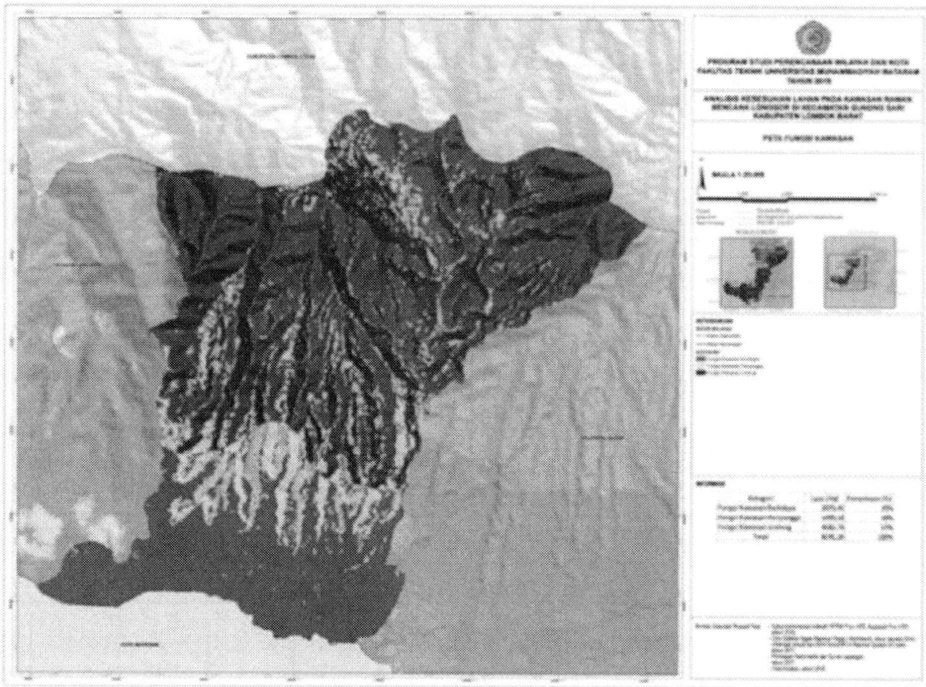

Figure 1. Land Function Map

3.2. Susceptibility Level of Landslide

Based on the result of landslide susceptibility analysis in Gunung Sari of West Lombok, it is found that there are four types of landslide susceptible potentials. They comprise a very low susceptible zone, low susceptible zone, moderate susceptible zone, and highly susceptible zone as shown in Figure 2 and Figure 3.

Figure 2. Landslides Susceptibility Mapping

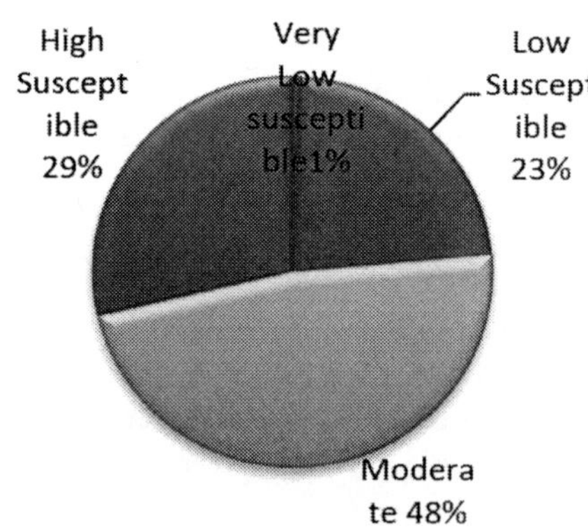

Figure 3. Landslides Susceptibility Analysis.

The area of very low susceptible is indicated in green with the area of 51.40 Ha. The low susceptible zone is indicated in blue with the area of 1874.66 Ha. The moderate susceptible zone is

IOP Conf. Series: Materials Science and Engineering **674** (2019) 012024 doi:10.1088/1757-899X/674/1/012024

indicated in yellow with the area of 3917.32 Ha. The high susceptible zone is illustrated in red on the map above with an area of 2351.88 Ha.

From the susceptibility analysis results, it can be seen that Gunung Sari is dominated by the range of susceptible to moderate potential areas which are equal to 48% of the total area. While the smallest category is a very low susceptible which is equal to 1% of the total area of the district.

3.3. Land Use

Land use in Gunung Sari covers plantation, paddy planting, field, settlement, production forest, and open land. The dominant land use is production forest, with approximately 4677.29 Ha or 57% of the total area and plantation, with 1931.42 Ha or 24% of the total land area in Gunung Sari. Table 3 illustrates the land use based on landslides susceptibility.

Table 3. Land Use based on Landslides Susceptibility

Susceptibility	Land Use	Area (Ha)	Location
Very Low	Plantation	51.40	Gunung sari, Dopang, Membalan and DasanGea
Low	Plantation	607.73	Jatisela, Sesela, Midang, Dopang, Taman Sari, Gunung sari, Guntur Macan, Mambalan, Mekarsari, Penimbung, Kekeri, DasanGeia
	Open space	1.57	
	Paddy planting	1034.20	
	Field	19.63	
	Settlement	200.71	
	production forest	10.82	
Moderate	Plantation	1077.27	Gunung Sari, Taman Sari, Guntur Macan,Membalan, Mekar Sari, Penimbung
	Paddy planting	132.48	
	Field	11.51	
	Settlement	54.07	
	production forest	2641.99	
High	Plantation	195.01	Gunung Sari, Taman Sari, Guntur Macan, Membalan, Mekar Sari, Penimbung
	Paddy planting	64.03	
	Field	56.31	
	Settlement	12.04	
	production forest	2024.48	
Total Area		8195.26	

3.4. Land Suitability Analysis

The land suitability map illustrated from the result of overlaying the function of the area, the potential for landslide susceptibility, and land-use. It can, therefore, be categorized into suitable, conditional and unsuitable areas as presented in Table 4.

IOP Conf. Series: Materials Science and Engineering **674** (2019) 012024 doi:10.1088/1757-899X/674/1/012024

Table 4. Suitability Analysis based on Area Function

Area Function	Susceptibility	Land Use	Land Suitability	Area (Ha)
Cultivation Zone	Very Low	Plantation	Suitable	51.40
	Low	Plantation	Suitable	334.28
		Paddy planting	Suitable	1026.56
		Field	Suitable	13.00
		Settlement	Suitable	199.21
		production forest	Suitable	0.75
	Moderate	Plantation	Conditional	155.24
		Paddy planting	Conditional	80.82
		Field	Conditional	4.27
		Settlement	Conditional	29.59
		production forest	Suitable	161.20
	High	Paddy planting	Unsuitable	15.35
		Settlement	Unsuitable	1.66
Cultivation Zone				**2073.33**
Buffer Zone	Low	Plantation	Conditional	233.93
		Open Land	Conditional	1.57
		Paddy planting	Conditional	7.64
		Field	Conditional	6.63
		Settlement	Conditional	1.50
		Production forest	Conditional	5.23
	Moderate	Plantation	Conditional	444.73
		Paddy planting	Conditional	51.01
		Field	Conditional	7.24
		Settlement	Unsuitable	23.50
		Production forest	Conditional	569.84
	High	Plantation	Unsuitable	36.39
		Paddy planting	Unsuitable	24.06
		Field	Unsuitable	18.60
		Settlement	Unsuitable	6.61
		Production forest	Unsuitable	0.71
Buffer Zone				**1439.21**
Conservatory Zone	Low	Plantation	Conditional	39.52
		Forestry	Conditional	4.83
	Moderate	Plantation	Conditional	477.30
		Paddy field	Conditional	0.64
		Settlement	Unsuitable	0.98
		Production forest	Conditional	1910.95
	High	Plantation	Unsuitable	158.62
		Paddy planting	Unsuitable	24.62
		Field	Unsuitable	37.71
		Settlement	Unsuitable	3.78
		Production forest	Unsuitable	2023.77
Conservatory Zone				**4682.72**
Total Area				**8195.26**

IOP Conf. Series: Materials Science and Engineering **674** (2019) 012024 doi:10.1088/1757-899X/674/1/012024

Based on the level of susceptibility illustrated above, it is apparent that the land use in Gunung sari is not all necessarily in accordance with its provisions. The appropriate land based on the analysis results in Gunung Sari is 1786.40 Ha or 22% of the total land, the conditional land area is 4032.49 Ha or 49% of total land, and the unsuitable land is 2376,36 Ha or 29% of the total land. Based on its function, the unsuitable land-use area covers approximately2249.48 Ha and the largest area of conditional land use which is located in the conservation zone covers approximately 2433,24 Ha. A more detailed explanation will be described in the following map. Yellow represents the conditional land use, red illustrates the unsuitable land use and the green shows suitable land use as shown in Figure 4 and Figure 5.

Figure 4. Land Suitability Analysis in the function area

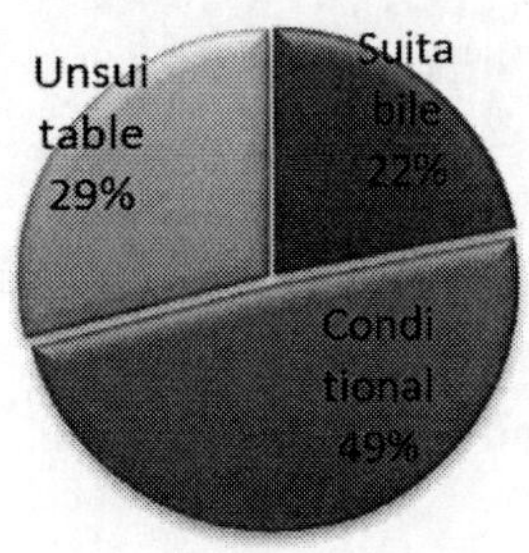

Figure 5. Percentage Land Suitability Analysis

Figure 6 explains the analysis of land suitability in cultivation area which is dominated by yellow, and that is a suitable land with 86% of the total cultivation area (Figure 7).

Figure 6. Land Suitability Analysis in Cultivation zone

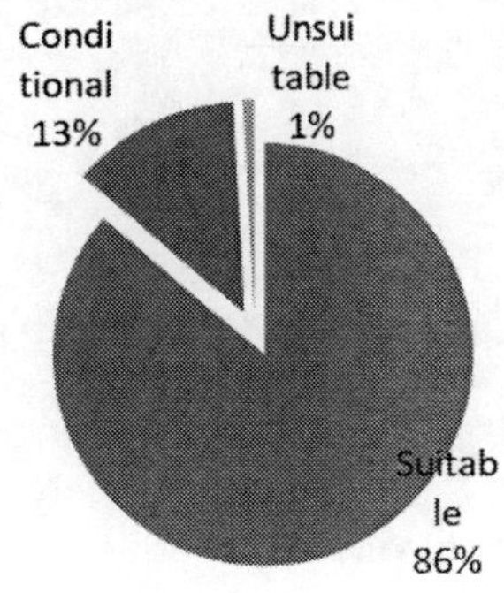

Figure 7. Percentage Land Suitability Analysis in caltivation zone

The map in Figure 8 explains the analysis of the land suitability buffer area which is dominated by red and that is a conditional land with 92% of the total buffer area (Figure 9).

Figure 8. Land Suitability Analysis in the buffer zone

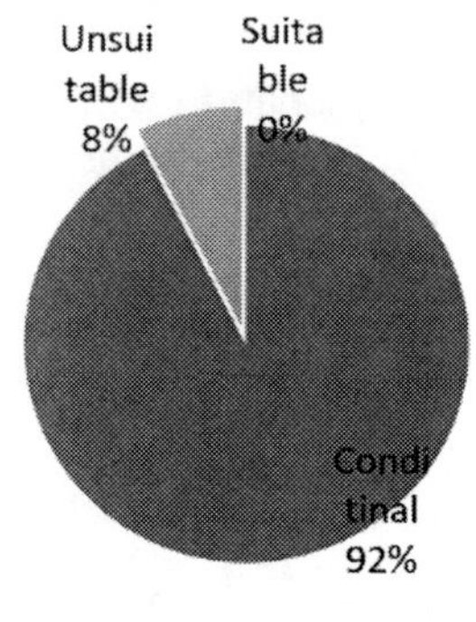

Figure 9. Percentage Land Suitability Analysis in buffer zone

Figure 10 explains the analysis of land suitability in the conservation zone. From the map shows red and blue are not much different in area, a red is a conditional land, and a blue is an unsuitable land, when seen in pie chat is dominated by conditional land with 52% while unsuitable land with 48% of the total area of the conservation area (Figure 11).

Figure 10. Land Suitability Analysis in the Conservation zone

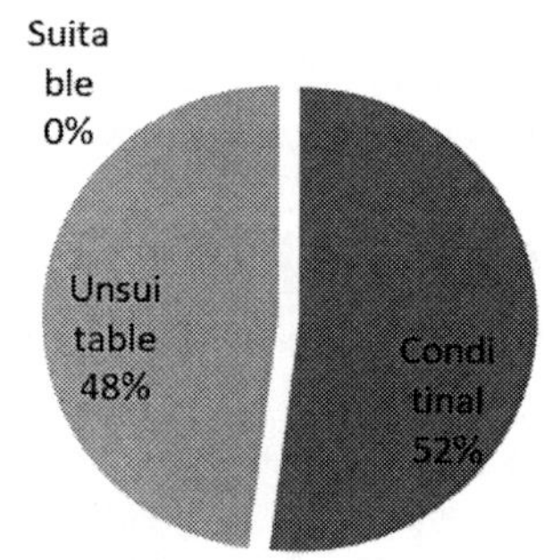

Figure 11. Percentage Land Suitability Analysis in conservation zone

4. Conclusions

The dominant function of the area in Gunung Sari sub-district is a protected area with a percentage of 57% of the total area. Dominant land in the Gunung Sari sub-district has a moderate level of vulnerability with a percentage of 48% of the total area. Dominant land use in Gunung Sari is production forest, The largest unsuitable land is in the conservation area with 48% of the total area

IOP Conf. Series: Materials Science and Engineering **674** (2019) 012024 doi:10.1088/1757-899X/674/1/012024

References

[1] B.H. Widayanti, A Yuniarman, and F Susanti, "Faktor Pemilihan Lokasi Bermukim pada Kawasan RawanBencana Longsor di Desa Guntur Macan Kabupaten Lombok Barat," *Jurnal of Regional and Rural Development Planning*, vol. 2 no.1, pp. 34-44, 2018.

[2] Rencana Penanggulangan Bencana Daerah Kabupaten Lombok Barat 2013-2017 revisi tahun 2015, 2015.

[3] Khoslah and A Ariani, "Tingkat Kerawanan Tanah Longsor di Dusun Landungan Desa Guntur Macan Kecamatan Gunung Sari Kabupaten Lombok Barat," *JIME ISSN 2442-9511*, vol. 3 No.1, 2017.

[4] D Winarti, Cristady Hardiyatmo, and Dwikorita Karnawati, "Gis-Based Landslides Susceptibility Mapping by Using Analytical Hierarchy Proses in The Hydrothermally Altered at Southern Mountain of Lombok Island Indonesia," *Jurnal Kurvatek*, vol. 1(1), pp. 84-91, 2016.

[5] Paimin, Sukresno, and Irfan Budi Pramono, *Teknik Mitigasi Banjir dan Tanah Longsor*. Balik Papan: Tropenbos International Indonesia Programme, 2009.

[6] Surat Keputusan Menteri Pertanian No.837/kpts/um/II/1980, 1980.

Analysis of drainage dimension at the density of housing environment (case study: urban drainage at Gampong Keuramat, Kuta Alam, Banda Aceh, Indonesia)

Akmal

Civil Engineering Department, Faculty of Engineering, UniversitasMuhammadiyah Aceh, 23245 Banda Aceh, Indonesia

Email: akmal@unmuha.ac.id

Abstract. In the relatively flat area such as Banda Aceh, the effort to prevent flooding could be made by monitoring the drainage system. This study was conducted at Gampong Keuramat, Kuta Alam subdistrict, Banda Aceh, in which the area has been densely populated yet having uneven population distribution and experience the continuous increasing of less-planned residential area coverage as the rise in the number of immigration to the region. If not anticipated further, the condition could lead to flooding since the amount of water discharge to the drainage could be larger than the capacity of existing drainage. Thus, this study was conducted to analyze the capacity of existing drainage and to identify the cause of drainage inundation in the area. The study was focus on the part of the drainage that was not functioned normally, while Rational Method was used to analyze the channel discharge data. The result shows that the E-F channel and F-G channel are not in the safe condition since the channel discharge design is broader (0.06 m3/sec and 0.08 m3/sec, respectively) than the actual channel discharge (0.05 m3/sec and 0.07 m3/sec, respectively). Furthermore, those channels can be improved by increasing the slope and enlarging the channel's dimension.

Keywords: flooding, drainage dimension, Banda Aceh

1. Introduction

The city of Banda Aceh has relatively flat topography, so it needs to pay attention to the drainage network system in preventing flooding, drainage system is one of the government's efforts to overcome the flood disaster that often struck the city of Banda Aceh. One of the areas that often occur flood the area Gampong Keuramat, Kuta Alam subdistrict. The drainage problem in this village is caused by uneven distribution of population, and the volume of the immigrant population is quite large; this resulted in the development of less planned settlements, resulting in drainage disposal system is not well coordinated. Based on the above problems, an analysis of the capacity of drainage channel dimensions can be known, so that the capacity of the drainage services is adequate or not, without causing puddles or floods.

The purpose of this paper is to analyze the capacity of channel dimension and to identify the cause of the inundation that occurred in the densely populated residential area of Gampong Keuramat, to give an illustration of the drainage condition. The method used in analyzing the discharge is the rational method.

2. Urban Drainage

The infrastructure of an area consists of several groups, namely: water infrastructure group, road groups, transportation equipment groups, waste management groups, urban building groups, energy groups, and telecommunications groups [1]. The drainage system is part of the water infrastructure group.

In order to support urban development, the need to improve and to have additional drainage facilities, especially rainfall discharge systems, are required. This is because rainfall discharge systems are set to prevent harmfully or disturbing effects on the environment resulted from rain streams by discharging them as quickly as possible to the receiving water bodies [2]. Thus, the condition of rainfall drainage system facilities should be the main concern in the flooding prevention, especially for areas that always experience flood every rainy season. Furthermore,the safety and reliability of the drainage system, as well as the problems related to that, should be taken into account and must be handled seriously.

3. Rainfall Design

The calculation of rainfall design can be done by statistical analysis, that is by calculating statistical parameters of data analyzed. The statistical parameters are an average value (ỹ), standard deviation (SD), asymmetric coefficient (Cs), and coefficient of variation (Cv) [3]. The value of asymmetric coefficient (Cs) determines the type of data distribution. If the coefficient asymmetric (C_S) = 0, the distribution is normal, if (C_S) = 3 C_V + C_V^3, the distribution is log-normal, if the (C_S) = 1.14 the distribution is Gumbel.If rainfall data do not follow either normal distribution, normal log or Gumbel, the data is suspected not to follow distribution [1].

4. Channel Dimension

The channel dimension must be capable of flowing the discharge plan rate(Qs), which has to be equal to or greater than the discharge design (Q_T) [4]. This relation is shown as follows:

$$Q_S \geq Q_T \tag{1}$$

where Q_S is measured discharge (m³/sec.), and Q_T for discharge design(m³/sec.).

The measured discharge (Qs) can be obtained using the following formula:

$$Q_S = A_S.V \tag{2}$$

where A_S for area section(m²), and V is velocity (m/sec.).

5. Rainfall and Runoff

There are many empiric-rational formulas that can explain the relation between rainfall and its runoff. Urban drainage design often uses the following rational modification formula [5]:

$$Q_T = 0{,}278.C.C_S.I.A \tag{3}$$

where Q_T discharge design(m³/sec.), C as a runoff coefficient, C_S for the storage coefficient, I is rainfall intensity during the time on concentration (mm/hour), and A for area (km²).

6. Research Method

The data used in this research are primary data and secondary data. The primary data are photo of water puddles and channel dimension measurement data. The secondary data consists of rainfall data [3], population data [2] and topographic maps. Equipment used to analyze the dimensions of drainage channels is tape measure, wood measuring, stationery, and spray paint.

The stages of analysis conducted based on the data obtained are as follows; calculate the rainfall plan, calculate the flow coefficient, calculate the channel sectional discharge, calculate the concentration-time, calculate the coefficient of the shelter, calculate the rain intensity, calculate the rain and runoff. The research flows diagram is shown in Figure 1.

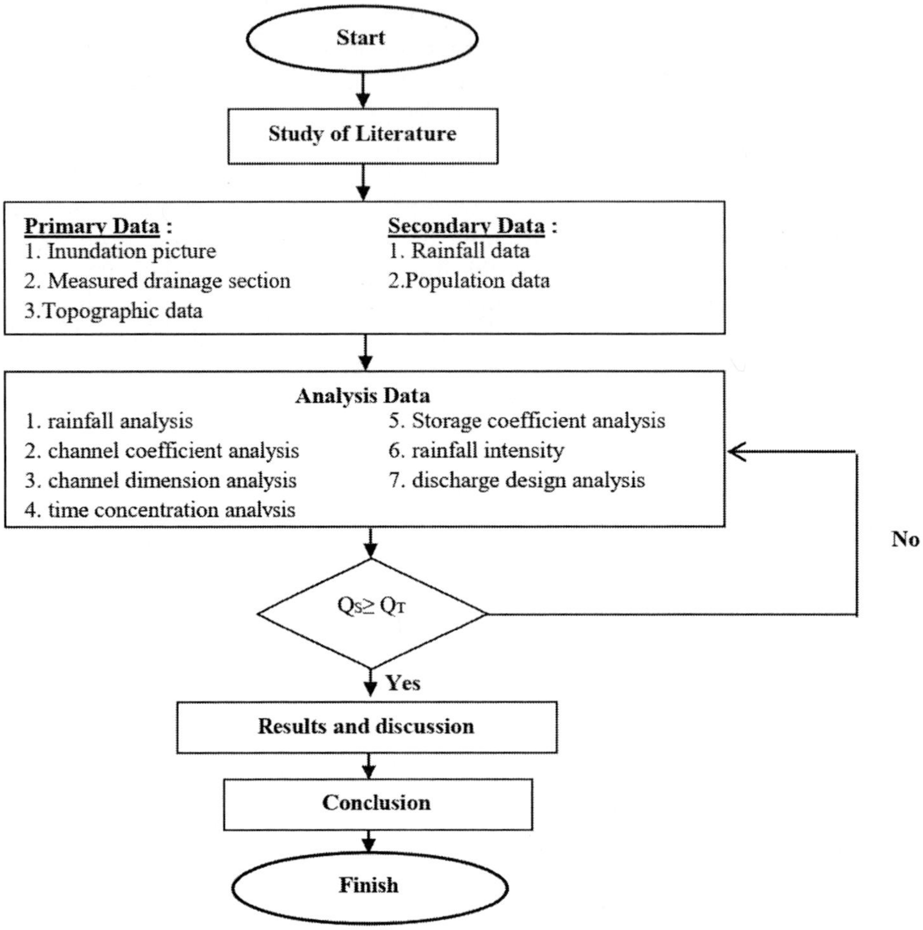

Figure 1. Research flow diagram

7. Results and Discussion

The discharge flow in the channel is influenced by the cross-sectional area, and the measured discharge in the channel, which is by using Formula 2. The calculation result can be seen in Table 1. Discharge design is calculated for each channel using Formula 3 which is influenced by flow coefficient, rain during concentration time. The results of the calculation of flood discharge can be seen in Table 2.

Table 1. Measured Discharge

Channel ID	A_s (m^3)	V (m/sec.)	Q_s (m^3/sec.)	Channel ID	A_s (m^3)	V (m/sec.)	Q_s (m^3/sec.)
A-B	0.16	0.80	0.13	I-J	0.17	0.24	0.04
B-C	0.19	0.19	0.03	J-K	0.15	0.44	0.07
C-D	0.14	019	0.03	K-L	0.19	0.74	0.14
D-E	0.14	0.51	0.07	L-M	0.42	0.60	0.25
E-F	0.18	0.25	0.05	N-O	0.30	0.12	0.04
F-G	0.23	0.32	0.07	O-P	0.30	1.17	0.35
G-H	0.51	0.20	0.10	P-Q	0.59	1.35	0.79
A-I	0.10	1.20	0.12	Q-R	0.22	1.34	0.30

Channel ID	A_s (m^3)	V (m/sec.)	Q_s (m^3/sec.)	Channel ID	A_s (m^3)	V (m/sec.)	Q_s (m^3/sec.)
R-S	0.34	0.50	0.17	E-Y	0.19	0.27	0.05
S-T	0.35	1.45	0.50	Z-M	0.16	0.86	0.14
T-U	0.34	0.47	0.16	F-Z	0.14	0.80	0.11
U-V	0.37	0.19	0.07	1-N	0.17	0.80	0.13
V-W	0.37	1.89	0.70	G-1	0.20	1.21	0.24
W-H	0.76	0.51	0.38	2-O	0.15	0.58	0.09
X-J	0.16	0.52	0.08	H-Q	0.77	2.10	0.60
D-X	0.22	0.98	0.22	W-R	0.40	0.81	0.32
Y-K	0.22	0.90	0.20	V-S	0.35	0.39	0.14

Table 2. Discharge Design

Channel ID	Q_S A(m^3)	A (Km^2)	C	CS	I (mm/hr)	Q_T (m^3/sec.)
A-B	907	0.00091	0.34	0.994	63.86	0.01
B-C	9671	0.00967	0.31	0.864	45.30	0.03
C-D	5254	0.00525	0.36	0.959	51.74	0.03
D-E	6017	0.00602	0.37	0.945	63.99	0.04
E-F	9555	0.00956	0.35	0.887	68.17	0.06
F-G	11667	0.01167	0.33	0.808	92.46	0.08
G-H	6695	0.00670	0.44	0.869	87.46	0.06
A-I	5369	0.00537	0.41	0.982	58.85	0.04
I-J	2740	0.00274	0.37	0.972	64.18	0.02
J-K	4178	0.00418	0.34	0.963	68.11	0.03
K-L	4127	0.00413	0.34	0.975	74.58	0.03
L-M	6196	0.00620	0.43	0.970	54.29	0.04
N-O	5443	0.00544	0.37	0.797	65.65	0.03
O-P	4209	0.00421	0.36	0.926	210.47	0.08
P-Q	7210	0.00721	0.40	0.926	77.61	0.,06
Q-R	2873	0.00287	0.36	0.993	63.64	0.02
R-S	3215	0.00322	0.36	0.911	163.64	0.05
S-T	4045	0.00405	0.35	0.932	256.32	0.09
T-U	5469	0.00547	0.36	0.907	89.60	0.04
U-V	5469	0.00547	0.36	0.837	76.04	0.03
V-W	2854	0.00285	0.36	0.973	189.27	0.05
W-H	4035	0.00404	0.35	0.938	119.93	0.04
X-J	6038	0.00604	0.37	0.938	66.83	0.04
D-X	4915	0.00492	0.36	0.916	148.99	0.07
Y-K	6081	0.00608	0.37	0.969	59.29	0.04
E-Y	6819	0.00682	0.37	0.879	63.81	0.04
Z-M	7477	0.00748	0.38	0.965	54.25	0.04
F-Z	5357	0.00536	0.35	0.900	137.03	0.07
1-N	5952	0.00595	0.38	0.941	90.17	0.05
G-1	6607	0.00661	0.37	0.908	145.41	0.09
2-O	6640	0.00664	0.37	0.890	100.36	0.06
H-Q	7117	0.00712	0.46	0.967	126.06	0.11
W-R	4907	0.00491	0.36	0.917	128.24	0.06
V-S	5343	0.00534	0.35	0.904	84.53	0.04

Conditions of channel discharge value must be greater than or equal to the design discharge ($Q_S \geq Q_T$), meaning the channel can accommodate the existing capacity. The results of the calculation of channel discharge for all channels have different values. The summary result of flood discharge control can be seen in Table 3.

Tabel 3. Discharge Control

Channel ID	Q_S (m^3/sec.)	Q_T (m^3/sec.)	Note	Channel ID	Q_S (m^3/sec.)	Q_T (m^3/sec.)	Note
A-B	0.13	0.01	safe	S-T	0.50	0.09	safe
B-C	0.03	0.03	safe	T-U	0.16	0.04	safe
C-D	0.03	003	safe	U-V	0.07	0.03	safe
D-E	0.07	0.04	safe	V-W	0.70	0.05	safe
E-F	0.05	0.06	unsafe	W-H	0.38	0.04	safe
F-G	0.07	0.08	unsafe	X-J	0.08	0.04	safe
G-H	0.10	0.06	safe	D-X	0.22	0.07	safe
A-I	0.12	0.04	safe	Y-K	0.20	0.04	safe
I-J	0.04	0.02	safe	E-Y	0.05	0.04	safe
J-K	0.07	0.03	safe	Z-M	0.14	0.04	safe
K-L	0.14	0.03	safe	F-Z	0.11	0.07	safe
L-M	0.25	0.04	safe	1-N	0.13	0.05	safe
N-O	0.04	0.03	safe	G-1	0.24	0.09	safe
O-P	0.35	0.08	safe	2-O	0.09	0.06	safe
P-Q	0.79	0.06	safe	H-Q	0.60	0.11	safe
Q-R	0.30	0.02	safe	W-R	0.32	0.06	safe
R-S	0.17	0.05	safe	V-S	0.14	0.04	safe

Based on results in Table 3, two channels are not secure, namely, channel E-F and F-G.

8. Conclusion

Based on the result, the physical condition of the drain in Gampong Keuramat is good, only E-F channel and F-G channel is not secure (where $Q_S<Q_T$) from thirty-four measurements. In this channel it requires slope repair and expansion of the channel dimension so that it can accelerate flow and not create puddles.

References

[1] Suripin, 2004. *Sistem Drainase Perkotaan yang Berkelanjutan*, Penerbit Andi, Yogyakarta
[2] Hendrasarie, N. 2005. *Evaluasi banjir pada area drainase Kali Kepiting dan KaliKenjeran Surabaya Timur.* J. Rekayasa Perencanaan 2(1):1-17
[3] Triatmojo, 1998. *Hidrologi Terapan*, Yogyakarta
[4] Chow, V. T, 1989. *Hidrolika Saluran Terbuka*. Terjemah Suyetman, dkk, Penerbit Erlangga, Jakarta.
[5] Wesli. 2008. *DrainasePerkotaan*.GrahaIlmu, Yogyakarta
[6] Data Curah Hujan Banda Aceh, 2015
[7] Data Penduduk Banda Aceh, 2012.

Acknowledgments

This research was supported by Universitas Muhammadiyah Aceh. We appreciate and thank to all colleagues and all those who have contributed from research conducted until this paper was published.

IOP Conf. Series: Materials Science and Engineering **674** (2019) 012026 doi:10.1088/1757-899X/674/1/012026

Comparison of cantabro test using *Arenga pinnata* fiber and coconut fiber in porous asphalt

T Syammaun* and F Rachman

Faculty of Engineering, University Muhammadiyah Aceh, Indonesia

* Email: tamalkhani@unmuha.ac.id

Abstract. This paper presents the results of a preliminary investigation on using *Arenga pinnata* fiber (APF) and coconut fiber (CCF) in porous asphalt (PA). The samples were prepared with various percentages of fibers (0%, 1%, 3%, and 5%) by weight of bitumen and evaluated. The laboratory tests included air abrasion loss and water abrasion loss. The results indicate that the addition of fibers provides a positive effect on the properties of PA. A certain amount of fibers as an additive appears to improve resistance to raveling and stripping. However, adding too much fiber would result in higher abrasion loss value and increase the air void.

1. Introduction

Porous asphalt (PA) is widely used for noise reduction and water drainage in order to improve traffic safety and comfort for both drivers and the residents living in the vicinity of roads [1]. Porous asphalt was developed in the early 50s in Europe and the United States. It is composed predominantly of narrowly graded crushed coarse aggregate without significant fines, which results in increased surface friction and permeability and noise reduction [2]. This kind of asphalt has also been called drainage asphalt, whispering asphalt, popcorn asphalt, open friction course, and porous friction course.

In general, there are several advantages, and also limitations of PA include the elimination of glare, splash, and spray, improvement in visibility, increased skid resistance of the road surface [3][4]. Nevertheless, PA the limitations of PA, such as air void can quickly be filled with detritus where pressure washing does not clear the voids and the only alternative is to remove the contaminated surface and relay new material, the unit cost of porous asphalt is far higher than conventional dense wearing coarse and predicted lifetime for porous asphalt is about 8 years against 20 years for conventional method [2]-[4]

Various kinds of fibers have been utilized as fillers in asphalt mixes to improve performances of certain properties of the mix. There are several types of fibers that could be used in asphalt mixtures, such as polyester fiber, cellulose fiber, glass fiber and mineral fiber [5][6]. This exploratory study aims to use *Arenga pinnata* fiber (APF) and coconut fiber (CCF) in its natural form in bituminous mixes primarily to help increase binder film thickness around aggregates without excessive binder drainage. As fiber may be considered as waste material its successful utilization in pavement mixes helps address the issue of waste management and environmental conservation. In this study, the samples were prepared and tested to evaluate cantabro on air cured samples and cantabro on water cured sample of porous asphalt.

2. Experimental program

2.1. Materials

In this study, aggregates employed were from locally crushed granite aggregates. The aggregates were selected based on the selected gradations which have been obtained from Australian Asphalt Pavement Association (AAPA) as shown in Table 1 to Table 4. The conventional bitumen penetration-grade 80/100 was used. Meanwhile, *Arenga Pinnata* Fiber (APF) and coconut fiber (CCF) are used as natural fiber (Figure 1). The fiber is washed thoroughly in clean water and dried at room temperature of 80°C for an hour until a constant percentage value of water loss was obtained.

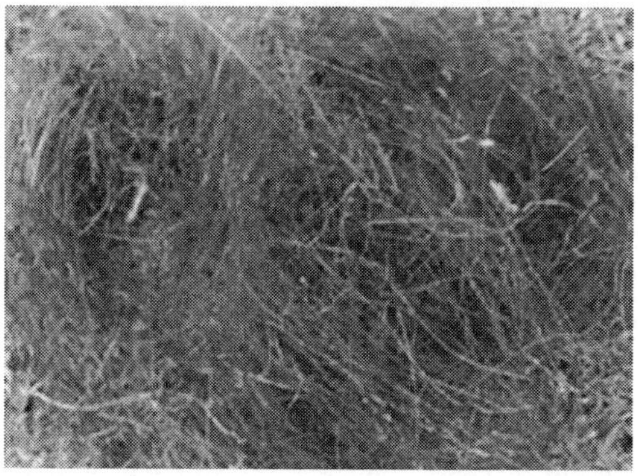

(a)

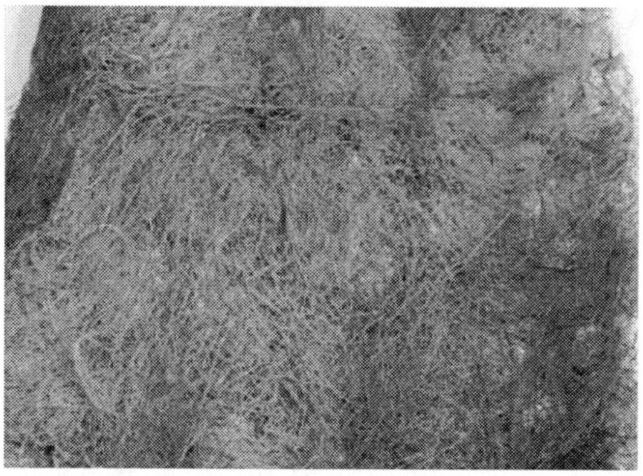

(b)

Figure 1. *Arenga pinnata* fiber (a) and coconut fiber (b)

Table 1. Gradation of PA based on the Australian Asphalt Pavement Association

Sieve size (mm)	Specification passing range (%)	Used passing (%)
19	100	100
13.2	85-100	92.5
6.7	45-70	57.5
4.75	25-45	35
2.36	10-25	17.5
1.18	7-15	11
0.6	6-12	9
0.3	5-10	7.5
0.15	4-8	6
0.075	2-5	3.5

Table 2. Physical properties of crushed granite aggregate.

Properties	Test Value (%)
Flakiness index	21.8
Elongation index	23.2
Impact value	26
Los Angeles abrasion value	16.9
Soundness	7
Aggregate crushing value	17

Table 3. Physical properties of asphalt binder.

Properties	Unit	Test Value (%)
Penetration at 25°C	1/10 mm	64
Ductility at 25°C	cm	>120
Softening point	°C	48

Table 4. Chemical properties of Fibers

Properties	*Arenga pinnata* fiber (%)	Coconut fiber (%)
Alpha Cellulose	-	37.36
Ash	2.54	2.87
Cellulose	51.54	-
Hemi Cellulose	15.88	27.68
Holo Cellulose	-	65.04
Lignin	43.09	48.21
Water	8.9	18.2

2.2. *Specimen preparation*
The amount of binder contents (BC) used in the mixes is 4.5%, 5%, 5.5%, and 6% by weight of aggregates. Various percentages of APF and CCF (1%, 2%, and 3%) were used and compared with control sample (no fibers). The fibers were clean for other materials and sieved through a 500 μm sieve. The samples were prepared using Marshall Compactor by applying 50 blows per-face and tested for raveling, and water damage.

2.3. *Test Method*

2.3.1. *Cantabro Test on Air Cured Samples*
The Cantabro test is a special test for porous asphalt to evaluate the resistance to particle loss by abrasion and impact. The Cantabro loss is used as an important indicator of bonding properties between aggregate and bitumen [7]. In this study, the samples were tested at temperature of 25°C and the weight of the Marshall sample was recorded before placing it in a Los Angeles drum without the steel balls. The drum goes through 300 rotations with a speed of 30 to 33 rpm and then the weight loss was calculated as illustrated in equation (1).

$$\text{Weight Loss (\%)} = [\,(Mi - Mr\,) / Mi\,] * 100 \quad (1)$$

Where; *Mi and Mr* are respectively the initial mass of sample and mass of the sample after rotation.

2.3.2. *Cantabro Test on Water Cured Samples*
The Cantabro test on water cured samples were used to measure the resistance of mixes to stripping and water damage. In this study, the initial weight of the Marshall sample was recorded and then samples were placed in a water bath for four days at 49±1°C. The samples were taken out of the bath on the fifth day and leave for 18 hours. The Cantabro test was then conducted according to the same procedure, as explained earlier.

3. Results and Discussion

3.1. *Air Abrasion Loss (Air Cured Samples)*
The effects of various percentages of binder on air cured samples are summarized in Figure 2. The maximum permissible abrasion loss should be no more than 18%. The results show that the abrasion loss decreases with increase binder content. It was also observed that the specimens made with APC

content had higher abrasion loss than the specimens containing CCF content. Moreover, sample made by 5%, 5.5% and 6% binder content shows the best resistance to abrasion in porous asphalt. It clearly shows that the increasing amount of binder is sufficient to coat aggregates and fibers.

Figure 3 shows that at any particular fiber content, the air abrasion loss will decrease as the binder content is increased. As such, the use of fiber in porous asphalt requires a bit more binder in order to avoid abrasion loss. This was due to the greater adhesion between the binder and aggregates because of the increased availability of binder to coat both the aggregates and fibers. The fluctuation in abrasion loss across the range of fibers content was found to be more pronounced as the binder content decreased. As the binder content increased, this fluctuation in abrasion loss across the range of fibers contents diminished (Figure 3). As such, there was a critical link between the number of fibers used in porous asphalt and the binder content in order to ensure that the durability of the porous asphalt pavement was not compromised.

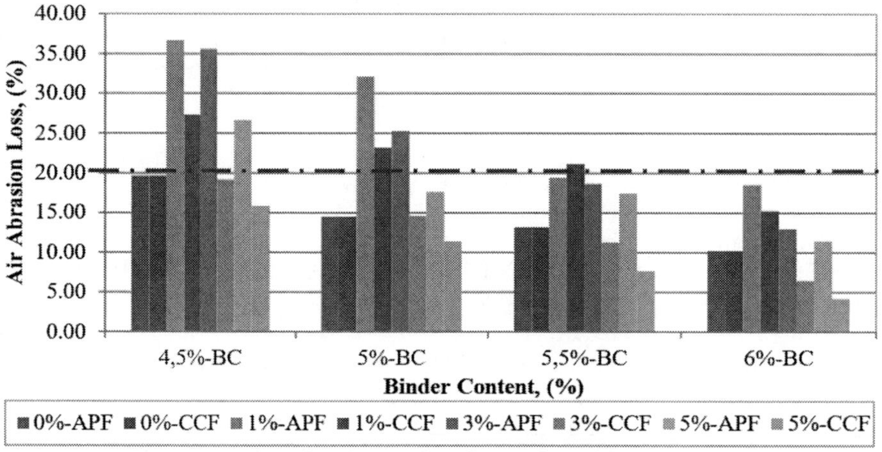

Figure 2. Abrasion loss value for various binder content

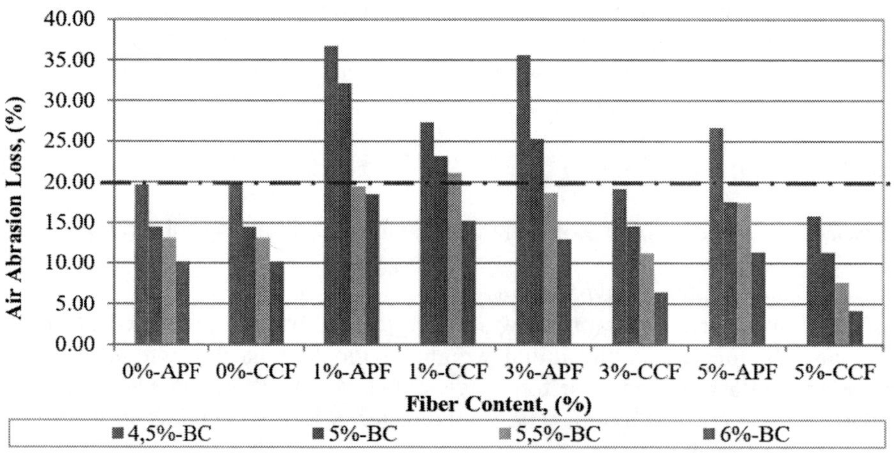

Figure 3. Abrasion loss value for various fiber content

3.2. Water Abrasion Loss (Water Cured Samples)

The standard tests on water cured specimens were conducted according to ASTM C131-96. The Cantabro test on water cured specimens were used to measure the resistance of mixes to stripping and water damage. The maximum permissible water abrasion loss should be no more than 40%. The values of water abrasion losses are given in Figure 4. The results revealed that in general, only specimens prepared with 5%, 5.5% and 6% binder content satisfy the recommended limit value. The content weight loss for water cured specimens was greater than that of air-cured specimens. This may be due to water influence, the abrasion resistance and also the binder-aggregate bonding in the porous asphalt. This also indicated that water is a serious factor in affecting the stripping.

The effects of the different fiber's contents on water cured specimens are summarized in Figure 5. The available results displayed the same trend as the revolution results of air abrasion loss with different fiber content. However, the abrasion loos value on water cured specimens was higher than the air-cured specimens. The general trend from Figure 5 shows that the addition of fiber decreased loss abrasion values when compared to specimens with no fiber (0%-APF & 0%-CCF), but the mix with more fiber content gave the abrasion loss of specimen's increases back. The sample made CCF content has better resistance to water abrasion loss than APF content. The percentage weight loss for water cured samples is greater than that of air-cured samples. This indicated that water did influence the abrasion resistance and also the binder-aggregate bonding in the porous asphalt.

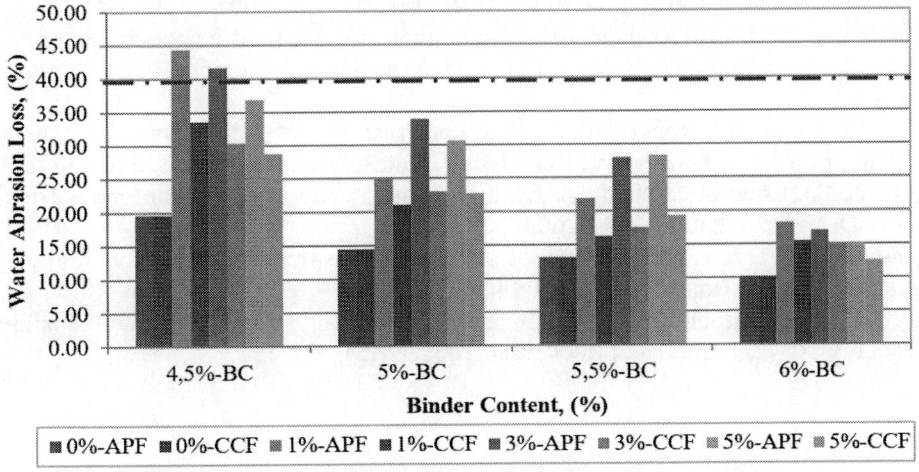

Figure 4. Abrasion loss value for various binder content

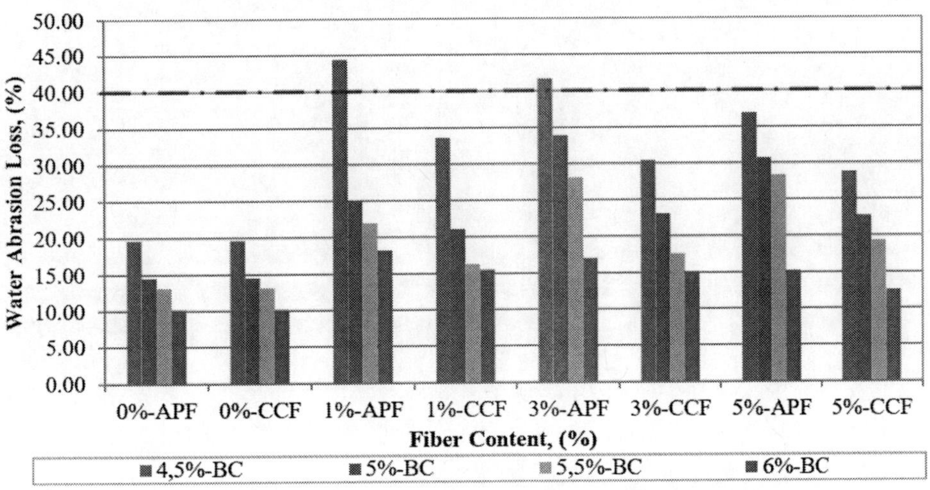

Figure 5. Abrasion loss value for various fiber content

IOP Conf. Series: Materials Science and Engineering **674** (2019) 012026 doi:10.1088/1757-899X/674/1/012026

4. Conclusion

From the exploratory laboratory investigations, the use of fibers affects the properties of the porous asphalt mix. For any particular binder content, the abrasion loss of the porous mix improves as the amount of fibers is increased. As the amount of fibers to be used increases, more binder is needed to avoid excessive abrasion loss. However, as the amount of binder to be used increases, the binder began to clog the air voids. As such, there is a need to strike a balance between the two opposing requirements if fibers (APF & CCF) are to be used as an additive in porous asphalt. Nevertheless, this initial study would hopefully open the door to more in-depth study on this matter.

References

[1] Takahashi S, Poulikakus LD, Partl MN. Evaluation of improved porous asphalt by various test methods, Proceedings of the 6th International RILEM Symposium, Performance Testing and Evaluation of Bituminous Material. Zurich; 2003. Pp. 23–236.

[2] Hossam F.H. and Khalifa S.A. Effect of organic fibers on open-graded friction course mixture properties, International Journal of Pavement Engineering. 2005; 6:1, 67-75

[3] Cooley Jr LA, Brown ER,Watson DE. Evaluation of OGFC mixtures containing cellulose fibers. Transportation Research Record 1723, Transportation Research Board, National Research Council; 2000.

[4] Moore LM, Hicks RG, Rogge DF. Design, construction and maintenance guidelines for porous asphalt pavements. Transportat. Res. Rec.: J. Transportat. Res. Board. 2001;1778:91-99.

[5] Chen H, Xu Q, Chen S, & Zhang Z. Evaluation and design of fiber-reinforced asphalt mixtures. Mater Design 2009;30(7):2595-2603.

[6] Moghaddam TB, Dynamic properties of stone mastic asphalt mixtures containing waste plastic bottles. Construct Build Mater 2013;34:236-242.

[7] Chui-Te, C. Use of ground tire rubber in asphalt pavements: Field trial and evaluation in Taiwan. Resources, Conservation and Recycling. 2008; 52(3), 522-532.

IOP Conf. Series: Materials Science and Engineering **674** (2019) 012027 doi:10.1088/1757-899X/674/1/012027

Evaluation of maintenance system in emergency installation of Meuraxa Hospital Banda Aceh, Indonesia

A Fatimah * **and J Amin**

Civil Engineering Department, Faculty of Engineering, Universitas Muhammadiyah Aceh, 23245 Banda Aceh, Indonesia

* Email: aldina.fatimah@unmuha.ac.id

Abstract. Hospital building and all its equipment require regular maintenance in order to avoid heavier damage. The implementation of continuity of optimal hospital building maintenance is expected to reduce the incidence of heavier damage and minimize the cost of existing repair and ensure the readiness of supporting facilities for the implementation of health services to the public in the hospital. The purpose of this research is to know the maintenance system that has been applied and to identify the condition of building emergency installation in Meuraxa Hospital. This research was conducted by distributing questionnaires to 91 respondents who are employees, nurses, doctors, and families of patients who are in the installation of Meuraxa hospital in Banda Aceh. Based on descriptive analysis, obtained the highest mean value is at component of roof maintenance at Building emergency installation (X1) for 3,934. This shows that respondents assess the roof maintenance variable is good. And the lowest mean is on ceramic maintenance component (X2) for 2,894. This indicates that respondents assess ceramic maintenance are less well. The observation results show the same thing with the opinion of the respondents which is the roof conditions on emergency installation Meuraxa is categorized as mild damage and still functioning well but the color of the roof cover has faded. Similarly, in the paint maintenance (X3), window and door frames maintenance (X4), ceiling maintenance (X5), water installation maintenance (X6), electrical installation maintenance (X7), air conditioning installation maintenance (X8), sanitary maintenance (X9)) is categorized as mild damage but still function well. On the results of these observations by researchers in general about the building emergency installation, it can be concluded that the implementation of building emergency installation maintenance has been done well.

1. Introduction

The building is a construction that plays an important role in all aspects. After the building is completed then the building will be used and maintained. Maintenance (maintenance) will make the age of the building becomes longer, several types of Maintenance Management Terms in Terotechnology is planned maintenance, preventive maintenance, corrective maintenance [2].

The age of the building plan is the time frame of the building used to the building is severely damaged. To avoid heavy damage is required routine maintenance system [8]. Hospital building and all its equipment require regular maintenance in order to avoid heavier damage. The occurrence of damage to hospital buildings and completeness can disrupt health services to the community.

The building of the hospital and all its equipment require regular maintenance in order to avoid heavier damage. The occurrence of damage to hospital buildings and completeness can disrupt health services to the community [7].

Meuraxa Hospital is a public building whose maintenance and maintenance activities should refer to the requirements of maintenance and maintenance of the reliability of the state-building [7]. The terms of maintenance and maintenance include safety, health, comfort, and ease of building. The purpose of this research is to know the maintenance system that has been applied and to identify the condition of building IGD Meuraxa Hospital. Maintenance components reviewed include architecture such as roofing, floor coverings, ceiling coverings and utilities such as air conditioning and sanitation installations.

2. Building Construction Maintenance Components

The components of building maintenance [3] :

1) Architectural building. Architectural buildings, including floors and stairs, walls and partitions, doors and windows, roofs and gutters, and ceilings.
2) Utilities. Building utilities include electricity, plumbing, air conditioning (AC), indoor communications, fire and elevator, wastewater treatment plants.
3) Structure. Building structures include columns, beams, and walls of buildings.

2.1. Classification of Damage Type for Building

The classification type of damage for each observation of building components are grouped into 3 conditions that are lightly damaged, moderately damaged, and heavily damaged. Limits on the three types of damage are defined as follows [3] :

1) Structure Damage Category:
   - minor damage is damage to structural components that do not reduce the function of the service (strength, stiffness, and ductility) of the overall structure, ie small cracks in beams, columns, and walls having a gap width between 0.075 to 0, 6 cm;
   - moderate damage is damage to structural components that can reduce strength but overall service capacity is in safe condition, ie large cracks in beams, columns, and walls with a gap width greater than 0.6 cm;
   - Heavily damaged is the damage to structural components that can reduce its strength so that the service capacity of the part or all of the structure is insecure, ie when the load-bearing wall is split and collapsed, the building is separated by the failure of the binder element and 50% of the main element is damaged or unfit for habitation.
2) Categories of Architectural Damage:
   - minor damage is damage that does not disturb the function of the building in terms of architecture, such as damage to the finishing work, namely peeling paint that does not cause disruption of function and aesthetics and does not pose a hazard at all, to residents;
   - medium damage is damage that can disrupt the function of the building in terms of architecture (function, comfort, aesthetics), such as damage to the building that is broken glass on windows and doors that can reduce the aesthetics of buildings and reduce comfort to residents; and
   - severe damage is very detrimental damage to the function and aesthetics of the building and result in the loss of comfort and can cause harm to the inhabitants.
3) Categories of Utility Damage:
   - minor damage is minor damage or malfunction of the utility sub-component that will not cause interference or diminish the function of utility components, for example on electrical installations, ie the extinguished one of the lights in the room;
   - moderate damage is the malfunctioning or non-functioning sub-component of the utility that causes interference or diminishes the functionality of the utility component, for example on

the installation of a disturbed telephone in one of the rooms causing the death of the telephone line; and

- heavily damaged is the malfunctioning or malfunctioning sub-component of the utility that may cause severe disruption or result in total utility malfunctioning.

3. Data Collection

The object of the study is indicators that support the implementation of the maintenance system at the installation of Meuraxa hospital in Banda Aceh. The subjects of this study are the staff, nurses, doctors, and patient families who are in the installation of Meuraxa hospital in Banda Aceh.

The required data in this study are primary and secondary data. The samples were taken approximately 10% of the total population [1] The total population is 914, and this study has 91 respondents. In this study have independent variables (X) and variable dependent (Y). There are 9 independent variables, roof maintenance (X1), ceramic maintenance (X2), paint maintenance (X3), window and door frames maintenance (X4), ceiling maintenance (X5), water installation maintenance (X6), electrical installation maintenance (X7), air conditioning installation maintenance (X8), sanitary maintenance (X9), and variable dependent is maintenance system (Y).

4. Result and Discussion

4.1. Respondent Characteristic

Based on a questionnaire distributed to 91 respondents. Characteristics of respondents in this study are gender, age, occupation and frequency of visits to Meuraxa Hospital within a month. Male respondents were 33 people (33.86%) and female respondents (58.64%). The characteristic of the age shows respondents who have age <17 years as many as 13 (14%), age 17 - 26 years as many as 76 people (84%), and respondents aged 27-35 Years as many as 2 people (2%).

Based on the work, the respondents who work as an employee amounted to 28 people (31%), doctors are 11 people (12%), nurses are 40 people (44%), and families of patients are 12 people (13%). The characteristic of the frequency of visitors who visited the hospital 1 to 3 times a month were 15 people (16%), and who visited the hospital more than 7 times a month were 76 people (84%).

4.2. Building Maintenance

Based on the respondent's answer, we get the mean value for each variable. The distribution of mean values for each variable can be seen in Table 1.

Table 1. Mean Value of Maintenance Building

Number	Variable	Mean	Rating
X1	roof maintenance	3,934	1
X2	ceramic maintenance	2,894	9
X3	paint maintenance	3,868	7
X4	window and door frames maintenance	3.901	4
X5	ceiling maintenance	3,853	8
X6	water installation maintenance	3,885	6
X7	electrical installation maintenance	3,886	5
X8	air conditioning installation maintenance	3,926	2
X9	sanitary maintenance	3,908	3

Based on descriptive analysis, obtained mean roof maintenance (X1) of 3.934. This shows that respondents rate good for roof maintenance variables. Furthermore, in the ceramic maintenance (X2) obtained value of 2.894 it shows that ceramic maintenance (X2) has good enough value based on respondent's opinion. In paint maintenance variables (X3) got a mean value of 3.868 respondents assessed the maintenance of paint goes well.

Furthermore, window and door frames maintenance (X4) got a mean of 3.901. means maintenance of window and door frames running well, then on the ceiling maintenance variable (X5), got a mean value of 3.853, maintenance of water installation (X6) got a mean value of 3.885, maintenance of electrical installation (X7) got a mean value of 3.886, maintenance of air conditioning (X8) gets a mean of 3.926, and sanitation maintenance (X9) got a mean value of 3.908. So it can be concluded the dominance of respondents argues that the variables X6, X7, X8, and X9 running well. Among other variables, the X1 variable that is roof maintenance in Building IGD (X1) has the highest mean value. And for the lowest is in the variable X2 is 2.894 has a mean value lower than other variables and in general this variable is running well.

4.3. Condition of emergency installation building Meuraxa Hospital

From the observation about the condition of damage to the IGD Meuraxa Hospital building on the components of the Architecture include Roof, Ceiling, walls, ceramics, and paint. While the utility components include AC, Piping, and Lift. Condition of the roof looks good only the color of the roof that has faded and does not disturb the function of the roof so categorized minor damage. On the ceiling, both the first floor and the second-floor type of damage that is the existence of holes and mossy so that damage to the ceiling is categorized minor damage.

On the wall seen that the existence of damage in the form of small cracks and can be categorized minor damage due to damage that does not disrupt the function of the wall and does not pose a hazard to the user at all. In ceramics seen the number of ceramics is broken and categorized as heavy damage because it has disrupted the function and aesthetics of the building and resulted in the loss of comfort and can cause harm to the user of the building.

In the paint looks damaged to the peeling paint, this can be categorized as minor damage because it is a disruption of function and aesthetics and does not pose a hazard to the user of the building. On the elevator seen damage to one of the available lifts are dead and can not be used, this can be categorized as moderate damage due to malfunction of one of the elevators causing interference or reduce the function of the existing elevator in the room. In air conditioner damage in the form of a loud sound, this can be categorized as moderate damage because it can cause interference to the user room. In the installation of Meuraxa hospital two lights are not lit; this can be categorized as moderate damage due to the non-functioning of some lights that can reduce the lighting in the room.

5. Conclusion

Based on the results of 91 respondents who responded to questionnaires distributed, the highest mean value is at component of roof maintenance at Building IGD (X1) for 3,934. This shows that respondents assess the roof maintenance variable is good. And the lowest mean is on ceramic maintenance component (X2) for 2,894. This indicates that respondents who assess ceramic maintenance are less well.

On the results of these observations by researchers in general about the building emergency installation, it can be concluded that the implementation of building emergency installation maintenance has been done well.

References

[1] Arikunto S., 2006, *Prosedur Penelitian (Suatu Pendekatan Praktik)*, Rineka Cipta, Jakarta.

[2] British Standard Institute BS 3811,1984, Glossary of Maintenance Management Terms in Terotechnology

[3] Kristianto, Usman, Winandi R, 2009, *Kajian Manajemen Pemeliharaan Gedung (BuildingMaintenance),* Jurnal Rekayasa, Vol 13, No.2,pp 157-166

[4] Kyle, R.C., and Baird, F.M. 1995. "Property Management" Dearborn Financial Publishing Inc. USA.

[5] Lee, R. 1987. "Building Maintenance Management". Collins Profesional and Technical Books.
[6] Oravetz, J. 1996. "Building Maintenance". Howard W. Sams & Co., Inc. Indianapolis.
[7] Peraturan Menteri Pekerjaan Umum Nomor: 24/PRT/M/2008, *Pedoman Pemeliharaan dan Perawatan Bangunan*
[8] Rudini, 2014, *Pengertian Bangunan Gedung*, http://rudiniacel.blogspot.co.id/2012/05pengertian-bangunan-gedung.html, access 9 June 2017

Acknowledgments
The Author would like to thank you for Dr. Hj. Hafnidar A.Rani, ST, MM (Dean of Engineering Faculty,Universitas Muhammadiyah Aceh). This research was supported by Universitas Muhammadiyah Aceh, Indonesia.

Development of construction dispute resolution process through arbitration (Indonesian National Board of Arbitration (BANI)

K Hayati[1,2,*], Y Latief[2] and A J Santos[3]

[1] Civil Engineering Department, Universitas Indonesia, Depok, Indonesia
[2] Civil Engineering Department, Universitas Muhammadiyah Aceh, Indonesia
[3] Universitas Juanda, Indonesia

* Email: kemalahayati1982@gmail.com

Abstract: The construction industry has various levels of complexity and competitiveness that may lead to conflicts among different principle and perspective. In case these conflicts are not managed properly, soon it will become disputes. One of the institutions in the field of arbitration for settlement of construction disputes in Indonesia is the Indonesian National Board of Arbitration (BANI). This study shows the construction project arbitration process and its risk factors in the Indonesian National Board of Arbitration (BANI). The study was conducted with structured interviews and surveys for data collection, followed by statistical and qualitative risk analysis. Current construction dispute resolution processes applied in the Indonesian National Board of Arbitration (BANI) is then evaluated by reviewing its processes followed by suggestion for a faster procedure of dispute resolution through the arbitration. This research will develop the dispute resolution process using arbitrage method in order to optimize the necessary time for dispute resolution.

Keywords: Arbitration, Construction, Dispute Settlement, Risk Management, Indonesian National Board of Arbitration

1. Introduction

The dispute is one of the main factors that prevent the success rate of construction project completion. Hence, it is important to consider the causes of disputes in order to complete the project at the expected time, budget, and quality [1]. As a form of dispute resolutions, arbitration fulfills few functions in conflict theory perspective. Despite the actual arbitrage work to proceed and resolve the conflict, the importance of arbitrage procedure is seen as a solution, no matter the result, by arranging the conflict to be resolved quietly or forcibly [2]. Indonesian National Board for Arbitration (BANI) is an independent institution that provides various services in the area of arbitrage, mediation, and other forms of dispute resolution outside the judiciary. For Indonesia case, dispute resolution through BANI is one of the best. With its 30 years of experience, BANI has standard procedure, various arbitrageurs from different disciplines and nationality with proven integrity to assure fairness of the verdict [3]. Dispute resolution through arbitrage in BANI shall be based on goodwill of the respective parties in such a cooperative way. Thus, arbitration is recommended as a form of better dispute resolution than other methods.

Nevertheless the intractable party in arbitrage can stop the process hence increase the arbitrage fee, but with proper planning and project management technique (includes avoidance and dispute mitigation), there should be no reason for the party to consider the possibility of resolution through arbitrage will impact in higher cost. With that consideration, arbitration can lead ahead compared to judiciary process in terms of efficiency [4]. By analyzing the arbitration process applied in the Indonesian National Board of Arbitration (BANI), the risk that can cause failure in dispute resolution process will be more efficient in terms of time. The performed evaluation is assessment of dispute resolution processes which is conducted through arbitrage. Furthermore, identifying hence to further anticipate the risks which will negatively affect the arbitrage processes and its achievement means the performance itself, nevertheless, can be improved.

2. Claim, Dispute and Risk Management

The claim can be defined as a demand or statement by one of the parties in seeking justification, whether for the rights, adjustment or interpretation in contract, payment, time extension, and other work-related to contract [5]. The unfulfilled claim or demand in construction projects can change to dispute. Construction dispute is a dispute related to implementation of construction service between related parties mentioned in the construction contract. Construction dispute is part of dispute in Civil Law. According to Act No. 30/1999 Clause 5, construction project is one of allowable disputes to be resolved through arbitrage or alternative way for dispute resolution. Risk management is a systematic process from planning, identification, analysis, response, and control the risk that may possibly lead to disadvantages. The objective of risk management is to maximize opportunity and consequence from positive events and also to minimize opportunity and consequence from negative events in response a goal. Risk management planning is a process to identify how risk management activity can be conducted. This risk management planning is urgent to communicate and formulate and also support from related parties in the project in order to held project life cycle effectively. The key success of this step is by convincing the created risk management is appropriate with the possible risk in project [6]. The risk aspect has a major impact on the performance of time in arbitrage process. In this research, the duration of time of arbitrage process is minimized to the possible level. Compared to other disputes, construction dispute is more exposed to risk since it is strongly related to time, complexity of the process, and no agreement between parties.

3. Methodology

The design in this study involved the selection of research strategy, survey process, research variables, research instruments, data collection, and method analysis. It used descriptive and parametric statistical correlation analysis. The research analysis for questionnaire result is done gradually, which is initialized by interview and continued by descriptive analysis, validity and reliability test to identify and extract most important or dominant factors. A homogeneity test is implemented to see the understanding of respondents on risk variables. Two used homogeneity tests are Kruskal-Wallis and Mann-Whitney test. The objective of descriptive analysis in this research is to present certain characteristics of data in particular sample to find out value of mean and median. Meanwhile, the validity test is a test for the accuracy of an instrument. A reliability test is used to determine the consistency of the measuring tool. In next step, qualitative risk analysis, level of risk is established in order to be set for further action. Cause and effect analysis is applied as well for any major or dominant risk to develop risk response strategy includes the implementation of preventive and corrective actions. Validation of research results is performed by experts to gain type of action input which is expected to be utilized in improving the performance of the arbitration.

4. Results and Discussions

4.1. Risk Identification

Identification of risk rating includes response and controls to the potential risks that considered as the obstacle in dispute resolution process in BANI are obtained in result of analysis (Table 1). This also includes improvement of the construction dispute resolution process through arbitrage based on previous result of analysis. The intention of the improvement is in preventive action and further in analysis of time impact to the process. There are three dominant risk factors which are divided into few process variable factors. Below are the dominant risk factors.

Table 1. Risk Identification.

Process	Risk Event	Weight	Risk Level
Arbitrage verification	Lack of evidence or documentation of the pleading from respective mentioned parties in contract whether the dispute issue can affect the arbitrage verdict	15,9	High
Arbitrage pleading	The defendant has no willingness to pay the arbitration fee, so that the applicant shall pay the amount before the trial is started.	14,4	High

4.2. Cause and Effect Analysis of the Arbitration Verdict

Based on described matrix in the previous section, all causes and effects from each mapped risk are dominant (Table 2). After the dominant risks identified and analysis of cause is performed for respective risk, it is then followed by arranging risk response strategy. As the result, recommendation for improvement to the construction dispute resolution through the Indonesian National Board for Arbitration (BANI) which is referred to risk handling can be formulated.

Table 2 Risk Factor, Cause, Preventive Action and Development Process.

Risk rank	Risk factor	Cause	Preventive action	Development process
1	Arbitrage verdict very rely on the arbitrageur competency	Arbitrageur qualification is not relevant to the area of dispute issue	BANI is expected to develop and provide access to list of trial groups and arbitrageur that meet the area of dispute complete with a competency profile that can be found through website.	Add a selection of arbitrageur steps with required competency in the construction project dispute.
2	Lack of evidence/documentation of pleading from mentioned parties in contract whether the disputed issue can affect the arbitration verdict.	The service provider neglects the documentation of events in the project, whether addendum or work changes.	Each documentation and evidence related to contract and claims shall be arranged in sequence and complete in order to speed up the arbitration verification process.	Add notification that in the project arbitration verification, the required documentation shall be arranged in sequence and complete.
3	The defendant party has no willingness to pay the arbitration fee, so that the applicant shall pay the amount.	For plea submission, one of the parties (the applicant) shall pay the arbitration fee since most of the defendant has no willingness to pay in early arbitration process.	Both parties have to prepare the necessary fees to resolve the construction dispute resolution once the construction contract is agreed and included in clause of arbitration registration fee responsibility.	Both parties shall prepare the necessary fee for construction dispute resolution once the construction contract is agreed and included in a clause of arbitration registration fee responsibility.

From the perspective of preventive action, the possible objective to eliminate necessary time for construction dispute resolution through arbitration is by developing the following processes:

IOP Conf. Series: Materials Science and Engineering **674** (2019) 012028 doi:10.1088/1757-899X/674/1/012028

a. Improving the process of plea and selection of arbitrageur through the addition of arbitrageur selection step as per required competency in construction project dispute in the area of the case and arbitrageur competency and experience profile in the website.
b. Erasing rejection process to an arbitrageur since the person is already appointed earlier according to the area of the case and known by the parties so the rejection process is no longer necessary.
c. Speeding up arbitration administration fee payment by the parties since the resolution fee is already prepared by the applicant and the parties have also created a written agreement in a contract stating their willingness to pay the necessary fee for dispute resolution.
d. Speeding up arbitration verification process through arbitrageur selection that has a strong competency in the area of dispute and with complete and well-arranged documentation from the mentioned parties since the first arbitration verification process. This step will ease the verification process that helps to speed up the verification time.

4.3. Improvement Process for Construction Dispute Resolution through the Indonesian National Board for Arbitration (BANI)

There are 14 steps in the dispute resolution process through BANI (Figure 1) and these can be reduced to 13 steps. The process with dominant risk value is arbitration verdict, arbitration verification, and arbitration pleading. By risk response from expert judgment, duration of time for construction dispute resolution process through the Indonesian National Board for Arbitration (BANI) can be reduced. The more efficient arbitration process can be achieved base on risk potential that has passed the identification process and risk response. Therefore, the improvement step of the arbitrage process is suggested as follows: (1) BANI Secretariat verifies the submitted plea file. Scopes of verification file are: *persona standi judicio*, *fundamentum petendi (posita)*, *petitum*, and the expected resolution complete with the selection of an arbitrageur by the applicant. (2) BANI appoints Arbitrageur Secretary internally. Arbitrageur Secretary responds by handed-over one copy of the arbitration plea and its attachment to the applicant. (3) Defendant responses by providing argument or inserting encountered lawsuit. (4) The head of BANI determines the arbitration board. (5) Calculation and announcement of administration fee bill to related parties by the Arbitration Secretary. (6) Payment of administration fee by the parties which can be paid directly by one party, the applicant or the defendant. (7) Payment confirmation and distribution of first trial invitation. (8) The first trial invitation is received by the parties. (9) Arbitration verification process (trial process). (10) Trial closing by Chief of Board. (11) Verdict signing by the arbitrageurs and final verdict announcement. (12) Administrative correction request, if any. (13) BANI register the arbitration verdict to State Court.

4.4. Effect of Time to Improve Dispute Resolution Process

Based on the research result, the total time for dispute resolution through the arbitrage process in BANI after the suggested improvement applied is 202 calendar days. It is way shorter compared to the maximum of 334 calendar days stated on the Procedure of BANI as in Act No. 30/1999 about Arbitrage and Alternative of Dispute Resolution [7]. The improvement reduces necessary time to close the construction dispute resolution at about 39.52%. The improved steps are:

1) The selection of arbitrageur in response to the defendant is more efficient at 15 days from 30 days in the current existing process. Profile of arbitrageur can be seen online and there is no rejection process for the selected arbitrageur.
2) Verification of arbitrage is more accurate and efficient. Previously, the verification issue of the arbitrage resolution can be completed in less than 90 days, with the conditions: complete and well-managed evidences as well as the commitment between both parties.
3) Payment process of administration fee by related parties. In existing condition, the maximum duration is 14 days, while it can be improved to 1 day. The total indicated time to be saved is 13 days.

IOP Conf. Series: Materials Science and Engineering **674** (2019) 012028 doi:10.1088/1757-899X/674/1/012028

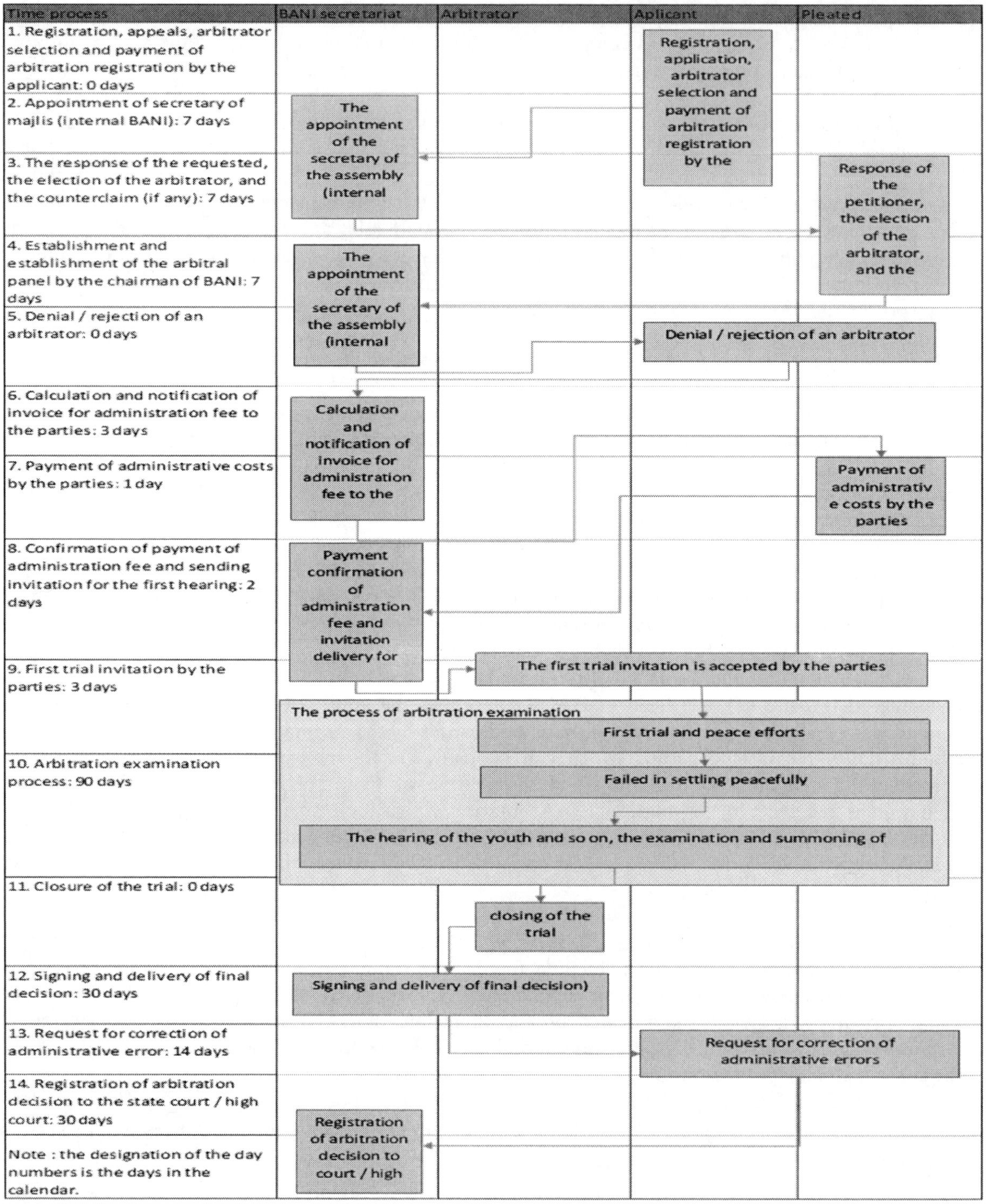

Figure 1. Improvement of Completion Time in BANI.

5. Conclusion

As described in Figure 1, there are 14 steps in the arbitrage process practiced by the Indonesian National Board for Arbitration (BANI). From the questionnaire, 39 risk factors that considered potential to be generated in the process of arbitrage in BANI are derived. The qualitative analysis result suggested three dominant risk factors in dispute resolution in BANI: (i) arbitration verdict rely on competency of the arbitrageur, (ii) lack of evidence or pleading from respective parties in contract whether the disputed issue can change the arbitration result, (iii) the defendant party has no goodwill to pay the arbitration fee so that the applicant shall pay the fee before the trial is started.

References

[1] E Cakmak Cakmak and Irlayici P 2014 *An analysis of causes of disputes in the construction industry using analytical network process* (Science Direct) p 183 – 187.
[2] Behrens 2007 *Conflict, arbitration, and dispute resolution in the German workplace.*
[3] N Yasin 2003 *Mengenal Kontrak Konstruksi di Indonesia* (Jakarta: PT. Gramedia Pustaka Utama).
[4] A Tanielian 2013 *Arbitration Still Best Road to Binding Dispute Resolution* (American Society of Civil) vol 5(2) p 90-96.
[5] V Chovichien and Tochaiwat K 2006 *Information System for Managing Employer's Construction Claims* p 87–91.
[6] PMI 2013 *A Guide to Project Management Body of Knowledge In (PMBOK Guide), fifth edition* (Pennsylvania: Project Management Institute).
[7] BANI 2016 *BANI Rules* (Jakarta: BANI Arbitration Center).

Physical properties of reclaimed asphalt pavement

S Sunarjono * **and N Hidayati**

Centre for Transportation Studies, Civil Engineering, Faculty of Engineering, Universitas Muhammadiyah Surakarta, Jl. A. Yani Pabelan Kartasura Tromol Pos I Surakarta 57102, Indonesia.

* Email: Sri.Sunarjono@ums.ac.id

Abstract. Sustainable construction is becoming an important issue in pavement engineering. The use of reclaimed asphalt pavement (RAP) for road reconstruction gives both technical and environmental benefits. Unfortunately, the properties of cold asphalt mixture using RAP is still behind of hot asphalt mixture using fresh aggregate. The objective of research is to explore the physical properties of RAP in order to understand the factors affecting the lack of cold asphalt mixture performance. The laboratory physical tests on RAP and fresh aggregate were performed to investigate particle shape, specific gravity, gradation, and bulk density. It is found that the presence of aged bitumen in the RAP affects the shape formation, specific gravity, absorption, and gradation of RAP. All those factors cause the bulk density of compacted RAP decrease. This finding opens the knowledge of why cold asphalt mixture using RAP is worse than using fresh aggregate.

Keywords: Physical Properties, Road Construction, Reclaimed Asphalt.

1. Introduction

The use of reclaimed asphalt pavement (RAP) materials for road construction become increasingly in demand due to environmental issues. It can reduce emissions gas production, decrease source of natural materials exploitation, and also solve road geometric problems. The RAP material which is the results of the damaged road milling can be reprocessed by cold mixing method to produce the new pavement. This technology can decrease the need for fresh aggregate and bitumen, reduce fuel consumption and CO2 production, and overcome the addition of road surface level when it is conducted by overlay method. It is noted that RAP materials can replace the fresh aggregate for road pavement construction up to 40%-60% with satisfied performance [1]-[2]. These benefits push the use of RAP as the prospective of road construction materials. The application of RAP in Indonesia using cold-recycling system was started in 2007 at Pantura road [3]. The indirect tensile strength and the stiffness modulus of cold asphalt mixture using RAP is found to be satisfied and meet the specification for road base course materials [3]-[4].

However the properties of asphalt mixture using RAP are still behind using fresh aggregate, the cold mix asphalt is also still behind the hot mix asphalt [5]. In the RAP cold mix, the high air void causes the low value of density [6] and stability [2]. This lack of asphalt mixture using RAP materials pushes the need to investigate intensively RAP materials. This paper reports the results of research exploring the physical properties of RAP. The objective of research is to open understanding factors affecting the mixture performance using RAP. This research is designed to open the knowledge of why asphalt mixture using RAP is worse than using fresh aggregate.

2. Research Method

The research method is conducted by analysis the results of physical tests on RAP, extracted RAP (RAP aggregate), and fresh aggregate. The laboratory physical tests used were aggregate identity and shape, specific gravity & absorption, sieving analysis, and loose & compacted unit weight. All test results on those materials were compared in order to get important information about factors' affecting the lack of asphalt mixture using RAP materials. The RAP material used in this research was collected from Tegal section of Java Pantura Road in Indonesia.

3. Result and Discussion

The research method is conducted by analysis the results of physical tests on RAP, extracted RAP (RAP aggregate), and fresh aggregate. The laboratory physical tests used were aggregate identity and shape, specific gravity & absorption, sieving analysis, and loose & compacted unit weight. All test results on those materials were compared in order to get important information about factors' affecting the lack of asphalt mixture using RAP materials. The RAP material used in this research was collected from Tegal section of Java Pantura Road in Indonesia.

3.1. Identity and Shape of RAP Materials

Figure 1 shows the RAP material used in this research that collected from Pantura road, Tegal, Indonesia. Figure 1 b and 1 c also show RAP materials collected from Cirebon [7] for comparison with RAP collected from Tegal. Color of RAP was grayish at dry condition or darker at wet condition, whereas Cirebon RAP looked browner. The color of RAP is not black because it was partially uncoated aggregates due to milling process. The RAP particles varied in size with nominal size was about 25 mm. This is an ideal size of asphalt concrete (AC) or hot rolled sheet (HRS), the popular asphalt mixture used in Indonesia. AC is a dense-graded mixture that normally used in rural road pavements, and HRS is a gap graded mixture that normally used in urban road pavements. The bitumen content of RAP was found to be about 4.16%. Bitumen content of RAP was found about 6.7% (Herman, 2004). It can be understood that the bitumen content will depend upon the type of mixture source. Normally bitumen content of AC will be lower than HRS. It is noted that RAP materials collected from stockpile can be combination of two or more mixture types. The water content of RAP was found to be about 1.3%. This water can come from trapped water in the pavement, or due to rainwater when stock in open area. Information on water content of RAP is important to understand the additional mixing water in the cold mix method. Property of aged bitumen of RAP was found to be low penetration i.e. 19.8, and it was found to be less ductile [8].

The result of observation on the RAP formation can be seen in Figure 2. It can be grouped at least into 6 formations, i.e. (1) single coarse or medium particle without mastic, (2) single coarse or medium particle coated partially by mastic, (3) single coarse or medium particle with coated fully by mastic, (4) coarse particles coated fully by mastic, (5) mastic formed combination of medium and fine aggregate, filler, and bitumen, and (6) mastic formed combination of fine aggregate, filler, and bitumen. Later these RAP formations are important to open understanding of RAP mixture properties.

Figure 1. Photograph of original RAP materials collected from (a) Tegal and (b,c) Cirebon (Widiyanto, 2013). Photographs in middle and bottom row are RAP particles retained in certain sieving size.

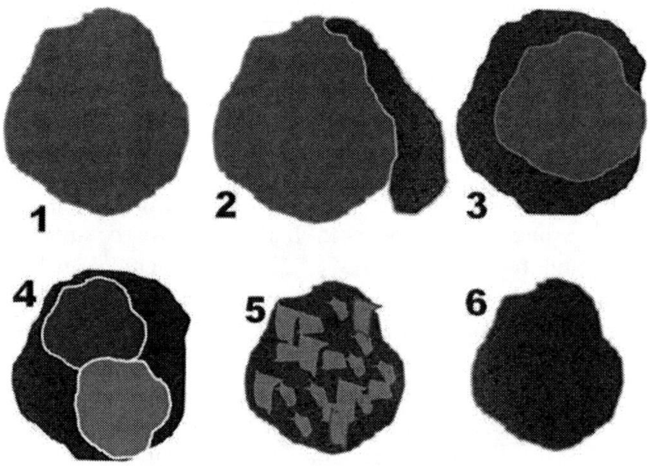

Figure 2. The Formation of RAP

Table 1 shows the flakiness and elongation test results of coarse aggregates. Three types of aggregate were tested i.e. RAP, RAP aggregate after extraction test, and fresh aggregate. All aggregate types were found to be not flaky and elongated, and then it was classified as uniform dimension. As seen in Table 1 that RAP aggregate is more flaky and elongated than RAP. It can be understood that the presence of aged bitumen on the aggregate surface makes RAP is more rounded. This fact is supported by flakiness and elongated index in which RAP aggregate values are higher than RAP values. It is noted that the higher the index the aggregate shape is more flaky and elongated. The flakiness and elongated index of all aggregates meet the specification (max 25%).

Table 1. Flakiness and elongation test results of RAP, RAP aggregate and fresh aggregate

Property	RAP	RAP Aggregate	Fresh Aggregate	Specification
Flakiness	0.743	0.285	0.286	0.593
Elongation	0.800	0.285	0.067	0.796
Classification	Uniform dimension	Uniform dimension	Uniform dimension	
Flakiness Index (%)	18.50	20.73	17.89	max 25
Elongated Index (%)	19.92	23.58	18.13	max 25

3.2. Specific Gravity and Absorption of RAP Materials

Table 2 shows the results of specific gravity (SG) tests of RAP, RAP aggregate and fresh aggregate (FA) materials. RAP aggregate is aggregate resulted from extraction test. Fresh aggregate properties are included for benchmark value. Aggregate specific gravity can be defined as a weight ratio between aggregate and water at similar volume. SG value of aggregate will affect value of density. In term of RAP materials which contain aggregate and aged bitumen, the density of RAP will be affected by their both component SG and their composition. It can be understood that the presence of aged bitumen will reduce RAP density. SG test is usually followed by absorption test i.e. the aggregate ability to absorb water into their pores. The absorption value of aggregate represents ability to absorb bitumen and adhesive strength of aggregate.

There are three types of aggregate SG i.e. bulk, saturated surface dry (SSD), and apparent SG. The difference between these three types of SG is in terms of weight and volume assumption used. Bulk and apparent SG use dry weight, while SSD SG uses weight of saturated surface dry aggregate condition. Bulk and SSD SG use total aggregate volume included solid, impermeable, and permeable part, while apparent SG do not account the permeable volume part. This assumption causes bulk SG value will be lowest, whereas apparent SG value will be highest.

In Table 2 show that the values of apparent SG are highest and bulk SG is lowest for all tested materials. These results meet the described theory. It was found that the SG of RAP aggregate was higher than RAP. It can be understood that the SG of RAP is combination SG of the old aggregate and the aged bitumen, in which the SG of aged bitumen is lower than aggregate. The SG of RAP aggregate was also found to be lower than the SG of FA. This finding can be described that it may the SG of FA is originally better than RAP aggregate, or there still presence the rest of aged bitumen trapped in the inside of RAP aggregate pores. Based upon this discussion it can be concluded that the SG of RAP or RAP aggregate will be potentially lower than fresh aggregate. It is known that SG will affect to the dry weight, and then it will affect density.

SG of RAP was also found different character than fresh aggregate. It is normally SG of aggregate will increase with higher particle size, because the accretion of aggregate weight is higher than the accretion of water weight. SG of FA and RAP aggregate follow this rule as shown in Table 1. However RAP has different unique character, the highest SG values were found to be the medium size particle (5-10 mm), not coarse aggregate (10-20 mm). The SG of RAP will be affected by the RAP mastic clusters that contain combination of aged bitumen and fine aggregates. In this case, it may the RAP mastic clusters is more distributed and attached on the aggregate size of 10-20 mm.

Specific gravity and absorption test were conducted in one package. Values of aggregate absorption will be affected by the wide of aggregate surface. It can be understood that the finer aggregate will have wider aggregate surface. Again the RAP materials have different character compare to other aggregate. The absorption of RAP medium fraction is less than the coarse fraction. This fact can be described in similar way with the fact of RAP SG. The coarse fractions can be actually shaped as the

RAP mastic clusters that contain combination of aged bitumen and fine aggregates. These fine aggregate parts induce the coarse fraction having higher absorption value.

The absorption of RAP was also found to be lower than RAP aggregate and fresh aggregate. This fact is caused by the aged bitumen coat partially or fully aggregate surface. This aged bitumen obstructs the water absorption process to the RAP pores. The value of aggregate absorption can be used to estimate aggregate absorption to bitumen. There is a term to indicate aggregate absorption to bitumen, i.e. effective specific gravity which is calculated as average value of bulk and apparent SG. The low RAP absorption shows two aspects, i.e. (1) the need of RAP to bitumen is lower, and (2) adhesion (coating) of RAP materials is lower when compared to fresh aggregates.

Table 2. Specific gravity test results of RAP, RAP aggregate and fresh aggregate

Properties	Particle Size								
	10 – 20 mm			5 – 10 mm			< 5 mm		
	RAP	RAP Aggregate	FA	RAP	RAP Aggregate	FA	RAP	RAP Aggregate	FA
Bulk SG	2.060	2.243	2.937	2.109	2.192	2,762	1.937	1,968	2,565
SSD SG	2.090	2.277	3.014	2.133	2.227	2,841	1.976	2,016	2,688
Apparent SG	2.122	2.322	3.183	2.170	2.272	3,000	2.016	2,068	2,926
Absorption (%)	1.416	1.506	2.632	1.117	1.611	2,878	2,041	2,459	4,822

Note: RAP agg is aggregate from RAP after extracted, whereas FA is fresh aggregate

3.3. Gradation of RAP Materials

Gradation is particle size distribution of aggregate. In terms of RAP that having unique particle shapes, RAP gradation is very interesting material to be discussed. First, RAP is combination of two materials, i.e. aggregate and aged bitumen. It will come up with a question, "which one as a true gradation: RAP or RAP aggregate?" Second, RAP contains mastic cluster part or fine particles attached to the coarse aggregate surface. It will come up a problem that a coarse aggregate size will be greater due to the attached fine aggregate. Third, is RAP gradation used for hot and cold mix system similar? This question arises because in hot mix system, the aged bitumen will normally be dissolved, but in cold mix system, the aged bitumen will not be dissolved in early age. In the cold mix asphalt using RAP and additional fresh bitumen, there needs research to explore the process of integration between old and new bitumen.

As seen in Figure 3, the gradation of RAP was investigated in three versions, i.e. gradation produced by dry sieving and wet sieving, and RAP aggregate resulted from the extraction test. Wirtgen [9] suggested using wet sieving system for RAP materials due to the attached fine aggregates problem. It was found that the wet sieving is finer than the dry sieving. The difference between these two sieving methods looks not significant. However, the recommendation of Wirtgen [9] is true in terms of getting the real particle size distribution, not cluster size distribution.

It was also found that the RAP aggregate gradation was much finer than RAP gradation. It is noted that RAP aggregate was generated by extracting the RAP particles. This fact can be understood because the aged bitumen attached to the RAP aggregate has been dissolved. The RAP aggregate gradation curve can be utilized to understand aged bitumen or mastic fine cluster distribution on the RAP gradation. As example, it can be estimated that the aged bitumen or the mastic is approximately 10% attached to the 10mm RAP particle size and 6% on the 1mm RAP particle size. It is found that the RAP aggregate gradation meet the AC-BC (Asphalt Concrete – Binder Course) specification envelope. However the RAP gradation is little bit coarser than the lower envelope border of AC-BC specification. It can be than understood that the source of this RAP from asphalt concrete pavement types.

There has been well understood that aggregate gradation can be grouped as dense, gap, or uniform graded. The way to determine the aggregate gradation grouped is by calculates value of Cu (Coefficient of Uniformity) and Cc (Coefficient of Curvature). It will be dense-graded if Cu>15 and 1<Cc<3, or uniform graded if Cu<15 and 1<Cc<3, gap graded if Cu>15 and Cc<1.Calculation results of these Cu and Cc values for RAP and RAP aggregate can be seen in Figure 3. It can be concluded that both RAP and RAP aggregate are dense-graded because their coefficient of uniformity is included in Cu > 15 and their coefficient of curvature is included in 1<Cc<3. This finding confirms that the source of RAP used is asphalt concrete, not hot-rolled sheet.

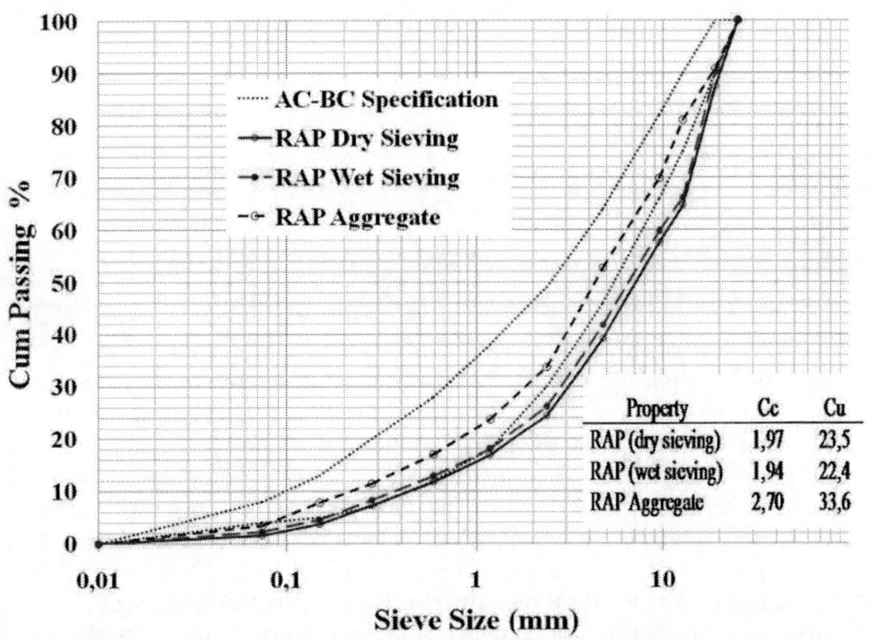

Note: AC-BC is asphalt concrete – binder course; Cc is coefficient of curvature = (D302)/(D10xD60); Cu is coefficient of uniformity =D10/D60; D10 is particle diameter at 10% cumulative passing

Figure 3. Gradation of RAP

3.4. Unit Weight of RAP Materials

Unit weight can be expressed as a ratio between weight and volume. It can be used to indicate the degree of density of a material. The higher the unit weight the higher the density. In this research the unit weight was investigated using fives conditions, i.e. loose, dense by stick, dense by shake, dense using standard Proctor, and dense using modified Proctor. The loose condition means the materials are introduced into the mould without any treatment. Dense by stick means the materials is pierced using stick before measured, whereas dense by shake means the materials are shake before measured. Unit weight using standard and modified Proctor means maximum dry density value after compaction process on the wet materials. The materials used for this test are RAP, modified gradation RAP, RAP aggregate, and fresh aggregate. As seen in Table 3 unfortunately not all data is available. However the data is enough to keep the results trend. It is supposed that the order of unit weight values for the particular material from low to high is loose, dense by stick, dense by shake, standard Proctor, and modified Proctor. It can be understood because the order represents the amount of the compaction energy given. In this case, RAP material has similar behavior with other materials. Whereas the order between those four materials from low to high is RAP, modified gradation RAP, RAP aggregate, and then fresh aggregate. RAP material has lowest value of unit weight. Modification of the RAP gradation can elevate the unit weight, but it is still lower than RAP aggregate or the aggregate.

Table 3. Unit weight test results of RAP, RAP aggregate, modified gradation RAP, and fresh Aggregate

Property (gr/cm^3)	RAP	Modified Gradation RAP	RAP Aggregate	Fresh Aggregate
Loose unit weight	1.418	NA	1.713	NA
Dense unit weight by pierced using stick	1.534	NA	1.854	NA
Dense unit weight by shake	1.568	NA	1.876	NA
Dense unit weight using standard Proctor	1.640	1.701	NA	2.110
Dense unit weight using standard Proctor	1.677	1.770	2.010	2.261

Based upon these unit weight test results, it is needed to understand the orientation of the particles during compaction for RAP materials. Understanding the effect of aged bitumen coating the surface of the particles is demanded in order to open the knowledge of compaction characteristics of RAP materials. Tests of particle orientation investigation during compaction process have been conducted. There were used two materials, i.e. RAP and extracted RAP. The material was compacted in a mould using vertical stress load. Three stone agents were put inside the material, i.e. on top, middle, and bottom of the mould. The position of stone agents before and after compaction was then monitored.

Figure 4 show the test results for RAP material (left) and extracted RAP (right). Blue point was the initial position of stone agent before the compaction process was started, whereas red point was the end position of stone agent after compaction process finish. As seen in the Figure 4 that for both materials the stone agent 3 located in the bottom of mould moved less dynamic if it is compared to agent 1 and 2. It can be understood that the effect of compaction load energy in the bottom of mould is lowest, and direction of mass movement is from the top to the bottom. It was recorded that the initial height of material was 14mm and the final height of material was 12.5mm (RAP) and 11mm (Extracted RAP). It means for RAP, the mass material on the top part go down about 1.5mm in average. The distance of mass movement decreases from top to bottom.

The discussion of aggregate orientation difference between RAP and extracted RAP is very interesting. The influence of aged bitumen will be identified from the provided difference. As seen in Figure 4, some important information can be explored. First, the orientation patterns of stone agents between RAP and extracted RAP are similar in which the stone agent 1 and 2 move more dynamic and opposite compared to stone agent 3. This fact gave evidence that the testing has been carried out correctly. Second, the movement distances of agent 1 and 2 in the extracted RAP mould were longer than in the RAP mould. Movement direction can be analyzed as horizontal, vertical, and their resultant. In this case, those three movement types of the extracted RAP were longer than the RAP.

This finding gives information that the presence of aged bitumen on the RAP material surface obstructs the particle movement. In the rule of the compaction process if the aggregate particles are difficult to move, it will cause the density of aggregate is low.

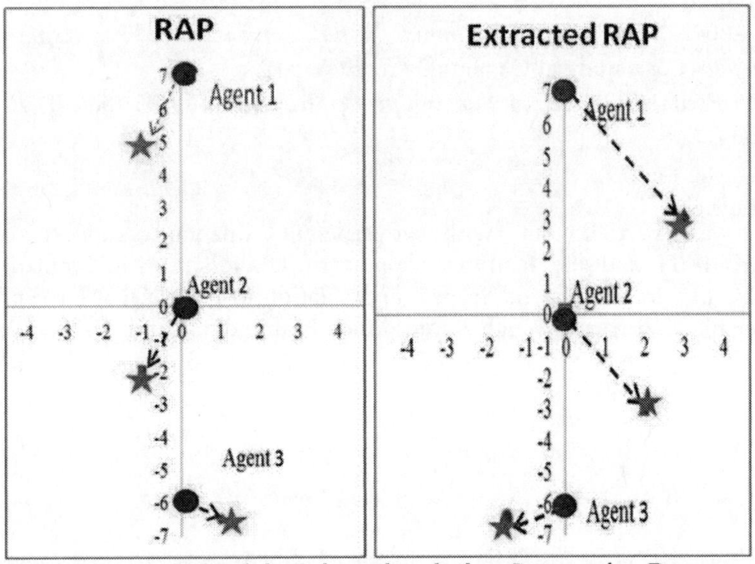

Figure 4. RAP Particle Orientation during Compaction Process

4. Conclusion

RAP is road material aggregate having combination component of old aggregate and aged bitumen with at least 6 shape formation combinations. Based upon flakiness and elongation test results the RAP material particles are classified as uniform dimensions. The presence of aged bitumen on the aggregate surface makes RAP is more rounded, but it cannot change the aggregate shape classification. The specific density and absorption value of RAP is potentially lower than fresh aggregate. The presence of aged bitumen affects the specific gravity combination between old aggregate and aged bitumen decrease. This presence of aged bitumen also obstructs water absorption process to the RAP pores. RAP can be grouped as dense-graded based upon their Cu and Cc values. RAP gradation is coarser than RAP aggregate gradation. The presence of aged bitumen does not change the RAP gradation type, but affect it coarser. The unit weight of RAP aggregate is higher than the unit weight of RAP and lower than the unit weight of fresh aggregate. The presence of aged bitumen on the RAP material causes the unit weight of RAP decrease. The unit weight of compacted RAP decrease due to 2 aspects, i.e. (1) the specific gravity of RAP is lower, and (2) the presence of aged bitumen obstructs the RAP particle movement.

References

[1] I C Van Winkle 2014 *Laboratory and Field Evaluation of Hot Mix Asphalt with High Content of Reclaimed Asphalt Pavement* Master Thesis The University of Iowa.

[2] S O Tambake, D N Kumar and K R Manjunath 2014 *International Journal of Research in Engineering and Technology* Vol 3 Issue 6 eISSN 2319-1163.

[3] R A Yamin and D Widajat 2008 *Jurnal Jalan-Jembatan* Vol 25 No 2 p 130-153.

[4] S Sunarjono 2007 Dinamika Teknik Sipil Vol 7 No 1 ISSN 1411-8904 p 1-10.

[5] KM Khan, M A Kamal, F Ali, S Ahmed and T Sultan 2012 IOSR Journal of Mechanical and Civil Engineering Vol 4 Issue 1 p 27-31.

[6] S Sunarjono 2010 *Compactability Study Of Foamed Asphalt Using Gyratory Compactor* The 13th FSTPT Symposium Unika Soegijapranata Semarang.
[7] A Widiyanto 2013 *The Use of Insitu Materials for Road Pavements and Green Road Technology in Indonesia*
[8] S Herman 2004 *The Effect of Reclaimed Asphalt Pavement on The Asphalt Concrete Physical Properties* (Case study at Gajahmada road Tegal)
[9] Wirtgen 2005 *Foamed bitumen mix design procedure using the Wirtgen WLB 10* (Wirtgen GmbH, Germany)

Acknowledgments

This research was carried out with government financial support of DITLITABMAS KEMENRISTEKDIKTI and the facilities support of Universitas Muhammadiyah Surakarta. The author would also like to express the deepest appreciation to people that have provided the data and managed the paper i.e. Widyorini Windi Astuti, Mada Pramindana, and Wahyu Setyawan.

IOP Conf. Series: Materials Science and Engineering **674** (2019) 012030 doi:10.1088/1757-899X/674/1/012030

Importance of evaluation for post-disaster assets: a preliminary study in Aceh, Indonesia

A Adamy* and A H A Bakar

Universitas Muhammadiyah Aceh, Indonesia

* Email: aulina.adamy@unmuha.ac.id

Abstract. Responding to the effects of the earthquake and tsunami in 2004, Badan Rehabilitasi dan Rekonstruksi Nanggroe Aceh Darussalam-Nias (BRR NAD-Nias) was assigned as the coordinating and executing agency for improving the region in Aceh province and Nias islands in Indonesia. The allocation for the projects directly implemented by the BRR was approximately USD 2.6 billion. Based on local regulation, these assets must be turned over to the Aceh government. Although most assets were already occupied, neither an appropriate handover nor external project evaluation was conducted. This preliminary study surveyed opinions from building users regarding building quality, the need for building evaluation, and the general opinion for occupying assets with no appropriate handover process in a post-disaster context. The samples included a school, health center, and a sub-district level government office in Banda Aceh city. This study used descriptive methods and a quantitative approach. Results showed the majority of respondents doubted the quality and condition of the assets and agreed with the importance of conducting asset evaluation. However, they are uncertain whether occupying assets before an appropriate handover was appropriate because the post-disaster situation required assets to be functional as soon as possible.

1. Introduction

In the history of disaster management in Indonesia, the central government first formed a temporary but powerful agency that aimed to respond to the effects of the massive earthquake and tsunami disaster on 26 December 2004. Aceh and Nias Rehabilitation and Reconstruction Agency, named Badan Rehabilitasi dan Rekonstruksi (BRR) Nanggroe Aceh Darussalam (NAD) or Nias, was formed by the Indonesian government on 16 April 2005. BRR NAD-Nias (BRR) was based in Banda Aceh (the capital city of Aceh province), to be “close to the action” and to be responsive to the local context. A total of USD 2.6 billion was given to the BRR for the implementation of programs or projects [1]. Four years later, in 2009, BRR completed 94.7% of the total key performance indicators (KPI) [2]. This achievement was considered a high record [3].

However, negative issues were raised regarding corruption and the delivery of poor reconstruction quality. GERAK (Anti-Corruption Movement) and Indonesian Corruption Watch (ICW) discovered Indonesian rupiah (IDR) 23.8 billion worth of the total project had problems [4]; included mark-up, unauthorized selection, corruption in at least five major BRR departments such as the publication of reports, appointment of staff, procurement of office equipment, illegal subcontractor practice, non-targeted donation, aid cutting, and non-specification-based project implementation [5]. Another

IOP Conf. Series: Materials Science and Engineering **674** (2019) 012030 doi:10.1088/1757-899X/674/1/012030

problem left by the BRR is that reconstructed assets directly built by the agency were not transferred "legally" to the Aceh government [6]. Problems in assets that were found included no documents, unclear ownership, unclear maintenance budget, damaged assets, no asset found, not functional, unfinished, and still ongoing projects.

The Aceh government already occupied most of the assets and faced problems alone because most donors have left and the BRR has been dissolved in April 2009. Therefore, this study aims to obtain the perspective of building users regarding the necessity of evaluation despite their occupying of those buildings, the condition of reconstruction assets during occupancy, and the decision of the Aceh government to occupy the assets even without a proper turnover. The importance of this preliminary study maps the state of mind of the beneficiaries and serves as basis in conducting proper post-disaster asset evaluation in the future.

2. Evaluation In Post-Disaster Project

Post-disaster projects have short periods of construction. On the one hand, the reality is more complex when there is a responsibility to return things back to pre-disaster state, and not to compromise survival on the other hand [7]. Practical action promotes a holistic and integrated approach to post-tsunami reconstruction, which demonstrates options for better building projects [8]. By integrating evaluation processes into post-disaster project designs, and by carrying them out both during and at the end of the project, the problems identified can be overcome earlier, and lessons can be learned for facing future post-disaster reconstruction. A focus on evaluation creates a legacy of useful lessons for future post-disaster projects.

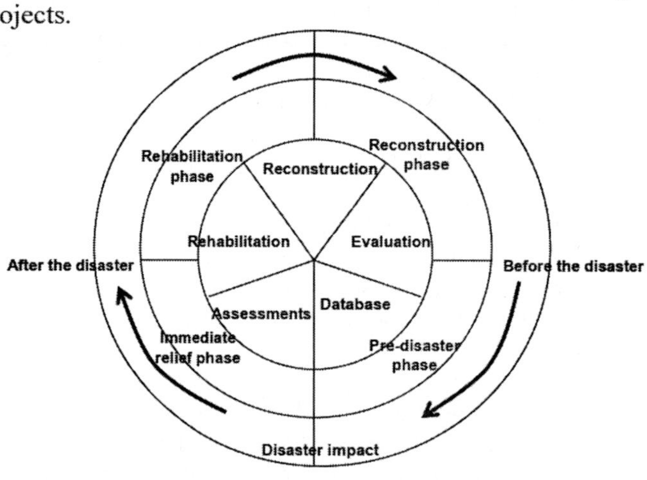

Figure 1. Post-Disaster Reconstruction Cycle [9]

The Organization for Economic Cooperation and Development (OECD) states that the aim of evaluation in post-disaster is to determine the relevance and fulfillment of objectives, developmental efficiency, effectiveness, impact, and sustainability [10]. An evaluation should provide credible and useful information that enables the incorporation of lessons learned into the decision-making process of both recipients and donors. One of the seven guiding principles for conducting a post-disaster project by the Habitat International coalition with the UN is the transparent and effective monitoring of the recovery process [11].

3. Point of view of building users

Direct feedback from building users captures information concerning areas where effective improvements can be made, and what measures can be used to lead to greater cost-efficient projects [12]. Therefore, the affected community and the local government should participate in their maximum capabilities. There is no accomplishment if a post-disaster project claimed to be successful did not benefit the target beneficiaries because of its poor results. The accountability for public buildings can only be claimed from the funds coming from aid grants if endorsements from both parties, the donors and the users, are obtained [13].

According to Kernohan et al. [14], people who really know about buildings are the people who use them as they are the experts in what buildings can do. They are true informants on questions of building use and serviceability. User requirements to be applied in buildings should be considered while finalizing the overall requirements. The users include people who will occupy the buildings on completion and people who are likely to occupy the buildings successively during the rest of the life of the building.

4. Methodology

This study used two stages: the first stage was cluster random sampling, which categorized under building types each sampling cluster as a government office, school, and community health center in sub-district level. The second stage was simple random sampling without replacement. The building that will be evaluated is a public building built by the BRR using government funds. The area of study was Banda Aceh, where there are 50 buildings. The building samples are:

- For the school, the evaluation object is Sekolah Dasar (SD) for elementary school level, Sekolah Menengah Pertama (SMP) for junior high school, and Sekolah Menengah Atas (SMA) for senior high school.
- For the health center, the evaluation object is Pusat Kesehatan Masyarakat (Puskesmas) or the community health center.
- For the government office, the evaluation object are kantor kecamatan or a sub-district government office, and kantor desa/lurah or a village office.

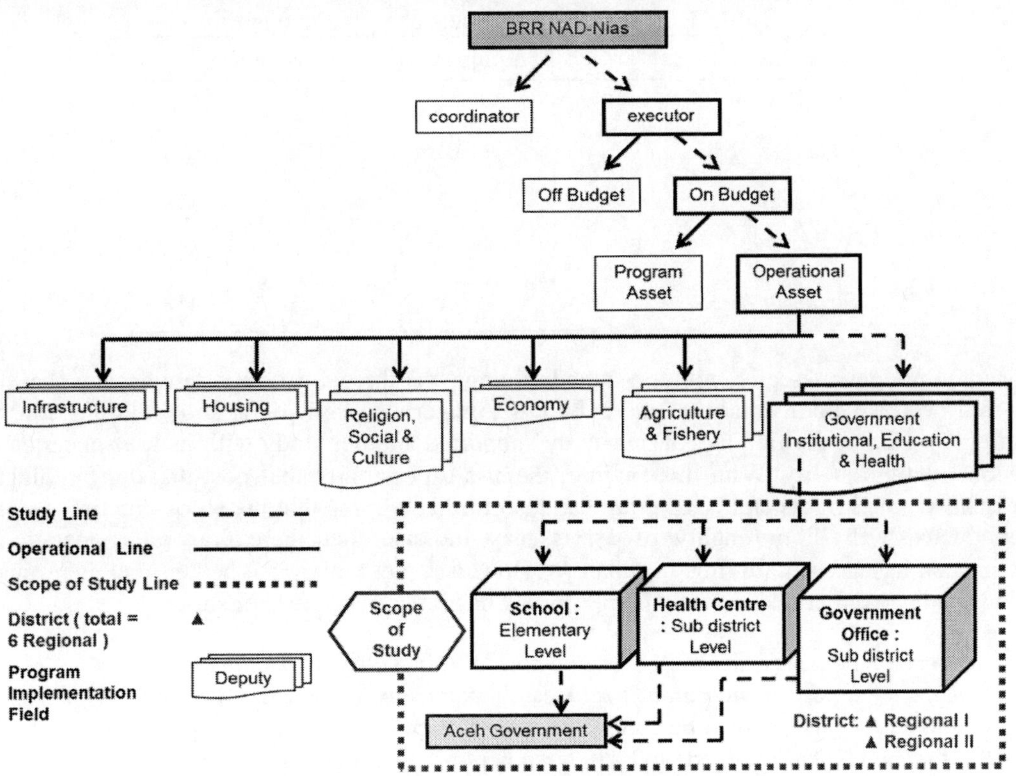

Figure 2. Scope of the Study

The research uses the quantitative approach and measures the evaluation objective based on respondent judgment. This study uses description methods and descriptive statistics, including median and mode, for each variable. Finally, the analysis was compiled from each variable for the final output. Descriptive measurement assesses the program participants and considers the ideas of building users regarding post-disaster asset evaluation. Frequencies are used to measure each variable.

5. Analysis

5.1. Background of Buildings and Respondents

In the first stage of cluster random sampling, from 50 buildings, this study managed to cover 45% of the buildings comprising the government office, school, and community health center built in Banda Aceh at a sub-district level. The government office represented 39% of the buildings, the school represented 67% of the buildings, and the community health center represented 60% of the buildings. The total value of the selected buildings is USD 1,727,954. For the building status, 88% of the buildings were completed and occupied. The research participants or respondents were building users. The majority of the respondents were above 25 years old, male, and Acehnese, which was defined as being born in Aceh and or staying in Aceh for more than 10 years. From 50 target respondents, only 27 returned the questionnaire for a response rate of 54%.

5.2. Doubt for the quality and condition of post-disaster assets in Aceh

Table 1 shows the percentage output for research statement number one. In total, 72% of respondents disagreed with the questionnaire statement. Only 20% of respondents agreed, and 8% of respondents remained neutral. Therefore, majority of building users doubted post-disaster building quality and condition.

Table 1. Reconstruction assets quality and condition

Description	Frequency	Percentage
Strongly Disagree	4	16%
Disagree	14	56%
Neutral	2	8%
Agree	4	16%
Strongly Agree	1	4%

Based on previous research with the same buildings, the three most important aspects that building users said were in poor condition were building materials, sanitation, and building construction strength [13]. Hence, the findings of this study supported another study with its user uncertainty with the reconstruction quality. With this finding, there is a possibility that post-disaster buildings could suffer quality issues in general. This point can also further assume that the reconstruction process was not conducted well. Poor quality of assets may indicate that there was poor operational and maintenance management during occupancy. However, determining whether the poor quality of buildings had any direct relation with inappropriate handover needs further study.

5.3. Need to conduct asset evaluation even the buildings are already occupied

Table 2 displays the percentage output for research statement number 2. Overall, 92% of respondents agreed that asset evaluation must be conducted even if a building has been occupied for several years. Only 4% of respondents disagreed, and 4% were neutral. This finding clearly shows that the majority of users endorsed building evaluation.

Table 2. Necessity conducting assets evaluation even Aceh government occupied the asset

Description	Frequency	Percentage
Disagree	1	4%
Neutral	1	4%
Agree	17	68%
Strongly Agree	6	24%

The Aceh Institute (AI) recommended an independent institution to assist in the transition process and to perform asset mapping, asset verification, and asset evaluation [6]. In 2008, the World Acehnese Association (WAA), with its headquarters in Denmark, demanded to the audit of the BRR [15]. However, BRR rejected the idea of external and independent evaluation but decided to do the handover process with only internal verification [1].

5.4. Occupying assets that have no proper handover and external project evaluation

Table 3 shows the finding for the last research statement. The results are divided into two big groups, and 44% respondents disagreed, and 48% of respondents agreed, while only 8% of respondents remained neutral.

Table 3. Aceh government-occupied assets where there was no evaluation by an independent committee during handover

Description	Frequency	Percentage
Strongly Disagree	6	24%
Disagree	5	20%
Neutral	2	8%
Agree	10	40%
Strongly Agree	2	8%

In the context of a state of emergency in post-disaster periods, there are doubts about whether the Aceh government should receive the assets even without an accurate handover and an evaluation by an independent committee. These measures must be implemented because project evaluation takes time even when assets are urgently needed to resume operations at Aceh after the disasters. Government assets are important for the basic needs of the communities as well as the presence of the government. Although time constraints will be always critical during post-disaster, the sustainability of the project should not be neglected. Based on the Ministry of Finance Regulation of Indonesia Numbers 62/2008 and 134/2009, these assets must be turned over to the Aceh government. Unfortunately, in the final report of transition team under the Aceh Governor Transformation Programme (AGTP) funded by the United Nations Development Programme (UNDP) in 2009, none of the Aceh provincial reconstruction assets was transferred to the Aceh provincial government [16].

6. Conclusion

As a preliminary study, we captured the point of view of building users regarding the quality of reconstruction assets, the necessity for conducting evaluation during occupancy, and the decision of the Aceh government to receive assets without proper handover. Most of the users doubt the quality of the buildings and support building evaluation even though they have already occupied the buildings. Doubts on building quality can be the starting point for an external and independent evaluation even during occupation periods. Meanwhile, the decision of the Aceh government to accept assets without an unclear handover process separated building users according to the users who agreed and disagreed. The consideration is the urgency to utilize these assets as soon as possible after post-disaster. Changing the past is impossible, but meaningful lessons could be learned from the enormous efforts on disaster reconstruction in Aceh. Perhaps conducting an external and independent evaluation during project handover can result in improvement in the future.

Further studies based on this survey need to focus on post-disaster evaluation during the occupancy period. These studies should concentrate on what type of evaluation is suitable, and what procedure should be followed during post-rehabilitation and reconstruction in Aceh.

References

[1] BRR NAD-Nias, "Asset Management and Administering Results for Rehabilitation and Reconstruction (On-Budget)," Center for Accounting and Asset Management, Ed., ed. Banda Aceh: BRR NAD-Nias, 2008.

[2] Badan Kesinambungan Rekonstruksi Aceh, "Kerangka Kerja Kebijakan dan Rencana Aksi Kesinambungan Rekonstruksi dan Percepatan Aceh 2010-2012," Badan Perencanaan Pembangunan Daerah Aceh, Ed., ed. Banda Aceh: Pemerintah Aceh, 2009.

[3] IFRC-RCS, "The Tsunami Legacy," Tsunami Global Lessons Learned Project Steering Committee2009.

[4] Bambang and Firdaus, *Budget Tracking: Aceh Rehabilitation and Reconstruction*. Banda Aceh: GERAK Aceh, 2006.

[5] Guerin and Bill. (2006, After the tsunami, waves of corruption in *Southeast Asia*. Available: http://www.atimes.com/atimes/Southeast_Asia/HI20Ae01.html

[6] The Aceh Institute, "Post 2009 completion of reconstruction work in Aceh," The Aceh Institute, Banda Aceh2010.

[7] M. Aaronson, "Designing for disasters: create long-lasting change," *RSA Journal*, vol. 152, pp. 40-7, 2005.

[8] P. V. Hidellage and V. Pullenayegem, "Adapting Strategies to Meet Needs in Post-Disaster Reconstruction," in *4th International i-Rec Conference*, 2008.

[9] B. Baradan, "Analysis of The Post-Disaster Reconstruction Process Following Turkish Earthquakes, 1999," in *Proceeding of 3th i-Rec International Conference on Post-Disaster Reconstruction "Meeting stakeholder interests, analysis of post-disaster reconstruction"*, Florence, Italy, 2006.

[10] E. U. Austrian Development Agency. (2009). *Guidelines for Project and Programme Evaluations* Available: https://www.oecd.org/development/evaluation/dcdndep/47069197.pdf

[11] B. J. Batra and S. Chaudhry, "International Human Rights Standards on Post-disaster Resettlement and Rehabilitation," Habitat International Coalition – Housing and Land Rights Network (HIC-HLRN) – People's Movement for Human Rights Learning (PDHRE)2005.

[12] G. K. Thampi, "Ensuring Effective Project Monitoring and Evaluation," presented at the Curbing Corruption in Tsunami Relief Operations, Jakarta, Indonesia, 2005.

[13] A. Adamy and A. H. Abu Bakar, "Asset Evaluation for Post-Disaster Project Built by Central Government in Banda Aceh Indonesia," *Asian Journal of Environment and Disaster Management*, vol. 3, pp. 405–423, 2011.

[14] D. Kernohan, J. Gray, J. Daish, and D. Joiner, *User Participation in Building Design and Management*. Oxford: Butterworth-Heinemann Ltd, 1992.

[15] S. Mukarram, "Acehnese People In Denmark Ask For Audit of BRR," in *The Hague* ed: ANTARA News, 2008.

[16] Transition Sector, "Inventory report asset rehabilitation and reconstruction of Aceh," Assistance Team for Aceh Governor Banda Aceh 2009.

Structural characterization of mixed local material for a ceramic porcelain insulator

H Aripin [1,*], I Usrah[1], N Busaeri [1], R Priyadi [2], Y Hermawan [3], I N Sudiana [4], I M Joni [5] and S Sabchevski [6]

[1] Department of Electrical Engineering, Faculty of Engineering, Siliwangi University, Jl. Siliwangi 24, Tasikmalaya 46115, Indonesia.
[2] Department of Agrotechnology, Faculty of Agriculture, Siliwangi University, Jl. Siliwangi 24, Tasikmalaya 46115, Indonesia.
[3] Faculty of Education, Siliwangi University, Jl. Siliwangi 24, Tasikmalaya 46115, Indonesia.
[4] Department of Physics, Faculty of Mathematics and Natural Sciences, University of Haluoleo, Kendari, Southeast Sulawesi 93132, Indonesia
[5] Nano Technology and Graphene Research Center (NTGRC), Padjadjaran University, Bandung, West Java 45363, Indonesia
[6] Lab. Plasma Physics and Engineering, Institute of Electronics of the Bulgarian Academy of Sciences, 72 Tzarigradsko Shose Blvd., Sofia 1784, Bulgaria.

[*] Email: aripin@unsil.ac.id

Abstract. A novel type of ceramic porcelain insulator material was produced by mixing kaolin with zeolite. The composition was prepared by adding a controllable amount of kaolin ranging from 30 wt% to 100 wt% into the zeolite. Then a series of samples were sintered at 1200°C. The influence of the composition on the structural properties was studied in detail using X-ray diffraction (XRD), Fourier transform infrared (FTIR) spectroscopy and Scanning Electron Microscopy (SEM). The results show that the microstructure of porcelain body is characterized by the presence of elongated mullite grains. It was found also that the decrease of the kaolin content in the porcelain body leads to a decrease in the crystalline phase of mullite. The analysis of the structural properties of such ceramics indicates that zeolite could not be recommended as an appropriate material for manufacturing of high-voltage ceramic porcelain insulators by mixing it with kaolin.

1. Introduction

Porcelain is being considered as a ceramic material, which has vitreous characteristics, i.e a high degree of melting on firing, porosity lower than 0.5%, and glass content greater than 40% [1]. Additionally, the porcelain possesses high hardness, low electrical and thermal conductivities, and brittle fracture [2]. It is known as a primary material for high-quality vases [3], tableware [4], decorative objects [5] and electrical insulators [6]. A promising and actively studied variety of the latter is the insulator material used in electrical power transmission systems due to the high stability of its electrical, mechanical and thermal properties in the presence of harsh environments.

Many studies have been carried out in order to investigate the properties of different porcelain materials used for electrical insulators and their dependence on the technological parameters, such as the temperature and time of the thermal treatment, chemical composition, and starting materials. Ezenwabude and Madueme [7] have prepared porcelain insulators with clay, feldspar, and quartz as starting materials. It has been found that the composition of 50% clay, 20% feldspar and 30% quartz with an appreciable insulation breakdown voltage of 45 V is appropriate for the manufacturing of electrical porcelain insulators. Ece and Nakagawa [8] have investigated the influence of various factors on the strength of porcelain insulators. Among them are the acicular mullite interlocking as well as mullite amount and size. Such observation is supported by the results of Arman Sedghi et al [9]. They have reported that the increase in the alumina content in porcelain material increases the mullite phase which, in turn, increases the bending strength. In the research carried out by Dunia and Al-Nasrawy [6], the acceptable electrical performance of the porcelain bodies has been achieved decreasing the glassy phase. These studies have demonstrated that the formation of the mullite and limitation of the glassy phase and unresolved quartz particles could effectively increase the strength and reliability as well as the electrical performance of the porcelain insulators.

The intention to use kaolin as a raw material for ceramic porcelain insulators is motivated not only by its abundance and low price but also by its plasticity during firing, which provides adequate mechanical strength for both the green and the dried body pieces. At the same time, a blend of clay and quartz still has some technical issues when it is used for fabricating ceramic porcelain insulators. The detailed analysis [10-11] of the experimental data shows that excessive plasticity can cause problems at the forming stage and make it difficult to achieve the required body permeability. Therefore, the proper selection of the raw material is very important for the further improvement of the technology for production of high-quality ceramic insulators. In this paper, we present the results of our study on the structural properties of a ceramic porcelain insulator derived from locally available materials (kaolin and zeolite). The purpose of this study was to investigate the effects of the replacement of quartz by zeolite on the properties of kaolin ceramic and on the underlying structural changes.

2. Experimental Procedure

2.1 Raw material preparation.

The raw materials used to prepare electric porcelain are kaolin and zeolite. They were obtained from Tasikmalaya (Indonesia) and selected manually. Then the raw materials were cleaned by washing with water, dried in the sun and grinded using a jar mill to pass through a 100 μm sieve. The samples were prepared according to the porcelain body formulations presented in Table 1. For example, the porcelain body sample K30Z was prepared by mixing 30 wt.% of kaolin in 100 mL deionized water, and then stirring at 80 °C for one hour. Next, 70 wt.% of the zeolite powder was added gradually into the mixture under continuous stirring at 50 °C for 6 hours. The resultant porcelain body powder was dried in an oven at a temperature of 120 °C for 2 hours and then stored in a desiccator for further treatment. The samples K50Z, K70Z, and K100Z were prepared following the same procedure.

Tabel 1. The porcelain body formulations

Sample	Mass percentages (wt.%)	
	Kaolin	Zeolite
30 wt% K (K30Z)	30	70
50 wt% K (K50Z)	50	50
70 wt% K (K70Z)	70	30
100 wt% K (K100Z)	100	0

2.2 *Sintering the synthesized porcelain body*

The powder samples were placed into a porcelain crucible and then inserted into a horizontal furnace, where they were sintered at a controlled heating rate of 10°C/min up to the temperature of 1200°C in air. Then the temperature was kept constant at 1200 °C for 2 hours. The cooling was performed by natural convection after turning off the electric furnace and leaving the samples inside.

2.3 *Method for characterization of the synthesized porcelain body*

The XRF measurements were carried out on a Philips PW2400 wavelength-dispersive spectrometer. The calibration was performed following the prescribed standard procedure. In addition, some correction factors for determining the influences of other elements were calculated. The measurement conditions were 24 kV – 100 mA, collimator mask 27 mm and Rh-X ray tube as an excitation source. A gas flow detector was used for controlling the rate of the gas flow. Furthermore, the experiments on the X-ray diffraction (XRD) of the samples have been carried out using a Smartlab X-ray diffractometer with filtered Cu Kα radiation at a wavelength of 0.15418 nm. The accelerating voltage and the applied currents were 40 kV and 30 mA, respectively. The diffraction patterns were registered over the range from 10° to 90° at a scan rate of 0.01 °/s. Fourier transform infrared (FTIR) spectra were obtained using a Varian 800 FTIR spectrometer (Scimitar Series model) in the wavenumber range of 400–4000 cm^{-1} with a spectral resolution of 4 cm^{-1}. All FTIR measurements were carried out at room temperature in a specular reflectance mode using the KBr pellet technique. The SEM photos of fracture surface of the samples were taken by using a scanning electron microscope (SEM) at a magnification of 10,000 times and at accelerating voltage of 15 kV. The detailed electrical strength test procedures are described elsewhere [12]. It was measured using a spherical and a plat brass electrode in liquid dielectric in the isolated chamber.

3. Results and Discussion

Table 2 shows the chemical composition of oxide contents for kaolin and zeolite from Karangnunggal. Both of the kaolin and zeolite contain SiO2 as the main constituents at about 42.20 wt.% and 59.30 wt.%, respectively. It is also accompanied by a significant amount of Al_2O_3 and Na_2O for kaolin and ZrO_2, Al_2O_3, Fe_2O_3, K_2O and CaO for zeolite. It can also be seen that kaolin and zeolite contain a small amount of TiO_2. The loss on ignition is associated with the presence of hydroxides, organic matter, and volatile components.

Table 2. Chemical composition of kaolin and zeolite from Karangnunggal

Chemical composition	Kaolin (%)	Zeolite (%)
SiO_2	42.20	59.30
TiO_2	1.10	0.70
ZrO_2	0.46	5.02
MgO	-	0.12
Al_2O_3	24.80	5.02
Na_2O	30.10	-
Fe_2O_3	1.11	8.47
K_2O	-	12.30
CaO	-	7.81
LOI	0.23	1.27

Figure 1a shows the XRD pattern of porcelain with a varying amount of kaolin and zeolite and sintered at 1200°C. In accordance with XRD data from Figure 1a, the sample of 100 wt.% consists of kaolin ($Al_2Si_2O_5(OH)_4$) at 2θ = 20.9° and 24.9° [13-14], cristobalite (SiO_2) at 2θ = 22.9° [15] and mullite ($3Al_2O_32SiO_2$) at 2θ = 27.9° [16]. The peak intensity of the kaolin and the mullite for the sample with a kaolin content 70 wt.% and below decreases gradually with the decrease of the kaolin content. The decrease of the peak is attributed to the formation of crystalline zeolite ($Mx/n[(AlO_2)_x(SiO_2)_y]mH_2O$) at 2θ = 27.2° and 30°, according to data from other studies [17]. The intensities of the zeolite peaks in the pattern increase with decreasing the kaolin content. From the above-mentioned data, we can conclude that the bending strength of porcelain decreases with decreasing the kaolin content since the latter correlates with a decreased amount of mullite formation and an increase of the cristobalite phase.

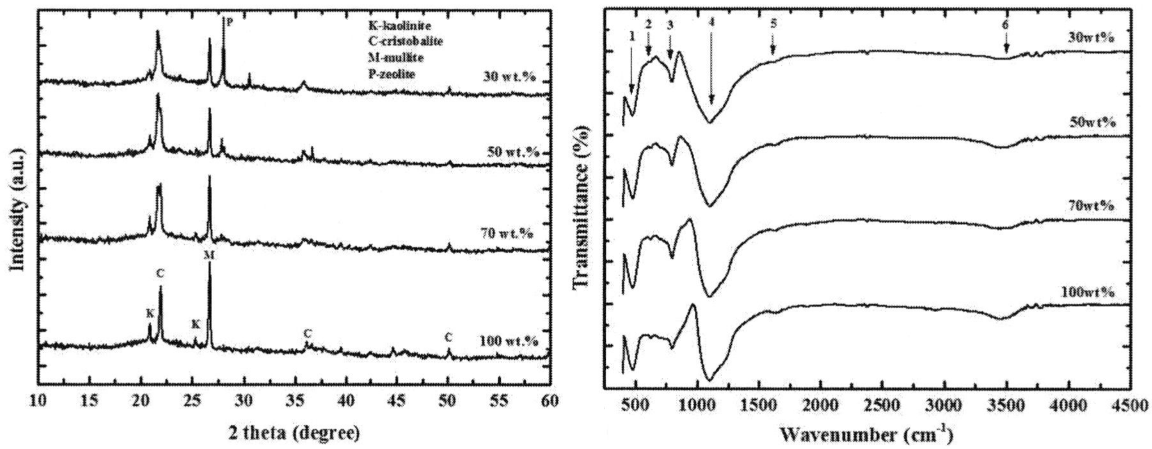

(a) (b)

Figure 1. (a) XRD patterns and (b) FTIR spectra for porcelain with a varying kaolin content and sintered at 1200°C.

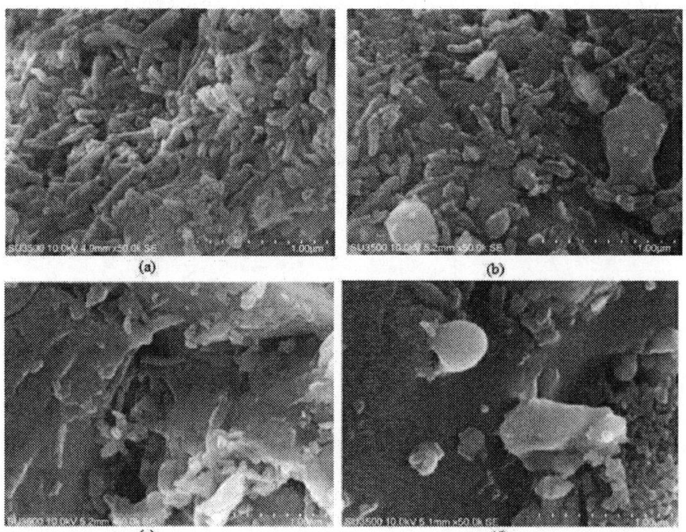

Figure 2. SEM images for porcelain with (a) 100wt%, (b) 70wt%, (c) 50wt%, and (d) 30 wt% kaolin and sintered at 1200°C.

Figure 1b shows the FTIR spectra for porcelain with varying kaolin content and sintered at 1200°C. For all samples, the band (peak 1) located at 464 cm^{-1} is assigned to tetrahedral T–O (T= Si or Al) bending vibrations, is common to the starting metakaolin [18]. Its intensity does not change substantially in the course of the zeolitization. The peak 2 that appears at 626 cm^{-1} corresponds to crystalline cristobalite [19]. The peak 3 at 800 cm^{-1} and the peak 4 at 1100 cm^{-1} can be assigned to the symmetric stretching vibrations of Si–O–Si and the anti-symmetric stretching vibrations of Si–O–Si [20], respectively. The absorption bands (peak 5) observed at about 1645 cm^{-1} was attributed to the bending H–O–H bond groups of adsorbed water molecules. The broadband (peak 6), at 3452 cm^{-1}, is attributed to the O–H stretching of hydroxyl groups that are present on the surface of the material [21]. It has been found that the integrated intensity of the symmetric stretching vibrations of Si–O–Si increases with decreasing the amount of kaolin. The big quantity of Si–O–Si bonds in the samples with smaller amounts of kaolin indicates that more zeolite particles facilitate the elimination of the SiO_2-rich phase so that the zeolite decreases the mullite grain growth in the porcelain body.

Figure 2 shows SEM images of porcelain with varying content of kaolin and sintered at 1200°C. The SEM image for the sample with 100 wt% kaolin illustrates that the micrographs are characterized by more elongated mullite grains surrounded by unsolvable cristobalite particles. In this sample, the elongated mullite grains are seen as major crystalline phases that are formed by using kaolin as the starting material. It may be due to the presence of a large amount of liquid phase during sintering. At 70 wt% kaolin, it can bee seen that the addition of zeolite provided the formation of a smaller amount of mullite grains in the sample. At 50 wt% kaolin, the elongated mullite grains gradually appeared. At 30 wt% kaolin, a compact microstructure with glassy phase was seen to be uniformly distributed in the sample. Then, the excess zeolite eliminated the SiO_2-rich phase then retarded the mullite grain growth at 1200°C. The results are supported by the detected results of the XRD patterns and FTIR spectrum.

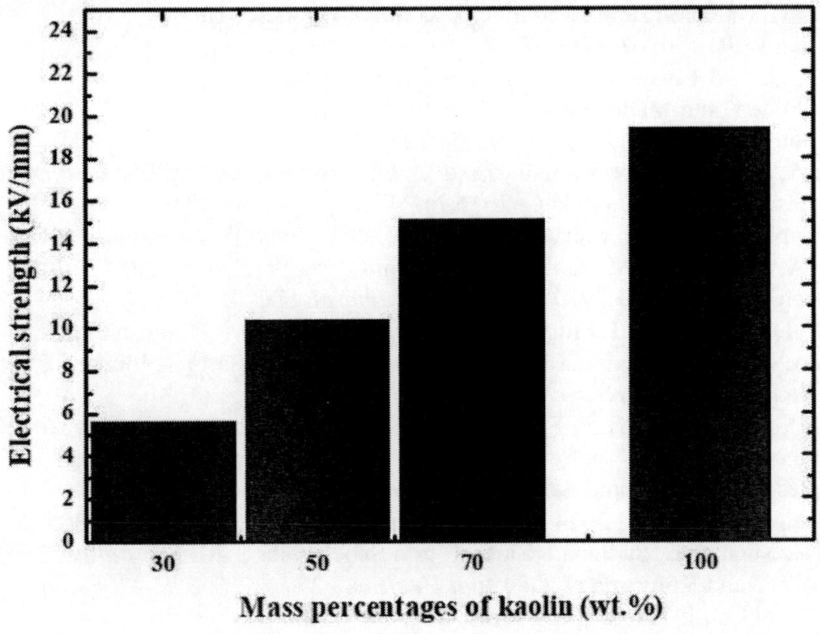

Figure 3. Electrical strength for porcelain with a varying kaolin content and sintered at 1200°C.

Figure 3 shows electrical strength for porcelain with varying kaolin content and sintered at 1200°C. Base on this figure, electrical strength increases with increasing kaolin content. There are at least two occurrences that result in increasing the electrical strength. The first one arises from an open pores reduction when the kaolin content increases in the sample. This behavior was attributed to the presence of calcium oxide (CaO) in zeolite. In this case, under sintering conditions it takes place the decomposition reaction between CaO and CO_2 gas from $CaCO_3$ [22] and then the open pores are ascribed to the empty sites left the released CO_2 gas. The amount of liquid phase produced by sintering SiO_2 does not fully fill the open pores, then it is responsible for decrease of electrical strength in the sample for higher content of zeolite. Another one stems from a decrease of the mullite phase and an increase of the cristobalite phase as detected by the XRD pattern. In the case, the presence of larger amount of cristobalite and smaller amount of mullite lead to a low electrical strength [6, 9].

4. Conclusion

A novel ceramic material for porcelain electric insulators has been produced successfully from kaolin and zeolite by controlling the amount of kaolin at 1200°C. The results presented in this paper show that incorporation of zeolite into the kaolin gives an appreciable effect on the microstructure of porcelain body. An interesting phenomenon has been observed analyzing the XRD pattern of samples. It has been found that the crystalline phase of mullite decreases as the zeolite content in the sample increased from 30 wt.% to 70 wt.%. Such interpretation is supported by the observed microstructure of the elongated mullite in the sample. Nonetheless, the overall results suggest that the replacement of kaolin by zeolite as a raw material could not be considered as a prospective route for the production of an appropriate material for high-voltage ceramic porcelain insulators.

References

[1] Perez J, Rincon, J and Romero M 2012 *Ceram. Int.* **38** 317.
[2] Callister D and Rethwich D 2008 *Fundamentals of Materials Science and Engineering* (Singapore: John Wiley & Sons, Inc.,)
[3] Jeoung-Ah K 2006 *J. Eur. Ceram. Soc.* **26** 1023.
[4] Boussak H, Chemani H and Serier A 2015 *Int. J. Phys. Sci.***10** 38.
[5] Colomban P 2013 *Arts* **2** 77.
[6] Dunia K, and Al-Nasrawy M 2016 *Iraqi J. Sci.* **57** 404.
[7] Ezenwabude E and Madueme T 2015 *Int. J. Mult. Sci. Eng.* **6** 28.
[8] Ece O, and Nakagawa Z 2002 *Ceram. Int.* **28** 131.
[9] Sedghi A, Riahi-Noori N, Hamidnezhad M and Reza Salmani M, 2014 *Bull. Mater. Sci.* **37** 321.
[10] Ajakor E, Anih L, and Ogwata C, 2015 *Int. J. Sci. Res. Publ.* **5** 1.
[11] Al-Araji N, Aal Dhabi L, and Alshammari W, 2014 *Int. J. Sci. Eng. Res.* **5** 1348.
[12] Sedghi A, Riahi-Noori N, Hamidnezhad N, and Reza Salmani M 2014 *Bull. Mater. Sci.* **37** 321.
[13] Gougazeh M, and Buhl J 2010 *Clay Miner. Mineral. Soc.* **45** 281.
[14] Zhao H, Deng Y, Harsh J, Flury M, Boyle J 2004 *Clay Clay Miner.* **52** 1.
[15] Aripin A, Mitsudo S, Sudiana I, Tani S, Sako K, Fujii Y, Saito T, Idehara T, Sabchevski S 2011 *J. Infrared Milli. Terahz Waves* **32** 867.
[16] Aripin H, Mitsudo S, Prima E, Sudiana I, Kikuchi H, Sano S, Sabchevski S 2015 Ceram. Int. **141** 6488.
[17] Rodrigues M, Souza A, and Santos I 2016 *Am. Chem. Sci. J.* **12** 1.
[18] Gougazeh M, and Buhl J 2014 *J. Assoc. Arab Univ. Basic Appl. Sci.* **15** 35.
[19] Aripin H, Mitsudo S, Sudiana I, Saito T, and Sabchevski S 2015 *T. Indian Ceram. Soc.* **74** 1.
[20] Shinohara Y and Kohyama N 2004 *Ind. Health* **42** 277.
[21] Yang S, and Gao L 2006 *J Am. Ceram. Soc.* **89** 1742.
[22] Kaur H, Kumar Bulasara V, and Kumar Gupta R 2015 *Desal. Water Treat.* **57** 15154.

Acknowledgments

This research was supported by a fund of the Siliwangi University through the Project of Research for Dosen Pembina in 2017 with contract number: 1140/D3/PL/2017.

IOP Conf. Series: Materials Science and Engineering **674** (2019) 012032 doi:10.1088/1757-899X/674/1/012032

Visualization of measurement of heart rate, oxygen saturation in blood and body temperature with a non-invasive method

M T Tamam[1,*], A J Taufiq[2], A Kusumawati[3] and D Setiawan[4]

[1, 2]Electrical Engineering Department of Universitas Muhammadiyah Purwokerto
[3]Medical Faculty of Universitas Muhammadiyah Purwokerto
[4]Pharmacy Faculty of Universitas Muhammadiyah Purwokerto
Jl. Raya Dukuhwaluh Kembaran, Banyumas, Jawa Tengah, Indonesia 53182

* Email: tamam@ump.ac.id

Abstract. Visualization of the measurement results of magnitude is a matter that will facilitate us in understanding the measurement results. Visual media is more interesting to most people than text media. This study aims to design and create an application system based on GUI (Graphical User Interface) that can display the form of heart rate waves, oxygen saturation in the blood, and body temperature. The measurement information is computed to produce a decision stored in the database. From the test results, the system can display the waveform of heart rate, oxygen saturation in blood and body temperature and diagnosis appears.

1. Introduction

The human body consists of various vital organs that are interconnected and have a unique and complex function. The heart as a vital organ that serves to provide oxygen throughout the body and cleanse the body of the metabolism (carbon dioxide). Heart disease is the number one cause of death in the world, so the need for early detection of this disease. The pulse can be measured at any point on the body where the vibrations in the arteries are sent to the surface frequently when it is pressed against a structure that lies beneath such a bone [1]. Pulse oximetry is a non-invasive method for measuring the percentage rate of hemoglobin saturation. This method uses the different wavelengths of red light (660 nm) and infrared light (940 nm) captured by the detection sensors after passing through the veins and capillaries at the tip of the index finger [2]. Pulse oximetry serves to manage long-term patients with oxygen therapy and to identify patients with congenital heart disease. Pulse oximetry has a high accuracy for detecting congenital heart disease in newborns [3].

Photoplethysmography is an optical technique of detecting cardiovascular pulse waves (heart) from the fingertips. By exploiting optical sensor reflection then warming the area around the skin, this is evidenced by the increasing beats component of photoplethysmography [4]. The design of monitoring of integrated signal monitoring system on a platform of the system on chip (SoC) and wireless technology in hospital with monitored parameter is ECG and body temperature then processed using microcontroller. Results from data processing are sent to the webserver using wireless communications [5]. Detection of beat pressure (pulse) is an integration of most analytical techniques for arterial blood pressure (ABP), intracranial pressure (ICP) and pulse oximetry signals (SpO_2). Pulse detection is used to estimate heart rate in ABP signals, to classify ICP morphology and to estimate blood pressure using pulse oximeter wave [6]. The use of sensors in e-health platforms connected to Matlab can display visually the quantities measured by the sensor, such as body temperature [7]. The

IOP Conf. Series: Materials Science and Engineering **674** (2019) 012032 doi:10.1088/1757-899X/674/1/012032

voltage changes or fluctuations detected by infrared LED mounted on the fingertips can be used to determine the heart rate [8]. Therefore, medical record media of heartbeat, oxygen saturation in blood and body temperature are computationally computed and study of observation approach of health condition of body without any surgery or physical or non-invasive, and is expected to prevent human death due to abnormalities heart early because of the less than optimal medical record and responsive.

2. Materials and Methods

2.1. Materials

Heart rate sensor and oxygen saturation in blood, body temperature sensor, Arduino Board Due, PC (Personal Computer), MULTISIM® software, BASCOM® SIMULATOR software, Matlab software, MySQL and Database Application, and Visual Basic 6.0.

2.2. Methods

Acquisition of data from heartbeat sensors, oxygen saturation in blood and body temperature placed on the fingers, presentation of measurement data on computer media that apply visual orientation and storage of database system, then by analyzing work characteristics and suitability of each component supporters of the design idea obtained the initial design of a stable and optimal in terms of mechanical, electronic and ergonomic (Figure 1 and 2).

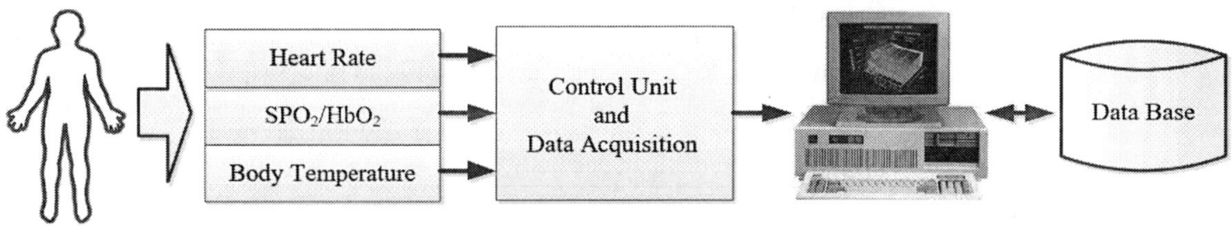

Figure 1. Block diagram of the system

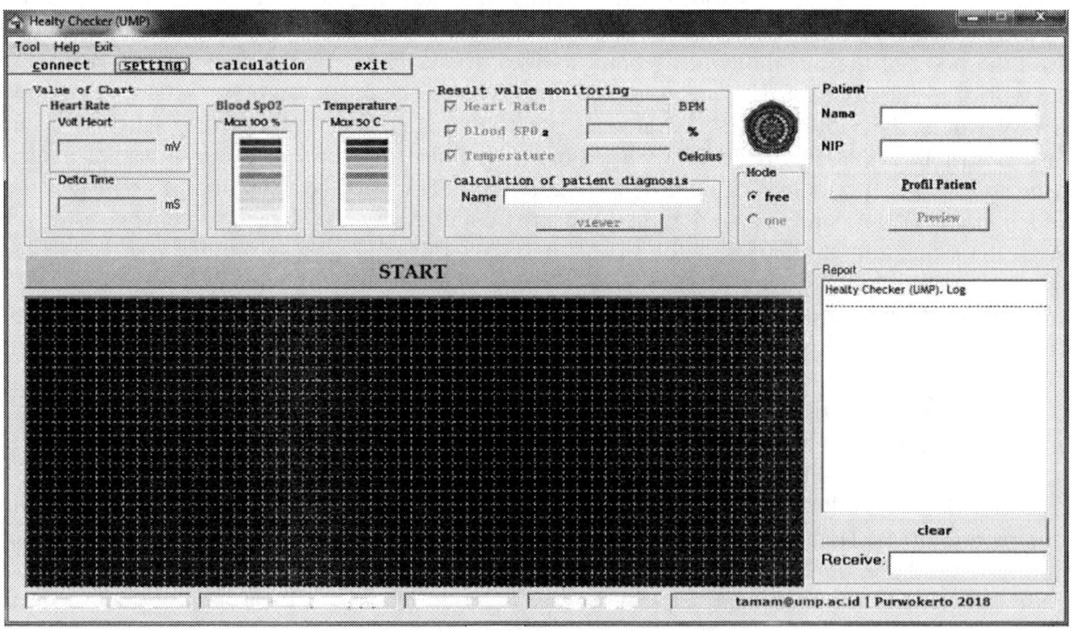

Figure 2. Visual design

3. Results and Discussion

A system is a form of visualization of the results of measurement of heart rate, oxygen saturation in blood and body temperature. This tool consists of heartbeat sensors, oxygen saturation in blood and body temperature mounted on the fingers. The signals that the sensor captures are further amplified and processed by the microcontroller.

Furthermore, with the device GUI (Graphical User Interface) Matlab data measurement results are displayed on the monitor screen. In addition, the data measurement results are also compared with the database that has been stored, so that from the measurement data results appear diagnosis. Table 1 shows the results of system testing of twelve respondents with sex and age varying.

Table 1. System test results

No.	Respondent	Sex	Age (years)	Heart Rate (BPM)	SPO_2 (%)	Temperature (°C)
1	Respondent 1	Male	48	82	98	32
2	Respondent 2	Female	12	81	99	34
3	Respondent 3	Female	54	82	97	35
4	Respondent 4	Male	59	81	94	35
5	Respondent 5	Female	5	94	98	35
6	Respondent 6	Male	45	80	98	30
7	Respondent 7	Female	34	93	97	34
8	Respondent 8	Mele	6	84	97	35
9	Respondent 9	Male	9	84	97	35
10	Respondent 10	Female	16	83	97	35
11	Respondent 11	Male	25	85	96	35
12	Respondent 12	Female	25	82	97	36

Figure 3 shows the view of a patient's biodata whose registration number, name, date of birth, sex, address, age, and other information are required. Biodata is stored in the database and one day can be changed if there are data changes.

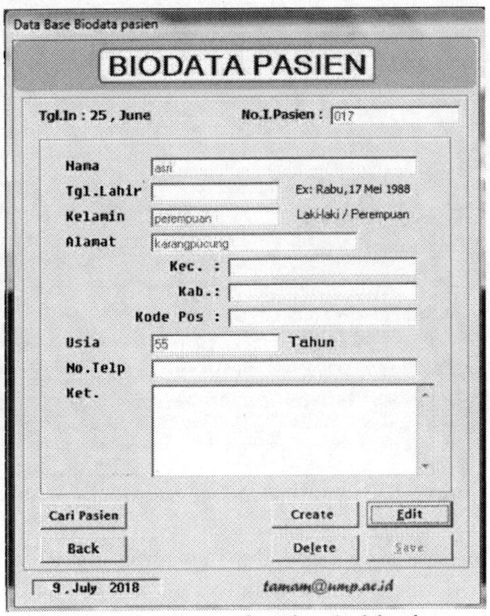

Figure 3. View of patient's bio data

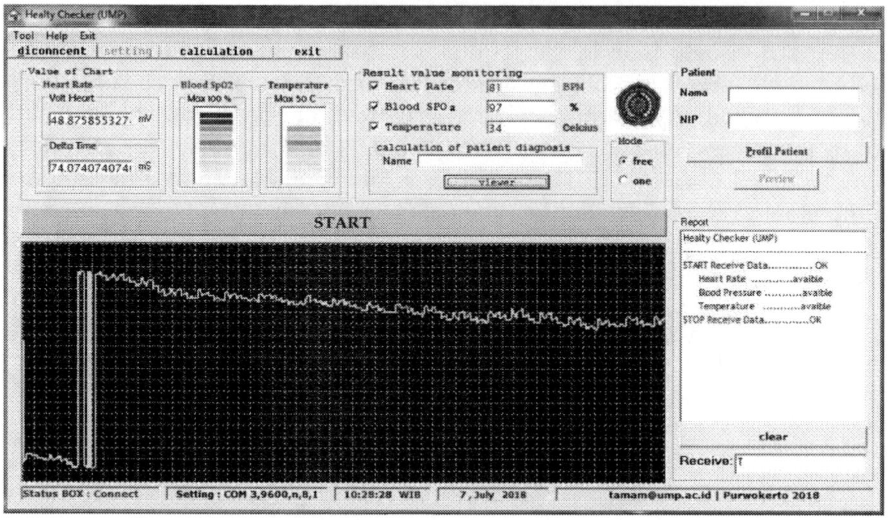

Figure 4. Visualization of measurement results

Figure 4 shows the visualization of measurement results of heart rate, oxygen saturation in blood, and body temperature. The apparent heartbeat waveform shows a voltage of 48.876 mV and a time of 74.07 mS. If converted to frequency scale to 13.52 Hz.

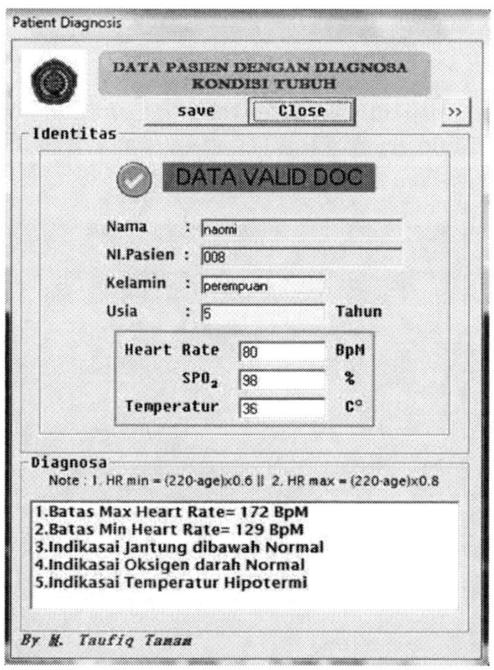

Figure 5. Visualization of diagnostic results

Figure 5 shows the diagnosis arising from the measurement of heart rate, oxygen saturation in blood and body temperature. The result of the diagnosis is a reference for medical personnel (doctors) to determine further treatment.

4. Conclusion

From the test results can be concluded that the system is able to work in accordance with the planned, the system can display the form of heart rate waves, oxygen saturation in blood and body temperature. The measurement data is stored in the database so that it will produce an appropriate diagnosis.

Acknowledgments

The authors are grateful to the Directorate of Higher Education, Kementrian Riset Teknologi dan Pendidikan Tinggi Republik Indonesia in providing a grant to conduct this research.

References

[1] Smeltzer, S.C. and Bare, B.G. (2005). *Buku Ajar Keperawatan Medisal Bedah.* Edisi 8 Vol.2. Jakarta: ECG. Harold S. Koplewich.

[2] Tavakoli, M., L. Turicchia, and R. Sarpeshkar. (2010) *"An Ultra-Low-Power Pulse Oximeter Implemented With an Energy-Efficient Transimpedance Amplifier."* Biomedical Circuits and Systems, IEEE Transactions On 4.1 : 27-38

[3] Bimbaum, S. (2009). *Pulse oximetry: identifying its applications, coding, and reimbursement.* American College of Chest Physicians; 135:838-841.

[4] Mendelson, Y. (2002). *Non-invasive pulse oximetry utilizing skin reflectance photoplethysmography.* Biomedical Engineering, IEEE Engineering in Medicine and Biology Society, Vol: 35, No: 10, pp:798-805.

[5] Lin, J.S., Huang, S. Y., Pan, K. W. and Liu, S. H. (2009), *A physiological signal monitoring system based on an SoC platform and wireless network technologies in homecare technology*, Journal of Medical and Biological Engineering, 29: 47-51.

[6] Beattie, Z.T. (2008), *Algorithm for automatic beat detection of cardiovascular pressure signals*, Engineering in Medicine and Biology Society, 30th Annual International Conference of the IEEE, pp:2594 – 2597.

[7] Mečiak, V. Blaho, M. Mrafko, L. Mudráková, T. (2015). *Sensor-Based Platform E-Health Connection With Mathlab.* Actas del 23 Technical Computing, Praga.

[8] Tamam, M. T. Taufiq, A. J. Kusumawati, A. (2017). *Design A System Of Measurement of Heart Rate, Oxygen Saturation in Blood and Body Temperature with Non-invasive Method.* International Conference on Engineering and Applied Technology (ICEAT), Lombok.

IOP Conf. Series: Materials Science and Engineering **674** (2019) 012033 doi:10.1088/1757-899X/674/1/012033

The current and voltage curves simulation solar module using IV tracer networks

Cholish[1,*], I D Sara[2], Y Away[3], A Azis[4] and Noorly E [5]

[1,4,5] Departement Electrical Engineering University of Muhammadiyah Sumatera Utara, Jl. Kapten Mukhtar Basri,BA No. 3 Postcode: 20238, Medan, Indonesia
[2,3] Magister of Electrical Engineering University of Syiah Kuala, Jl. TGK. Shaykh Abdul Rauf No. 7 Postcode: 23111, Banda Aceh, Indonesia

* Email: cholish@umsu.ac.id

Abstract. Simulation is done to see that the output parameter value of a solar module. Output produced a solar module as measured manually does not provide information that solar modules working at optimal power. Information on the performance of a solar module can be done with a piece of equipment called IV Tracer. IV Tracer work to collect data by varying the value of the working voltage of solar modules ranging from short circuit to open relations. IV Tracer circuits using MOSFETs IRF540 with snubber circuits produce 52 samples of data for 506 mS. The shape of the characteristic curve of current and voltage (IV) produced in accordance with the datasheet PSIM with the fill factor of 0.71.IV Tracer circuit can generate output parameters according to the theory of power changes to variations in temperature values of 0.05% and variations in light intensity of 0.14%. IV Tracer circuit with MOSFET switching system and may dampen stepping snubber circuit voltage of -0.2 V in accordance with the solar module to be tested.

Keywords: Solar Module, Current and Voltage Characteristic Curve (IV), Tracer IV, PSIM

1. Introduction

The potential of solar energy is strongly influenced by climate change that occurred. Knowing the characteristics of a solar module needs to be tested. Parameters that can affect the results of the above output is the light factor, the temperature of the solar module, the angle of incidence of light and solar radiation [1][2].

Testing solar modules with a variety of conditions in a given environment is performed to determine the nature of the solar modules produce electrical energy. Measurement of the value of the output of a solar module that is done manually will not give information about the families, the solar module if the parameter is a maximum performance value. Energy produced by a solar module illustrated in a characteristic curve shape current and voltage (IV) [3,4]. The curve contains information about important parameters that are owned by a solar module. the parameters contained in the current and voltage characteristic curve (IV), among others; The maximum power (Pmax), maximum voltage (Vmax), the maximum current (Imax), the open-circuit voltage (Voc), short circuit current (Isc) and a fill factor (fill factor / FF) [1].Current and voltage characteristic curve (IV) a solar module is obtained by tracking the value of output using a piece of equipment called IV Tracer [5,6,7,8].

IV Tracer works by varying the working voltage of solar modules ranging short circuits to open circuit relationships [4].The circuit loading on the previous study using two (2) pieces of the MOSFET to produce forms of current and voltage characteristic curve (IV) a solar module [4,5,9,10]. In this study, the simplification of the circuit using a single MOSFET as a switching system with a snubber circuit as a load for varying the voltage. IV Tracer circuit is simulated in the PSIM software to show the form of current and voltage characteristic curve (IV) on the module according to the solar module datasheet.

Simulations carried out with a view to obtaining the conditions according to standard tests (STC) is the light intensity of 1000 W / m2 and a temperature of 25 0C [1,6,8]. Some environmental conditions are tested in simulations carried out to demonstrate the influence of the characteristic curves of current and voltage (IV) on the solar module. Circuit simulation performed by using a polycrystalline solar module types with a small capacity is expected to provide comprehensive information about the shape of the characteristic curve of current and voltage (IV) according to the datasheet.

2. Literature Review

2.1. Characteristics of Solar Modules

There are some important points of the characteristic curve of current and voltage (IV) which shows the performance of solar modules on certain conditions (Figure 1), namely [9,11,12]:

1. Rated maximum power point (MPP)
2. Short-circuit current (Isc)
3. Open circuit voltage (Voc)

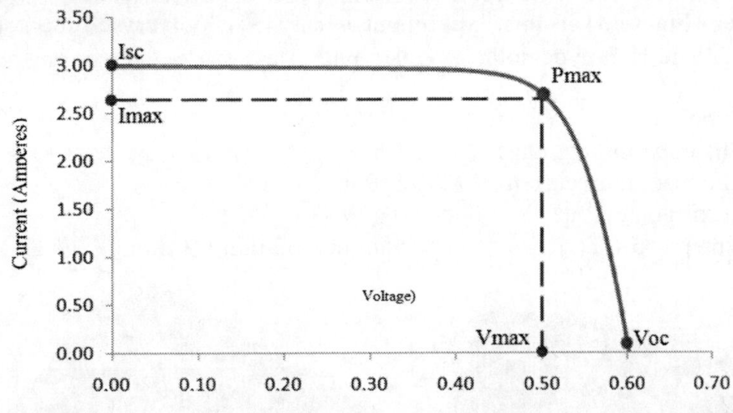

Figure 1. The characteristic curves I - V on solar modules [11,13]

2.2. Switching systems

Snubber circuit (Figure 2) needed to reduce interference when switching occurs. In the DC-DC switching circuit is needed once a snubber circuit to cut the VDS voltage having a voltage ripple (spike) is high on the MOSFET [3,14,15,18,19]. Supplied switching circuit after the voltage on the input side, the voltage spike caused by leakage inductance (inductance leak) is high enough so that the required snubber circuit which serves to reduce the voltage spike. In the snubber circuit greater the value of the capacitors used, the greater the voltage is cut [3,16,17,19]. But the value of the capacitors used must conform with the series resistor with a capacitor and diode are mounted parallel fast.

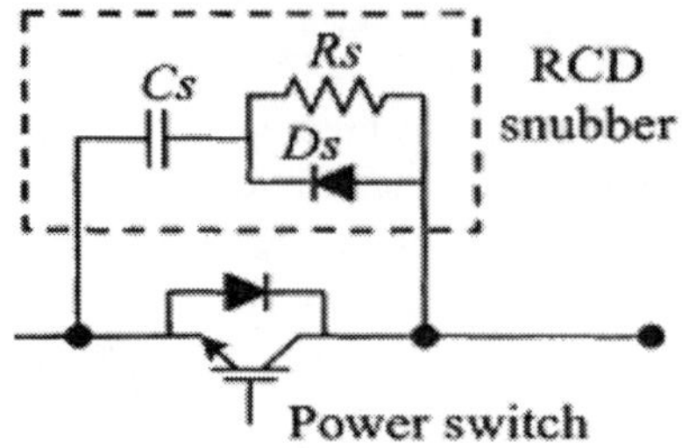

Figure 2. Snubber circuit [3,18]

The value of Rs and Cs values can be calculated with the following equation [3].

$$R_S = \frac{t_{on}}{5C} \quad (2)$$

$$Cs = \frac{I_L t_f}{2V_f} \quad (3)$$

Rs = custody snubber (Ω)
Vf = Voltage MOSFET (V)
ton = The lifetime of the MOSFET (S)
Cs = snubber capacitor (F)
IL = Current MOSFET Expenses (A)
tf = Time Fall MOSFET (S)

3. Method

In this research, the circuit simulation using Power Simulation software (PSIM). PSIM is a software that is used to determine the condition of the circuit in detail the parameters that can be adjusted. The results of the current and voltage characteristic curve (IV) produced is expected to show the value of the parameter characteristics of solar modules according to the datasheet.

The solar modules (Figure 3) in this experiment is a type of polycrystalline solar modules with little power is 5VDC. The Prototype solar modules under test can be shown on the specification as follows [20]:

Type	: GH65x65
The maximum voltage	: 5.5 V
The maximum current	: 110 mA
maximum output	: 0.4 W
dimension	: 65 mm x 65 mm x 3 mm.

Figure 3. The solar modules are tested

Modules such as the above parameters will be included in the conditioning based on simulation software PSIM input. Further by simplifying the design using MOSFETs and snubber circuits. The IV Tracer schematic drawing of the solar module with a series of short circuits is shown in figure 4.

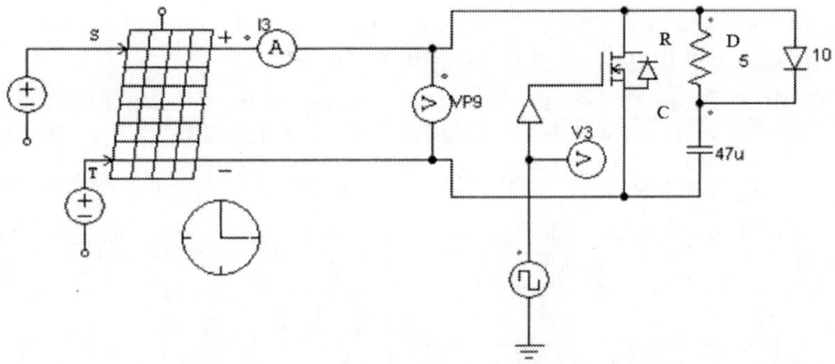

Figure 4. The series of solar modules IV Tracer loading

Design series solar modules using n-type MOSFET IRF540N the snubber circuit. The circuit is applied is a system of short circuit of solar modules. The process of switching voltage spikes that cause snubber designed to minimize disturbance to the combination of a high-frequency diode with the type of FR-207. A snubber circuit is designed based on the value of the component datasheet IRF540 MOSFET ie IL = 16A ton = 11 ms, tf = 35 ms Vt = 6V so that it can be determined as a matter of the following; [3]. Based on the calculations above, the selected value component approach these results with 47uF capacitor and resistor values 5Ω.

4. Results and Discussion

4.1. The circuit simulation Tracer IV

Circuit simulation applied to the program PSIM to see the shape of the current and voltage curves of solar modules produced by the circuit in accordance with the parameters contained in the datasheet. IV Tracer circuit testing is done with a few variations of the environmental conditions that the light intensity and temperature.

In testing simulated by testing solar module parameters to variations in the environment that affect. Results obtained through simulation to temperature variation showed a decrease in the voltage parameter value is quite large. Increasing the value of testing temperature between 25 0C, 45 0C and 65 0C with a fixed light intensity conditions.

The power parameters showed an average decline of 0.05% to the value of the voltage drop of 0.59%. The overall value can be shown in table 1.

Table 1. Parameters of test results based on changes in the value of the temperature

No.	Parameter	Condition 1	Condition 2	Condition 3	The average change
1	Temperature (0C)	25	45	65	20
2	Irradiant (W / m2)	1000	1000	1000	0
3	Pmax (W)	0.430	0.380	0.330	0.050
4	Vmax (V)	4.500	3.930	3.330	0.590
5	Imax (A)	0.095	0.097	0.100	0.002
6	Voc (V)	5.500	4.910	4.320	0.590
7	Isc (A)	0.110	0.113	0.116	0.003

In the form of current and voltage characteristic curve (IV) showed a large enough voltage drop, in this case, is different from the current which shows very little improvement. This occurs in accordance with that contained in the theoretical study with results in line with the solar module to be tested.

In testing the variations in light intensity showed a decrease in current parameter values is quite large. Impairment testing of light intensity between 1000 W / m2, 700 W / m2 and 400 W / m2 with fixed temperature conditions. Power parameters showed an average decline of 0.14% to the value of the current decline of 0.03%. The decrease is not too large due to the current capacity of solar modules tested specifications. The overall value can be shown in table 2.

Table 2. Parameter results based on changes in light intensity value

No.	Parameter	Condition 1	Condition 2	Condition 3	The average change
1	Temperature (0C)	25	25	25	0
2	Irradiant (W / m2)	1000	700	400	300
3	Pmax (W)	0.430	0.280	0.150	0.140
4	Vmax (V)	4.500	4.360	4.090	0.210
5	Imax (A)	0.095	0.065	0.035	0.030
6	Voc (V)	5.500	5.350	5.100	0.200
7	Isc (A)	0.110	0.077	0.045	0.032

The current scale of impairment in accordance with the capacity of solar modules listed on the dataset. In the voltage values also occur down outside generated so that the parameter changes in light intensity changes substantially to the power generated.

4.2. Switching circuit perform

The capacitor current at the time of switching showing the conditions of charge and discharge of the load coming from the solar module. Rated current and voltage that occur in solar modules, capacitors, and resistors show the same value. MOSFET in that it has an important role when performing switching current shows rapid reduction in the figure below. The parameter performs output current in the circuit can be shown in the following figure.

Based on the current output parameters of the solar module and circuit IV Tracer shows the same value in the short-circuit current is 0.110 A with switching times for 506 mS. Based on the results of the comparison between the circuit and the IV Tracer happens to solar modules provide the same value. Factors charging the capacitor showed a very fast along with the MOSFET switching process is done.

The snubber circuit and MOSFET show the condition of the properties as inverting the value that these disorders can be mitigated by a combination of a series resistor, a diode with a capacitor. Inverting the value given by the circuit at 5.48 V. Figure 5 shown a graphical form output voltage snubber circuit.

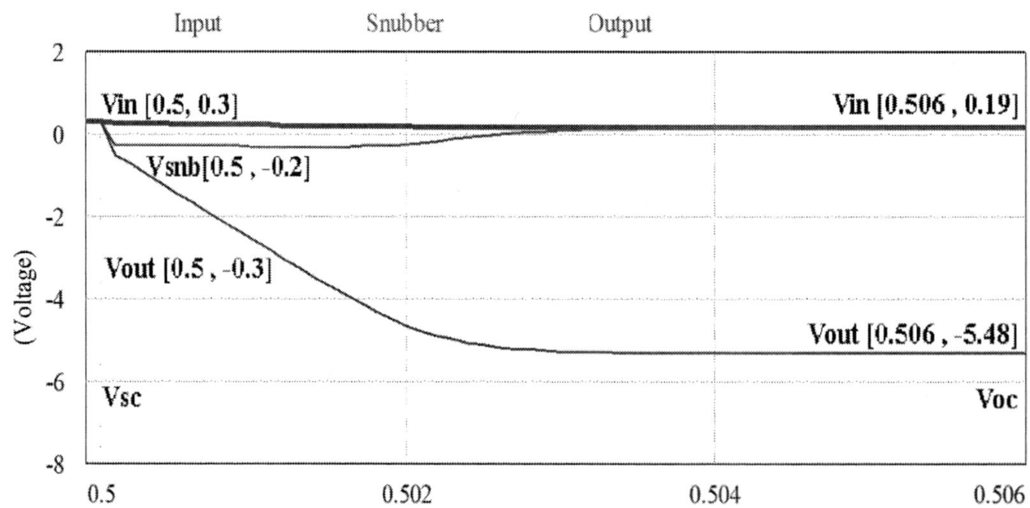

Figure 5. On circuit switching voltage curve Switching

Snubber circuit to provide voltage inverting so that interference can be reduced. A voltage spike on a graph form a solar module provides a voltage drop when the switching process of -0.2 V as a voltage inverter, which causes an interference voltage value.

5. Conclusion

Simulations using a series conducted IV Tracer shows the curve shape of solar modules in accordance with the solar module datasheet with the fill factor was 0.71. IV Tracer circuit can generate output parameters according to the theory of power changes to variations in temperature values of 0.05% and variations in light intensity of 0.14%. System circuit used to mitigate disruptions such as that shown in the graph of -0.2 volts corresponding to the solar modules are tested. Based on the results, researchers can provide advice on further research to be carried out analyzing the factor loadings combined with other types of snubber circuits.

References

[1] Tom Markvart and Luis Castafier, 2003, "Practical Handbook of Photovoltaics: Fundamentals and Applications" Elsevier Ltd., *Copyright*.

[2] KJR, Slren Bkhlj, 2005, "Design And Control Of An Inverter For Photovoltaic Applications", Aalborg University, Denmark Institute of Energy Technology.

[3] Muhammad H. Rashid, 2011, "*Power Electronics Handbook Devices, Circuits, And Applications* Third Edition ", *Electrical and Computer Engineering University of West Florida,* Elsevier Inc.Copyright c.

[4] Gradella Marcelo Villalva, Jonas Rafael Gazoli, and Ernesto Ruppert Filho, 2009, "Comprehensive Approach to Modeling and Simulation of Photovoltaic Arrays", IEEE Transactions On Power Electronics, Vol. 24.

[5] Matic Herman, Jankovec Marko, Marko Topic, 2013, "Optimization of the I-V measurement scan time through the dynamic modeling of solar cells", Faculty of Electrical Engineering, Laboratory of Photovoltaics and Optoelectronics, University of Ljubljana, Slovenia, IET Renew. Power Gener.

[6] Shuren Wang, Wei Jiang, and Zhengyu Lin, July 2015, "Practical Photovoltaic Simulator with a Cross Tackling Control Strategy Based on the First-hand Duty Cycle Processing", School of Hydraulic, Energy and Power Engineering, Yangzhou University, Yangzhou, China, Journal of Power Electronics, Vol. 15, No. 4, pp. 1018-1025.

[7] John Bird, 2010, "Electrical Circuit Theory and Technology Fourth Edition", Elsevier

[8] Emily S. Kopp, 2012, "IV Analysis Of Photovoltaic Modules deployed Outdoors At Tucson Electric Power Solar Test Yard", The University of Arizona.

[9] Slamet Riyadi, 2011, "Alternative Energy Solar Panel as" Subject Matter, Soegi pranata Catholic University of Semarang.

[10] RA Sinton and A. Cuevas, 2000, "A Quasi-Steady-State Open-Circuit Voltage Methode For Solar Cell characterization", 16th European Photovoltaic Solar Energy Conference, Glasgow.

[11] Satwiko S, April 2012, "Characteristics of Solar Cell Test System On 24 Volt Dc For Catudaya In Hybrid Power Systems", Department of Physics, Mathematics and Natural Sciences Scientific Meeting Proceedings XXVI UNJ HFI Central Java and Yogyakarta, Purworejo.

[12] Tetsuya Nakamura, Mitsuru Imaizumi, Shin-Ichiro Sato and Takeshi Ohshima, 2011, "Change in the I-V Characteristics of Subcells in a Multi-junction Solar Cell due to Radiation Irradiation", 978-1-4673-0066-7 IEEE, Japan.

[13] Jaharap Situmorang, Linus Ampang Pasasa, June 2011, "Utilization Characteristics of Solar Cells For Dynamic Power Physics Learning Media", Proceedings of the National Symposium on Innovation Education and Science 2011, Bandung, Indonesia.

[14] Karen Wilson, Denis De Ceuster, Ronald A. Sinton, 2006, "The Effect Of Cell Measuring Mismatch On Module Output", IEEE 1-4244-0016-3, SunPower Corp. USA.

[15] Rudito Prayogo, December 2012, "Setting the PWM (Pulse Width Modulation) with PLC", ruditoprayogo@rocketmail.com, the Ministry of Education and Culture of the University of Brawijaya.
[16] Sardiyanto, 2012, "Making Module 3-Phase Inverter Sinusoidal Pulse Width Modulation For Induction Motor Speed Settings 3 phase 220 Volt Connected Triangle", Department of Electrical Engineering, Faculty of Engineering, University of Diponegoro in Semarang.
[17] CK Huang, HH Nien, SK Changchien, CH Chan, and CK Chen, December 2008, "An Optimal Designed RCD snubber for DC-DC Converters", Department of Industrial Education and Technology, NCUE, Changhua 500, Taiwan, 10th Intl. Conf. on Control, Automation, Robotics and Vision Hanoi, Vietnam.
[18] Sam Ben-Yaakov and Gregory Ivensky, March 1999, "Passive Lossless Snubbers For High Frequency PWM Converters", Power Electronics Laboratory Department of Electrical and Computer Engineering, Ben-Gurion University of the Negev, ISRAEL.
[19] Semiconductor at www.onsemi.com, "Switcher Efficiency & Snubber Design" portfolio of power management products from ON
[20] 0,4W solar cells _ 5.5V 65x65x3mm - Shop roboticists - BOTLAND.com.pl.html

Efficiency analysis on the inverter using the energy-saving lamp

N Evalina *, A Azis H, Rimbawati and Cholish

Department of Electrical Engineering, University of Muhammadiyah Sumatera Utara Jl. Kapten Muchtar Basri postcode 20238, Medan Indonesia

* Email: noorlyevalina@umsu.ac.id

Abstract. Electric energy savings are influenced by magnitude of reactive power used by electric customers, one of the tools that can minimize reactive power is the inverter, which is an electronic equipment, that can change the direct current of electricity to alternating current, The method used is by measuring along with analysis of calculation each time the inverter is given the burden of energy-saving lamps with varying power to determine the change in the electrical quantities to be analyzed. Inverter efficiency in doing the convert direct current to alternating current during low load (5W) amounted to 94.66%, and the highest load current (203W) amounted to 90.55%.

Keywords: Inverter, Energy Saving, Efficiency

1. Introduction

Indonesia is the largest consumption energy in Southeast Asia, thus the strategy of saving energy, costs and reducing emissions is important issue in this country. The subsidies have been taken off cause the electricity tariffs increased [1], especially for household customers. Therefore, the reactive power resulting in the loss of active power must be reduced so that electricity will be used by customers to be better, Electric energy savings are influenced by the magnitude of reactive power used by electric customers, one of the tools that can minimize reactive power is the inverter [2], which is an electronic equipment that can change the direct current of electricity from devices such as batteries, accumulators or solar cells into electric current alternating current. This tool can also work to back up electricity at home within a certain time (temporarily) when there is no power supply.

To cope with power outages, especially in households, it is necessary to have emergency equipment capable of working to distribute electrical power while as a substitute for the supply of electricity. The tool is an inverter, which is an electronic device that can convert the direct current emanating from devices such as batteries, accumulators or solar cells into alternating current. This tool can work for backup electricity in the house within a certain time (temporarily) when the supply of electricity goes out and over the input voltage of the battery is not discharged. Source voltage of the inverter is the battery or batteries and is only used when needed or as a temporary replacement of National Electrical Power, electricity when subjected to power outages.

2. Literature Review

2.1. Inverter

The inverter is a circuit that functions to convert a DC input to AC output voltage and frequency can be set as desired. Alternating voltage square waveform and in particular the use of the necessary filters to produce a sinusoidal waveform. Great setting voltage can be done in 2 ways. Firstly by regulating the DC input voltage from the outside but wide fixed ignition timing. Second, set the width of the time to connect with the DC input voltage remains.

The inverter is composed of the main circuit formed by the rectifier circuit whether controlled or not to convert the alternating current (AC) into direct current (DC) and eliminates the ripple contained on direct current. The rectifier serves as a charger voltage on the battery/accumulator when the voltage source inverter from the battery has been depleted. To prevent damage to the battery due to overcharger, then you should add automatic circuit rectifier circuit that will decide the filling process when the voltage on the battery is full [3][4]. Chart inverters generally have a shape like in Figure 1.

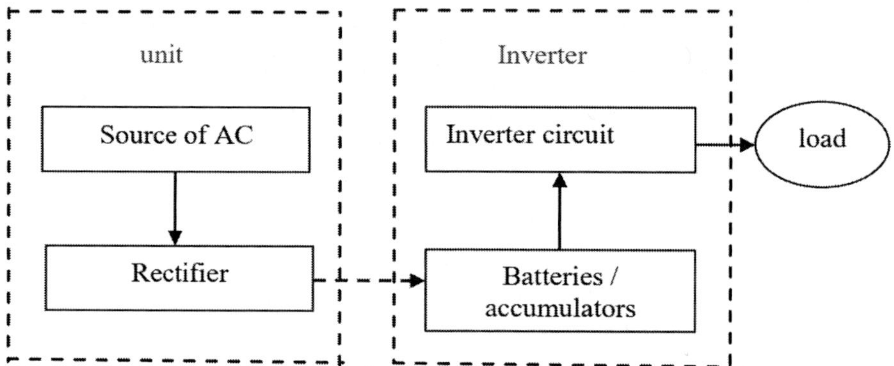

Figure 1. Chart Inverter

2.2. *Inverter structure*

The structure of the inverter (Figure 2)shows that the inverter with transistors that produces an alternating current (AC) with the frequency of commercial sources, (50 Hz or 60 Hz). The first part of the circuit converter (which converts a source of commercial AC into DC source and eliminates the ripple on the DC output. The second circuit is an inverter that converts the direct current into alternating current three-phase with frequency diverse (adjustable); the circuit is called the primary circuit. The third part is a control circuit that serves as a controller of the main circuit. The combined total of the circuits is called an inverter.

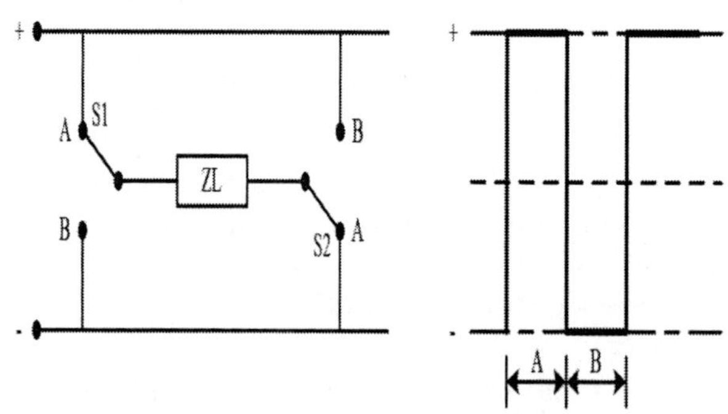

Figure 2. Structure simple inverter

When the position of S1 and S2 in A, the burden of getting a positive voltage, while a negative voltage is obtained when S1 and S2 in position B. Thus the removal of the switch (S1 and S2) in turn will generate alternating voltage is a square which is determined by the source , and the frequency is determined by the switching speed switch. Based on inverter configuration can be differentiated into two types namely single-phase half-bridge inverter and single inverter phase full bridge [5]. Meanwhile, based on the number phase, the inverter can be divided into two types as well that inverters single-phase and three-phase inverter.

2.3. Energy Saving Light Bulb
The energy-saving lamp is a type of fluorescent lamp with the development of a compact form. Energy-saving lamps have the same working principle with fluorescent lighting, which is casting the gas in the lamp tube causing ultraviolet rays due to electrical energy flow. The type often used is a model of soft light (SL) and power light (PL). In the SL models electric component consisting of ballasts, capacitors, and integrated starter in unity in the lamp, while the PL models for electric component separate from the lamp. LHE fundamental differences with standard fluorescent lamps are lamps that have been designed to the shape of the cap and are made the same as at the foot of incandescent lamps is by the screw system [6]. The advantages of energy-saving lamps are the use of electric power is more efficient than incandescent bulbs and TL; and It has a service life span is longer, which is about 8 times the service life of incandescent lamps.

2.4. *Inverter* efficiency
Inverter efficiency illustrates the effectiveness of the conversion of the inverter to convert direct current into alternating current, efficiency is influenced by the amount of output power and magnitude of reactive power, reactive power is the real power losses in system and loss as much as possible in the system, in the system, in order to shrink the more efficient. Large reactive power is influenced by the power factor.

3. Methodology
As for the equipment used in this study are as follows:
1) Single-phase inverter, as transformers DC to AC that will be examined use.
2) Energy Saving Lamp types SL, as an electrical load that will weigh on the inverter which these lights are arranged in parallel and power consumption varies greatly.
3) Accumulators, as a DC voltage source to power the inverter.
4) Rectifier (charger), used as auxiliary equipment when the voltage needed for charging the accumulator.
5) Ammeter AC and DC used to measure the flow of inputs and outputs on the circuit.
6) The multimeter is used to measure the voltage on the circuit.
7) Electronic energy meter, used as a measure of power and cos phi load.
8) Jumper cables, as the connecting terminals on the circuit.
9) Camera (mobile), used to perpetuate research on the indicators to be studied.

The study was conducted through several stages of testing, measurement and measurement are as follows:
1) Prepare all the equipment that will be used in this study.
2) Ensuring that the accumulator would be used in full condition and ready for use.
3) Connecting the DC ammeter in series with the positive terminal of the accumulator and the positive input terminal of the inverter.
4) Connecting in parallel a DC voltmeter (multimeter) to the accumulator terminal and inverter input terminal.
5) Large measure voltages and currents that occur in large measure input voltage inverter and the inverter output when the inverter in case of no load.
6) Connect in series ammeters AC (can be replaced by a power meter) to one inverter output terminals and one terminal (cable) load energy-saving lamps.
7) Connect in parallel AC voltmeter (multimeter) and frequency meter with inverter output terminals and terminal (cable) load energy-saving lamps.
8) Turning on the inverter with the state of load is a load with a large energy-saving light bulbs predetermined power (starting from the smallest lamp power to the largest).
9) Measuring the large input voltage inverter, inverter input current, the output voltage of the inverter, inverter output current, frequency and power factor of the inverter in case of the load.

IOP Conf. Series: Materials Science and Engineering **674** (2019) 012034 doi:10.1088/1757-899X/674/1/012034

10) Doing repetitive experiments (8) and (9) to load a large energy-saving lamp with the other lamp power (lamp power is raised varies to the extent desired).
11) Noting the results of measurements of indicators that have been obtained at the study table.
12) When it has finished, then measure the whole load on the inverter, inverter shutoff, releasing all the test series that have been assembled in advance and store the equipment back in place of the original.
13) Conduct analysis of the calculation of some indicators that can not be obtained through measurement, such as counting much voltage ignition inverters based on the signal waveform, substantial average voltage, the effective voltage, power input and power output of the inverter and the inverter efficiency at light load with various power, did calculation of the duration of the charging current for the accumulator and the duration of use of the accumulator on the inverter.
14) Record all calculation results obtained in the chapter on data analysis and discussion.

4. Results and Discussion

In testing with this load, several energy-saving bulbs are used starting from 5W to 203W to overload a 12V accumulator with a maximum current output rate of 65Ah. The schematic form of mounting measuring tools, accumulators, and lights on the test can be seen in Figure 3 and Figure 4.

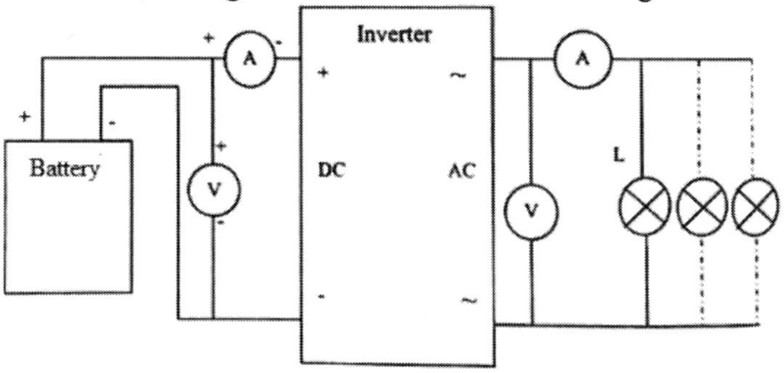

Figure 3. Schematic Inverter with Load Tests

Inverter with load testing aims to determine the magnitude of the voltage and current input from the accumulator as well as to determine the amount of voltage, current consumption, power factor, and frequency generated by the inverter after getting a load of energy-saving lamps[2][7].

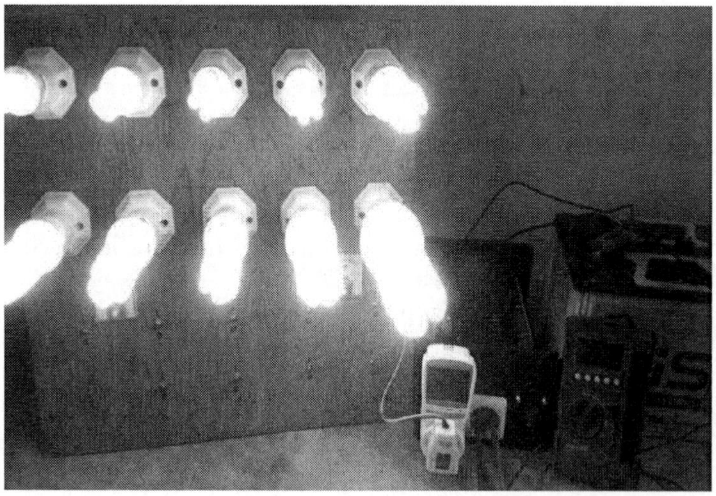

Figure 4. Load Testing Inverter with Energy Saving Light Bulb

IOP Conf. Series: Materials Science and Engineering **674** (2019) 012034 doi:10.1088/1757-899X/674/1/012034

The comparison of power factor saving lamps based on inverter brings much difference to the result, the load using an inverter much better, compared with power factor the light load without using an inverter [2][8] (Figure 5).

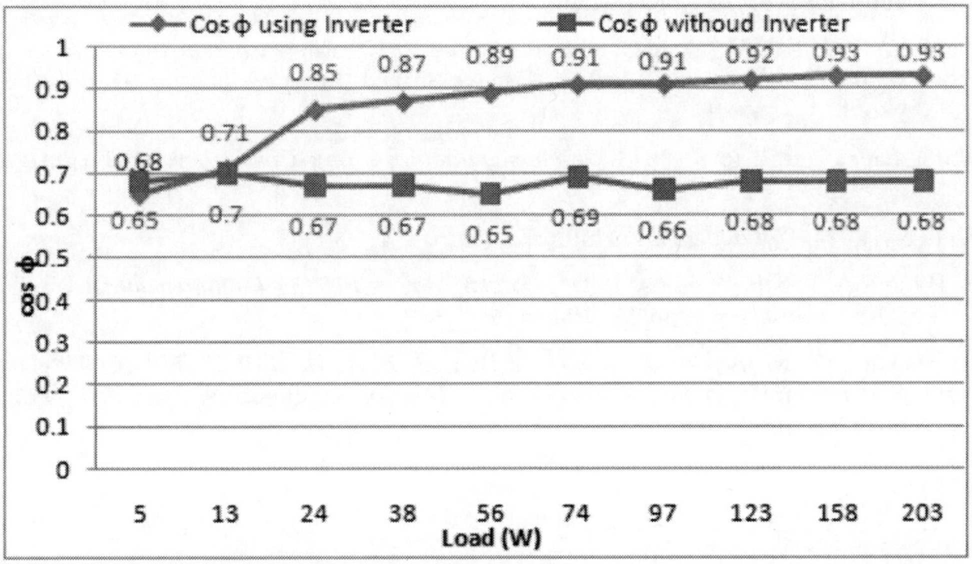

Figure 5. Comparison of the chart again power factor using an inverter and without inverter

From data testing of Inverter (Figure 5), we can calculation, input power, the power consumed by energy-saving lamps, out power, reactive power, and efficiency we will get the results as in Table 1.

Table 1. Calculation Data Power and Efficiency Inverter

Load (W)	P supply (W)	S load (VA)	P load (W)	Q load (VAR)	Eff (%)
5	5.06	7.38	4.79	5.61	94.66
13	12.81	17.06	12.11	11.94	94.53
24	23.86	26.52	22.54	14.06	94.46
38	35.18	38.08	33.13	18.66	94.17
56	50.87	53.66	47.75	24.68	93.86
74	62.34	64.01	58.24	26.24	93.42
97	78.61	80.32	73.08	32.93	92.96
123	97.24	97.47	89.67	38.01	92.21
158	123.17	121.09	112.59	44.79	91.41
203	149.78	145.84	135.63	53.96	90.55

From the above calculation results obtained when the inverter efficiency energy-saving lamps burdened with a load varied from 5 watts to 203 watts was obtained efficiency of 94.66 - 90.55 %

5. Conclusion

The load using inverter it can be concluded reactive power smaller only 53,96 VAR, and power factor during the height load is 203 watts used power factor better 0.93, consumption active power lamps are very good 135.65 Watt, and inverter efficiency in doing the convert direct current to alternating current during low load (5W) amounted to 94.66% and the highest load current (203W) amounted to 90.55%., it means using inverter can increase the value of efficiency.

References

[1] International Energy Agency, 2017, *Energy Efficiency 2017 Laporan Khusus: Efisiensi Energi di Indonesia*

[2] Z. Tharo, A. P. U. Siahaan, and N. Evalina, 2016, *Improvisation Analysis of Reactive Power Energy Saving Lamps Based on Inverter*, Int. J. Eng. Tech., vol. **2**, no. 5, pp. 141–145

[3] S. N. Hutagalung, M. Panjaitan, and I. Pendahuluan, 2017, *Protype Rangkaian Inverter Dc Ke Ac 900 Watt*, vol. **16**, pp. 278–280

[4] Ankit, S. K. Sahoo, S. Sukchai, and F. F. Yanine, 2018, *Review and comparative study of single-stage inverters for a PV system,* Renew. Sustain. Energy Rev., vol. **91**, no. April, pp. 962–986

[5] N. Kashappa and R. R. K, 2013 , *Implementation of a Low Cost 5 Level PWM Inverter*, vol. **13**, no. 9, pp. 1–5,.

[6] K. P. Nv, 2014, *Soft and gentle*. Phillips

[7] S. Ha and A. Papalexopoulos, 1997 , *Reaktif Power Pricing Management*, IEEE Trans. Power Syst., vol. **12**, no. 1, pp. 95–104

[8] H. Muaelou, K. M. Abo-al-ez, and E. A. Badran, 2015, *Control Design of Grid-Connected PV Systems for Power Factor Correction in Distribution Power Systems Using PSCAD*, vol. **6**, no. 8, pp. 1092–1099,.

Comparative analysis between the switch mode power supply (SMPS) using IC Tl494cn transformer based on power supply linear

A Azis H[*], Cholish, Rimbawati and N Evalina

Department of Electrical Engineering University of Muhammadiyah Sumatera Utara
Jl. Kapt Muchtar Basri No. 3, postcode 20238, Medan, Indonesia

[*] Email: abdulazis@umsu.ac.id

Abstract. In general Audio Amplifier using linear transformer-based power supply as energy conversion. Such use may cause interference with electrical installation sourced from the substation circuit. The non-linear load would cause problems in the electrical installation system. This research shows that Switch Mode Power Supply (SMPS) has a minimum voltage Vdrop output load of about 100 to 33.33 ohms. On the burden of 25-20 ohms as the voltage Vdrop output, which is different from Linear Power Supply Transformer-based parameters, that its voltage Vdrop output was changed significantly.

Keyword: Power Supply Linear, Audio Amplifier Efficiency, SMPS

1. Preliminary

At present, the use of electrical energy is one of the most basic needs of people. It can be seen from the power consumer demand for electrical energy needs [1]. This condition is not far from people's dependence on electrical energy requirements as a staple. These needs include the need for electrical energy for household such as a motor on the machine water, washing machines, fans,*blender*Lamp as lighting, cooling in the refrigerator, air conditioner, heater (Heater) at an iron, rice cooker, and goods - other electronic goods that also serves as an entertainer entertainers such as televisions, audio players, radios, audio amplifiers, and others.

Texas Instruments Incorporated designing Switch Mode Power Supply (SMPS) [2]-[8] using MOSFET TL598. A control circuit that uses MOSFET TL598 is designed with the ability to avoid the need for additional output circuits. In addition, MOSFET TL598 here used as a control circuit pulse-width-modulation.

2. Literature Review

Subsequent research of Switch Mode Power Supply Using Boost For PFC Converter. This is done to eliminate the problems that occur in Switch Mode Power Supply (SMPS), which is not using power factor correction and, thus, requiring the addition of great value as a filter capacitor DC [2]-[5]. To solve this problem is the addition of power factor correction (PFC) in switch-mode power supply (SMPS)[6]-[9]. In this SMPS, power factor correction is placed at the output of the rectifier by using Boost converter. Boost converter working in discontinuous conditions, because the conditions are not continuous did not appear backflow (IRR) on the components of the boost converter diode, the diode so that less can be used. In addition to the continuous condition resulting in lower I2R losses and low

IOP Conf. Series: Materials Science and Engineering **674** (2019) 012035 doi:10.1088/1757-899X/674/1/012035

ripple current in the inductor resulting in lower core loss. The boost converter is connected in series with the load buck converter to supply 24V/60W [9]-[10]. Boost converter as a power factor correction (PFC) in the design of an output voltage of 50 V and a current of 3 A. While the buck converter design produces a voltage of 24 V with a current output 2.5 A.

3. Method

Before modifying an audio amplifier, it needs to know construction and part of it to simplify the design modifications. A series of Switch Mode Power Supply (SMPS) functions the same as a linear transformer-based power supply, supplying voltage and current on the audio amplifier, the only difference being its mechanism of action. For more details, can be seen one-line diagram in Switch Mode Power Supply (SMPS) (Figure 1) and Power Supply transformer-based linear (Figure 2).

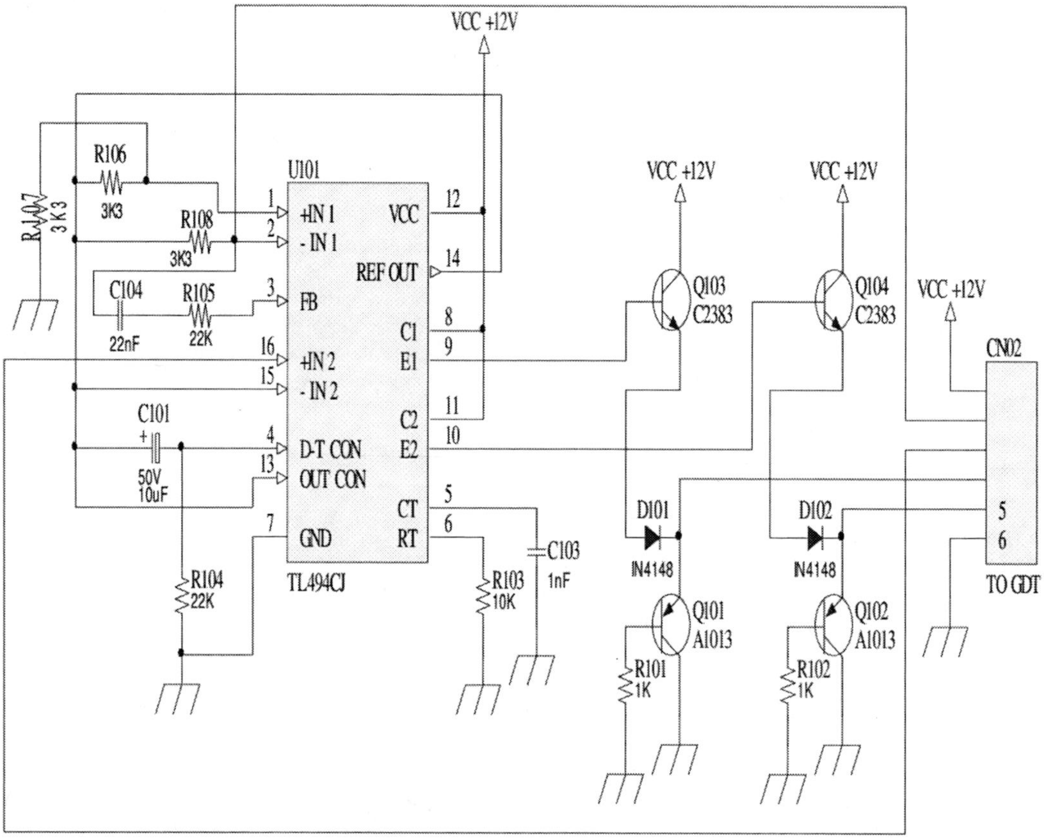

Figure 1. Schematic SMPS Power Control at +/- 35 VDC - 350W using IC TL494CN

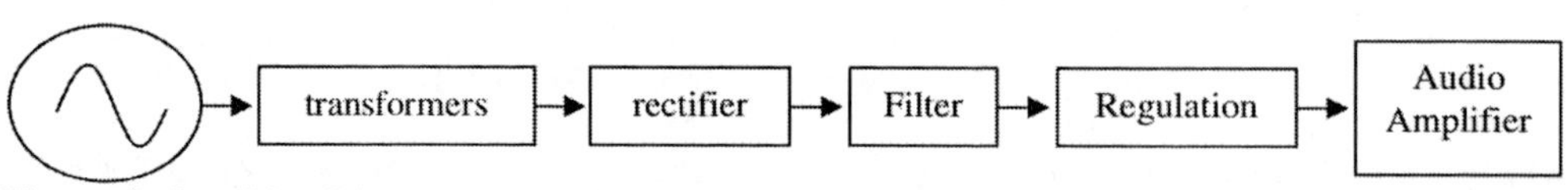

Figure 2. One-Line Diagram In Audio Amplifier using Linear Transformer-Based Power Supply

One-line diagram (Figure 2) is a condition when the audio amplifier using transformer-based Linear Power Supply. AC voltage is converted into AC voltage other smaller ones with the help of transformer, This voltage is then rectified by using a series rectifier voltage (Rectifier), and at the end added condenser as smoothing voltage (Filter) so that the DC voltage generated is not too bumpy. Then the DC voltage refines again regulator so that the voltage generated is better (no ripple on the waves). Linear transformer-based power supply produces a DC voltage that varies between 0-60Volt with a current between 0-10 ampere,

After a linear transformer-based power supply switched using the Switch Mode Power Supply, a different mechanism of action, which is directly rectified (Rectify) and filter (Filter) 220 V AC input voltage to obtain a DC voltage. The DC voltage is then Switch ON and OFF at a high frequency to high-frequency circuit that generates an AC that can pass through a high-frequency transformer. How to rectify the voltage is to use a high frequency between 10KHz to 1MHz, which is much higher frequency than the frequency of around 50Hz AC. Additionally In SMPS gave feedback circuit so that voltage and current are out of this circuit can be controlled either automatically using PWM Control.

4. Results and Discussion

From the tests performed on Switch Mode Power Supply (SMPS) and Power Supply Transformer Based Linear, then obtained the desired parameter data and analysis have a comparison (Figure 3 to Figure 5).

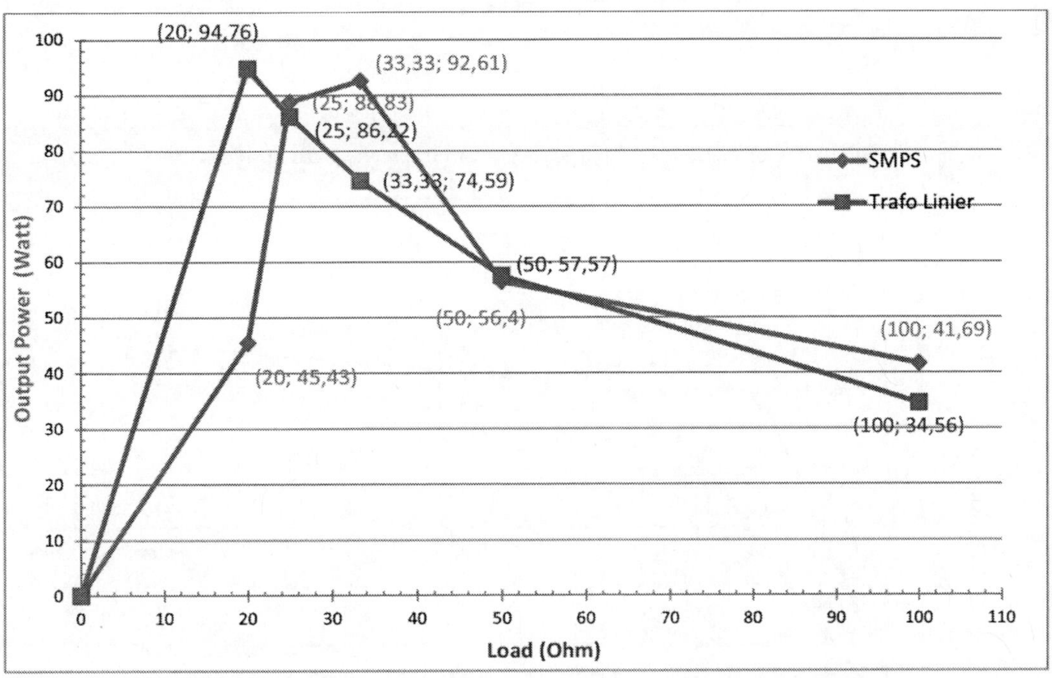

Figure 3. Expense ratio graph of the Output Power On Second Device

IOP Conf. Series: Materials Science and Engineering **674** (2019) 012035 doi:10.1088/1757-899X/674/1/012035

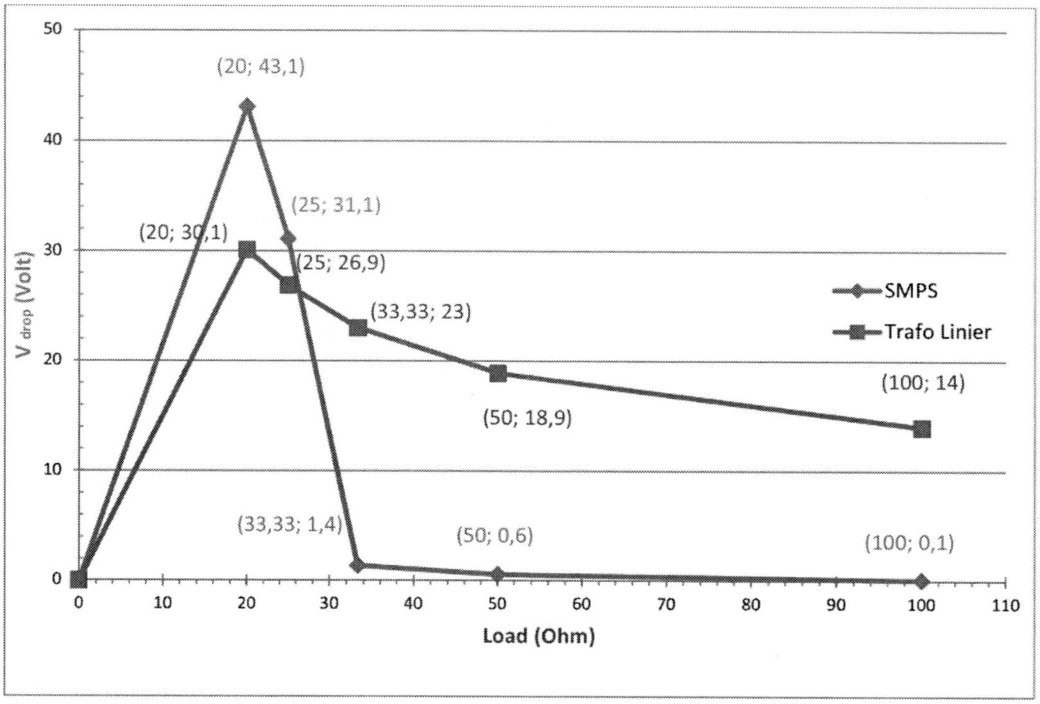

Figure 4. Comparison of Charges Against Graph V_{drop} On both devices

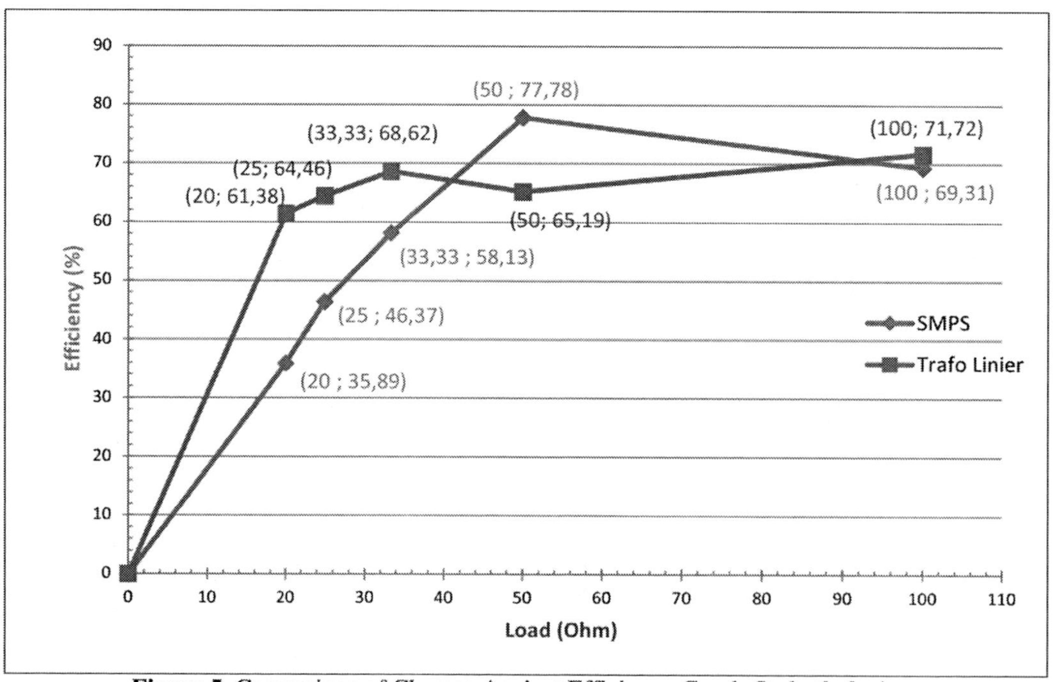

Figure 5. Comparison of Charges Against Efficiency Graph On both devices

5. Conclusion

From experiments conducted on Switch Mode Power Supply (SMPS) and Linear Transformer shows that the Switch Mode Power Supply (SMPS) experienced minimal V_{drop} output voltage to the load 100 to 33.33 ohms. On the burden of 25-20 ohms, the output voltage V_{drop} very much. Different from Linear Power Supply Transformer-based parameters - parameters such as output voltage V_{drop} changed significantly, as well as its efficiency. From these facts, it can not be said SMPS is perfect because it is not as expected at first. This is because the modified Switching Transformer not meet the quality standards, and there are several other factors so that the resulting efficiency is low. The results of this experiment can be made about improving the performance of the SMPS itself in order to obtain future results that perfect as expected.

References

[1] Fundamentals of Power Electronics "Chapter 15: *Transformer Design"*.

[2] EE IIT, "*Design of Transformer for Switched Mode Power Supply (SMPS) Circuits*", Version 2 EE IIT, Kharagpur.

[3] IEEE Canadian Review, 2009, "Switching Mode Power Supplies", Fall / Automne 2009, No.61, Canada.

[4] Kwon Su-Han, Yoo Doo-Hee, Jeong Gang-Youl, 2014, 2014, "*High-Efficiency AC-DC Switch-Mode Power Supply Using Full-Bridge Converter Circuits*", International Jurnal Of Control and Automation Vol.7. No.6.

[5] L Wuidart, 1999, "*Topologies For Switched Mode Power Supplies*", Application Note, STMicroelectronics, Italy.

[6] Manasa, 2014, "Variable Switched Mode Power Supply Using TL494", *Project Report Submitted In Partial Fulfilment Of The Requirements For The Award Of The Deegre Of Bacheloor Of Technology In Electrical And Electronics Engineering*, Gokaraju Rangaraju Institute Of Engineering And Technology Hyderabad, Andra Pradesh.

[7] Ngah Demon Siti Zulaikha, Tamuri Abdul Rahman, Bidin Noriah, 2008, *"High Voltage Switch Mode Power Supply For Laser System"*, Derpartment of Physics Centre of Foundation Studies, National Defense University Malaysia, 57000 Kuala Lumpur.

[8] SCILLC, 2014, "*Switch-Mode Power Supply Refrence Manual"*, On Semiconductor, USA.

[9] Texas Instruments Incorporated, 2001, "*Section 4 - Power Transformer Design"*, Dallas, Texas.

[10] Zhang. J Henry, 2013, "*Basic Concepts Of Linear Regulator And Switching Mode Power Supplies*", Linear Technologi, USA.

IOP Conf. Series: Materials Science and Engineering **674** (2019) 012036 doi:10.1088/1757-899X/674/1/012036

Enhancement of genetic algorithm optimization on distribution operation control by fuzzy inclusion

A Ulinuha[1,*] and G Ariyanto[2]

[1]Dept. of Electrical Engineering, Universitas Muhammadiyah Surakarta, A Yani ST, Po Box 1, Pabelan, Kartasura, Surakarta, Postal Code 57102, Central Java, Indonesia
[2] Dept. of Informatics, Universitas Muhammadiyah Surakarta, A Yani ST, Po Box 1, Pabelan, Kartasura, Surakarta, Postal Code 57102, Central Java, Indonesia

* Email: Agus.Ulinuha@ums.ac.id

Abstract. The optimal control of distribution system operation is presented in this paper. The objective of the control is minimum real power loss simultaneously achieved with voltage profile improvement. The distribution system equipped with Load Tap Changer (LTC) and switchable Shunt Capacitors is optimally controlled by determining the tap position of LTC and switching status of Shunt Capacitor at every hour for the 24-hour period. Two Genetic Algorithms (GAs) are developed for load curve partition and hourly components status determination. The performance of GA for components optimal scheduling is further enhanced by integrating Fuzzy Strategies in fitness evaluation and constraints satisfaction. The scheme of Fuzzy Convex Decision Making (FCDM) that provides flexible objective achievement and soft constraints is employed. This enables considering more extensive solutions that may lead to better optimal control. The proposed control strategy is implemented on 30-bus distribution system and the system improvements are observed. The enhancement of fuzzy integration is highlighted.

Keyword: Distribution System, Optimal Control, Genetic Algorithms, Fuzzy Strategy

1. Introduction

With constantly increasing demand of electricity, distribution systems have become more complicated. The growing demand for electricity may be satisfied in both technically adequate and reasonably economical way by careful distribution system planning. The objective of distribution system planning is to ensure that this can be satisfied in an optimum way [1]. Constantly changing load makes the planning very complicated. If not carefully managed, this may cause several problems, such as voltage violation and additional power loss.

An adaptable enhancement is required to restore the voltage and to reduce the power loss. A modern distribution system is normally equipped with devices such as Load Tap Changer (LTC) and Shunt Capacitors that may be controlled to minimize power loss and to improve the voltage profile. The aim is maintaining the system voltage and minimizing system energy loss. This can be practically achieved by finding the set of adjustments of the controllable devices for 24-hour period [2,3].

The optimal dispatch of LTC and Shunt Capacitors for 24-hour period is a multi-phase decision-making problem with discrete variables and nonlinear objective function [4-6]. The interdependence and interaction between these controllable devices lead to control complexity due to the switching effects of a device to the other devices [7]. This may cause oscillation of the controlled devices resulting in frequent switching that may reduce life expectancy and increase the maintenance cost. It is therefore preferable to achieve the control objectives in the least possible number of control steps. Simultaneous scheduling the controllable devices is the best approach to effectively solve the problem yet unfortunately, this will result in unavoidable complicated calculation [8].

The proposed method for the problem in hand is Genetic Algorithm (GA) [9,10]. In addition to its general features, GA is justified due mainly to its encoding ability that enables comprehensively considering the simultaneous dispatch of the switchable devices and confirming the switching limit before evaluating the possible dispatch schedules. The performance of GA will be further enhanced by including fuzzy approach. The scheme of Fuzzy Convex Decision Making [11] that provides flexible objective achievement and soft constraints enables the possibility of considering more extensive solutions leading to better optimal control. Implementation on a 30-bus distribution system confirms the enhancement the hybrid Fuzzy-GA method.

2. Genetic Algorithm

For solving optimization problem using Genetic Algorithm (GA), an initial population consisting of some chromosomes is initially constructed. Every chromosome represents a possible solution. This step is then followed by a cycle of three stages: fitness evaluation of each chromosome, selection of chromosomes for regeneration of new population and manipulation of chromosome by crossover and mutation [12]. Completing this cycle means that one generation has occurred. After some number of generations, the algorithm converges and the best chromosome represents an optimal solution. In this paper, 2 GAs are developed for determining load curve interval and optimal dispatch schedule.

In order to determine the intervals for entire load curve that result in an optimal LTC schedule, the number of interval in the load is initially assumed. GA is then employed to determine those intervals. The chromosome consists of a number of binary substrings according to the number of interval considered. Every substring corresponds to the value indicating the length of an interval. The sum of all substring values indicates the total length of the entire interval, which is 24 hours. Every chromosome is evaluated using the following fitness function [13]:

$$F \quad = F_{max} - min \sum_{l=1}^{L} \sum_{t=1}^{T} [(P_{tl} - PA_l)^2 + (Q_{tl} - QA_l)^2] \tag{1}$$

Subject to

$$\sum_{l=1}^{L} T = 24 \tag{2}$$

Where F_{max} is a constant converting fitness function to standard form, P_{tl} and Q_{tl} are active and reactive power at hour t and load interval l, PA_l and QA_l are average active and reactive power at load interval l, T is number of hour at l^{th} load interval, and L is number of assumed interval.

With the load intervals in hand, the schedule of LTC tap position can be effectively decided, where the tap position is fixed in the same interval. This will take into account the overall daily load change and easily satisfy the switching constraint. For Shunt Capacitors at substation, their switching operation is limited by a maximum number. Therefore, the chromosome representing this schedule consists of several substrings where, every substring denotes the switching status for each Shunt Capacitor for 24-hour period. The shunt capacitors installed at distribution feeders are normally allowed to be switched on and off once a day. Therefore, it requires determining the time for switching the capacitor on and the duration for keeping it on. The final eligible chromosome representing the 24-hour scheduling of LTC and shunt capacitors is consecutively constructed by the eligible chromosomes of LTC, substation shunt capacitors, and feeder shunt capacitors. Every chromosome at each generation is evaluated using fitness function derived from objective function and voltage constraint.

$$F = \max\left[F_{\max} - \left(w_{loss}\sum_{t=1}^{24} loss_t + w_V \sum_{t=1}^{24}\sum_{i=1}^{m} \Delta V_{it}\right)\right] \tag{3}$$

where $loss_t$ is per-unit real power loss of the system at hout t, ΔV_{it} is per-unit *rms* voltage violation at bus i at hour t, w_{loss} is weighting coefficient of real power loss, and w_V: weighting coefficient of *rms* voltage violation.

3. Fuzzy Inclusion

The previously developed GA is further enhanced by incorporating fuzzy into the existing GA forming hybrid GA-fuzzy method. The mode of fuzzy proposed is Fuzzy Convex Decision Making (FCDM) that facilitates a collaboration of objective function and constraints forming a fuzzy multi-objective optimization. The effectiveness of this approach is due to the ability of fuzzy providing soft restrictions for objective achievement and constraints satisfaction such that the modified GA enables maintaining the promising individuals while improving the solution.

For optimization using hybrid GA-fuzzy, membership functions that represent the objective and constraints are developed [14]. The membership functions are established to measure the accomplishment of objective function as well as fulfillment of constraints.

To minimize the energy loss, a membership function that rewards high membership value for minimum hourly power loss is used. A linear membership function with negative slope , $-(100/loss_0)$, declining from the maximum value of 100 is employed as represented by the following equation.

$$\tilde{\mu}_{loss} = \begin{cases} \dfrac{(loss_0 - loss)}{loss_0} \times 100; & loss \le loss_0 \\ 0; & loss > loss_0 \end{cases} \tag{4}$$

Where $\tilde{\mu}_{loss}$ is membership value of real power loss reduction, $loss_0$ is real power loss for uncompensated system (MW), and *loss* is real loss for compensated system (MW).

For voltage regulation purposes, a membership function that maintains voltage levels as close as possible to the preset value is employed. Exponential decreasing membership function is used in this paper as given in the following equation.

$$\tilde{\mu}_{\Delta V} = e^{-100\,\Delta V} \tag{5}$$

Where $\tilde{\mu}_{\Delta V}$ is membership of voltage regulation and ΔV is absolute deviation of *rms* voltage with respect to the preset voltage.

For the purpose of minimizing LTC tap movement, a membership function that keeps the switching number small is employed. The membership returns three values of membership as indicated in the following equation.

$$\tilde{\mu}_{TAP} = \begin{cases} 1; & 0 \le TAP \le 8 \\ (TAP_{max} - TAP)/(TAP_{max} - 8); & 8 < TAP \le TAP_{max} \\ 0; & TAP > TAP_{max} \end{cases} \tag{6}$$

Where $\tilde{\mu}_{TAP}$ is membership for LTC tap movement, *TAP* is LTC switching number, and TAP_{max} is maximum LTC switching allowed.

The abovementioned membership functions are clearly intended to reward the high membership values for the most preferred operating conditions. Therefore, the fuzzy fitness function of hybrid GA-fuzzy is constructed by simply maximizing the memberships rewarded by the membership functions. For optimal scheduling of the controllable devices, the fuzzy fitness function as given in the following equation is used.

$$F = \max\left[w_{loss} \sum_{t=1}^{24} \tilde{\mu}_{losst} + w_V \sum_{t=1}^{24} \sum_{i=1}^{m} \tilde{\mu}_{\Delta Vit} + w_S\, \tilde{\mu}_{TAP} \right] \tag{7}$$

Where $\tilde{\mu}_{losst}$ is membership for real power loss of the system at hour t, $\tilde{\mu}_{\Delta Vit}$ is membership for voltage regulation at bus i at hour t, $\tilde{\mu}_{TAP}$ is membership for LTC switching for the whole period, w_{loss} is weighting coefficient for real power loss, w_V is weighting coefficient for *rms* voltage regulation, and w_S is weighting coefficient of LTC switching minimization.

The calculation flowchart for optimal dispatch scheduling of LTC and Shunt Capacitor using GA and Hybrid Fuzzy-GA is given in Figure 1.

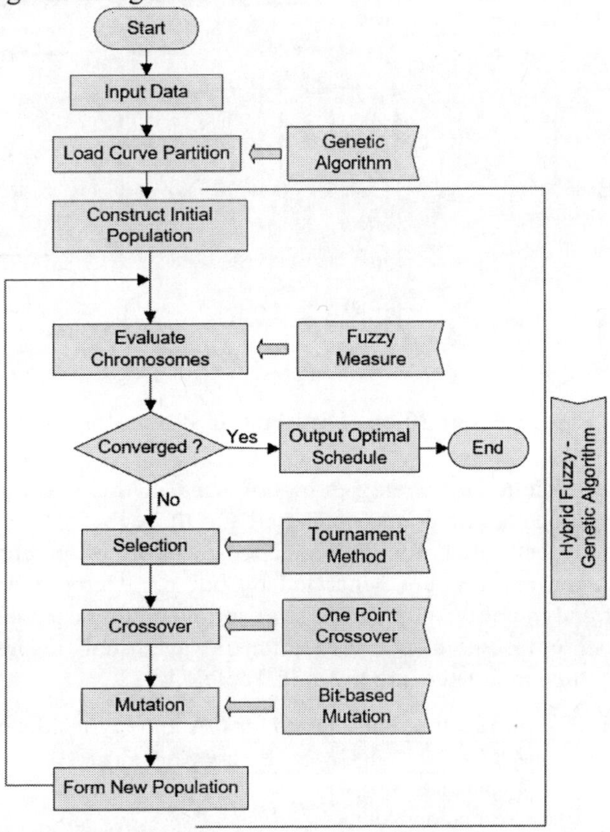

Figure 1. Optimization flowchart for optimal dispatch scheduling

Two GAs are developed, to determine optimal load curve partition and optimal components dispatch schedule, respectively. The second GA is further enhanced by Fuzzy inclusion. For comparison purpose, the calculation is initially run using GA and is then repeated using Fuzzy-GA. The improvement of the generated result is highlighted.

4. Results and Discussion

The proposed strategy is implemented on a 30-bus distribution system as shown in Figure 2. The system data is available in [13].

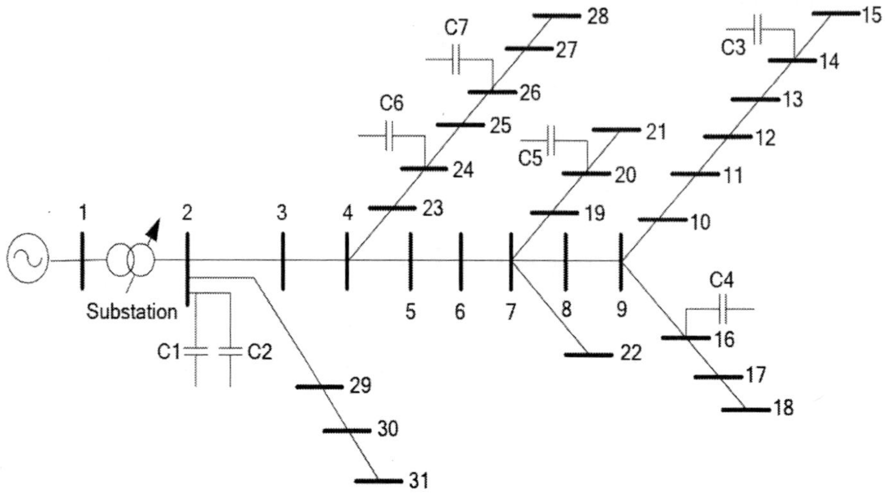

Figure 2. The 30-bus Distribution System for Simulation

The following is the evolutionary strategies of GA used in this study. The initial chromosomes representing possible solutions are randomly generated and those that satisfy the switching constraints are selected to construct the initial population. The selection of parent for crossover uses the tournament method and the children are generated by one-point crossover from their parents. The probability of crossover and mutation, population size, and weighting coefficients are fixed throughout the iteration. The best chromosome at every iteration is saved and transferred directly to the next population. The optimization parameters are given in Table 1.

Table 1. Optimization Parameters for GA and Hybrid Fuzzy-GA

Parameter	Value
Population Number	30
Generation Number	50
Probability of Crossover	60%
Probability of Mutation	1%
Weighting factor for GA	
Voltage Improvement	75%
Real Power Loss Minimization	25%
Weighting factor for Fuzzy - GA	
Voltage Improvement	60%
Real Power Loss Minimization	20%
LTC Switching Minimization	20%

The result of optimal dispatch scheduling of LTC and Shunt Capacitors may be represented in terms of voltage control and real power loss minimization. The complete operation schedule of the controllable devices cannot be displayed in this paper due to the space limitation. However the generated schedules have been checked to confirm that all switching constraints are satisfied.

For the purpose of controlling the voltage, the hourly voltage at every bus is observed before and after implementing the optimization strategy. It should be noted that the voltage at every bus is allowed to deviate at no higher than 5% from the nominal voltage. Since it is impossible to indicate the hourly voltage profile of all buses, the bus with worst voltage profile is detected. The improvement of voltage profile of the bus is then figured. It is to assure that the voltage profile of the entire buses is within the maximum deviation of 5%. For the simulated distribution system, the optimization result of voltage improvement is shown in Figure 3.

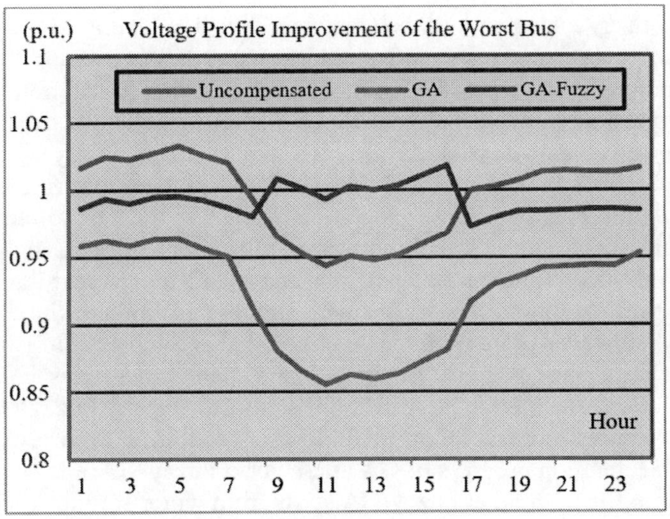

Figure 3. Voltage Profile Improvement of the Worst Bus

It may be observed from Figure 3 that for the worst bus, Fuzzy-GA may perform better by completely controlling the voltage to be within the deviation limit of 5%. Although GA may generally improve the voltage, in this case, the method may not entirely maintain the voltage to be within the limit. Detail inspection of the optimization results presented by GA, the voltage magnitude at hour 11 a.m. and 01 p.m. are slightly lower than 95%, which are 94.38% and 94.79%, respectively.

To further observe the performance of the proposed methods, the reduction of real power loss is presented. The hourly power loss for the system is presented in Figure 4. It may be observed from the figure that in general, the assigned optimization methods enable reducing the real power loss of the system. However, Hybrid Fuzzy-GA presents the better losses reduction. The energy saving provided by GA and Fuzzy-GA is 177.834 MWh and 205.023 MWh, respectively.

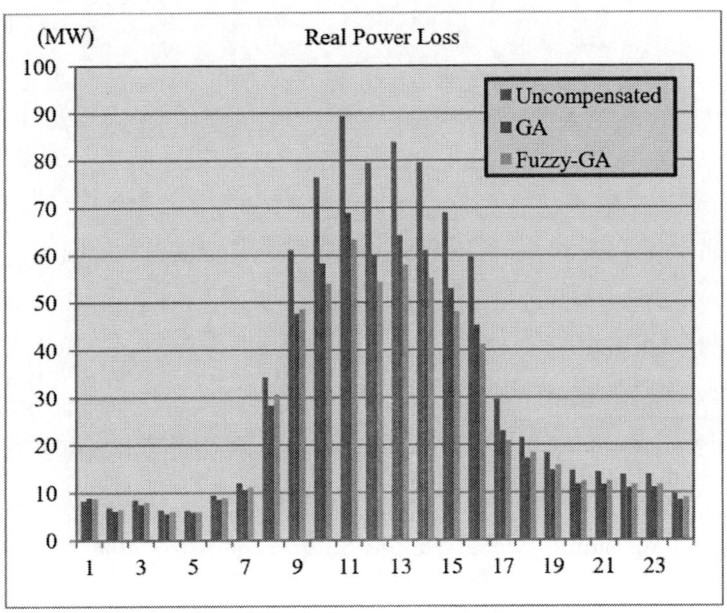

Figure 4. Hourly Real Power Loss

The advantage of fuzzy inclusion into the existing GA is due mainly to the possibility of the hybrid method providing soft restriction of objective achievement and constraints satisfaction. This feature enables exploring more extensive possible solution leading to better solution. Every possible solution represented as chromosome is evaluated based on its level of objective achievement and constraints satisfaction. The objective and constraints are all represented as Fuzzy Membership Function that accommodates the potential solutions to still be considered in the optimization process. These will be further refined in the following iterations (generations) to arrive the best solution.

Genetic Algorithm strictly checks the constraints fulfillment prior to include the chromosome in fitness evaluation and some following steps. This may lead to disregarding the potential solutions resulting in less optimal solution. The refinement process is limited since the number of eligible solution is not many. However GA ensures the solution that is totally eligible for the problem in hand.

In term of calculation process, GA is simpler and, sometimes, gets convergence rapidly. It is because of the variation of solution is small. In contrast, Hybrid Fuzzy-GA that considers more solutions normally requires more iteration to find best solution or even oscillates before converges. Figure 5 illustrates the iteration progress of GA and Hybrid Fuzzy-GA.

It may be seen from the figure that Fuzzy-GA starts from wider solution area may even cause lower best-fitness value. But it leads to a better solution. It also encounters some oscillations before getting the convergence. Therefore, the computation time required by Hybrid Fuzzy-GA is longer than that required by GA. For the optimization problem in hand, the computation time needed by GA and Hybrid Fuzzy-GA is 46.8 sec and 53.6 sec, respectively.

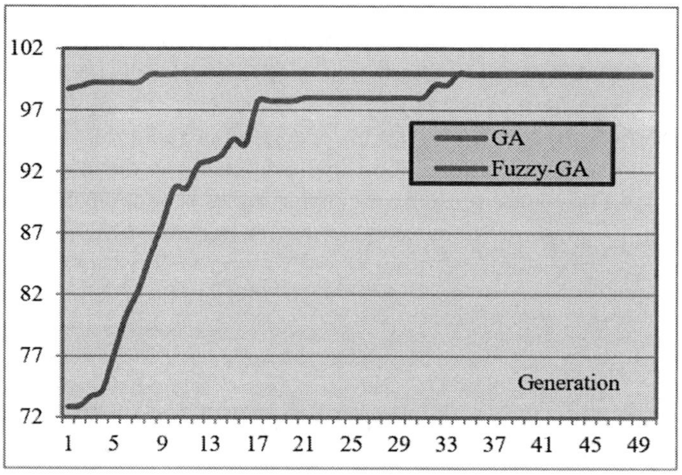

Figure 5. Iteration Progress of Optimization Calculation

5. Conclusions

Optimization methods, GA and Hybrid Fuzzy-GA are developed and implemented for the optimal dispatch scheduling problem of controllable devices. The followings are the main conclusions from the study:

1. The optimal dispatch scheduling of LTC and Shunt Capacitor may improve the voltage profile and reduce the real power loss,
2. Genetic Algorithm is considered to be suitable for the optimal dispatch problem due to the ability of the method simultaneously taking into account the objective and constraints in the calculation,
3. Inclusion of Fuzzy into the existing GA may extend the solution space leading to better solution,
4. While Hybrid GA-Fuzzy enables finding better solution, it causes higher computation charge and longer computation time.

References

[1] C.-T. Su and C.-T. Lin, "A new fuzzy control approach to voltage profile enhancement for power systems," *IEEE Transactions on Power Systems,* vol. 11, pp. 1654-1659, 1996.

[2] M. M. A. Salama and A. Y. Chikhani, "An expert system for reactive power control of a distribution system. Part I: System configuration," *IEEE Transactions on Power Delivery,* vol. 7, pp. 940-945, 1992

[3] J. R. P.-R. Laframboise, G. Ferland, A. Y. Chikhani, and M. M. A. Salama, "An expert system for reactive power control of a distribution system. Part 2: system implementation," *IEEE Transactions on Power Systems,* vol. 10, pp. 1433-1441, 1995.

[4] I. Roytelman, B. K. Wee, and R. L. Lugtu, "Volt/var control algorithm for modern distribution management system," IEEE Transactions on Power Systems, vol. 10, no. 3, pp. 1454-1460, 1995

[5] G. Ramakrishna and N. D. Rao, "Adaptive neuro-fuzzy inference system for volt/var control in distribution systems," *Electric Power Systems Research,* vol. 49, pp. 87-97, 1999/3/1 1999.

[6] Y. Deng, X. Ren, C. Zhao, and D. Zhao, "A heuristic and algorithmic combined approach for reactive power optimization with time-varying load demand in distribution systems," *IEEE Transactions on Power Systems,* vol. 17, pp. 1068-1072, 2002.

[7] M. E. Baran and M.-Y. Hsu, "Volt/VAr control at distribution substations," *IEEE Transactions on Power Systems,* vol. 14, pp. 312-318, 1999.

[8] Z. Gu and D. T. Rizy, "Neural networks for combined control of capacitor banks and voltage regulators in distribution systems," *IEEE Transactions on Power Delivery,* vol. 11, pp. 1921-1928, 1996.

[9] J. H. Holland, "Adaptation in Natural and Artificial Systems," University of Michigan Press, Ann Arbor 1975.

[10] D. E. Goldberg, *Genetic Algorithms in search, optimization, and machine learning*. Massachusetts: Addison-Wesley Publishing Company, Inc., 1953.

[11] A. Trebi-Ollennu and B. A. White, "Multiobjective fuzzy genetic algorithm optimisation approach to nonlinear control system design," in *Control Theory and Applications Conference*, 1997, pp. 137 - 142.

[12] Z. Michalewics, *Genetic Algorithms + Data Structures = Evolution Program* 3ed. New York: Springer-Verlag Berlin Heidelberg, 1996.

[13] Z. Hu, X. Wang, H. Chen, and G. A. Taylor, "Volt/VAr control in distribution systems using a time-interval based approach," IEE Proceedings-Generation, Transmission and Distribution, vol. 150, no. 5, pp. 548-554, 2003

[14] H. J. Zimmerman, Fuzzy Set Theory and Its Applications, 3 ed. Boston: Kluwer Academic Publishers, 1996.

Acknowledgement

The Authors would like to acknowledge that this research is supported by The Directorate of Research and Community Service (DRPM), The Ministry of Research and Higher Education (Kemenristekdikti) under the scheme of International Research Collaboration and Scientific Publication for the Funding Year of 2017. The official contract number is 211.17/A.3-III/LPPM/V/2017

IOP Conf. Series: Materials Science and Engineering **674** (2019) 012037 doi:10.1088/1757-899X/674/1/012037

Optimization of overcurrent relay coordination using cuckoo search algorithm in electronic induction systems

Z S Rahmania, L Syafa'ah, I Pakaya and Zulfatman*

Electrical Engineering Department
University of Muhammadiyah Malang
Jl. Raya Tlogomas No. 246, Malang-Indonesia 65144

* Email: zulfatman@umm.a.cid

Abstract - Overcurrent relay protection is one of the electrical power supply protection mechanisms that serve to prevent the system from overcurrent. The main issue that occurs when developing the coordination of overcurrent relays is the timing determination of the primary and back-up relay operation. The objective of this study is to develop an optimization method to result a faster operating time of relay coordination. The optimization method was developed and implemented in case of PT Semen Padang (Indarung VI) at 150 kV PLN's source, which was operated with conventional calculation method. The Cuckoo Search Algorithm (CSA) method was proposed as optimization method for calculation. The optimization was used to set the relay time and the secondary current (I_{pickup}) according to the British standard to obtain the optimal Coordination Time Interval (CTI). The CSA method was calculated by MATLAB R2016 software, while the relay testing was done on the ETAP 12.6.0. Results of simulation indicate that the CSA method more capable and faster than the manual calculations in calculating relay setting coordination since almost all of the CTI values were reached in 2.0 to 4.0 seconds.

1. Introduction

The overcurrent relay is a relay which used to protect the system from the overcurrent noise [1]. The overcurrent relay is connected to the circuit breaker, which will trigger the circuit breaker to open if there is interference with certain trip times. Overcurrent relays can be used to protect almost any part of the electrical system such as the transmission lines, transformers, generators, or motors [2]. To coordinate the overcurrent relay, there are many parameter values that must be reached, one of those parameters is used to get the ideal setting value. The ideal setting value must be accompanied by accurate analysis, with the current and time setting values calculation.

One of the obstacles that occur when doing the coordination studies of overcurrent relays is the timing determination of the primary and back-up relay operation. The desired operating time is as fast as possible by the time setting parameters consideration. This is done to minimize the interference effects on the electrical equipment from the arc flash. This exposure explains that failures in overcurrent relay coordination can be caused by the operating time of the back-up relay on the primary relay have not the right timing values [3].

To solve the problem of overcurrent coordination, the various methods and techniques that have been done are classified into 3 types i.e. curve fitting technique, graph theory technique, and optimization technique. Optimization techniques are also divided into 3 types i.e. conventional optimization techniques, interior point optimization techniques, and optimization techniques using artificial intelligence [4].

The Cuckoo Search Algorithm (CSA) is one of the optimization techniques. CSA is a new heuristic algorithm based on the previous studies results using twelve different stochastic functions which shows better search performance than the other heuristic algorithms such as Particle Swarm Optimization (PSO) algorithms or Genetic Algorithms. The advantage of CSA is using the Levy Flights (the concept of random food searching in animals while flying) as the random step, which allows the search for optimal values in a shorter time [5].

Cuckoo Search Algorithm (CSA) is used to analyze the overcurrent relay coordination. This algorithm shows better performance value because the CSA evaluates the parameter values 2 times, so it will get more optimal value for the overcurrent relay coordination if compared with the manual calculations method.

2. Method

In general, this study was started by designing a single line diagram that explains the data of short circuit current disturbance calculation, load data, and data calculation of maximum load current. The CSA algorithm was developed based on these data to obtain list of relay setting values. These values then test on the relay system by using as software. The development of CSA algorithm and calculation of the relay setting values by the CSA were done by MATLAB R2016 software, while the system testing was done by ETAP 12.6.0 software.

2.1. Indarung VI Electrical System PT. Semen Padang.

The electrical system in Indarung factory is divided into two major parts i.e. the electrical system in VI factory & 2 SS existing and one plant to PLN in connection to VI factory. The electricity load at Indarung VI Plant will be supplied by PLN through 3 units of the 150 / 6.3 kV power transformer with the capacity on each transformer is 30/35 MVA (ONAN / ONAF). The loads in Indarung VI are divided into 3 groups according to the amount of existing substations, for the data loading of the transformer 1 substation can be seen in Table 1. In Indarung VI Plant there are many loads used to support the production process. One of them is a motor i.e. the small and large capacity motor.

Table 1. The burden of Transformer on Substation 1 Indarung Factory.

Substation and Transformer ID	**Voltage (kV)**	**Capacity (MVA)**	**Input Loading**		**Output Loading**	
			MVA	**%**	**MVA**	**%**
Trafo TRF 1	150/6.3	30	20.955	59.9	20.487	58.5
SS 158 LSC, SILICA-1:						
6 A1Q211T1	6.3/0.4	2	1.277	63.8	1.271	63.6
6C1Q211T1	6.3/0.4	1.6	1.229	61.4	1.214	60.7
SS 348 RAW MILL DRIVE:						
6R1Q212T1	6.3/0.4	1.6	1.266	79.1	1.252	78.2
SS 348 RAW MILL (ID FAN):						
6R1S01T1	6.3/0.7	0.8	0.309	38.7	0.307	38.4

2.2. Single Line Diagram of Indarung VI Factory.

The single line diagrams are designed using ETAP 12.6.0 using the IEC (International Electrotechnical Commission) standard. In this final project, the data of *the* single line diagram is as expressed in Figure 1 which will be used as an analysis material.

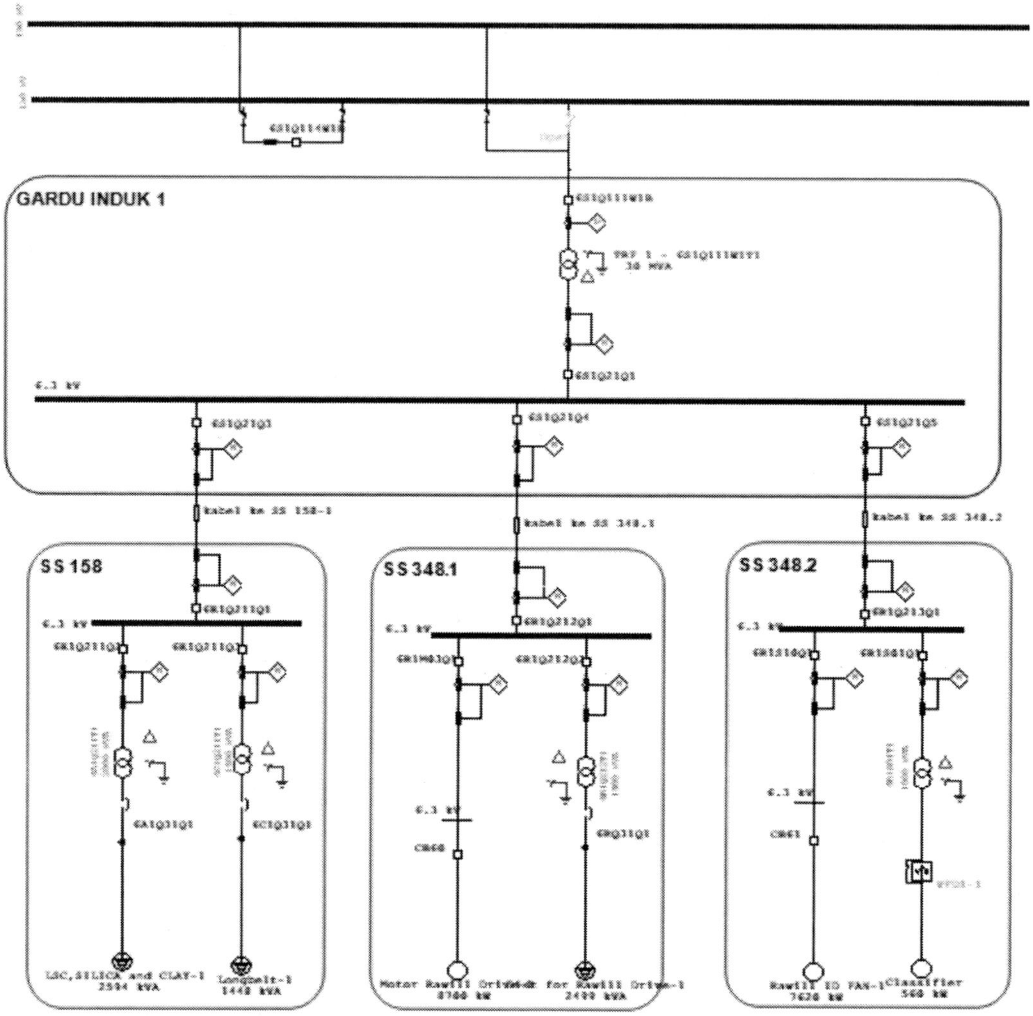

Figure 1. Single Line Diagram Substation Stations 1.

2.3. Method of Calculation of Over Flow Relay Coordinate Settings

There are several values needed to perform the overcurrent relay settings, such as Full Load Ampere (FLA) value and minimum Short Circuit Current (I_{sc} min). These values are used to determine the Current Setting (I_{set}), Time Dial Setting (TDS) and the current pickup (tap currents).

- **Full Load Current (FLA)**

  To use the *Very Inverse* standard relay, it must be calculated first the full load current value. The full load current value is used to calculate the setting current on the primary side (Iset).

$$I\ FLA = \frac{Va}{V \times \sqrt{3}} \quad (1)$$

- **Current setting on the primary side (Iset)**

$$Iset = \frac{kfk}{kd} \times I\ FLA \quad (2)$$

The Iset value above is the current setting on the primary side, while the value to be set in the relay is the secondary side setting value. Therefore, the setting value on the secondary side should be calculated using the CT ratio.

- **Current setting on the secondary side (Ipickup)**

$$I\ pickup = I\ sett \times \frac{1}{Rasio\ Ct} \quad (3)$$

- **Minimal interrupt current (Min Isc)**

$$Iscmin = \frac{Vll}{Z} \quad (4)$$

The calculated noise current value is used to determine the time dial setting.

- **Time dial (Td)**

$$Td = \frac{((\frac{Isc\ minimum}{I\ set})^{\alpha} - 1) \times t)}{k} + C \quad (5)$$

Where K and α are coefficients of the curve type, can be seen in Table 2.

Table 2. The coefficient of the curve type.

Curve Type	Coeficient	
	K	α
Standard Inverse	0.14	0.02
Very Inverse	13.5	1.00
Extremely Inverse	80	2.00

2.4. Design of Optimization Relay Setting using Cuckoo Search Algorithm (CSA)

In terms of optimization, Indarung VI electrical system relay will be optimized using the CSA method. The optimized value is the time setting (td) and current setting value (I_{set}), and for optimization using CSA requires some parameters to be referenced such as the relay working time (t), primary relay td value (t_p), secondary relay td value (t_b).

The CSA optimization design starts with the initialization of td value with restrictions 0,05 <td <1,2 and the I_{set} limit value i.e. (1,05 × I FLA) <I_{set} <(1,3 × I FLA). So the initial random value of the CSA produces the constraint value, then in the second iterations that exist in the script in the MATLAB software will be evaluated by the objective function of the overcurrent relay coordination on the mathematical equation as follows:

$$min \sum_{f=1}^{fmax}((\sum_i(t_i + t_{ij})) + \text{PEN} \times \sum_{K=1}^{P}(\Delta t_{Pbk} - |\Delta t_{Pbk}|)^2 \quad (6)$$

PEN value is the *penalty constraint* value of CSA algorithm, and P is the *iteration* value in constraint with the constraint value = Δt_{Pbk}. The value of Δt_{Pbk} can be searched with the following mathematical equations:

$$\Delta t_{Pbk} = t_{pk} - t_{bk} - CTI \quad (7)$$

The optimization calculation with the CSA is done using MATLAB. After obtaining the value of time dial setting, it is necessary to create a table of primary relay and backup relay in order to know the value of Coordination Time Interval (CTI). CTI value is obtained from the t backup value minus by the t primary value and said to be fulfilled if the CTI value greater than 0.2s.

3. Results and Discussion

3.1. Conventional Calculations for More Flow Relay Settings A subsection

Table 3 is result of conventional calculation on CTI that calculate by using above calculation method on overcurrent coordination setting.

IOP Conf. Series: Materials Science and Engineering **674** (2019) 012037 doi:10.1088/1757-899X/674/1/012037

Table 3. The CTI value result of the conventional calculation.

Primary Relay	tPrimary	Backup Relay	tBackup	CTI	Condition
Line 158.1	0.3	Incoming 158	0.4	0.1	Not fulfilled
Line 158.2	0.3	Incoming 158	0.4	0.1	Not fulfilled
Line 348.1.1	0.34	Incoming 348.1	0.4	0.06	Not fulfilled
Line 348.1.2	0.3	Incoming 348.1	0.4	0.1	Not fulfilled
Line 348.2.1	0.3	Incoming 348.2	0.4	0.1	Not fulfilled
Line 348.2.1	0.3	Incoming 348.2	0.4	0.1	Not fulfilled
Incoming 158	0.4	Outgoing 158	0.6	0.2	fulfilled
Incoming 348.1	0.4	Outgoing 348.1	0.6	0.2	fulfilled
Incoming 348.2	0.4	Outgoing 348.2	0.6	0.2	fulfilled

From Table 3, it can be seen that there are several CTI conditions for all unresolved primary relay and backup relay. This condition is due to the interval value was less than 0.2 seconds.

3.2. Optimization of Relay Settings with Cuckoo Search Algorithm (CSA)

The optimization calculation with the CSA was done using MATLAB. After obtaining the value of time dial setting, it is necessary to create a table of primary relay and backup relay such as have been detailed in Table 4, in order to know the value of CTI.

Table 4. The CTI value result of the optimization calculation.

Primary Relay	tPrimary	Backup Relay	tBackup	CTI	Condition
Line 158.1	0.45	Incoming 158	0.87	0.42	fulfilled
Line 158.2	0.48	Incoming 158	0.87	0.39	fulfilled
Line 348.1.1	0.48	Incoming 348.1	0.87	0.39	fulfilled
Line 348.1.2	0.15	Incoming 348.1	0.87	0.72	fulfilled
Line 348.2.1	0.49	Incoming 348.2	0.89	0.4	fulfilled
Line 348.2.2	0.49	Incoming 348.2	0.89	0.4	fulfilled
Incoming 158	0.87	Outgoing 158	1.18	0.31	fulfilled
Incoming 348.1	0.87	Outgoing 348.1	1.11	0.24	fulfilled
Incoming 348.2	0.89	Outgoing 348.2	1.13	0.24	fulfilled

From Table 4, it can be seen that the CTI condition for all primary relay and the backup relay has been reached because the interval value is more than 0.2 seconds.

3.3. Relation Coordination Simulation Testing of Substation (SS) 158 LSC, SILICA

Substation (SS) 158 received the power supply from the substation 1 transformer with a capacity of 30/35 MVA (ONAN / ONAF) and a voltage of 150 / 6.3 kV. SS 158 was divided into 2 groups i.e. 158.1 LSC, Silica Storage & Clay Crusher and 158.2 Longbelt as can be seen in Figure 2 below.

IOP Conf. Series: Materials Science and Engineering **674** (2019) 012037 doi:10.1088/1757-899X/674/1/012037

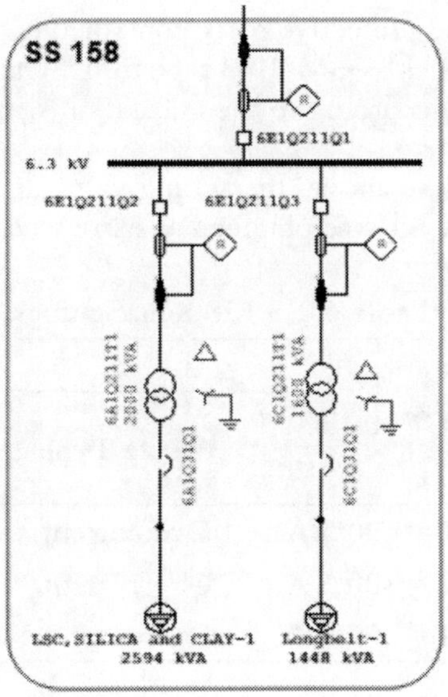

Figure 2. Single Line SS 158.

In SS 158 there were 2 groups i.e. 158.1 LSC, Silica Storage & Clay Crusher and 158.2 Longbelt where each group has the different load types as can be seen in Table 5 of SS 158.1.

Table 5. Load Data of SS 158.1 LSC, Silica Storage & Clay Crusher.

Code	Description	kW	Amp	Volt
6C1L03Q1	Bucket Chain Reclaimer Clay	80.75	250	380
6E1L02Q1	Side Scrapper Reclaimer Silica	80.75	250	380
6A1L02Q1	Bridge Reclaimer Limestone	323	1000	380
6C1Q31SP1	Spare	323	1000	380
6C1Q31Q2	Feeder to MCC 6C1Q311	664.02	1250	380
6C1Q31Q3	Feeder to NDB 6C1Q319	396.98	1250	380
6A1U01M1	Reclaimer Conveyor	90	240	380
6A1U02M1	Belt Conveyor	90	240	380
6A1U02M2	Belt Conveyor	90	240	380
6C1Q31SP2	Spare	110	290	380
6C1M01M1	Clay Crusher	110	290	380
6C1M01M2	Clay Crusher	110	290	380
	Total kW	**2,468.50**		
	Total Kw Static Load	**1,472.57**		
	Total Kw Dynamic Load	**995.93**		
	% of Static Load	**59.65**		
	% of Dynamic Load	**40.35**		

IOP Conf. Series: Materials Science and Engineering **674** (2019) 012037 doi:10.1088/1757-899X/674/1/012037

Each group has a Siemens 7SJ62 protective relay contained in the feeder and for incoming on each substation using a MICOM relay of ALSTOM P143 type from Schneider for the detect if there is a phase noise. The name of the relay was according to the substation's name and coupled with the sequence number of relay.

From the load data and single line above, the setting value analyzed result of I_{pickup}, time dial and $I_{setting}$ in MATLAB software using CSA was obtained as expressed in Table 6.

Table 6. SS 158 Rele Settings.

No	Relay's Name	CT Ratio	Relay			
			Curve Type	Setting		
				Ipickup	Time Dial	Isetting
1	INCOMING 158	CT = 600/5 A	Overcurrent	3.51	0.87	421.2
2	158.1	CT = 300/5 A	Overcurrent	3.74	0.45	224.4
3	158.2	CT = 200/5 A	Overcurrent	4.14	0.48	165.6

The setting value as Table 7 will be applied to each relay in substation 158. If those setting values have been applied to the relays, it will be simulated by giving the interference to each end repeater as expressed in Figure 4 below which was simulated at ETAP 12.6.0.

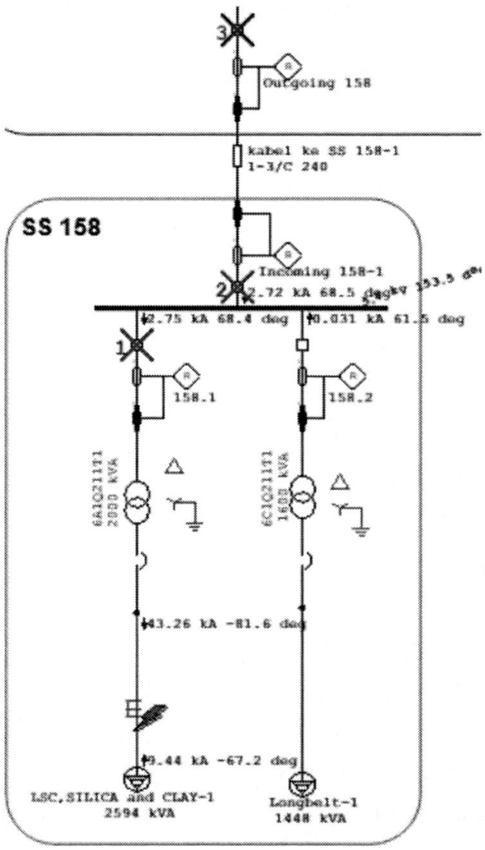

Figure 3. The sequence of Operation SS 158.

IOP Conf. Series: Materials Science and Engineering **674** (2019) 012037 doi:10.1088/1757-899X/674/1/012037

The disturbance was given to the ended repeater and the simulation was done gradually. The first was the disturbance given to group 158.1, then will show the first breaker or power breaker which works faster. From Figure 3, it can be seen that the main safeguard works first (indicated by the cross number 1), followed by backup safety (indicated by number 2 and 3).

The primary relay of group 158.1 works in 0.05 s, with 43.26 kA and SS 158.2 interruption current working at 0.05 s, with 35.69 kA interruption current. The simulation result also expressed with a curve as shown in Figure 4.

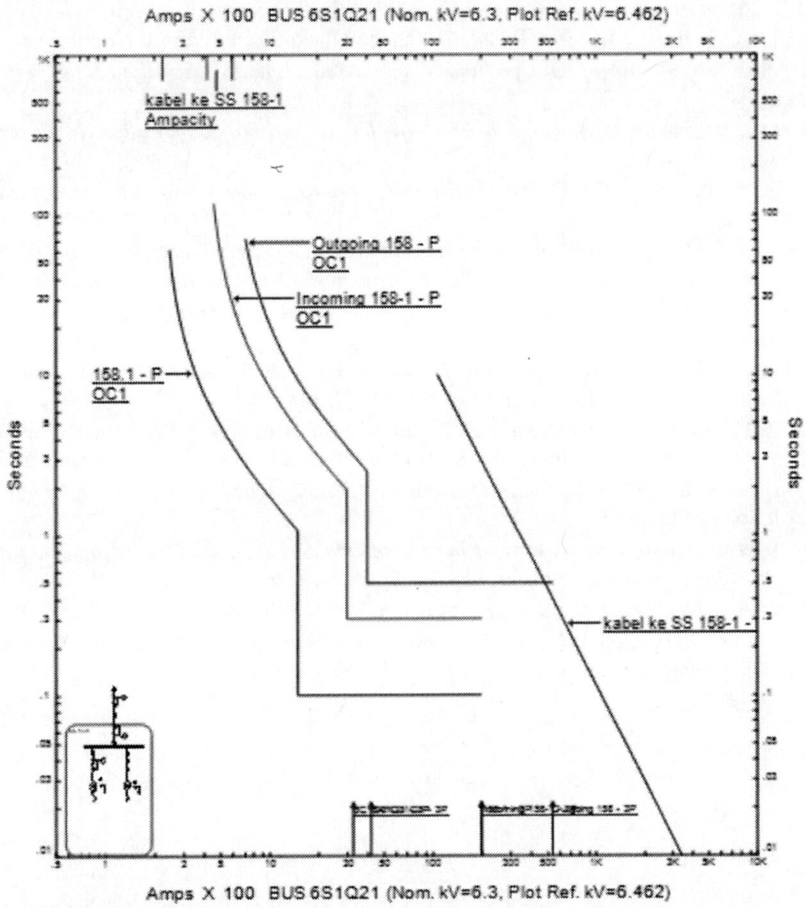

Figure 4. Characteristics of the SS Phase Disorder Curve 158.1.

Figure 4 shows the relay 158.1 as the primary relay, incoming relay 158 and outgoing relay 158 as the backup relays. The 158.1 relays coordinate to incoming 158 and the outgoing 158 relay curve has no overlapping; this condition indicates that the relay settings are correct because the first trip should occur at the primary relay.

4. Conclusion

The Cuckoo Search Algorithm (CSA) method was developed successfully to obtain the ideal setting value of the primary relay (t_p) and the backup relay (t_b). The test results of the CSA method show that all Coordination Time Interval (CTI) values are in accordance with the existing standards i.e. 0.2 seconds up to 0.4 seconds. The implementation of the CSA method in the calculation of coordination relay setting as the comparison with the manual calculation results shows that the CSA method has more capable.

References

[1] Bedekar, P P, et al., 2009 *Optimum Time Coordination of Overcurrent Relays in Distribution System Using Big-M (Penalty) Method* Visvesvaraya National Institue of Technology Nagpur (Maharashtra)

[2] Novi, A, 2012 *Koordinasi Rele Arus Lebih pada Sistem Kelistrikan PT. Pertamina Ubep Tanjung setelah Penambahan Beban dan Pembangkit Baru* Institut Teknologi Bandung

[3] Aditya, D P, et al., 2016 *Optimisasi Koordinasi Directional Over Current Relay (DOCR) pada Sistem Distribusi Mesh Menggunakan Modified Adaptive Particle Swarm Optimization (MAPSO) dengan Pembangkit Tersebar* Institut Teknologi Sepuluh Nopember

[4] Hakimi, M K, 2016 *Optimasi Koordinasi Rele Arus Lebih pada Sistem Distribusi Mesh dengan Pembangkit Tersebar Menggunakan Algoritma Genetika* Institut Teknologi Sepuluh Nopember

[5] Suwardi, D, et al., 2013 *Penerapan Algoritma Cuckoo Search pada Travelling Salesman Problem* STMIK Mikroskil

[6] Darji, G U, et al., 2015 *A tuned cuckoo search algorithm for optimal coordination of directional overcurrent relays* IEEE

[7] Gokhale, S S, et al., 2015 *Time overcurrent relay coordination using the Levy flight Cuckoo search algorithm* IEEE

[8] Guo, F, et al., 2014 Overcurrent Protection Coordination in a Power Distribution Network With the Active Superconductive Fault Current Limiter *IEEE Transactions on Applied Superconductivity* **24** 5

[9] Ghose, R, et al., 2014 *Cuckoo Search Algorithm for Speech Recognition* Institute of Engineering & Management Kolkata India

[10] Judanto, H P, 2017 *Studi Analisa Koordinasi Rele Proteksi pada Sistem Kelistrikan Indarung VI PT. Semen Padang-Sumatera Barat* Universitas Muhammadiyah Malang

[11] Bariq, J F, 2016 *Analisis Perencanaan Koordinasi Sistem Proteksi Relay Arus Lebih pada Jaringan Distribusi Tenaga Listrik di Pusdiklat Migas Cepu* Universitas Muhammadiyah Surakarta

[12] Ahmarinejad, A, et al., 2016 *Optimal Overcurrent Relays Coordination in Microgrid Using Cuckoo Algorithm* Science Direct

[13] Ahmadi, S A, et al., 2016. *Application of hyper-spherical search algorithm for optimal coordination of overcurrent relays considering different relay characteristics* Science Direct

[14] Rajabioun, R, 2011 *Cuckoo Optimization Algorithm* Science Direct

[15] Rathinam, A, et al., 2010 *Optimal Coordination of Directional Overcurrent Relays using Particle Swarm Optimization Technique* International Journal of Computer Applications

Improvement of maximum power point tracking (MPPT) efficiency using grey wolf optimization (GWO) algorithm in photovoltaic (PV) system

D Suhardi, L Syafaah, M Irfan, M Yusuf, M Effendy and I Pakaya*

Department of Electrical Engineering, University of Muhammadiyah Malang, Jl. Raya Tlogomas No. 246, Malang 65144, Indonesia

* Email: ilhampakaya@gmail.com

Abstract. Photovoltaics are becoming very popular, because the system does not produce pollution and can be installed anywhere, including in remote areas. However, in the use of photovoltaic found some common problems, it is difficult to get maximum and stable power. Therefore, to overcome these problems using Maximum Power Point Tracking method. On a photovoltaic system it is necessary to determine which converter will be used to increase the power output of the photovoltaic. The design of MPPT using gray wolf optimization algorithm that can track the output power quickly and reduce the oscillation in photovoltaic system. The result of the maximum power tracker using the gray wolf optimization algorithm is better than the incremental conductance algorithm of up to 0,4 s. Converters are used using soft - switching buck converter method to overcome the power losses that often arise in PV systems. The result of power output using soft-switching buck converter is greater than using buck converter. When comparing the efficiency of photovoltaic systems using the gray wolf optimization algorithm increases from using the incremental conductance algorithm.

Keywords: photovoltaic, MPPT, converter, power

1. Introduction

PV has become popular, because this system does not produce pollution and capable installed anywhere, including in remote areas. However, in the use of PV, some common problems are found, such as the difficulty of getting the stable and the maximum power when cloudy conduction or the panels are covered by trees. The power produced by PV depends on light intensity and temperature. Therefore, to overcome the problems which often occur in the PV system, a control system which capable to track the maximum power point of PV is needed, this system is known as the Maximum Power Point Tracker (MPPT).

Several research papers have been published on the developing MPPT algorithm. Jamil's study shows the hill-climbing algorithm works better than conventional algorithms on MPPT systems [1]. However, the output of the system using hill-climbing algorithm has a large noise, so another algorithm that more efficient is developed, such as the Incremental Conductance (IC) algorithm which can increase the power output of the PV compared with the Perturb & Observe (P & O) algorithm [2].

Kumar also proposed the fuzzy algorithm is better than the conventional algorithm, and fuzzy algorithm showed more effective and efficient [3]. Fuzzy algorithm is the basis for the development of new algorithm which inspired by the lives of animals such as Artificial Bee Colony (ABC) which has better Maximum Power Point (MPP) tracking capability on partial shading [4]. The MPPT algorithm functions used to find the maximum power point and maintain at the work point so the desired output is obtained. The purpose of developing algorithm is to get an MPPT system which should work efficiently, simply and accurately. So Gray Wolf Optimization (GWO) algorithm was made which inspired by the life of a gray wolf.

In PV system other than the determination of the algorithm used, it should also be noted the converter will be used, namely the DC-DC converter which has several basic circuits such as buck converter, boost converter, buck-boost converter, cuck converter and SEPIC converter. DC-DC converter serves to convert direct current into another direct current whose output value could be lower or greater than the input current. The growing development of existing technologies has emerged several novel buck type converter designs, such as soft-switching DC-DC converters which could improve the efficiency of MPPT [5]. Soft-switching is a switching method which functions to store temporary the input energy and release it to the output so increase the efficiency in the system.

2. Method

2.1. Photovoltaic design

The system used in this study uses PV ADT24200. In general, the design of this system is obtained from the equations based on the characteristics of the PV (Figure 1). The replacement circuit and the characteristics of the PV are shown in Fig. 1 and Table 1, respectively.

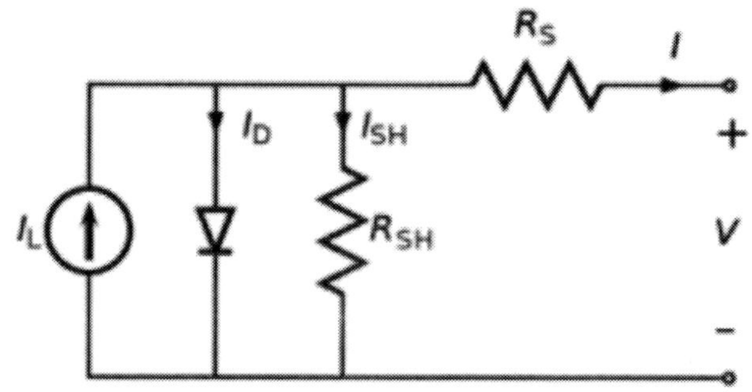

Figure 1. Replacement circuit of photovoltaic

Table 1. Characteristics of PV ADT24200

Characteristics	Amount	Unit
Maximum Power (Pmax)	200	W
Maximum Power Voltage (Vmp)	35	V
Maximum Power Current (Imp)	5,71	A
Open Circuit (Voc)	43	V
Short Circuit Current (Isc)	6,11	A
Jumlah Sell	25	Cel

From the reference and mathematical modelling parameters found in Table 1, the equation 1-5 is obtained [6]:

$$I_{pv} = I_{ph} + I_R \qquad (1)$$

$$I_{ph} = (I_{sc} + K_i(T_k - T_{ref}))\frac{G}{G_{ref}} \qquad (2)$$

IOP Conf. Series: Materials Science and Engineering **674** (2019) 012038 doi:10.1088/1757-899X/674/1/012038

$$I_{rs} = \left(\frac{I_{scr}}{\left[\exp\left(\frac{qV_{OC}}{N_s KAT} \right) - 1 \right]} \right) \tag{3}$$

$$I_s = I_{rs} \left(\frac{T_c}{T_{ref}} \right)^{3} \left[exp\left(\frac{qE_g\left(\frac{1}{T_{ref}} - \frac{1}{T_k} \right)}{KA} \right) \right] \tag{4}$$

$$I = N_p I_o \left[exp\left(\frac{q(vpv + I_{pv}R_s)}{N_s AKT} \right) - 1 \right] - V + \frac{IR_s}{R_{sh}} \tag{5}$$

2.2 Converter Design

DC-DC converter has many forms according to their respective functions, each converter has advantages and disadvantages. In this study, the converter used is soft-switching buck converter which used to reduce the voltage and the resulting power losses are less than the usual buck converter [7]. Determination of components in the soft-switching buck converter is the same as determining the value of the components in the buck converter which distinguishes only the soft switching. By determining the parameters to be used, it can perform calculations using equations 6-7 as follows:

$$D = \frac{V_{out}}{V_{in}} \tag{6}$$

$$L = \frac{V_{out}(V_{in} - V_{out})}{\Delta I\, x\, f\, x\, V_{in}} \tag{7}$$

$$C = \frac{\Delta I}{8\, x\, \Delta V\, x\, f} \tag{8}$$

In terms 6-8 it is shown the calculation of the value of the components to be used, as desired. The values of each component can be determined according to the parameters in table 2 as follows:

Table 2. Data concept on PV parameters

Characteristics	Amount
Power	200 W
Voltage Input	43 V
Voltage Output	34 V
Frequency	10 kHz
Forward Voltage	0. V
Current Output	5,7 A
R	6 Ohm

2.3 MPPT Design

MPPT serves to find maximum power and keep the power produced always at maximum and stable. There are several algorithms used and developed in the MPPT method. Efficient algorithms are needed, fast-tracking and resulting oscillations are more stable [8].

Gray Wolf Optimization (GWO) is one of the new algorithms developed to improve fast-tracking efficiency and the resulting oscillation is more stable. This algorithm is inspired by grey wolves to attack prey for hunting purposes [9]. The negligible value is the best position, in terms of tracking prey. In applying GWO to track maximum power [9]. Figure 2 shows the GWO suitability flowchart when tracking maximum power.

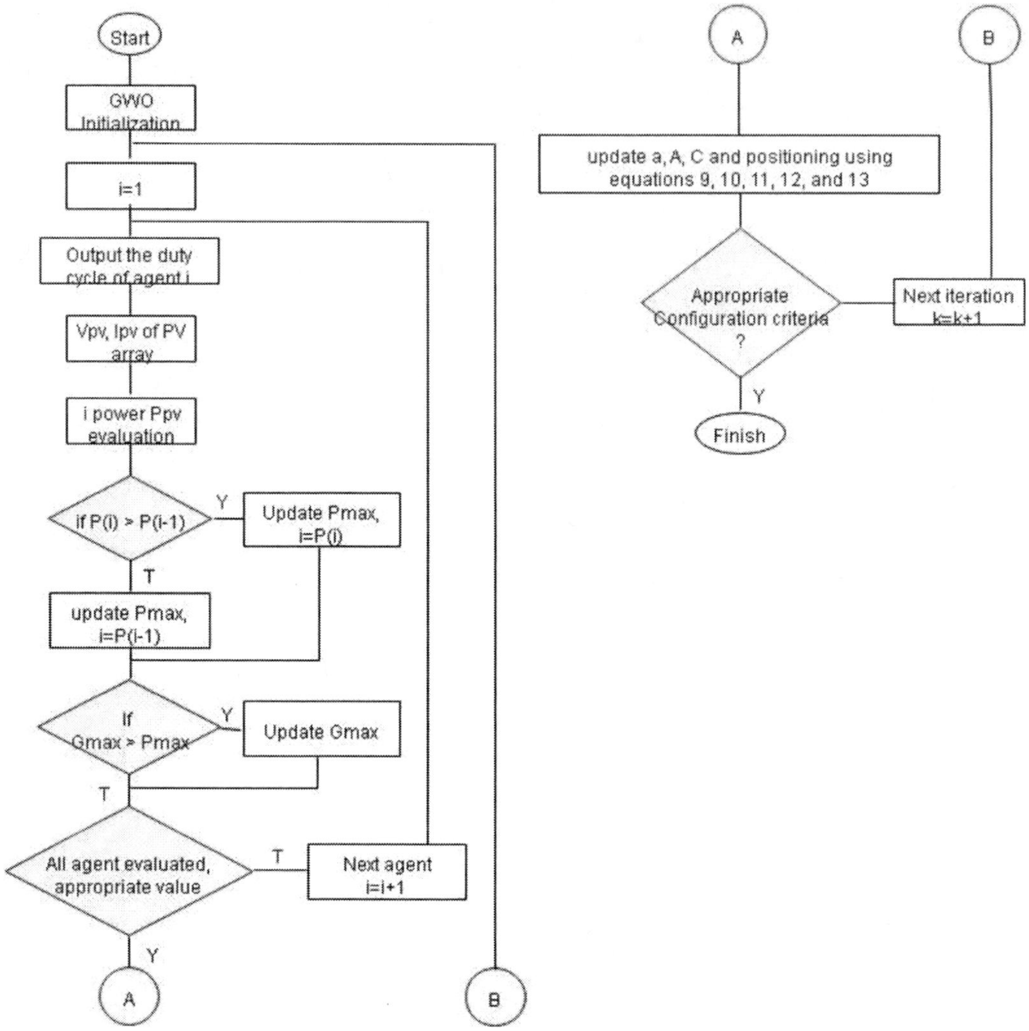

Figure 2. Flowchart of Grey Wolf Optimization (GWO)

In determining the value of the best position in tracking the maximum point of power in a photovoltaic system can be shown in equations 9-10.

$$\bar{D} = |\bar{C}.\overline{Xp}(t)||\overline{Xp}(t)| \quad (9)$$

$$\bar{X}(t+1) = \overline{Xp}(t) - \bar{A}.\, \bar{D} \quad (10)$$

Where t is the repetition of the current, D, A, and C represent the coefficient of vector, Xp is the position vector of the prey, and X shows the position vector of the gray wolf. Vector A and C are calculated using equations 11-12 as follows:

$$\bar{A} = \overline{2a}.\overline{r1} - \bar{a} \quad (11)$$

$$\bar{C} = 2.\overline{r2} \quad (12)$$

Where the α component decreases linearly from 2 to 0 during the loop time and r1, r2 is a random vector in [0,1].

In GWO method on MPP, the duty cycle is continued at a constant value. The output value using the GWO oscillation algorithm is reduced rather than using conventional algorithms, reduced power loss and results higher efficiency. To implement MPPT based on GWO, the duty cycle that is considered as a gray wolf. Therefore, it can be changed into equation 13 as follows:

$$D_i(k+1) = D_i(k) - A\,.\,D \quad (13)$$

IOP Conf. Series: Materials Science and Engineering **674** (2019) 012038 doi:10.1088/1757-899X/674/1/012038

3. Results and Discussion

The results of the simulation of the system have been made and tested to find out the system has been desired. Testing is divided into 3 stages, which are carried out separately.

3.1. Analysis of Photovoltaic

In the first test, the input of the intensity of the sun and temperature is set under normal conditions, so that the I - V and P - V curves can be shown in Figure 3.

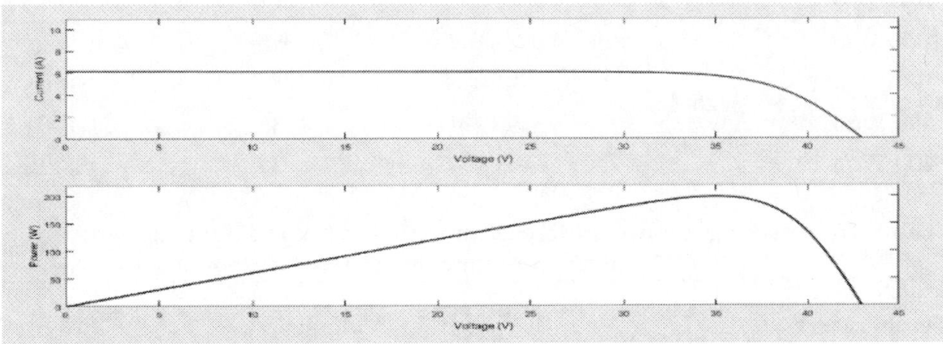

Figure 3. I-V and P-V curves at 25 ° C and light intensity of 1000 W / m2

Figure 3 shows when using a temperature parameter of 25 ° C and Light Intensity of 1000 W / m2, the power reaches 200 Watts at a voltage of 35 V and currents 6 - 6.2 A.

At the second test, the input of the intensity of the sun is set to change and the temperature is set at 25 ° C, so that the P-V curve can be shown in Figure 4.

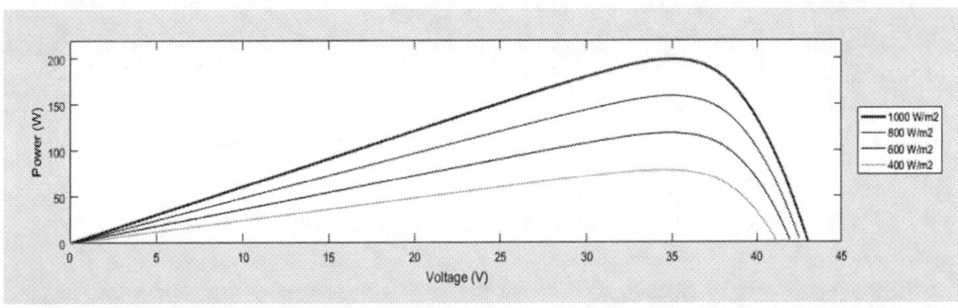

Figure 4. P-V curves at a temperature of 25 ° C and the intensity of light changing

Figure 4 shows the results of the test, if the light intensity received is greater then the power produced is also large. Because the resulting current is getting bigger.

In the third test, the input intensity of the sun is set at 1000W / m2 and the sun's temperature is in changing conditions so that the P-V curve can be shown in Figure 5.

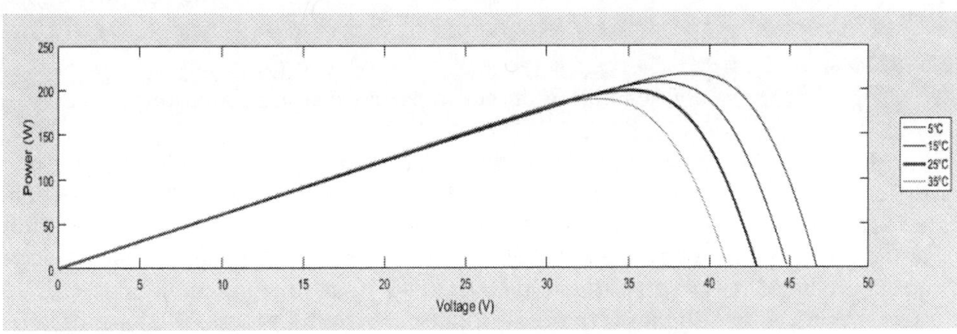

Figure 5. P-V curve at a light intensity of 1000 W / m2 and temperature change

Figure 5 can be analyzed that the test results show that if the temperature received is getting smaller, the resulting power increases, this is inversely proportional to the intensity of light. This test are shown in tables 3 and 4.

Table 3. The maximum difference in power produced by PV at changing light intensity and temperature 25 ° C

Light intensity	Temperature	Current (A)	Voltage (V)	Power (W)
1000 W/m^2	25°C	6 – 6,2	35	200
800 W/m^2		4,5 – 5	35	150 - 180
600 W/m^2		3,2 – 4	35	100 - 120
400 W/m^2		2 – 3	35	60 – 80

Table 4 The maximum power difference produced by PV at a light intensity of 1000 W / m2 and temperature that changes

Temperature	Light intensity	Current (A)	Voltage (V)	Power (W)
5°C	1000 W/m^2	6 – 6,1	41 – 42	240 – 242
15°C		6 – 6,1	39 – 40	224 - 226
25°C		6,1 – 6,2	35	200
35°C		6,1 – 6,2	34	188 – 190

The next test is conditioned when partial shading, so that the I - V and P - V curves can be shown in Figure 6.

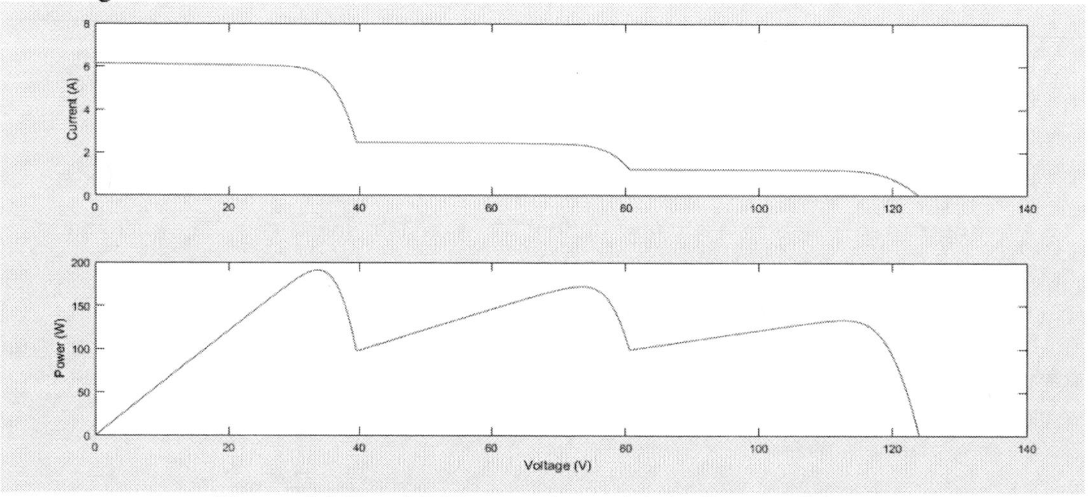

Figure 6. I-V and P-V curves in partial shading conditions

Figure 6 the condition made using 3 panels arranged in series, but the intensity of sunlight is arranged differently from panel 1 to panel 3: 1000 W/m2, 400 W/m2, 200 W/m2. The first test produces a current of 6A and a voltage of 30-40 V produces a maximum power of 200 W. In the second test, the intensity of sunlight received is only 800 W/m2 causing a decrease in electric current. Electric current produced is 4.5 - 5 A, the voltage is 30-40 V, and the maximum power is 150 -180 W. In the last test, the intensity of light decreases, the electric current becomes 2 - 3 A, voltage 30 - 40 V, and maximum power 60 - 80 W.

3.2. *Test of Converter*

In testing the converter can be seen in Figure 7.

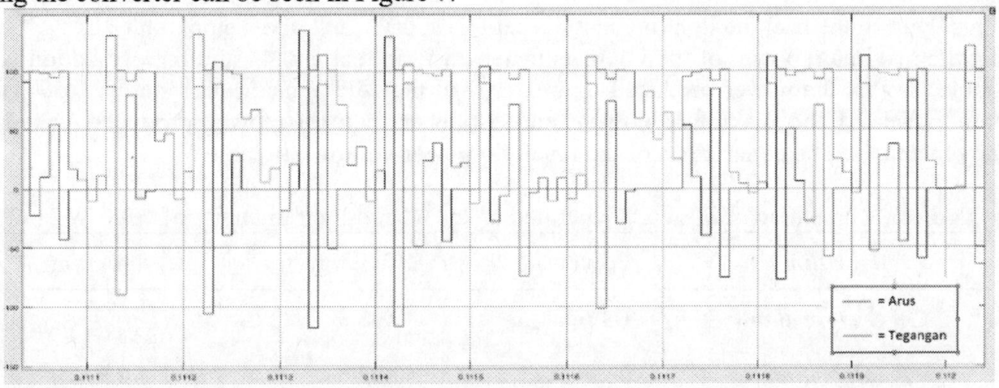

Figure 7. Output Current and Voltage on MOSFET Using Buck Switch Soft Switching.

The converter experiment uses an input voltage of 100V with a duty cycle of 50%, so that the expected output is 50% of input. In Figure 7 shows there is no overlap between current and voltage so that the power produced is greater when using a normal buck converter. In Table 5 below, it will show the output value of each converter which is carried out for 3 s.

Table 5 Soft Switching Comparison Simulation Results Buck Converter and Buck Converter

Converter	Vin (V)	D(%)	Iout (A)	Vout (V)	P (W)
Soft Switching Buck Converter	100	50	8,609	50,64	431,6
Buck Converter	100	50	8,32	50,01	416,2,4

3.3 *Test of MPPT*

In the first test, each MPPT was connected to PV and the converter used. The power generated can be seen in Figure 8 and 9.

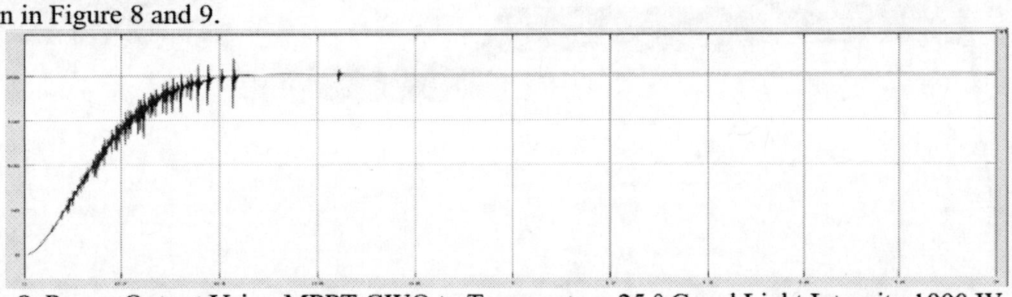

Figure 8. Power Output Using MPPT GWO to Temperature 25 ° C and Light Intensity 1000 W / m2

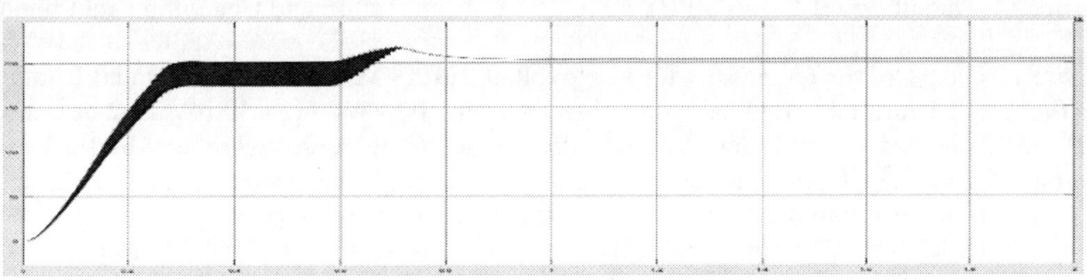

Figure 9. Output Power Using MPPT IC to Temperature 25 ° C and Light Intensity 1000 W / m2

Figures 8 and 9 show the results of testing for Photovoltaics using MPPT using GWO and IC algorithms. During the trial the light intensity is set to 1000 W / m2 and Temperature 25 ° C. It can be seen that the maximum power of the GWO is faster tracking that is 0.44 and the oscillation caused is more stable, so that the power produced is greater than the MPPT using IC requires 0.68 s to reach maximum power. if the maximum power track is faster then the power produced is longer and efficiency increases. The comparison of the resulting power is shown in table 6.

Table 6. Comparison of Power Generated at 25 ° C and light intensity of 1000 W / m2

Algorithm	**Power**	**Efficiency**	*Tracking*
GWO algorithm	203.6 W	0.98 %	0.55 s
IC algorithm	202.2 W	0.97 %	0.95 s

In the second test, each MPPT was connected to PV and the converter used. but in changing light intensity conditions, the power produced can be seen in Figures 10 and 11.

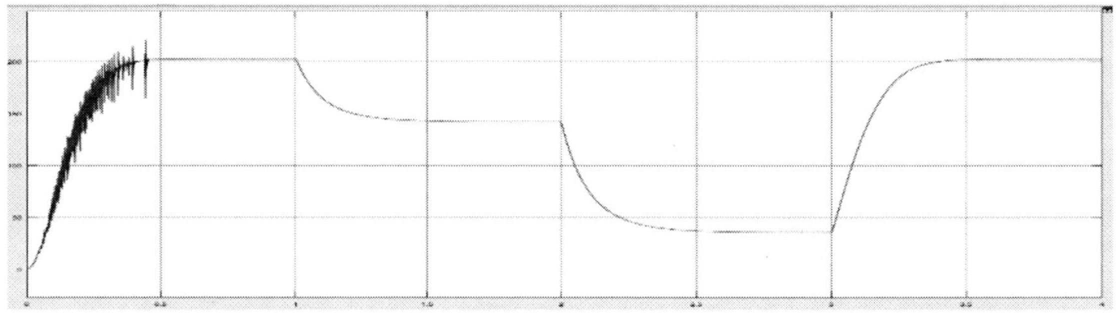

Figure 10. Output Power Using MPPT GWO to 25 ° C Temperature and Changing Light Intensity

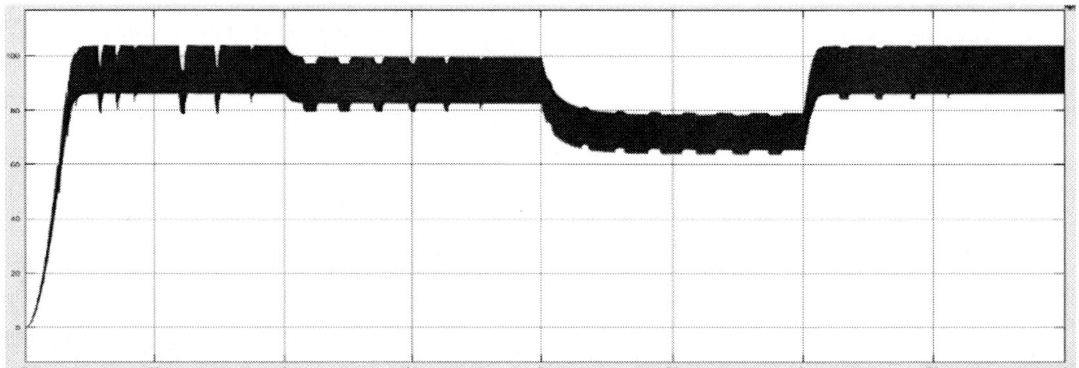

Figure 11. Output Power Using MPPT IC to 25 ° C Temperature and Changing Light Intensity

Figure 10 and 11 of the test results for Photovoltaics using MPPT using GWO and IC algorithms. During the test, the adjustable light intensity changes from 1000 W / m2 - 400 W / m2 decreases every second and at the last second the light intensity is set to increase again to 1000 W / m2, the temperature is set to 25 ° C. It is proven when tracking the maximum power of GWO is faster to track that is 0.55 and the resulting oscillation is more stable, so that the power produced is greater than the MPPT that uses IC. The IC fails to track the maximum power that is caused because the resulting losses are very large. The comparison of the resulting power is shown in table 7.

Table 7 Comparison of Power Generated at 25 ° C and Light Intensity Changed

Algorithm	Light intensity	Maximum power	Efficiency	*Tracking*
GWO algorithm	1000 W/m^2	203,2 W	98 %	0,55 s
	800 W/m^2	142,2W	98 %	
	400 W/m^2	35,9 W	98 %	
	1000 W/m^2	203,2 W	98 %	
IC algorithm	1000 W/m^2	95,76 W	92 %	-
	800 W/m^2	91,88 W	93 %	
	400 W/m^2	72,86 W	95 %	
	1000 W/m^2	81,92 W	92 %	

In the third test, each MPPT is connected to PV and the converter is used in changing temperature conditions, the power produced can be seen in figures 12 and 13.

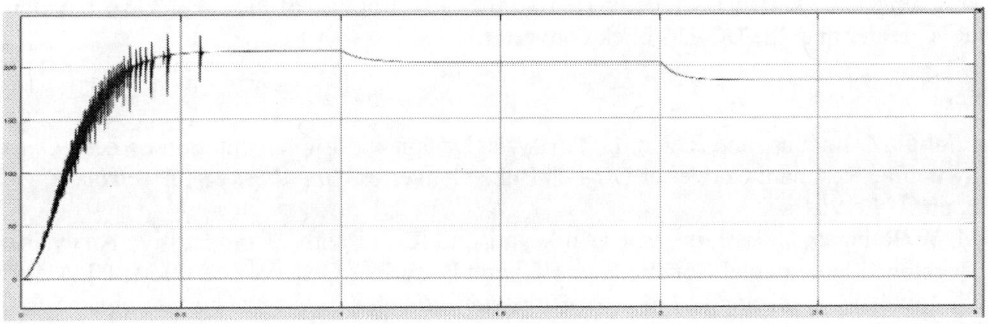

Figure 12. Output Power Using MPPT GWO to Light Intensity of 1000 W / m2 and Changing Temperature

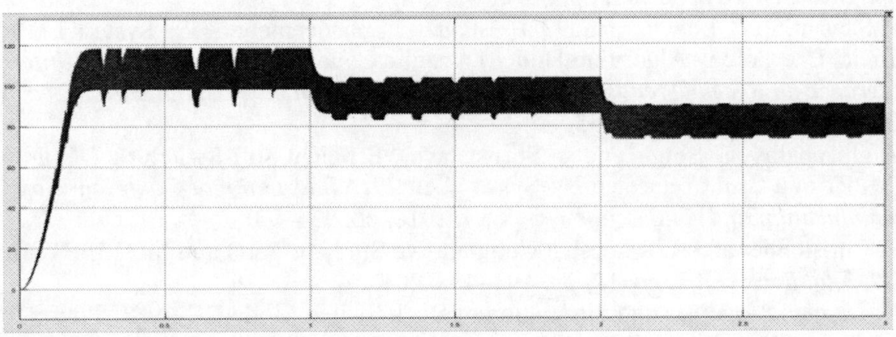

Figure 13. Output Power Using MPPT IC to Light Intensity of 1000 W / m2 and Changing Temperature

Figures 12 and 13 are the results of testing for Photovoltaics using MPPT using GWO and IC algorithms. During the test, the light intensity is set from 1000 W / m2 and the regulated temperature changes from 15 ° C, 25 ° C, 35 ° C every second. From the two results that have been done, it looks like GWO only takes 0.55 s to track its maximum power, while the IC fails to track its maximum power due to the greater losses. The comparison of the resulting power is shown in table 8.

Table 8 Comparison of Power Generated at Light Intensity of 1000 W / m2 and Temperature Changing

Algorithm	Temperature	Maximum power	Efficiency	*Tracking*
GWO algorithm	15 °C	214,8 W	98 %	0,55 s
	25 °C	203,2 W	98 %	
	35 °C	185,8 W	98 %	
IC algorithm	15 °C	109,1 W	92 %	-
	25 °C	95,76 W	92 %	
	35 °C	83,42 W	92 %	

4. Conclusion

Maximum Power Point Tracker (MPPT) using GWO algorithm is better in tracker compared to MPPT using IC algorithm, GWO tracker time only requires 0.44s. IC requires a longer time of 0.68 s and efficiency increase. Losses are caused when using a DC-DC soft switching buck converter lower than when using a normal DC-DC buck converter, so the output power of the DC-DC soft switching buck converter is greater than the DC-DC buck converter.

References

[1] I. Jamil, Z. Jinquan, and R. Jamil, "Analysis, Design and Implementation of Zero-Current-Switching Resonant Converter DC-DC Buck Converter," *Int. J. Electr. Electron. Eng. Vol*, vol. 2, pp. 1–12, 2014.

[2] M. W. Rahman, C. Bathina, V. Karthikeyan, and R. Prasanth, "Comparative Analysis of Developed Incremental Conductance (IC) and Perturb & Observe (P&O) MPPT Algorithm for Photovoltaic Applications," in *2016 10th International Conference on Intelligent Systems and Control (ISCO)*, 2016, pp. 1–6.

[3] A. Kumar, P. Chaudhary, and M. Rizwan, "Development of Fuzzy Logic Based MPPT Controller for PV System at Varying Meteorological Parameters," in *2015 Annual IEEE India Conference (INDICON)*, 2015, pp. 1–6.

[4] P. T. Sawant, P. C. Lbhattar, and C. L. Bhattar, "Enhancement of PV System Based on Artificial Bee Colony Algorithm Under Dynamic Conditions," in *2016 IEEE International Conference on Recent Trends in Electronics, Information & Communication Technology (RTEICT)*, 2016, pp. 1251–1255.

[5] A. Ambikapathy, G. Singh, and A. Shrivastava, "Efficient Soft-Switching DC-DC Converter for MPPT of a Grid Connected PV System," in *2016 International Conference on Computing, Communication and Automation (ICCCA)*, 2016, pp. 934–938.

[6] I. W. Christopher and R. Ramesh, "Comparative Study of P&O and InC MPPT Algorithms," *Am. J. Eng. Res.*, vol. 2, no. 12, pp. 402–408, 2013.

[7] G. C. Diyoke, "Analysis and Simulation of Buck Switch Mode DC to DC Power Regulator," *Int. J. Tech. Res. Appl.*, vol. 3, no. 1, pp. 97–103, 2015.

[8] S. Mohanty, B. Subudhi, and P. K. Ray, "A Grey Wolf Optimization Based MPPT for PV System Under Changing Insolation Level," in *2016 IEEE Students' Technology Symposium (TechSym)*, 2016, pp. 175–179.

[9] S. Mohanty, B. Subudhi, and P. K. Ray, "A New MPPT Design Using Grey Wolf Optimization Technique for Photovoltaic System Under Partial Shading Conditions," *IEEE Trans. Sustain. Energy*, vol. 7, no. 1, pp. 181–188, 2015.

IOP Conf. Series: Materials Science and Engineering **674** (2019) 012038 doi:10.1088/1757-899X/674/1/012038

Appendix

Nomenclature:

K_i	: Short-circuit current/Temperature Coefficient (0.0017A/K)
K	: Boltzman Constant (1.3805 x 10^{-23}J/K)
T_{ref}	: Temperature(K)
G	: Intensity
I_{sc}	: Short Circuit Current
I_s	: Saturation Current
T_k	: Actual Temperature (K)
T_{ref}	: Reference Temperature
q	: Electron charge (1.6 x 10^{-19}C)
E_g	: Bandgap energy (E_{g0} = 1.1 eV)
V_{oc}	: Open Circuit Voltage (V)

SBVI: A low-cost wearable device to determine location of the visually-impaired safely

S A Akbar, T Nurrohman, M I F Hatta and I A Kusnadi

Department of Electrical Engineering, Universitas Ahmad Dahlan,
Jalan Prof. Dr. Soepomo, SH, Yogyakarta, Indonesia

Email: sonali@ee.uad.ac.id, tofik216nurochman@gmail.com, hattafaiq@gmail.com, atmakusnadi@gmail.com

Abstract. Visually Impaired (VI) is blind people that low vision. They are difficult to recognize an environment like an indoor building or pedestrian. So, that needs a tool which suitable and smart for them. Many research and products have been developed assistive device increasing with smart technology, but it requires a high cost. For our purpose, its to develop a low-cost device and connected to the smart mobile. We are consisting to develop a smart prototype, it can use to determine the localization of blind people when lost or need support. By utilizing mobile apps on smart mobile, it provides a map-based message for the Visually Impaired safely. The device made small size and accurately.

1. Introduction

Blind or we called the Visually Impaired people, they have health disorders due to major difficulties for mobility access in the environment. So, they will get risk and complexity for obtaining the environment like room building or the pedestrian's spot [1], [2]. Those viewed when moving around pedestrian and unknown spots without guiding assistance, it gets lost or disturbed by irresponsible people. Actually, some people can be helping them, but that not giving more solutions if they stayed out of place. As for the Visually Impaired usually they utilize instincts in recognizing the environment but not all can be known or memorized. We know around in environment it not only available a signpost for blind people but also mostly not appropriated for them. So, much research is done developed and built more devices or may special line by various technologies implementation [3]–[5]. As we all use smart devices but it has a complexity to hard or expensive when used for blind caused their needs practices, adapted in several times as well as needs much money. According to blind people, they need some wearable tools to accompany and protect them in activities safely.

So, from that background, this research has developed a small smart device to give simplicity and also inexpensive. By using the Internet of Thing (IoT) which is currently very developed by various researches [6]. Our project name is SBVI (Smart Bracelet for Visually Impaired), which can work to help the blind when lost and need supports, also protecting them. Its made by low-cost parts and material. Furthermore, this study only focuses on Indonesian blind people. In addition, this research was supported by a university research grant and Indonesian blind private institution.

2. Low-Cost Design

When the Visually Impaired needs assistive devices. This research develops worthy technology for the deaf, especially to protect the Visually Impaired safely. So, we design a device with prioritizing suitable functions. Our concept is to help the Visually Impaired when get lost Our concept is to help the Visually Impaired when get lost in various locations. This system has developed utilize low-cost material part which uses the electronics component. Moreover, that design uses a minimum budget of more than $100, each part available to buy in the electronic market. The purpose idea is facilitating the assistive technologies for the Visually Impaired in their activities. The overall cost spent to build the prototype is $47; its shown detail presented in Table 1.

Table 1. Parts of device

Component	Unit Price ($)	Quantity needed	Subtotal Cost
(GPS) U-Blox Neo-6M	18	1	18
(GSM) SIM 900-a Module	14	1	14
Arduino Micro	7	1	7
Button	1	4	4
Li-ion Batteray	2	1	2
Additional Material	2	1	2

3. Research Work

3.1. SBVSI architecture

The system section of the design mainly using GPS sensor and GSM module installed, it used to detect and to send satellite data and convert as a message. GPS has a unique ability that can transmit data through real-time and mostly smart device used it for obtaining a position [7]. We used a microcontroller that can be controlling the system also proceed the data. In message text, its included the support message when they are needs a help. In other that, we used the google map apps to display the location map in the smartphone. To activate the device and transmits a notification, the user must be pressed the buttons it. There is two button which is the first button for activation and the second button for sending. Our system only uses minimum supply and long life caused design system its smaller and worthy. In another case, we use the cheap materials and wearable parts like a elastics bracelet and 3D printing its for PCB design. Figure 1 and Figure 2 both are shown the system design and the system work.

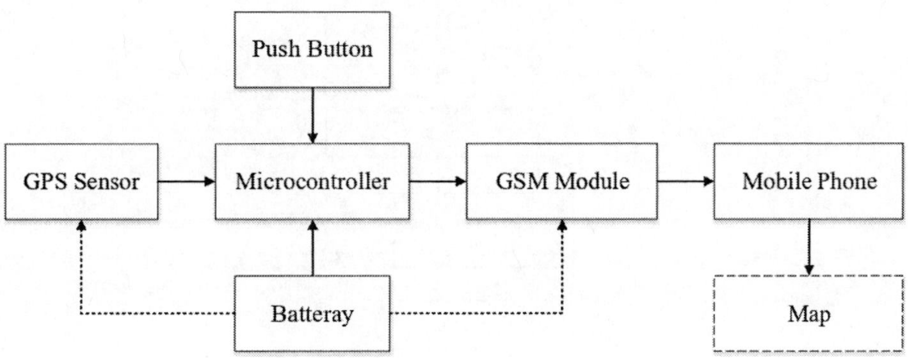

Figure 1. System Design

3.2. SBVI Hardware and software

Our device is used electronic sensor and part. The main sensor is the GPS U-Blox Neo 6M series that important for our device its to obtaining the land coordinate data (i.e. latitude, longitude, speed Mph). After getting the coordinate, the system will convert it using Arduino pro-mini (microcontroller series) as a message notification. For send the data to a smartphone, the system used the SIM 900a-GSM module type which needs an Indonesian SIM card model. To build the device it uses additional software for drawing schematics system [8]. Two push button mounts in the device are used to process start/stop and transmit the message data. Due to several part needs and limited supply, so we utilized the step-up/down module it can be floating of voltage. Furthermore, to switch on the system, we used a Li-ion battery of 9 Volt. To adjust the minimize the wearable device, also we design neatness the cable tagging.

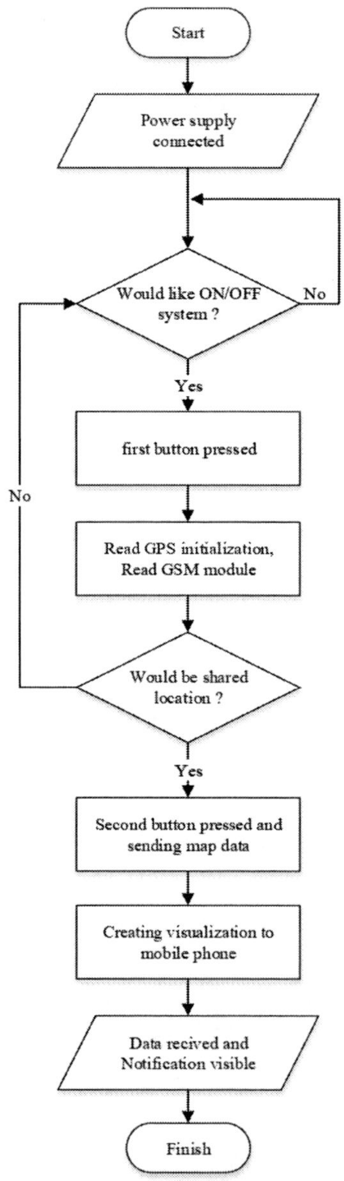

Figure 2. Flowchart

4. Results and Discussion

After a developed prototype, next stage is implemented to subject and trained by user exercise and get around in several locations. The system has been tested in five locations of four campus buildings and two public areas. Performance measurement that is determined accuracy of location recognized and message received

4.1. User Experiences
The prototype was small size designed it can be mounted on the arm or wrist. The tutorial installing process for the blind is shown in Figure 3. The procedure of test and train the device as follows:
1) We choose the visually impaired as the user and the trainer as the personal operator.
2) To introduce the system setting and start through an additional native speaker.
3) A user activates the system by the first button and next, they must be walking around into nearby spot
4) After that, they must button pressed of the second button in the random spot
5) Before the Personal operator activates the internet access in the smartphone. while they will get a message from a user device, they can be displayed through an online map and take it.
6) The personal operator also observes and check the accuracy of the direction parameter of user location

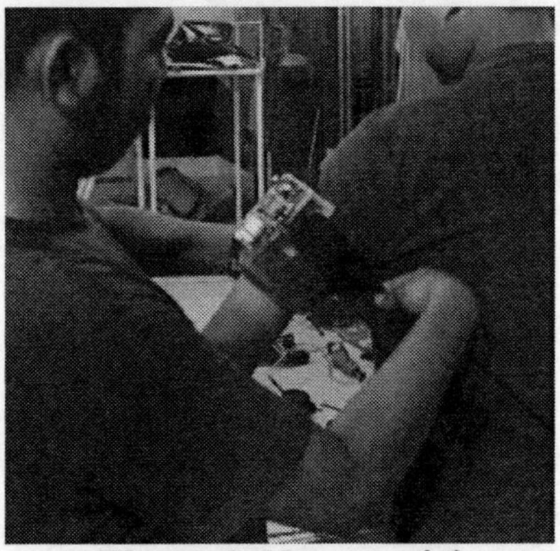

Figure 3. User tutorial

We were tasted by five subjects and get around six locations. we still semi-guided training and testing by walking or using the vehicle.

4.2. Message Notification
A notification message it provides for the personal operator. That included two messages which a help message and location map. The aiding message has three options, first about the information needs of directions, disruptions, traps of strangers. Figure 4 shown the message notification.

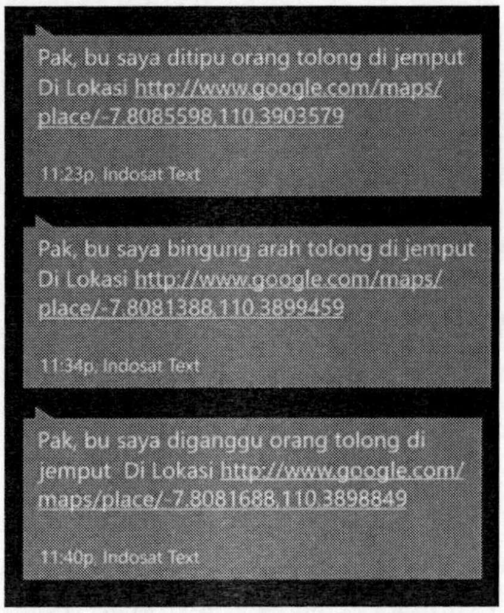

Figure 4. Three optional message and link of map for knowing the user need

IOP Conf. Series: Materials Science and Engineering **674** (2019) 012039 doi:10.1088/1757-899X/674/1/012039

4.3. Map-based localization

The prototype device is utilized the online map through mobile apps by google map database. That has the advantage to recognize and detect the location within accurate of position coordinates using satellite database also its free access. To display the map, the operator used a smartphone and must connected internet network. One of visualization of map-based localization its present in Figure 5. That is the result of user test it conducted between 3rd campus to the 5th campus of Universitas Ahmad Dahlan.

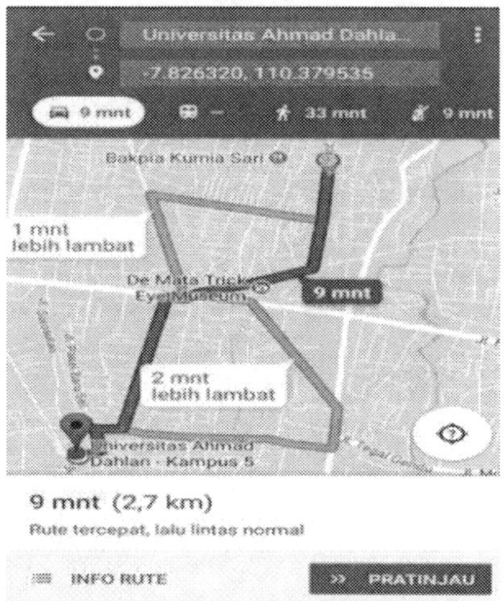

Figure 5. Tracking user using online map

4.4. Performance Correction

This study is done tested that finished by walking around in six locations. We conducted to recognize four campuses spot and two random spots both happen in Yogyakarta, Indonesia. The campus spot is chosen at Universitas Ahmad Dahlan which have five campuses building in several locations also conducted city centre spot and airport in Yogyakarta. We used two parameters to measure its obtained by distance and accuracy of system recognize. Eq (1) its use to performance calculate of the system. Table 2 is present the system ability through user experience.

Table 2. Accuracy average of localization.

Operator Spot	**User Location**	**Distance Rate (KM)**	**Miss Distance (Meter)**	**Accuracy (%)**
3rd Campus of UAD	1th Campus of UAD	1.3	10	99.23
3rd Campus of UAD	2rd Campus of UAD	1.4	20	98.57
3rd Campus of UAD	4th Campus of UAD	2.9	21	99.27
3rd Campus of UAD	5th Campus of UAD	2.7	50	97.82
3rd Campus of UAD	City Centre of Yogyakarta	2.8	20	99.28
3rd Campus of UAD	Yogyakarta Airport	5.2	15	99.71

$$\text{Accuracy avarage as follows} = \frac{(Distance\ rate - Miss\ distance)}{(Distance\ rate)}\ x\ 100\% \qquad (1)$$

5. Conclusion

We purposed to build a low-cost wearable device it appropriate for blind. By using low-cost parts and material which provided to develop a simple device and can be mounted on the arm or wrist. The device can be accompanied and protect the visually impaired safely it provided user tracking and support assist through the real-time message. Our device has generated an ideal concept and suitable

IOP Conf. Series: Materials Science and Engineering **674** (2019) 012039 doi:10.1088/1757-899X/674/1/012039

prototype for ease of use. In other studies, we will build a different smart device by medical approachment to blind and disability people.

References

[1] A. Riazi, F. Riazi, R. Yoosfi, and F. Bahmeei, "Outdoor difficulties experienced by a group of visually impaired Iranian people," *J. Curr. Ophthalmol.*, vol. 28, no. 2, pp. 85–90, 2016.

[2] V. Mothiravally, S. Ang, G. M. Baloch, T. T. Kulampallil, and S. Geetha, "Attitude and Perception of Visually Impaired Travelers: A Case of Klang Valley, Malaysia," *Procedia - Soc. Behav. Sci.*, vol. 144, pp. 366–377, 2014.

[3] R. Tapu, B. Mocanu, and E. Tapu, "A survey on wearable devices used to assist the visual impaired user navigation in outdoor environments," 2014.

[4] P. T. Mahida, S. Shahrestani, and H. Cheung, "Localization techniques in indoor navigation system for visually impaired people," *2017 17th Int. Symp. Commun. Inf. Technol.*, no. 28, pp. 1–6, 2017.

[5] G. Yang and J. Saniie, "Indoor navigation for visually impaired using AR markers," *IEEE Int. Conf. Electro Inf. Technol.*, pp. 1–5, 2017.

[6] N. S. Mala, S. S. Thushara, and S. Subbiah, "Navigation gadget for visually impaired based on IoT," *Proc. 2017 2nd Int. Conf. Comput. Commun. Technol. ICCCT 2017*, pp. 334–338, 2017.

[7] A. Mulla, J. Baviskar, A. Baviskar, and A. Bhovad, "GPS assisted Standard Positioning Service for navigation and tracking: Review & implementation," *2015 Int. Conf. Pervasive Comput. Adv. Commun. Technol. Appl. Soc. ICPC 2015*, vol. 00, no. M, 2015.

[8] "Eagle PCB Design," *Autodesk Inc.*, 2018. [Online]. Available: https://www.autodesk.com/products/eagle/.

Acknowledgments

This research is realized with collaborated and supported by Universitas Ahmad Dahlan research grant and Indonesia Blind Private Institution of Yaketunis Yogyakarta

IOP Conf. Series: Materials Science and Engineering **674** (2019) 012040 doi:10.1088/1757-899X/674/1/012040

DC motor simulation transfer function estimation: case study Proteus Ver. 7

S A Wahyu and D P Riky

Electrical Engineering. Ahmad Dahlan University, Yogyakarta, Indonesia

E-mail: wahyusaji@gmail.com, rikydp@gmail.com

Abstract: Computing technology development today has a significant influence on the design process of a product or system even including the learning process in the field concerned. The electronics and control field is a field that greatly helped by these developments. Developments of electronic products or systems become faster and more efficient with the availability of CAD (Computer-Aided Design) software. Some Electronic CAD software provides DC motor model, as Proteus 7.0. Availability of this model is beneficial in control system designing and learning but the availability of a DC motor models are often constrained because are not accompanied with adequate information about internal parameters of the dc motor model or dc motor modeling done in programming so it is not possible to derive dc motor transfer function based on internal parameters. This study successfully showed transfer function derivation of dc motor in Proteus CAD simulation with a black-box approach for linear systems.

1. Introduction

Understanding the character of a system is a prerequisite for the analysis of systems[1,7]. Often this knowledge is also needed in designing a new system composed of several well-known systems before. For systems that are known in detail we can derive system model from the system components mathematics model that has been developed previously For a system with unknown subsystem, then the system identification can be done by observing the system's response to a particular input. This approach is often referred to as a black box approach [2,3].

Within time domain, the black-box approach includes applied several alternative signals into the system; some of the signals are step signal, ramp signal, and impulse signal.

Proposals to use the unit step system response for the system transfer function determination started by Kupmuller [3] and this method is commonly used in industry [4,9].

In this study, we use the system response of step signal input to estimate system transfer function using MatLab identification toolbox. The system in this study is a dc motor model in Proteus ECAD tools ver7. A comparison between estimated transfer functions with real transfer function is carried out to judge validity of estimated transfer function. The contribution of this study is that it allows performing dc motor modeling on ECAD tools without having to obtain prior internal parameters of the dc motor.

2. Method

2.1 DC motor modeling

For a dc motor open-loop dynamic equation is[7,8,9] :

IOP Conf. Series: Materials Science and Engineering **674** (2019) 012040 doi:10.1088/1757-899X/674/1/012040

$$L\frac{di}{dt} + iR + k_b\varpi = V \tag{1}$$

$$J\frac{d\omega}{dt} = k_t i + T_d \tag{2}$$

where L is armature inductance (h), R is armature resistance (ohm), V is input voltage (volt), I for armature current (ampere), k_b for back emf constants, and K_t reveals as torque constants.

2.2 DC motor model in ECAD Proteus ver 7.0

Proteus ver 7.0 has DC motor model. DC motor model accommodates dc motor with inertia, and loading while the dc motor model with encoder is a model that accommodates dc motor and load inertia and equipped with encoder position. This study used a model of dc motors. The schematic model for dc motors as shown in figure 1.

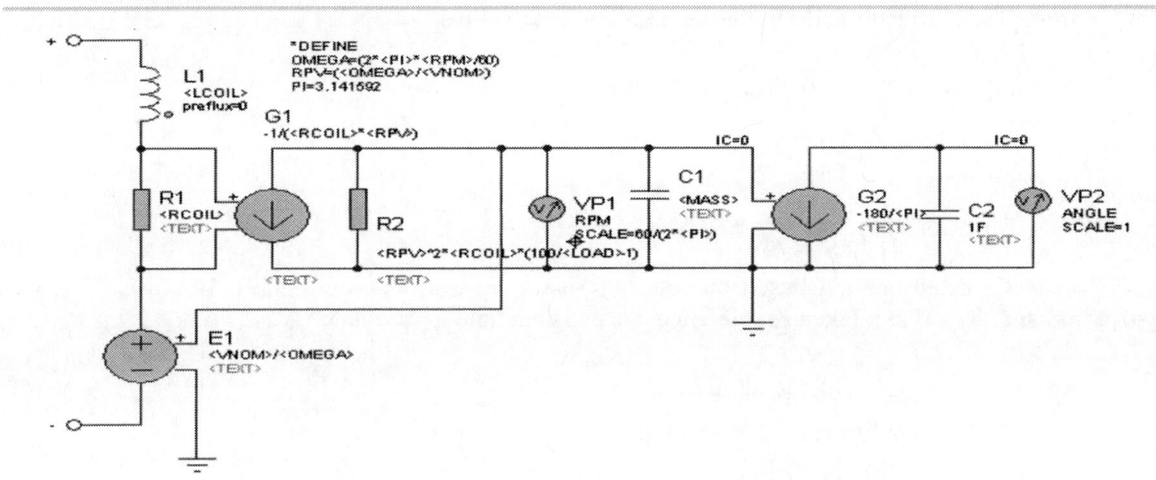

Figure 1. DC motor schematic model in Proteus [10]

2.3 DC motor model used in this study

In this study, DC motor with encoder is used. The motor model parameters are shown in Table 1

Table 1. DC motor model parameter being studied

Parameters	Nominal
R_a ,(Armature Resistance)	22 ohm
L_a (Armatur Inductance)	100 mH
Nominal Voltage	5V

The output of Proteus dc motor model is in the form of pulses. These pulses are generated by an internal rotary encoder. Pulse duty cycle and frequency can be set. Due to output is not in the voltage form, but in form of pulse, it is necessary to use a frequency-to-voltage converter circuit so that the block diagram of the system shown in Figure 2.

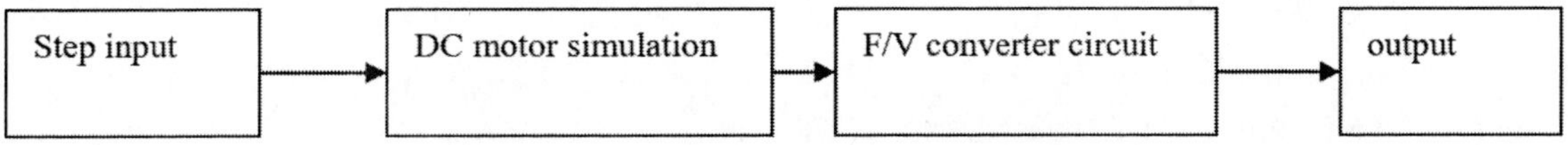

Figure 2. System experiment blog diagram

2.4 Frequency to Voltage Converter Circuit Design

In this F/V converter circuit, LM131 IC is used as active component F/V converter output at maximal input is designed at 10 V, with a maximum of 10% ripple voltage at a speed of 100 rpm. According to table 1, the maximum frequency is:

$$f_{max} = \frac{20}{rev}\frac{900}{min}\frac{1}{60}hz$$

$$f_{max} = 300hz$$

From this maximal frequency, we can derive a minimal time period (T_{min})) as:

$$T_{min} = 1/f_{maks} = 1/300 = 3.3ms$$

Based on T_{min}, we get $T_{pulse\ min}$: 0.5 T_{min} = 1,67ms; In the *datasheet* IC LM131, $T_{outhigh}$ is suggested at 0.8 from T_{min} ([6]Jacob,J.,1989) or 0.8 x3,33 ms = 2,664 ms.

In this study, we use Ct = 0.33μF, so Rt becomes:

$$R_t = \frac{t_{outhigh}}{1.1C_t} \quad (3)$$

$$R_t = \frac{2{,}664ms}{1.1 \times 0.33\,\mu F} = 7{,}3K\Omega$$

R_d and C_d determination based on rule 5 R_d $Cd \ll T_{pulsemin}$ (Jacob,J.,1989). Because $T_{pulsmin}$ is =1,67mS and R_d= 10kΩ, taken as reference, we can conclude to calculate C_d as :

$$5R_DC_D \le (0.1)(1{,}67ms) \quad (4)$$

Using formula 6 above, we get :

$$C_D \le \frac{(0.1)(1{,}67ms)}{(5)(10k\Omega)} = 03{,}3nF$$

$$V_{ave} = \frac{(2v)(1.1R_tC_t)R_Lf_{in}}{R_s}$$

V_{ave} is used to determining Rs value. With V_{ave}=10 v, R_s has value (Jacob,J,1989):

$$R_s = \frac{(2V)(1.1R_tC_t)R_Lf_{in}}{V_{ave}} \quad (5)$$

$$R_s = 14{,}8k\Omega$$

As part of F/V converter circuit, a filter is added. This filter is consists of R_L C_F, filter circuit has time constant,τ, at $\tau = R_L C_F$. V_{out} equation becomes [5] :

In order to get ripple voltage at 10% then :

$$\frac{V_{out}}{V_{pk}} = 90\% = e^{-\tau/T}$$

$$\ln(0.9) = -\frac{t}{R_LC}$$

$$C_F \ge 0{,}72\mu,$$

In this filter circuit, C_F=1μF is used.

3. Results and Discussion

3.1 Characteristic and Stability test of DC motor simulation and In Proteus

In order to obtain the characteristics of the system, the system is feed with several step voltage. These step voltages have magnitude at 14V and 15 V. These step voltage response graphs shown in Figure 3 and Figure 4.

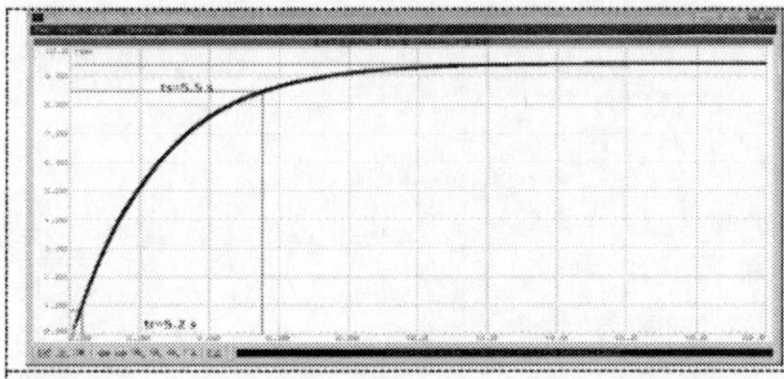

Figure 3. Graph Response for 14 V step input

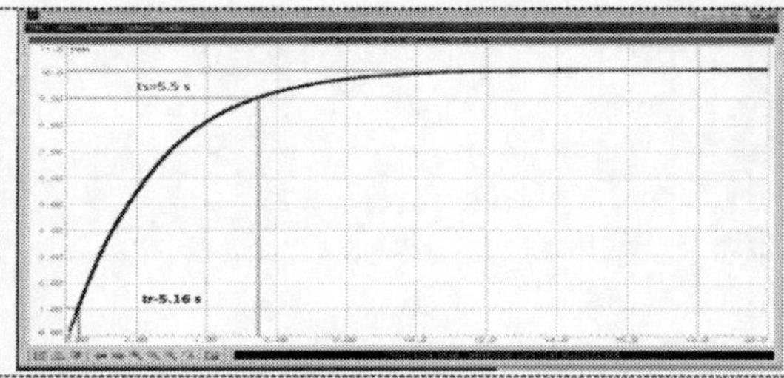

Figure 4. Graph Response for 15 V step input

Figure 4 to 5 visual examination, shows the system is an overdamped system, its means that system has a dumping ratio greater than 1. Also based on graphics examination, proteus simulation results give consistent results, the system has a rise time value and setting a relatively constant as shown in the following table 2.

Table 2. Settling time, t_r and t_s of DC motor model for various step input

V_{step},V	$N_{setling}$,rpm	t_r ,s	t_s ,s
14	940	5.2	5.5
15	1100	5.2	5.5
16	1080	5.2	5.6

3.2 Sampled data from the DC motor model step response.

In order to get step response data sample, motor dc model is feed with 15 V step input and then data is sampled in 0.25 second. Table 3 shown this sampling result.

Table 3. Data sampled at 0.25 s interval of dc motor model speed when feeding with 15 v step signal

Time, s	N, rpm	No.	N, rpm	No	N, rpm	No	N, rpm	No	N, rpm
0	0	1	6.4900	2	8.8200	3	9.7100	4	9.99
0.25	0.6900	1.25	6.8200	2.25	8.9600	3.25	9.7600	4.25	9.99
0.50	1.5900	1.50	7.1100	2.50	9.0800	3.50	9.7900	4.50	9.99
0.75	2.3400	1.75	7.3600	2.75	9.2000	3.75	9.8100	4.75	10

Data in Table 3 is used for transfer function estimation. Transfer function estimation obtained using Matlab, through the identification tool application. The model used is a process model for a second-order system, without delay and overdamped. The transfer function estimation results are:

$$\textit{Estimated transfer function} = \frac{0.676}{0.583s^2 + 2.629s + 1}$$

where Wn equals to 1, K is 0.676, and Dumping ratio is 1.31.

3.3 Transfer function confirmation

Transfer function confirmation test is done by comparing the dc motor response to a step input its included response from estimated transfer function and response from dc motor model. From both step response graph, graphically its seen that the value of t_r t_s and settling consistency between the two graphics, as we can see in figure 5 and figure 6.

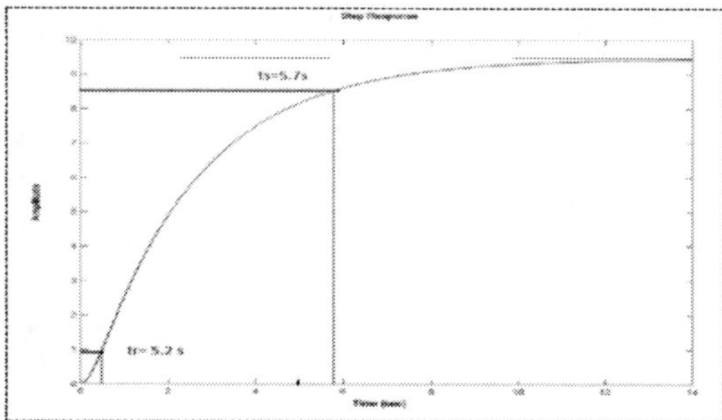

Figure 5. T System response when feeding with 14v step voltage

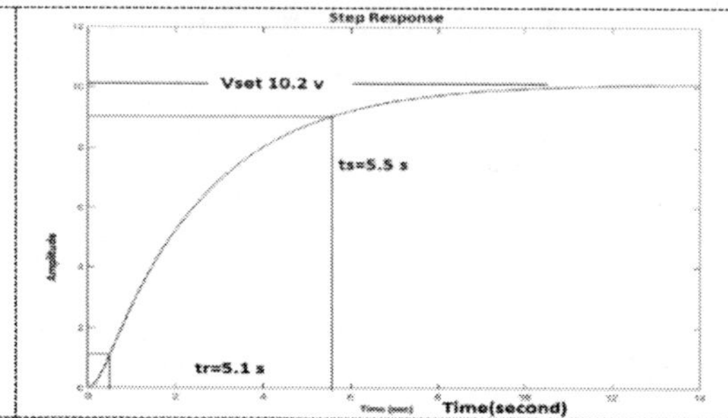

Figure 6. System response when feeding with 15v step voltage

Table 4 shows the comparison of t_r,t_s and settling time between estimated transfer function and dc motor when it is fed with 14V.

Table 4. Comparison table when the system is feed with 14V step voltage

Parameter	Proteus DC motor model	Matlab Transfer function	Deviation
$N_{settling}$,rpm	9.4	8.6	-8.5%
t_r ,s	5.2	5.2	0%
t_s ,s	5.5	5.7	3,6%

Based on the value of t_s and t_r from the three tables above, it can be concluded that the resulting estimates MatLab function approach with the actual transfer function from a dc motor in proteus simulation deviation values for t_r, , t_s and $N_{setting}$ quite small at less than 1.5% when it was feed with 15 v input.

Confirmation of the transfer function estimation by transfer functions are also performed by comparing the actual value of a point per point for the sample time 0.25 seconds when 14 v was used as input (Table 5).

Table 5. Result when system is feed with 14V step voltage

Time(s)	Simulation	Estimation	$\sqrt{(Err^2)}$ (%)	Time(s)	Simulation	Estimation	$\sqrt{(Err^2)}$ (%)
0.25	0.497	0.36	27.57	6	8.59	9.23	7.45
0.5	1.48	1.06	28.38	6.25	8.74	9.34	6.86
0.75	2.02	1.82	9.90	6.5	8.81	9.48	7.60
1	2.96	2.55	13.85	6.75	8.86	9.47	6.88
1.25	3.65	3.23	11.51	7	8.91	9.58	7.52
1.5	4.26	3.86	9.39	7.25	8.99	9.65	7.34
1.75	4.64	4.05	12.72	7.75	9.03	9.71	7.53

The average error between real value and estimated value from table 7 above is 8.8%. It means that estimated transfer function close enough approach real transfer function.

4. Conclusion
Based on the description above, we can conclude that the dc motor model contained in Proteus 7 can be modeled mathematically using the black-box approach.

References
[1] Ahmed S, Buang, Biao, Shah, Shirish May 2007 Novel Identification Method from step Response *Journal Of Control Engineering Practice* Vol.15

[2] Jacob J *1988 Industrial Control Electronics: Application and Design* Prentice-Hall New Jersey

[3] Franklin G F, Powell J D, and Workman M L 1990 *Digital Control of Dynamic Systems* Addison Wesley 2nd edition

[4] Unbehauen H and Rao G P 1998 A review of identification in continuous-time systems *Annual Reviews in Control* vol. 22 p. 145–171.

[5] Ljiung L 1999 *System Identification: Theory for the User* Prentice Hall, 2nd edition

[6] Ogata K,1990 *Modern Control Engineering* Prentice Hall Publisher New Jersey p. 57-58

[7] Saab S S and Kaed-Bey R A 2001 Parameter identification of a DC motor: an experimental approach *The 8th IEEE International Conference* on vol .2 p. 981-984

[8] Valdivia V, Barrado A, Laazaro, Zumel A, Raga P and Fernandez C 2009 Simple Modeling and Identification Procedures for "Black-Box" Behavioral Modeling of Power Converters Based on Transient Response Analysis *IEEE Transaction on Power Electronics*

[9] Wei Wu 2012 *DC Motor Parameter Identification Using Speed Step Responses* Modelling and Simulation in Engineering Hindawi

[10] Labcenter Electronic 2012 *Proteus VSM DC Motor Model* Manual LabCenter

Palmprint recognition using the cosine method

M Kusban*, B Hari P and A Budiman

Universitas Muhamamdiyah Surakarta, Indonesia

* Email: Muhammad.Kusban@ums.ac.id

Abstract. Palmprint Recognition Technology now requires breakthrough identification of diverse people. Palmprint recognition is the right choice of the system, namely the acceptance of biometrics that can be done quickly and cheaply because it has a significant enough media dimension that is difficult to manipulate. The four classifications of the biometric algorithm, the use of matching methods is still less attractive to researchers. In general, the part of the research concern is in the preprocessing and dimension reduction sections. Although researchers more widely use the Euclidean matching method, the selection of cosine methods is worth considering. The cosine method for the palmprint recognition matching process will have a significant effect on increasing the verification value when inserting the use of the contra-variance formulae in the equation. The selected amount of contra-variance is the data of the training. From the results of research that have been carried out, the rate of EER and verification are quite promising. The value of research results can compensate for other researchers in the same field.

1. Introduction

The palmprint recognition is part of the biometrics discipline used to get to know someone through the palm. As with the use of fingerprints, the palmprint recognition is increasingly in demand by researchers when in real life it can be used as a means of identifying evidence of a crime. In America, more than 30% of recent crime can be traced from the mark of the palm and not from fingerprints. Advantages use the palmprint recognition, the process of recognizing someone can run faster by merely waving. Besides that, the identification of palms can be made using low image resolution. As a recognition medium, the palms of the palms are used as feature matching obtained from three specifications: main lines, principal ridge, and bifurcation. Knowing the feature of hands can be done because no one has the same line pattern, even for twins. The different imprint of the hand pattern begin to form when the baby is still in the embryo for the 13th week, and it is correctly printed by the 18th week [1]. For the medical field, the recognition of the palms can be used to diagnose several inherited diseases such as down syndrome, aarskog syndrome, cohen syndrome and fetal alcohol syndrome [2].

There have been several researchers who have contributed to the palmprint biometrics field. Kang stated that the use of image filters was able to improve the performance of retinal biometric systems [3]. Shanmugapriya concluded that the use of the image filter was able to increase the local and global detail value of the palmprint [4]. Sharma proposed using the image filter method to reduce and even eliminate noise in nonlinear mapping [5]. After image normalization, it is necessary to homogenize the position of all images. Sung stated that the right method for normalizing the location of various images is to use Gabor [6]. Along with Sung's opinion, Wang noted that Gabor's use helped strengthen the information feature [7]. The adverse effect of using the Gabor method is the increasing number of images that must be processed. To overcome the Gabor method problem then the dimension reduction process is used. Badrinath prefers the use of the PCA method to dimensions reduction in his research; meanwhile, Aykut suggests using KPCA dimension reduction techniques [8]. After the remaining data is only valuable data, the analysis continues with the matching process. According to Ong, the matching method that

IOP Conf. Series: Materials Science and Engineering **674** (2019) 012041 doi:10.1088/1757-899X/674/1/012041

many researchers use in biometrics is the Euclidean technique [9]. The advantage of using the Euclidean technique is the simplicity of the function, the ability to find a different scope of resolution, and excel in completing different models [10]. Until now there are no researchers who claim to identify the object of research correctly without errors successfully. The best way to improve a biometric recognition system is to improve the performance of each stage of classification, starting from the data acquisition process, preprocessing, extracting information patterns, and last is using matching methods. According to Kong, to obtain a biometric system with high accuracy and fast processing, it is necessary to classify algorithmic methods in four stages, namely image acquisition, preprocessing, data extraction, and matching [2]. Zhang said that research that follows the four classifications could to accommodate five primary objectives of a biometric system that are to resist changes in illumination, rotation, translation, increasing insignificant the amount of data, and accurate in the matching process [11]. Aykut and Ekinci used the original PolyU database then combined it with the Gabor method [8 × 5], the KPCA dimension reduction technique, and Euclidean matching succeeded to get a verification value of 97.12% [12]. Badrinath and Gupta used morphology method for image enhancement followed by Gabor technique [8 × 5], PCA method, and the Hamming method for matching was able to produce a verification value of 95.409% [13]. Li [14] in a study using binary form data input, then using the Gabor method [6 × 3], PCA applications for dimensional reduction, and the use of weight sum matching techniques managed to obtain research values for EER of 0.03. Finally, Xu [15] used the skeleton method for image enhancement processes, continued the use of Gabor [6 × 6], and the multiple machers matching method succeeded to gain an error rate of 0.18.

Most researchers only offer algorithms in specific subsections of the overall biometric recognition classification. In this research, the author suggests using a 3W (wavelet, Wiener, weighting) method to improve the appearance of the image. The 3W technique is image filter algorithms that is a combined algorithm consisting of wavelet transforms, Wiener filters, and weighting methods. In general, researchers in the palmprint recognition divide the research subsection into four parts. The warkac method is a term for four algorithmic systems which includes the choice of 3W method, Gabor method, KPCA, and cosine matching. Expected research output is the capability of the system algorithm to obtain a low EER (equal error rate) rate and high verification value.

2. Method

In research, input data using images from PolyU as many as 600 images. Each image has a variation of 9 pieces with a division of 3 for training, three testing, and 3 for evaluating. The simulation process uses Matlab software. The output of the research is the value shown in the table and four biometric curves, namely CMC, ROC, DET, and EPC. As a comparison, at the end of the paper also included a comparison of test scores with other researchers. The sequence of the research was carried out in series as shown in Figure 1.

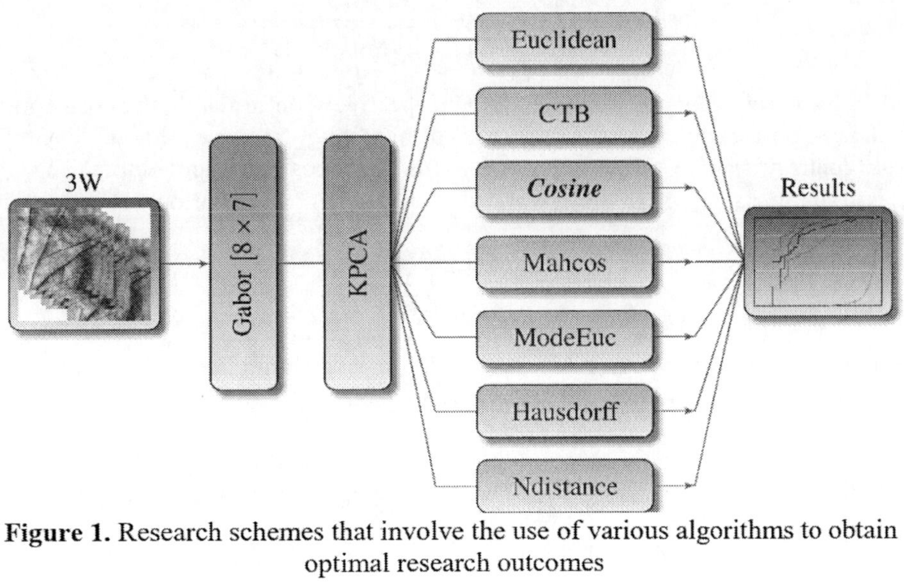

Figure 1. Research schemes that involve the use of various algorithms to obtain optimal research outcomes

After the image filter process using the 3W algorithm, the research process is continued using the Gabor method. The orientation parameter and the Gabor scale function to set the reference point for all images. The reference point is needed because of the thousands of images used in the study have different angles and scales so that a method is required to normalize the position of all images. Using the orientation and scale Gabor of [8 × 7] proved to have verification outputs that are suitable for the palmprint recognition [16]. The Gabor process has a negative impact, namely increasing the number of dimensions. The processed image increases by the multiplication between orientation and scale. The increase in the amount of data is identical to the increase in the computational process. A dimension reduction method is used to overcome the problem of large amounts of data which also means to reduce computational processing time. The dimensional reduction is used to eliminate as much unnecessary data as possible while maintaining only essential data. The presence of crucial data in the dimensional reduction process must be able to represent all existing data even though it has been deducted from unnecessary data. By Nibouche, the PCA method is often used by researchers in the palmprint recognition process [17]. In the study, the dimension reduction method used was the KPCA method according to the results of the study [1]. After the data that is left behind is only essential data, the research continues with the matching process. The Euclidean method is a commonly used matching method while in this study, the technique used is the Cosine method from the results of the study [1]. The distance method or matching technique in biometrics is a statistical technique used to evaluate the similarity of training information in impostor and genuine data. If no match is found, then the system is considered superior and vice versa if there is a lot of data compatibility between impostor and genuine then the system is deemed to be weak. According to Kang,, in the introduction of the palms, the matching method that is widely used by researchers is the Euclidean method [9]. The advantage of using the Euclidean method is the simplicity of the function, the ability to search for a unique, and robust resolution scope in completing far different models [10]. Although it has advantages, Velasquez stated that the use of the Euclidean method has the disadvantage of not including the scope of correlation attributes which results in no consistency when adding attributes [18]. To overcome the weaknesses of Euclidean, some researchers used other matching algorithms such as the cosine method [19]. Qian [20] stated that there is a similar function between the use of the Euclidean method and cosine. The Euclidean method with cosine angle distance or known as cosine has a comparable measurement value when used in data input less than 128 samples [21]. The cosine matching is used to measure the similarity of two non zero vectors. This measurement principle of cosine is when measures 0^0 or does not have an angle then it will have the highest value of 1, whereas if it has an angle it is said to have a value below 1. Two vectors have similarities when the two values are parallel or have an angle difference of 0^0, and are supposed to have no similarity when perpendicular. The cosine distance formula can be written in the following equation.

$$\cos\theta = \frac{A \cdot B}{|A||B|} = \frac{\sum_{i=1}^{n} A_i B_i}{\sqrt{\sum_{i=1}^{n} A_i^2}\sqrt{\sum_{i=1}^{n} B_i^2}} \tag{1}$$

With A_i and B_i are components of vector A and B respectively. In research, the cosine method is used to compare data impostor with genuine generated from testing data or evaluating data. The effect of using a covar (contra-variance) values impacts towards better cosine system output. The covar value in research uses training data (ξ). If the covar value is inserted in Equation (1), the cosine equation becomes the following form.

$$\cos\theta = \frac{A \cdot \mathrm{covar}(\xi) \cdot B}{|A||B|} \tag{2}$$

The covar calculations are not given to data testing or evaluating data. The output of the research will be worse when the covar value is allocated to both data.

3. Results and Discussion

The research that discusses matching methods using data from PolyU as many as 600 samples. Compared algorithms include the use of Euclidean techniques, City Block, cosine, Mahalanobis Cosine, Euclidean Modulation, Hausdorff, and Ndistance. The research results are shown in Table 1, and the biometric curves are shown in Figure 2.

Table 1. The research results that involved various distance methods for the palmprint recognition

Method	Time (s)	FRR	FAR	EER	Ver. (%)
Euclidean	0.73178	0.00364	0.00407	0.00385	99.63636
CTB	0.71109	0.01182	0.01190	0.01186	98.81818
Cosine	0.73379	0.00273	0.00272	0.00272	99.72727
Mahcos	0.74022	0.00364	0.00364	0.00364	99.63636
ModEuc	0.72987	0.00364	0.00407	0.00385	99.63636
Hausdorff	0.71819	0.50364	0.50385	0.50375	49.63636
Ndistance	0.73510	0.49727	0.49641	0.49684	50.27273

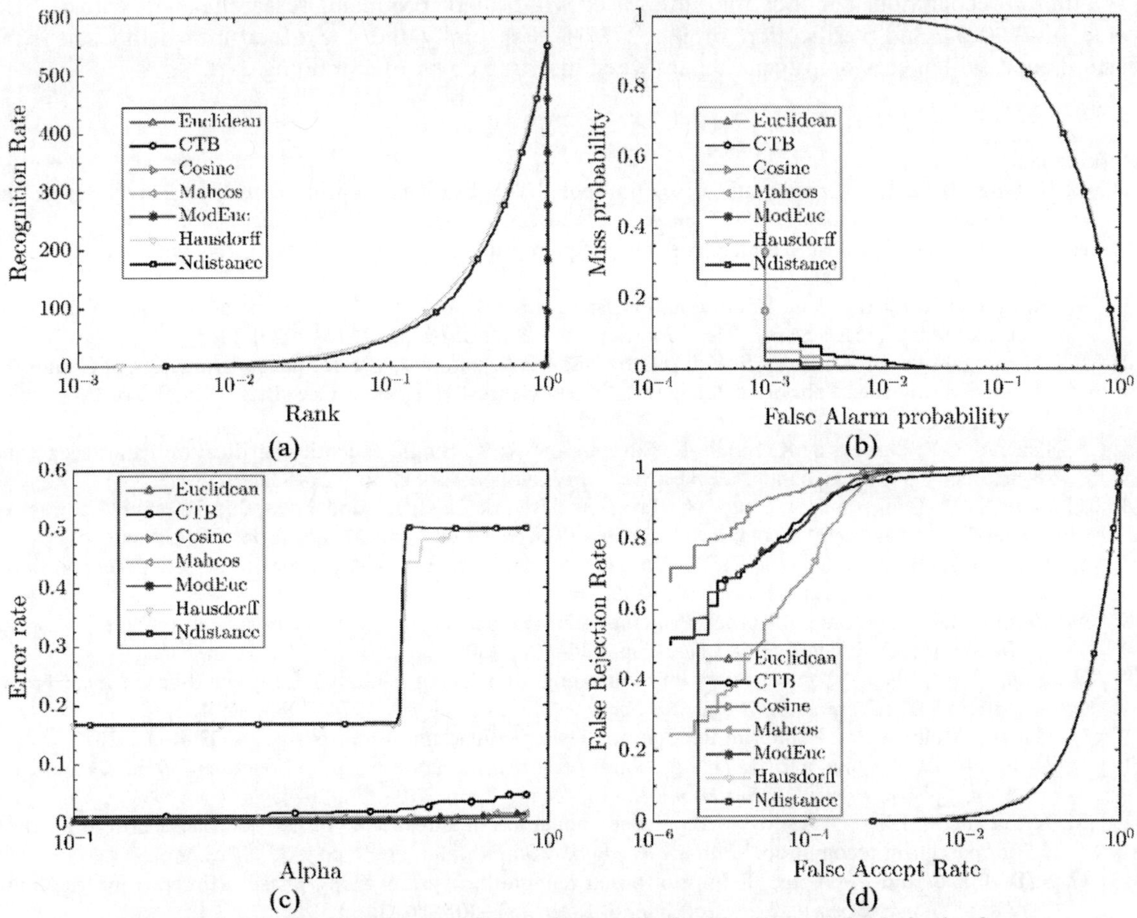

Figure 2. Comparison value of the cosine method with other methods in the form of four biometric curves: (a) CMC (b) DET (c) EPC (d) ROC

Of all the distance methods used, the cosine method has the most prominent research value with an EER rate of 0.00272 and a verification value of 99.72727%. Some other techniques that have verification values above 99% are the use of the Euclidean, MahCos, ModEuc methods. To find out the comparative value of the distance method, the four biometric curves were used to strengthen the research result shown in the Table 1. The curves will reinforce the advantages of the cosine compared other methods. The cosine method with the representation of the green line in Figure 2 shows the superiority method in the four curves. Figure 2 (c), the green line is in the lowest position which means having the smallest error value. This fact is strengthened by the ROC curve in Figure 2 (d) that the green line has the most dominant range of the false accept rate which is close to 1 for all research values. To strengthen the results research, Table 2 shows the results of a study for the use of different distance methods.

Table 2. Comparative value of matching methods between researchers in The palmprint recognition

Research	Distance	EER	Verification (%)
Aykut [8]	Original	0,00637	99,36364
Badrinath [13]	Hamming	0,02905	97,09091
Kusban *proposed method*	Cosine	0,00272	99,72727

4. Conclusions

The choice of 3W, Gabor [8 × 7], KPCA, and cosine methods can be used as an alternative algorithm in palmprint recognition. The four algorithms used will provide promising research results with an EER value of 0.00272 and verification of 99.72727%. Especially the use of cosine methods, system performance, will have a significant impact when inserting covar over training data.

References

[1] G. K. Ong Michael, T. Connie, and A. B. Jin Teoh, "Touch-less palm print biometrics: Novel design and implementation," *Image Vis. Comput.*, vol. 26, no. 12, pp. 1551–1560, Dec. 2008.
[2] A. Kong, D. Zhang, and M. Kamel, "A survey of palmprint recognition," *Pattern Recognit.*, vol. 42, no. 7, pp. 1408–1418, 2009.
[3] W. Kang and X. Chen, "Fast Representation Based on a Double Orientation Histogram for Local Image Descriptors," *IEEE Trans. Image Process.*, vol. 24, no. 10, pp. 2915–2927, 2015.
[4] K. Shanmugapriya, M. Karthika, S. Valarmathy, and M. Arunkumar, "Performance Evaluation of Contourlet Transform based Palmprint Recognition using Nearest Neighbour Classifier," vol. 3, no. 1, pp. 294–299, 2013.
[5] S. Sharma, S. R. Dubey, S. K. Singh, R. Saxena, and R. K. Singh, "Identity verification using shape and geometry of human hands," *Expert Syst. Appl.*, vol. 42, no. 2, pp. 821–832, 2015.
[6] J. Sung, S. Y. Bang, and S. Choi, "A Bayesian network classifier and hierarchical Gabor features for handwritten numeral recognition," *Pattern Recognit. Lett.*, vol. 27, no. 1, pp. 66–75, 2006.
[7] H. jun Wang, H. nian Qi, and X. F. Wang, "A new Gabor based approach for wood recognition," *Neurocomputing*, vol. 116, pp. 192–200, 2013.
[8] M. Aykut and M. Ekinci, "Kernel Principal Component Analysis of Gabor Features for Palmprint Recognition," *Icb*, vol. 5558 LNCS, pp. 685–694, 2009.
[9] W. Kang and Q. Wu, "Contactless palm vein recognition using a mutual foreground-based local binary pattern," *IEEE Trans. Inf. Forensics Secur.*, vol. 9, no. 11, pp. 1974–1985, 2014.
[10] Q. Matrix, "Multispectral Palmprint Recognition Using a Quaternion Matrix," pp. 4633–4647, 2012.
[11] B. Zhang, W. Li, P. Qing, and D. Zhang, "Palm-print classification by global features," *IEEE Trans. Syst. Man, Cybern. Part ASystems Humans*, vol. 43, no. 2, pp. 370–378, 2013.
[12] M. Aykut and M. Ekinci, "AAM-based palm segmentation in unrestricted backgrounds and various postures for palmprint recognition," *Pattern Recognit. Lett.*, vol. 34, no. 9, pp. 955–962, 2013.
[13] G. S. Badrinath and P. Gupta, "Palmprint based recognition system using phase-difference information," *Futur. Gener. Comput. Syst.*, vol. 28, no. 1, pp. 287–305, 2012.
[14] Q. Li, X. Li, Z. Guo, and J. You, "Online personal verification by palmvein image through palmprint-like and palmvein information," *Neurocomputing*, vol. 147, pp. 364–371, 2015.
[15] Y. Xu, L. Fei, and D. Zhang, "Combining Left and Right Palmprint Images for More Accurate Personal Identification," *Image Process. IEEE Trans.*, vol. 24, no. 2, pp. 549–559, 2015.
[16] M. Kusban, A. Susanto, and O. Wahyunggoro, "Excellent performance of palmprint recognition by using wavelet filter," *ICIC Express Lett.*, vol. 11, no. 8, 2017.
[17] O. Nibouche and J. Jiang, "Palmprint matching using feature points and SVD factorisation," *Digit. Signal Process. A Rev. J.*, vol. 23, no. 4, pp. 1154–1162, 2013.
[18] N. H. Zardari, K. Ahmed, S. M. Shirazi, and Z. Bin Yusop, "Weighting Methods and their Effects on Multi-Criteria Decision Making Model Outcomes in Water Resources Management," *Int. J. Oper. Res.*, vol. 10, no. 2, pp. 56–66, 2015.
[19] A. Meraoumia, S. Chitroub, and A. Bouridane, "Gaussian modeling and Discrete Cosine Transform for efficient and automatic palmprint identification," *2010 Int. Conf. Mach. Web Intell.*, pp. 121–125, Oct. 2010.
[20] G. Qian, S. Sural, Y. Gu, and S. Pramanik, "Similarity between Euclidean and cosine angle distance for nearest neighbor queries," *Proc. 2004 ACM Symp. Appl. Comput. - SAC '04*, p. 1232, 2004.

[21] E. Liu, A. K. Jain, and J. Tian, "A coarse to fine minutiae-based latent palmprint matching," *IEEE Trans. Pattern Anal. Mach. Intell.*, vol. 35, no. 10, pp. 2307–2322, Oct. 2013.

Acknowledgments
This research of the palmprint recognition field is a National Strategic Research program - Institution, with funding from the Directorate of Research and Community Service Directorate General of Strengthening Research and Development of the Ministry of Research, Technology and Higher Education in accordance with Research Contract number: 72.67 / A.3-III / LPPM / III / 2018.

A Smart Sustainable approach for waste management in post-natural disaster phase

Y A Fatimah[1], R Murniningsih[2], A Setiawan[1] and M Aman[1]

[1] Engineering Faculty, Universitas Muhammadiyah Magelang, Indonesia
[2] Economic Faculty, Universitas Muhamamdiyah Magelang, Indonesia

* Email: yun.fatimah@ummgl.ac.id

Abstract. As an archipelago country, Indonesia is lied on the Pacific ring of fire - the center of active volcano and earthquake, located in tsunami-prone regions and several tectonic plates and had intense tropical rain, that places Indonesia at regular risk of earthquake, volcano eruption, tsunami, flood, and other natural disasters. Instead of nature and severity destructions, these natural disasters create a large volume of debris and waste, about 5 to 10 times of the solid waste generated usually, that have negative impacts on environmental, social and public health if not managed carefully. This study proposes a smart, sustainable approach for waste management in post-natural disaster. Some natural disaster happened in Indonesia was collected from Indonesia's National Disaster Mitigation Agency to identify the global waste and debris potential from the disaster. Smart, sustainable municipal waste management framework based on four priority areas of governance, economy, social and environment, from the previous study, is adopted to develop a post-disaster smart, sustainable waste management framework and a pre-design system. The proposed framework is focused on the technical mechanism for collecting and converting generated waste and debris into valuable materials that are integrated into the waste management system.

Keywords: Indonesia, Disaster waste, Framework, Smart, Sustainable

1. Introduction

Indonesia is a marvelous archipelago country, lied between the Indian and Pacific oceans, rich with gorgeous mountains, beaches, tropical areas, green forest, and natural environment. On the other hand, Indonesia is also lied on the center of active volcano and earthquake, located in tsunami-prone regions and several tectonic plates and had intense tropical rain, which is placed Indonesia as the most vulnerable country prone magnifies natural disaster. Volcano eruption, tsunami, earthquake, floods, forest fires, and landslides ae frequently happen in Indonesia. Indonesia's National Disaster Mitigation Agency, stated that there were at least 2,342 natural disasters in 2016 and 2,213 natural disasters in 2017 [1] which jeopardized Indonesian people and environment. This natural disaster not only creates nature and severity destructions (i.e. death, injured, property damages, destroyed infrastructure, and building) but also generates unavoidable and tremendous waste and debris. Managing waste and debris in post-natural disaster becomes very critical as they can create negative impacts on environment, social and public health. Furthermore, the waste and debris generated from the catastrophe are possibly mixed and contaminated with hazardous and toxic materials, which are very dangerous for human health and environmental performance. It was estimated that tsunami in Aceh generated 780,000 tonnes of waste and debris, while a mountain of waste and debris created other health and environmental disasters. It was found that managing this waste and debris was very challenging due to technical and infrastructure issues.

This research aims to develop a post-disaster smart sustainable waste management framework and its design system based on four priority areas of governance, economy, social and environment, modified from municipal waste management framework in the previous municipal waste management study. It is expected that the contingency program improves disaster treatment in sustainable ways. A comprehensive literature review provides trend of waste management pyramid, circular economy as the baseline of purposed framework, ICT and IoT for disaster waste management system and smart-sustainable waste management system. A technical mechanism for handling, collecting and converting generated waste and debris into valuable materials that are integrated into the disaster recovery system is also proposed in this study. ICT and IoT are used for Information Computer Technology and Internet of Things to improve the efficiency and effectiveness of waste management.

2. Literature Review

2.1. The trend of waste management hierarchy

Waste has negative significant impacts not only on to environment but also human if it does not manage properly. The waste creates pollution, greenhouse gas (GHG) emission, dirty environment that can create serious economic, social and environmental issues. Therefore, proper waste management becomes essential need to minimize the waste negative impacts. Moving up the hierarchy of waste management (prevention, reuse, reduce, recycle, remanufacturing, energy recovery and disposal) [2] as presented in Figure 1 is urgently required to achieve more sustainable waste management.

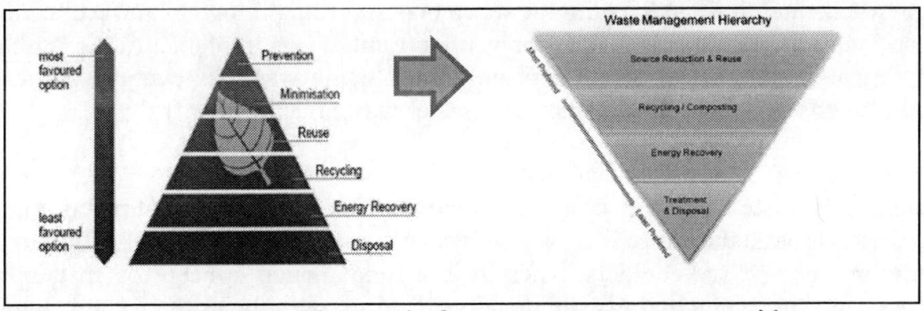

Figure 1. The trend of waste management pyramid

In Figure 1 The landfill is expected to be the last option with small portion or even zero portion in the hierarchy, while reduction activity is expected to be the top priority in waste management system. Therefore, a new concept of circular economy is expected to turn up the current WMS pyramid into sustainable WMS pyramid, in which the landfill will have the minimum portion of waste after reduction, recycling or composting energy recovery options. This trend of waste management considers achieving zero waste through the implementation of appropriate strategies related to consumer consumption behavior to treatment technology.

2.2. Circular Economy as the baseline for future disaster waste management system

A circular economy is an innovative system that has objectives to close the waste loop and to design waste out of the system. The circular economy offers transformations of waste management linear model into a more economic, social and environmentally responsible waste management system [3]. Circular economy facilitates the reduction of material depletion, waste landfilling, dangerous pollution and emission, environmental degradation, energy leakage and narrowing energy through long-life design, good maintenance, recondition, remanufacturing, recycling and remodification as presented in Figure 2.

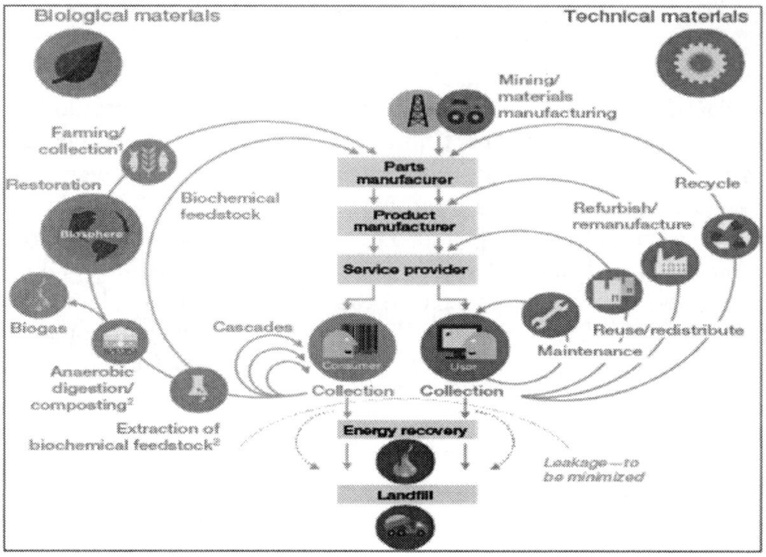

Figure 2. Circular Economy as the fundamental approach

The fundamental principal of the circular economy is cradle to cradle which is in contrast with the make-take-waste system. The circular economy is the future way to achieve sustainable economic, social and environmental through keeping the more material (i.e. biological and technical materials) in production and consumption circle. Two main opportunities for implementing circular economy in waste management are turning waste into resource and using waste as energy source. This circular concept can be an effective approach to sustain disaster waste management.

2.3. Smart and sustainable waste management system

The sustainability of waste management can be achieved if it is proactive (not reactive), has objectives to minimize waste through the entire life cycle of waste management (from collection to disposal), and need cooperation amongst stakeholders. Therefore, a fundamental concept of sustainable and smart waste management for municipal waste developed from previous research in Figure 3 can be considered to enrich the development the purposed post-disaster waste management system.

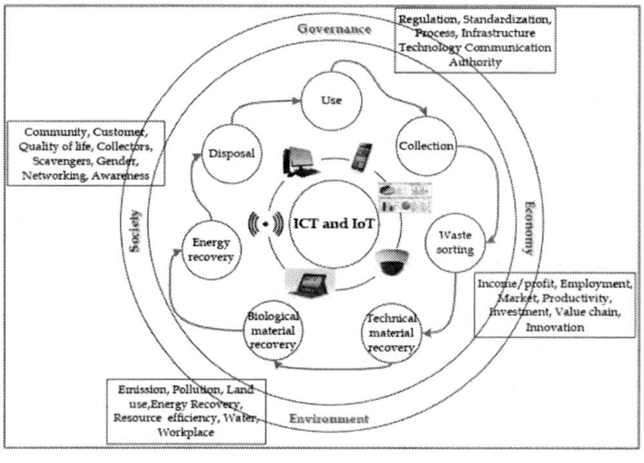

Figure 3. The fundamental concept of sustainable and smart waste management

Figure 3 illustrates the use of ICT and IOT as the core of the waste management system through the integration of four elements (i.e. governance, economy, society, and environment) which play important roles in achieving a sustainable disaster waste management system.

3. Methods

In-depth literature review and direct communication with Government agencies including Indonesia Indonesia national disaster management agency (BNPB), environmental department, and Non-Government Organisation (NGO) were carried out to gain data and information related to the post-disaster waste management in Aceh, Lombok, Yogyakarta, and Palu. The data is then used to identify the existing situation of the post-disaster waste management in those cities including the amount of the waste and debris, the collection strategy applied, the vehicle used, the temporary storages chosen, the appropriate treatment technologies conducted and landfill sites provided. An in-depth analysis was conducted for the existing models and approaches for disaster waste management. The result was then used as the fundamental framework for developing smart, sustainable municipal waste management framework based on four priority areas of governance, economy, social and environment. To help to formulate the framework, this research adopted municipal waste management framework conducted in the previous research. The framework is focused on the technical mechanism for handling, collecting and converting waste and debris into valuable materials that are integrated into the disaster recovery system. This framework considers the integration of ICT and IoT to waste management to improve the efficiency and effectiveness of waste management.

4. Result and discussion

4.1. Indonesia post-disaster waste stream and management

Earthquake and Tsunami predominantly create devastating impacts on Indonesia compared to the other natural disaster flood which is the most frequently hit Indonesia. The worst natural disaster was The Indian tsunami which hit Aceh in 2004. The tsunami created huge disaster waste in this area. The disaster waste was estimated to reach 780,000 tonnes, mainly containing C & D waste, vegetation, and soil waste. The waste including mixed building, households, woods, and vehicle wastes, was removed from the disaster area and sent to a landfill site to clean the area for the reconstruction process. The waste handling tools and transportation uses dump trucks belong to the municipal and contractors [4]. The handling, treatment, reuse and recycling processes of waste management mechanisms were applied to demolition waste. About 12,000 wood furniture was made from the tsunami [5]. However, there is very minimum information about how waste is exactly collected, stored, transported, and treated. The other natural disaster happened in Indonesia is earthquake in Lombok. During this catastrophe disaster, there were about 13,000 damage buildings. The waste and debris were estimated consist of C & D waste, vegetation, and soil waste. The collection strategy applied, the collection used very limited excavators thus the waste cleanup was very slow. Lack of heavy equipment limited the waste cleanup process. the cleanup process was conducted by Indonesian military and private companies using total 48 excavators. The community involvement was very limited as they have no knowledge on how to manage the disaster waste [6]. The other earthquake in Yogyakarta in 2006, created about 350.000 damage buildings and houses and produced about 2,9 million tones of waste and debris [7]. The disaster tsunami and earthquake recently also hit Palu in 2018, devastated environment and building, brought abundant waste and sweep people to death. Lack of an effective mechanism to warn people when the disaster-hit Palu, has caused massive destruction. The disaster has left complex issues including the massive volume of disaster waste (e.g. mixed soil, building materials, debris, trees, boats, household waste, and hazardous waste) which is poorly managed. There were about 68.451 house building destroyed, that created tons of mixed waste [8]. The waste was collected in the temporary waste boxes before they sent to final disposal place. Provincial Government of Palu provided conventional trucks, excavators and bulldozers to transport the waste from temporary waste collection to final disposal place. However, limited number of people, machines, and vehicles delayed the waste management process, thus the waste was mountain everywhere. Some of the organic waste was burned to reduce mounted waste in the disaster location. Not only humans who were negatively affected by the disaster, a lot of animals (i.e. cows and goats) were poorly neglected and eat the waste for their food.

4.2. Lessons learned on waste management

Well preparation for potential disasters and its impacts are urgently required to help community, city, and area to reduce the impacts created from disasters. Such planning for post-natural disaster waste and debris waste management systems will prevent the issues of unmanaged massive waste, speed recovery process and avoid high cost and dangerous risks for mistakenly handling the waste. Some experience of natural disaster waste management presents good practices in waste management plan [9]. Generating a long period waste management plan has become priority of California City to prepare disaster impacts potentials. The plan includes waste collection and handling strategy, temporary waste storage, waste treatment technology (recycling programs), disposal, hazardous waste categorization, and waste management dissemination. Therefore, some formal and informal agreements between communities and cities are also conducted to get quick access to urgent needs of personal and equipment. Implementation of an appropriate treatment technology such as recycling, reconditioning and remanufacturing program [10], [11] in a place is very important to quickly expand the existing waste management capacity when a disaster occurs. An updated solid waste management plan is also very essential as it will present the existing waste management capacity, location and types to supporting agencies (communities and contractors) thus will help in responding the appropriate action during the disaster. In addition, developing a special communication approach involving local communication media has also helped to direct the community to manage their disaster waste [9]. Preparing waste management staff involving local community people is a great source of labor to help government in managing disaster waste. Identifying waste management transportation, equipment, and supplies before the disaster has a great impact to reduce the lack of resources and to improve the waste transfers. Providing the information on the collection and storage locations are necessary to be available during the disaster to help in transforming the waste. Furthermore, segregating hazardous waste strategy is essential to be applied for controlling and avoiding hazardous contamination from waste stream [12].

4.3. Role of ICT and human in post-natural disaster waste management

To transform the conventional waste management system into a smart and sustainable waste management system in post disasters, three essential factors including the role of technology, information, and human are essentially involved. The technology used in the disaster area is varying depended on the disaster types (i.e. landslide, volcano eruption, floods, tsunami, and earthquake), topography position and location (i.e. urban, semi-urban, rural areas), country (i.e. developed, developing countries), etc. [13]. Appropriate technology (e.g. machinery, vehicles) is required for collecting variety of waste (i.e. debris, building waste, automobiles, furniture, and electronic products), transporting from disaster areas to disposal centers, and threating the waste. The waste management system in post-disaster is challenging and complex as the waste is often mixed and spread. However, the current waste management technology offered automatic waste collection system which changes the way waste collection system work with automatic technology by using ICT such as sensors for sorting process, GPS for waste transportation process monitor, logistic digital system, waste mobile apps, automatic compactor trucks etc. [14] Sources of waste information during post-disaster are important to be communicated and integrated through waste collection process, transportation and treatment. However, the most waste information during post-disaster is still collected by different organization (e.g. government and non-government organization) which has a lack communication and integration. Thus, there is missing communication and data integration that should be shared among the waste management stakeholders. Some research discussed the importance of waste data collection (i.e. big data), including to identify some issues during the waste management operations and to make appropriate decisions to provide best solutions to collect, transport and threat the waste [15].

4.4. A sustainable post-disaster waste management framework

Learn from a developed country (sophisticated technology), and adopt the local contents, a framework to achieve sustainable waste management in post-disaster is developed. The framework covers five important waste management stages as illustrated in Figure 4. First stage concerns the understanding and estimation of potential debris and waste together with its approximate amount and type of waste. The second stage focuses on determination of the collection strategy and centers. The third stage concerns the selection of effective transportation and storage. Fourth is determining the appropriate treatment. The fifth is determining the proper sites of landfills.

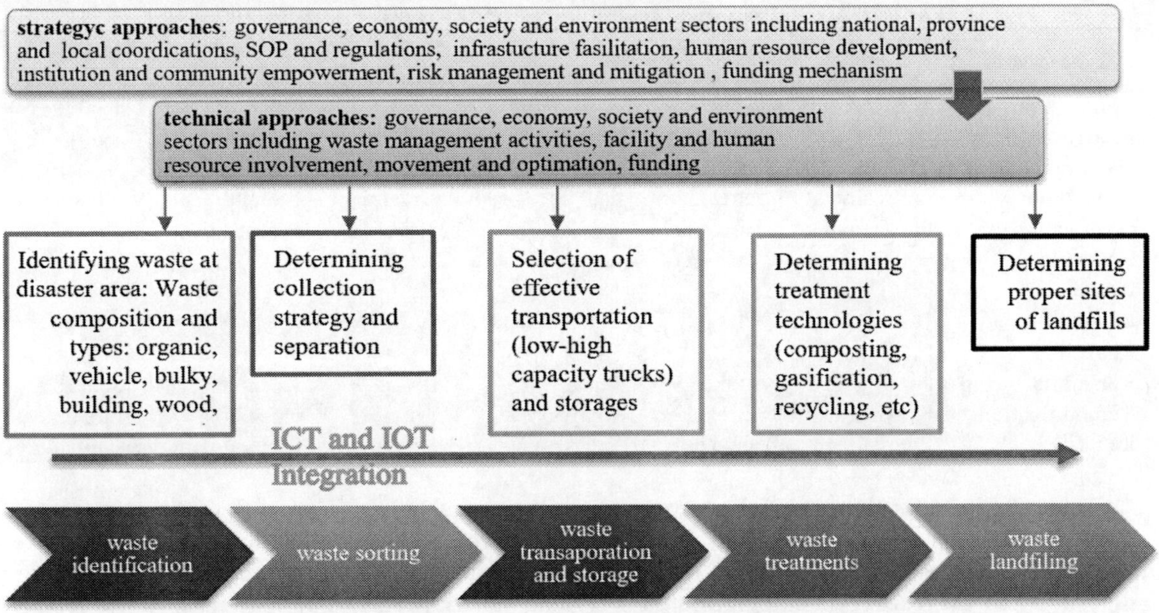

Figure 4. The sustainable post-disaster waste management framework

Waste composition and generation [4] post-disaster in this research are categorized into six groups, including organic waste, vehicle waste, bulky waste, building waste, wood waste, electronic waste, as presented in Table 1.

Table 1. waste composition and generation

Waste type	Composition
Organic	Green waste, trees, stumps, brush, leaf, animal
Vehicle	Automobile steel components, tires, plastic accessories
Bulky	Furniture (couches, table, etc.), oven, plumbing fixtures (toilets, sinks)
Building	Timber, concrete, brick, asphalt, plastic board, asbestos,
Wood	Wood particle, garden trees,
Electronic	Computer, refrigerators, dispensers, AC,

Generally, the volume of disaster waste reaches 5 to 15 times the annual waste generated. One house can generate 30 – 113 tons of waste, while amount of waste and debris can reach 4-15 million tons. The assessment and estimation of the waste volume, waste type, degree of mixed waste are specific factors used in this system. The assessment result is also used to determine the type of vehicles specifically. The second layer in the framework is determining collection strategy and separation. There are on-site and off-site separation strategies that can be used in this system [16]. The selection of the effective strategy depended on some factors, including time limitation, availability of the vehicles and human resources, waste composition and volume, public health risk. Determining effective transportation and storage. Effective transportation is determined based on waste volume and type. There are at least three types of vehicles to be considered in this system, including low-medium and high capacity trucks. In this research, sustainable waste treatment technologies are technologies that are expected to provide best economic and social performance for disaster waste management and a minimum negative impact on the environment, during the entire process of the waste treatment. This sustainable waste treatment technology can be used to help identify the most appropriate treatment technology of waste/ debris created from the disaster that is locally suited for the context of the disaster waste/materials type. The following Table 2 presents the waste treatment technology options offered in this framework [2], [17], [18].

Table 2. Waste treatment technology options based on materials type

Treatments	Organic materials	Vehicles materials	Bulky materials	Building materials	Woods, materials	Electronic, materials
Composting	Suitable	Unsuitable	Unsuitable	Unsuitable	Suitable	Unsuitable
Anaerobic digest	Suitable	Unsuitable	Unsuitable	Unsuitable	Suitable	Unsuitable
Gas recovery	Suitable	Unsuitable	Unsuitable	Unsuitable	Suitable	Unsuitable
Incineration	Pretreatment	pretreatment	Suitable	Pretreatment	Pretreatment	pretreatment
Gasification	Pretreatment	pretreatment	Unsuitable	Unsuitable	Pretreatment	pretreatment
Reduce	Suitable	Suitable	Suitable	Suitable	Suitable	Suitable
Reuse	Unsuitable	Suitable	Unsuitable	Suitable	Unsuitable	Suitable
Reconditioning	Unsuitable	Suitable	Pretreatment	Unsuitable	Unsuitable	Suitable
Recycling	Unsuitable	Suitable	Suitable	Suitable	Unsuitable	Suitable
Remanufacturing	Unsuitable	Suitable	Unsuitable	Unsuitable	Unsuitable	Suitable
Remodification	Unsuitable	Pretreatment	Unsuitable	Unsuitable	Unsuitable	Pretreatment
Landfill	Pretreatment	Unsuitable	Pretreatment	Pretreatment	Pretreatment	Unsuitable

Table 2 Illustrates the potential appropriate treatment technologies for different waste materials. For example, composting is suitable for most organic material including food waste, small trees, leaves, market waste, and is suitable mostly for wood materials from garden trees [19]. Anaerobic digestion and gas recovery are appropriate for organic materials and wood materials. Waste produced from electronic and household appliance are mostly suitable for reconditioning, recycling, remanufacturing and remodification [20]. The table is used for determining the appropriate treatment technology in system.

4.5. Pre-design of a smart post-disaster waste management system

Waste of post-disaster is categorized as unpredictable waste, not based on historical data or demand analysis. Thus, post-disaster waste management system requires a new approach which is able to provide accurate and real-time data information, appropriate technology, facility, and human skills. The use of ICT (e.g. sensors) and IoT (e.g. web) become key factors to integrate physical and cyber system sustainable waste management system through waste sorting, waste collection, waste transportation, and waste disposal. In order to create a smart disaster waste management system, accurate and reliable waste data identification, collection and analysis are required. A specific data collection system is developed to effectively manage the waste, as illustrated in Figure 5. The system is expected to be able to identify the type of waste, amount of waste, fraction of waste (i.e. grouping based on waste properties – paper, glass, fibers, automobile parts, furniture, etc.) and treatment of waste (i.e. recycling, waste to energy, remanufacturing, landfilling). The data collection system is expanded through the use of ICT and IoT, including the use of bins equipped by sensors, the GPS location of collected waste, and the type and contents of waste using automatic sorting machines. The integration among the technology, ICT, and IoT is designed to improve the global waste collection system.

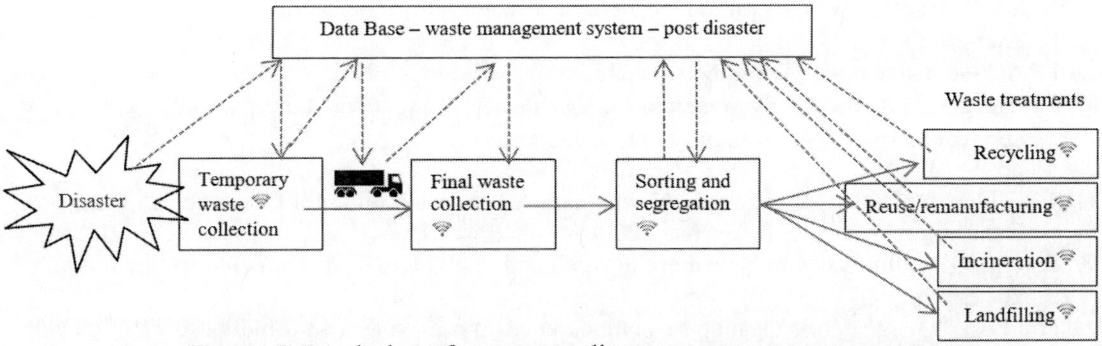

Figure 5. Pre-design of smart post disaster waste management system

Figure 5 illustrates a system which is able to estimate the different types of wastes generated from the disaster, the types of debris, the waste volume, the locations where debris are located, and the appropriate treatment technology can be applied. Integrated system which involves the use of high-quality technology such as sensors, big data, and other ICT infrastructures is essential factors in this system. A detailed design of the system is discussed in the future study.

4.6. Lessons for consideration

The sustainable smart post-disaster waste management (SSDWM) could be considered as important framework for an effective and efficient waste management practice post-disaster phase in Indonesia. The SSDWM offers number of benefits including reduction of waste, improvement of job creation, economic development, human quality life improvement, fast and effective decision making and better Information on waste management performance which can be real-time, accurate and reliable. However, the implementation of SSDWM also face number of challenges including unpredicted waste created from natural disaster, mixed waste which is highly contaminated with dangerous materials (pesticide, fuel, etc), and different waste impacts created from different natural disaster, thus need different approaches. On the other hand, there are also some opportunities offered by the purposed waste management, for example massive amount of waste means money, massive amount of waste means great resources, advanced waste collection and segregate technologies which can reduce the cost and time usages, and appropriate waste treatment based on the waste characteristics. Accordingly, to successfully implement the SSDWM, some technical approaches need to be conducted, including intensive education, and campaign of waste management, effective waste collection, transportation, treatment and landfill processes, appropriate ICT and IoT infrastructure election, and community, private sector, industry, and government participation.

5. Conclusion

Generally, developing countries like Indonesia suffer the most from natural disasters, as it appears that natural catastrophes hit that region more often. The difficult situation in this country comes from the fact that Indonesia is not prepared to face such crises, including waste management for disaster. In reality, Indonesia still has a big problem with its regular waste management as waste management in this country has not been adequate to meet the daily waste collection and handling. The situation will be even more difficult, in the event of a catastrophe. As a result, the lack of preparation will lead to an increase in losses in lives and properties. An efficient way, strategies, and plans must be prepared earlier to limit the potential destruction of the disaster. This research presents a smart, sustainable waste management framework as potential future solution. A further study is ongoing conducted to assess and evaluate the sustainability of the purposed disaster waste management system.

References

[1] BNPB, "2016 Was a Year Full of Natural Disasters in Indonesia | Indonesia Investments," 2018. [Online]. Available: https://www.indonesia-investments.com/id/news/todays-

headlines/2016-was-a-year-full-of-natural-disasters-in-indonesia/item7478. [Accessed: 08-Aug-2018].
[2] DEFRA, "Guidance on applying the waste hierarcy," London, 2011.
[3] A. Del Borghi, M. Gallo, C. Strazza, and M. Castagna, "Waste management in Smart Cities : the application of circular economy in Genoa," *Impresa Progett. 0 Electron. J. Manag.*, vol. 4, pp. 1–13, 2014.
[4] UNEP, "Disaster Waste Management Mechanism A Practical Guide for Construction and Demolition Wastes in Indonesia," 2008.
[5] Z. Zakiya, "Tsunami waste management in Aceh and resilience," *NAtional Geography Indonesia*, 2012. .
[6] Jakarta Post, "Quake debris cleanup in Lombok needs more excavators - Indonesia | ReliefWeb." [Online]. Available: https://reliefweb.int/report/indonesia/quake-debris-cleanup-lombok-needs-more-excavators. [Accessed: 21-Oct-2018].
[7] Bappenas, "Preliminary damage and losses assessment, Yogyakarta and CEntral Java," 2016.
[8] World Health Organization, "Situation Report #10 Sulawesi Earthquake and Tsunami, Indonesia," 2018.
[9] C. Brown and M. Milke, "PLANNING FOR DISASTER DEBRIS MANAGEMENT."
[10] A. King, S. Burgess, W. Ijomah, and C. McMahon, "Design for End-of-Life: Repair, Recondition, Remanufacture or Recycle?," 2009, pp. 745–754.
[11] Y. A. Fatimah and W. K. Biswas, "Remanufacturing as a means for achieving low-carbon SMEs in Indonesia," *Clean Technol. Environ. Policy*, vol. 18, no. 8, pp. 2363–2379, Dec. 2016.
[12] K. Joseph, A. Selvam, and J. W. C. Wong, "Waste Storage, Segregation, and Collection," in *Sustainable Solid Waste Management*, 2016, pp. 35–51.
[13] G. Karunasena, D. Amaratunga, R. Haigh, and I. Lill, "POST DISASTER WASTE MANAGEMENT STRATEGIES IN DEVELOPING COUNTRIES: CASE OF SRI LANKA," *Int. J. Strateg. Prop. Manag.*, vol. 13, no. 2, pp. 171–190, 2009.
[14] M.-C. Emmanouil, E. Stiakakis, M. Vlachopoulou, and V. Manthou, "An Analysis of Waste and Information Flows in an ICT Waste Management System," *Procedia Technol.*, vol. 8, pp. 157–164, 2014.
[15] D. Celaj, J. Desmond, A. Shah, S. Shirai, F. Carrera, and H. Lauer, "Preparing for the Use of Big Data in Denmark's Waste Management Sector," 2017.
[16] K. Asari, M., Sakai, S. I., Yoshioka, T., Tojo, Y., Tasaki, T., Takigami, H., & Watanabe, "Strategy for separation and treatment of disaster waste: a manual for earthquake and tsunami disaster waste management in Japan," *J. Mater. Cycles Waste Manag.*, vol. 15, no. 3, pp. 290–299, 2013.
[17] G. Tchobanoglous, "Solid Waste Management," *Environ. Eng.*, pp. 177–308, 2009.
[18] M. C. Bronson, "Treatment of solid wastes," in *Separation Techniques in Nuclear Waste Management (1995)*, 2017, p. 153.
[19] The University of Georgia Environmental Engineering Outreach, "Food Waste Composting: Institutional and Industrial Application | UGA Cooperative Extension." [Online]. Available: http://extension.uga.edu/publications/detail.html?number=B1189&title=Food Waste Composting: Institutional and Industrial Application. [Accessed: 12-Sep-2018].
[20] X. V. Wang and L. Wang, "Digital twin-based WEEE recycling, recovery and remanufacturing in the background of Industry 4.0," *Int. J. Prod. Res.*, 2018.

Acknowledgment
This study adopts a fundamental concept of sustainable smart waste management system developed from a research project titled an adaptive smart waste management system toward the sustainable circular economy, which was conducted in 2018.

A review of literature on types, stages of recovery and humanitarian logistics operations in the tsunami and earthquake disaster in Indonesia

I Masudin* **and F W Fernanda**

University of Muhammadiyah Malang, Jl. Raya Tlogomas 246 Malang, Indonesia

* Email: masudin@umm.ac.id

Abstract. Recovery after a disaster determines the survival of the population in the affected area. However, there is still little lecturer to discuss recovery compared to other processes. This literature review aims to develop findings from previous literature reviews with a focus on the area of Indonesia and the earthquake and tsunami disaster. The conclusions that can be drawn are the majority of journals discussing earthquakes and researching on physical recovery. Economy recovery being the least discussed them, and the most widely researched recovery phase is a long term recovery. In 2016-2017 there is now a majority of themed journals of recovery management, and 2010-2012 the most widely publicized long term reconstruction journal. This article also presents 5 gaps in the recovery themed literature.

1. Introduction

Indonesia is located on of the most vulnerable island nation. This is due to the position of Indonesia which is above the collision line between two continental plates or often called the ring of the fire. Disasters such as the 2004 tsunami that hit Aceh province were also caused by an earthquake resulting from collisions between the two continental plates. The effects of disasters are very wide. Two areas in Aceh province that suffered the most damage is the city of Banda Aceh and Maoulaboh. Hundreds of thousands of people have been killed by tsunami waves, almost all infrastructure destroyed, economic crippling, farmland destruction, and massive psychology impact on Acehnese survivors. Various humanitarian organizations and the Indonesia government continue providing continuous assistance and recovery to the affected Acehnese.

Given the recurrent cycles of earthquakes and tsunami occurring in Indonesia, government and communities must be prepared to reduce post-disaster impacts by having a good type and recovery phase. From a disaster like a tsunami that hit several countries in 2004, it can be concluded that some types of recovery are needed as well as the relief operations. Recovery types include psycho-social recovery, recovery economics, physical recovery, and environmental recovery [1], while recovery starts from disaster assessment, short term recovery, long term reconstruction, and recovery management [2]. In the category of each recovery, it is always related to the strategy how to distribute all the things required for the survivors and the recovery/reconstruction itself.

In addition to the many variations and volatile demand, humanitarian logistics also have to deal with different types of disasters. Natural disasters, wars, and epidemics are the most disastrous types of disasters [3]. However, with such complex problems, humanitarian logistics must remain concerned with performance (effectiveness and responsiveness), acting as a link between various phases and processes in disaster management, as a source of data primarily for post-disaster analysis, and determining the success of a disaster management operation [4].

The 2004 tsunami disaster was referred to as a turning point in humanitarian logistics because of the many judgments that came over the weaknesses of humanitarian logistics at the time [5]. But the various disasters that are still happening lately, then the humanitarian logistics must have changed in that time period.

Banomyong, Varadejsatitwong, and Oloruntoba [6] mentioned in his journal that the number of recovery studies during the years 2005-2016 only amounted to 8 studies. The least number of recovery studies are also confirmed by Altay and Green [7] and Kovács and Spens [8]. This paper aims to update the data and clarify the details of the previous literature review by specializing in the tsunami earthquakes in Indonesia. This paper reviews humanitarian logistics and the types and stages of recovery in research journals with the theme of recovery in the tsunami and earthquakes in Indonesia. The time period used for the limitation of the literature collection is 2005-2017.

2. Methodology

By filtering the literature in accordance with the subject area, it can be concluded state of the art of field [9]. The statement from Baker [10] and Hart [11] is also the same direction, adding another purpose of the review literature is to identify gaps and development of existing literature. Here are the steps are taken in writing this review literature :

- The first step in this review literature is to determine the criteria used to filter the literature, it's the subject field and time range.
- The second step is the screening stage of literature published by a particular publisher.
- The third step is to arrange and sort the literature that has been collected so that the gap and detail of each literature can be found
- The fourth step is to make a review and classification of the subject area of the literature that has been collected
- Finally, the fifth step is to find the gaps inferred from the review in step four.

Such measures have been used by Jabbour et al. [12] in the literature review on humanitarian logistics and previously tested on research from Jabbour et al. [12] and Mariana, Sobreiro, & Do Nascimento Rebelatto [13] [14]. As a restriction, reviews are conducted on published literature such as theses, proceedings, and literature from search engines such as Google Scholar, Science Direct, Emerald Insight, and others. Keywords used in the literature search are "recovery phase", "Indonesia's earthquake", "Indonesia's disaster" and "Indonesia's disaster and logistics operations".

3. Classification and Granting Code

Classification and coding are used to facilitate a review of the literature that has been selected. The methods used take references from Junior and Godinho Filho [15], Jabbour [16], and Jabbour et al. [17]. The classification contains 4 categories numbered from 1-4. Each section in the category is numbered with the A-5 alphabet. One article can have more than code so it will look a gap in all the articles collected. The categories and codes used are as follows:

- Classification 1- disaster type: there are two types of disasters taken in this research, namely earthquake and tsunami.
- Classification 2- type of recovery: type of recovery has four kinds of psycho-social recovery, Economy recovery, physical recovery, Environment recovery
- Classification 3- recovery phase: there are 4 different types of recovery steps: disaster assessment, short-term recovery, long-term reconstruction, and recovery management
- Sub-category long term reconstruction: there is a discussion of disaster risk reduction and no discussion of disaster risk reduction.
- Classification 4- year of journal issuance: the year of journal publishing begins in 2004-2017, coded A-E for a period of 2 years.

Table 1 classifying and coding of central structures disaster recovery phase literature

	Classification		Code
1	Kind of disaster	1A	Earthquake 1A1- There is humanitarian logistics discussion on earthquake disaster 1A2 - No humanitarian logistics discussion on earthquake disaster
		1B	Tsunami 1B1- There is humanitarian logistics discussion on earthquake disaster 1B2 - No humanitarian logistics discussion on earthquake disaster
2	Kind of Recovery	2A	Psycho-social Recovery
		2B	Economy Recovery
		2C	Physical Recovery
		2D	Environment Recovery
3	Step of Recovery	3A	Disaster Assessment
		3B	short term Recovery
		3C	long term reconstruction 3C1 – there is review/discussion on disaster risk reduction 3C2 – no review/discussion on disaster risk reduction
		3D	Recovery Management
4	Years of journal publication	4A	2004-2006
		4B	2007-2009
		4C	2010-2012
		4D	2013-2015
		4E	2016-2017

4. Basic Concept of Humanitarian Logistics, Type and Stage Recovery

Any potentially affected area should have an adequate humanitarian logistic system. This is intended to minimize damage and casualties when disaster strikes. The definition of humanitarian logistics according to Thomas and Mizushima [18] is: "*The process of planning, implementing, controlling, and storing efficient and low-cost good flows is accompanied by relevant information. Starting from the point of origin to the point of consumption in accordance with the needs of the recipient*".

In the event of a disaster, at every stage of emergency relief such as mitigation, preparedness, response, and recovery depends largely on the smoothness of humanitarian logistics [19]. At the pre-disaster stage (mitigation and preparedness), disaster victim will receive more assistance if disaster management equipment is served first [20]; [21]; [22]; [23]; [24]. Evacuation of the population will also be more coordinated if previously there has been planning a shelter position [25]. Humanitarian logistics in the post-disaster phase (response) include mass evacuation, aid delivery, victim evacuation, debris collection, road cleaning and local development [26]. Furthermore, in the recovery phase, humanitarian logistics includes activities such as temporary shelter construction, lifelines recovery and recovery of infrastructure, and provision of medicines for the prevention of common diseases (long-term recovery).

According to Baker et al., the recovery phase is divided into 4, namely disaster assessment, short term recovery, long-term reconstruction, and recovery management [27]. The disaster assessment phase runs in the direction of the assessment of the emergency response so that the physical impact of the disaster can be accurately assessed. After a disaster assessment is completed, short-term recovery is done with the main objective of securing affected areas, housing victims and creating conditions conducive to initiating the recovery process. Reconstruction of disaster-affected areas marks a long-term reconstruction phase. Reconstructed aspects can be such as psychological, economic, demographic and political. The last stage of recovery management is to coordinate and ensure all stages of recovery run smoothly [27].

Table 2 coding of the studies considered

No	Authors	Category				
1	Wimbardana & Sagala (2012) [28]	1A, 1A1	2A	3C	3c2	4C
2	Amijaya & Kloft (2013) [29]	1A, 1A1	2C	3C	3c1	4D
3	MacRae & Hodgkin (2011)	1A, 1A2	2C	3B		4C
4	Kusumasari & Alam (2012) [30]	1A, 1A2	2C	3C	3c1	4C
5	Boen (2006)	1A,1B, 1A2	2C	3C	3c2	4A
6	Resosudarmo et al. (2008)	1A, 1A1, 1A2	2B	3C	3c2	4B
7	Grundy (2010)	1A. 1A2	2C	3C	3c1	4C
8	Maly, Kondo, & Shiozaki (2012)	1A, 1A2	2C	3C	3c1	4C
9	Bayudono (2009)	1A, 1A2	2A, 2B, 2C, 2D	3D		4B
10	Steinberg (2007)	1A,1B, 1A2	2A, 2B, 2C	3C	3c1	4B
11	Ophiyandri et al. (2013)	1A, 1A2	2C	3C	3c1	4D
12	Andriansyah (2015)	1A. 1A2	2D	3D		4E
13	Joakim (2016)	1A, 1A2	2A, 2B, 2C, 2D	3D		4E
14	Wijayanto (2016)	1A,1B, 1A2	2C	3D		4E
15	Brown et al. (2015)	1A,1B, 1A2	2D	3D		4E
16	Joakim & Doberstein (2014)	1A, 1A2	2C	3C	3c1	4D
17	Scawthorn et al. (2006)	1A, 1A2	2C	3A		4A
18	Sofyan (2012)	1B, 1A1	2A, 2B, 2C, 2D	3D		4C
19	Dahuri (2006)	1B, 1A2	2A, 2C, 2D	3A		4A
20	Sugiyasu & Murao (2013)	1B, 1A2	2C	3D		4D
21	Tinning (2011)	1B, 1A2	2B, 2D	3C	3c2	4C
22	Brintnell et al. (2009)	1B, 1A2	2A	3A		4B
23	Meilianda et al. (2017)	1B, 1A2	2A 2B 2C 2D	3D		4E
24	Thorburn (2009)	1B, 1A2	2A 2B 2C 2D	3D		4B
25	Wong (2009)	1B, 1A2	2D	3A		4B

The types of recovery discussed in this study are psycho-social recovery, economic recovery, physical recovery, and environmental recovery. The effects of disaster on each aspect are explained as follow:

4.1. Psycho-social recovery

The real impact of psychosocial for the victims of the disaster has been investigated since 25 years ago and found in some literature such as [31]; [32]; [33]; [27]; [34]. Psycho-social impacts of the disaster are not only affecting mental aspect but also physically affected to disaster victims such as fatigue and gastrointestinal illness and musculoskeletal disorder. Moreover, Emotional signs of anxiety, depression, sadness, changing hours of sleep and appetite, and abuse of certain drugs are also common [2]. However, the impact of psycho social perceived by each person is different depending on the place where the victim suffered a disaster [35].

4.2. Economy Recovery

Economy recovery has an impact on the redistribution of economic value on a large scale. Proper financing arrangements will affect the number of successful economic players recovering or failing, but still depending on the source and terms of the fund [36]. All elements who are responsible for post-disaster planning should focus on key economic factors in order for economic recovery to proceed smoothly [37]. However, the thing to remember is that the economic conditions before the disaster can not be repeated on the economic conditions after recovery [38].

4.3. Physical Recovery

Physical damage is the worst damage after a disaster. The impact of disasters will change the urban physical environment even though development tends to be the same as it was before the disaster [36]. Proper physical recovery actually opens the opportunity to redevelop the damaged building with a better and stronger structure [39]. If the physical recovery management is not properly regulated, there will be a failure of poor quality of the building, lack of adequate housing stock, and loss of opportunity to reduce the risk of disaster protection in the building to be re-created [40].

4.4. Environment Recovery

Natural events such as storms and hurricanes cause floods, heavy rain, and high sea waves that have the potential for each hazard. But the potential for severe damage is also due to the vulnerability of people who are on the path of natural events [41]. The environmental damage experienced must depend on the characteristics of the area. For example, the damage caused by the tsunami resulted in a lot of solid waste and disaster debris [41]. Reports from United Nation Environment Program (UNEP) show severe damage to the infrastructure, building, and industrial premises, including water and systems [42].

5. Result and Discussion

In search of articles randomly using related keywords, 25 articles were taken that fit the topic of discussion. The code obtained will be displayed as a percentage diagram. Here is a discussion of gaps for each category.

5.1. Type of Disaster

Based on the percentage in Figure 1, 54% of journals chose the theme of recovery in earthquake disaster. 35% discussed recovery for both, most discussed the earthquake disaster followed by tsunami (Aceh), but some are not in one area. 11% of journals discuss tsunami disaster recovery. The most widely discussed earthquake disaster was the Yogyakarta earthquake of 2006. Whereas during the period of 2004-2017, in various areas of Indonesia there were frequent earthquakes, but the most attention highlighted was the Yogyakarta earthquake. Gaps that can be drawn from it are: G1: What can be applied from the Yogyakarta earthquake to manage earthquake recovery in areas that are difficult to reach and potentially repeat earthquakes?

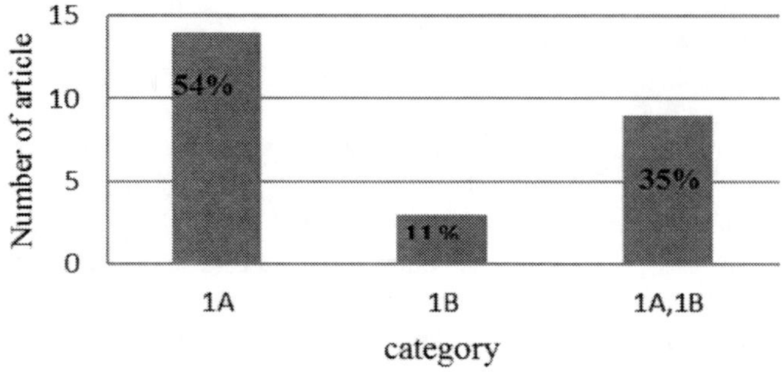

Figure 1. kind of disaster

5.2. *Recovery Type*

The second category maps the journal for recovery types (Figure 2 and Figure 3). 42% of journals chose physical recovery types. 19% of journals discuss all types of recovery that focus on 1 particular area. Environment recovery is the third most discussed theme with a 12% percentage, while psycho-social recovery gets 11% percentage. Journals with the theme of physical recovery are very much discussing housing recovery both from technical aspects and management, even differentiated by area and the value of local wisdom. The percentage the smallest is the type of recovery economy is only 4% of the total journal. Economy recovery is an important factor in long-term reconstruction to support the lives of local people, but not so much the subject of discussion. Considering this, the gap that appears is: G2: How is the sustainability of the current housing recovery in the affected area?; and G3: Does the number of housing-themed studies affect sustainability?

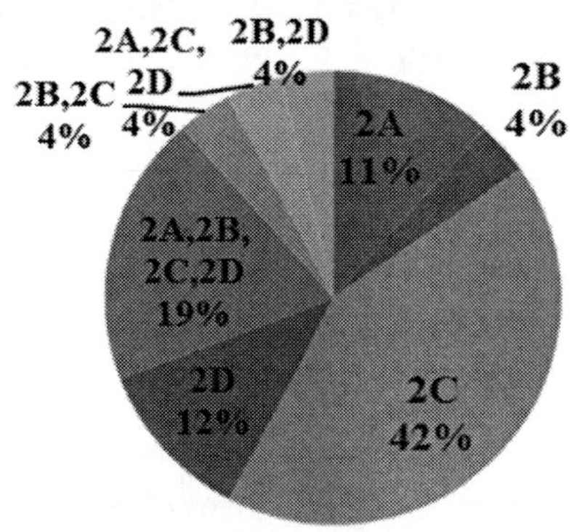

Figure 2 kind of recovery

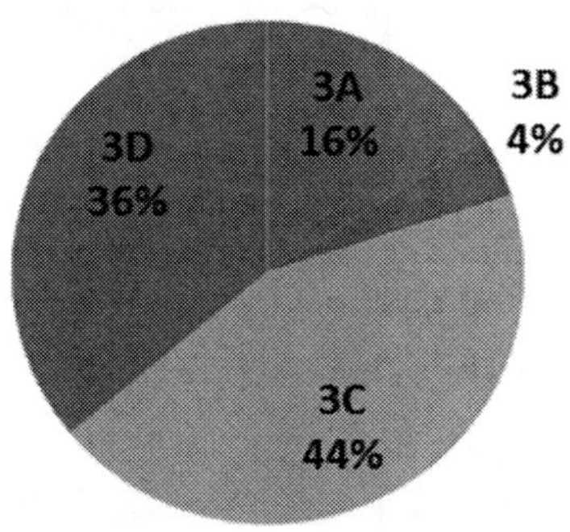

Figure 3 stage of recovery

5.3. *Recovery Phase*

Journals with the Recovery Phase category have only 1 in each journal. So there are 4 codes, namely 3A, 3B, 3C, and 3D (Figure 3). Code 3C or long-term reconstruction has a percentage of 44%. Code 3D or recovery management of 36%, then the code 3A or disaster assessment by 16%. The short term recovery theme is at least in percentage with 4%. The number of journals with the longest reconstruction theme is at most and is directly proportional to the 2C category or physical recovery which mostly contains the theme of long-term housing/reconstruction, while most recovery management journals discuss all types of recovery (psycho-social recovery, economic recovery, physical recovery, environment recovery). Long term recovery should contain planning aimed at addressing future risk disaster. For example, retrofitting of steel structural construction buildings that cost 3 times more expensive but more powerful than ordinary buildings. However, these protection actions should not only be applied to physical recovery but should be equitably planned in every aspect, reducing the impact of future disasters. In this collected journal, 7 of 11 long term reconstruction journals have discussed how preventive measures on long term recovery in areas with repeated catastrophic potential (Figure 4). Based on the value, then the gap that appears is: G4: Taking into account the vulnerability of repeatedly affected Indonesian areas in the short term, will long term recovery be able to reduce the risk of such damage? (Is there any research that has made a review about the effectiveness of disaster risk reduction on long term reconstruction?)

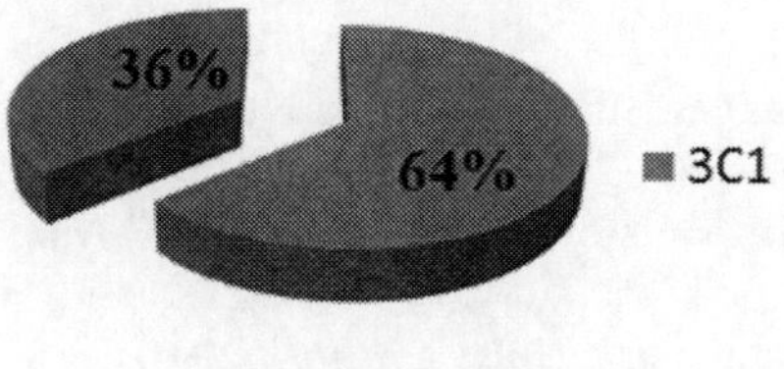

Figure 4. Long term recovery

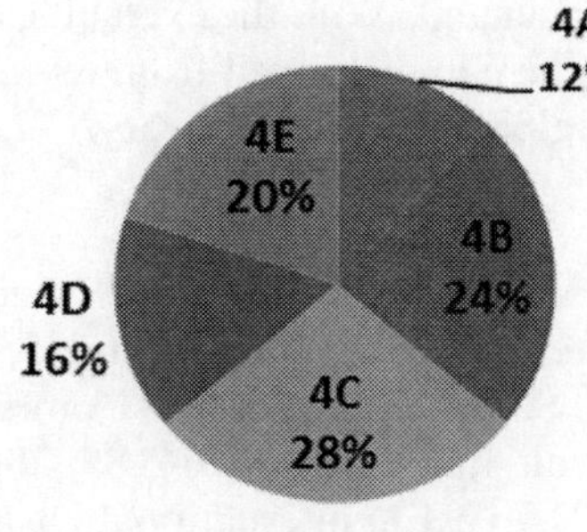

Figure 5. years of article publication

5.4. Years of Journal Publishing
In late December 2004, Indonesia gained much international attention for being the most severely affected region after the tsunami disaster. Followed after the earthquake in Yogyakarta in 2006. The percentage of journal issuance year started from 4C - 28%, then 4B - 24%, 4E - 20%, 4D - 16%, and last 4A - 12% (Figure 5). Code 4C is the year 2010-2012 to be the largest in number along with 3C code or long-term reconstruction. Judging from the year of the disaster (2004/2006), 2010-2012 is a time when many long-term reconstructions are being built in disaster areas. The percentage value does not vary much with the 4B code that is 2007-2009. While the 4E or 2016-2017 code all discuss recovery management. Judging from the gap that is drawn is: G5: In every evaluation and analysis of recovery management, what is the most essential for the stage and type of recovery when viewed from the time period per year?

5.5. Humanitarian Logistics
Previous literature collected from the databases related to the natural disaster in Aceh and Yogyakarta shows that there are only a few numbers of articles concerns with humanitarian logistics (see Code 1A1 in Table 2). For instance, a study by Wimbardana & Sagala mentions that during the process of recovery to build back survivors' house and temporary shelters, demographics and unstable disaster locations become the major problems in implementing logistics and distribution activities in accordance with its function [28]. Moreover, Kusumasari & Alam uses Yogyakarta earthquake as the location of their study to understand what the government can do to relief the disaster operations [30]. It is stated that the policy made by the government to handle the disaster operations should do managing the command with other unit coordinators regarding the logistics distribution, especially within the three-days after the disaster. Finally, Amijaya & Kloft believe that the key concerns for planning of temporary housing in the period of post-disaster are location decisions and time of delivery which are a type of humanitarian logistics activities [29]. Humanitarian logistics in the previous natural disaster in Yogyakarta earthquake and Aceh's tsunami literature concerns with rebuilding the survivor's house for the earthquake in Yogyakarta so judging from the gap that can be drawn is: G6: How is the implementation of humanitarian logistics in other types and phases of disaster especially in Tsunami disasters?

6. Conclusion
The essence of this article is to develop the detail of the gap in the previous literature, which states that the recovery phase theme is the least discussed. Indonesia becomes the chosen area for the boundary of the article, as it is more familiar and includes disaster-prone areas that require a mature recovery plan. From the results and discussion it can be concluded that the majority of recovery journals discuss

IOP Conf. Series: Materials Science and Engineering **674** (2019) 012043 doi:10.1088/1757-899X/674/1/012043

earthquakes in Indonesia, while tsunami recovery is mostly discussed globally or in conjunction with other tsunami-affected countries. Physical recovery is the type of recovery most discussed in journals, while economic recovery is at least in number. Thus, long-term reconstruction is also the most discussed recovery phase, followed by recovery management. The year of issuance of journal recovery is the most that are 2010-2012. The review results also show the most discussed theme in the year 2017 that is recovery management. With these results, obtained 5 gaps associated with the 4 categories that have been reviewed. The gap is derived from reviewing the collected literature and compared with other literature discussing global disaster recovery.

References

[1] C. M. Alfano and J. H. Rowland, "Recovery Issues in Cancer Survivorship: a New Challenge for Supportive Care," *Cancer J.*, vol. 12, no. 5, pp. 432–443, 2006.
[2] M. K. Lindell and C. S. Prater, "Assessing Community Impacts of Natural Disasters," *Nat. hazards Rev.*, vol. 4, no. 4, pp. 176–185, 2003.
[3] C. Em-dat, "The OFDA/CRED International Disaster Database," *Univ. Cathol.*, 2010.
[4] L. N. Van Wassenhove, "Humanitarian Aid Logistics: Supply Chain Management in High Gear," *J. Oper. Res. Soc.*, vol. 57, no. 5, pp. 475–489, 2006.
[5] M. Crum, R. Poist, G. Kovács, and K. M. Spens, "Trends and Developments in Humanitarian Logistics--a Gap Analysis," *Int. J. Phys. Distrib. Logist. Manag.*, 2011.
[6] R. Banomyong, P. Varadejsatitwong, and R. Oloruntoba, "A Systematic Review of Humanitarian Operations, Humanitarian Logistics and Humanitarian Supply Chain Performance Literature 2005 to 2016," *Ann. Oper. Res.*, pp. 1–16, 2017.
[7] N. Altay and W. G. Green III, "OR/MS Research in Disaster Operations Management," *Eur. J. Oper. Res.*, vol. 175, no. 1, pp. 475–493, 2006.
[8] M. Jahre, G. Persson, G. Kovács, and K. M. Spens, "Humanitarian Logistics in Disaster Relief Operations," *Int. J. Phys. Distrib. Logist. Manag.*, 2007.
[9] J. Rowley and F. Slack, "Conducting a Literature Review," *Manag. Res. news*, vol. 27, no. 6, pp. 31–39, 2004.
[10] D. A. Baker and J. L. Crompton, "Quality, Satisfaction and Behavioral Intentions," *Ann. Tour. Res.*, vol. 27, no. 3, pp. 785–804, 2000.
[11] C. Hart, *Doing a Literature Review: Releasing the Research Imagination*. Sage, 2018.
[12] C. J. C. Jabbour, V. A. Sobreiro, A. B. L. de Sousa Jabbour, L. M. de Souza Campos, E. B. Mariano, and D. W. S. Renwick, "An Analysis of the Literature on Humanitarian Logistics and Supply Chain Management: Paving the Way for Future Studies," *Ann. Oper. Res.*, pp. 1–19, 2017.
[13] V. R. Minciacchi *et al.*, "Large Oncosomes Contain Distinct Protein Cargo and Represent a Separate Functional Class of Tumor-Derived Extracellular Vesicles," *Oncotarget*, vol. 6, no. 13, p. 11327, 2015.
[14] E. B. Mariano, V. A. Sobreiro, and D. A. do Nascimento Rebelatto, "Human Development and Data Envelopment Analysis: A Structured Literature Review," *Omega*, vol. 54, pp. 33–49, 2015.
[15] M. L. Junior and M. Godinho Filho, "Variations of the Kanban System: Literature Review and Classification," *Int. J. Prod. Econ.*, vol. 125, no. 1, pp. 13–21, 2010.
[16] C. J. C. Jabbour, "Environmental Training in Organisations: From a Literature Review to a Framework for Future Research," *Resour. Conserv. Recycl.*, vol. 74, pp. 144–155, 2013.
[17] L. B. L. Amui, C. J. C. Jabbour, A. B. L. de Sousa Jabbour, and D. Kannan, "Sustainability as a Dynamic Organizational Capability: a Systematic Review and a Future Agenda Toward a Sustainable Transition," *J. Clean. Prod.*, vol. 142, pp. 308–322, 2017.
[18] A. Thomas and M. Mizushima, "Logistics Training: Necessity or Luxury," *Forced Migr. Rev.*, vol. 22, no. 22, pp. 60–61, 2005.
[19] N. G. A. C. for Policy Research and U. S. D. C. P. Agency, *Comprehensive Emergency Management: A Governor's Guide*. [Department of Defense], Defense Civil Preparedness Agency, 1979.

[20] B. Balcik, B. M. Beamon, and K. Smilowitz, "Last Mile Distribution in Humanitarian Relief," *J. Intell. Transp. Syst.*, vol. 12, no. 2, pp. 51–63, 2008.
[21] S. Duran, M. A. Gutierrez, and P. Keskinocak, "Pre-positioning of Emergency Items for CARE International," *Interfaces (Providence).*, vol. 41, no. 3, pp. 223–237, 2011.
[22] S. Ichoua, "Humanitarian Logistics Network Design for an Effective Disaster Response," 2010.
[23] H. O. Mete and Z. B. Zabinsky, "Stochastic Optimization of Medical Supply Location and Distribution in Disaster Management," *Int. J. Prod. Econ.*, vol. 126, no. 1, pp. 76–84, 2010.
[24] S. V Ukkusuri and W. F. Yushimito, "Location Routing Approach for the Humanitarian Prepositioning Problem," *Transp. Res. Rec.*, vol. 2089, no. 1, pp. 18–25, 2008.
[25] M. J. Widener and M. W. Horner, "A Hierarchical Approach to Modeling Hurricane Disaster Relief Goods Distribution," *J. Transp. Geogr.*, vol. 19, no. 4, pp. 821–828, 2011.
[26] L. Özdamar and M. A. Ertem, "Models, Solutions and Enabling Technologies in Humanitarian Logistics," *Eur. J. Oper. Res.*, vol. 244, no. 1, pp. 55–65, 2015.
[27] A. C. Baker, P. W. Glynn, and B. Riegl, "Climate Change and Coral Reef Bleaching: An Ecological Assessment of Long-Term Impacts, Recovery Trends and Future Outlook," *Estuar. Coast. Shelf Sci.*, vol. 80, no. 4, pp. 435–471, 2008.
[28] R. Wimbardana and S. Sagala, "Understanding Social Recovery Process in Pangalengan Community after the 2009 West Java Earthquake: Challenges to Post-Disaster Recovery Planning," in *The 2nd International Conference Sustanaibale Built Environment. Yogyakarta*, 2012, pp. 10–12.
[29] S. Y. Amijaya and H. Kloft, "Social and Environmental Aspects Significance on The Rebuilding After Natural Disaster in Indonesia Case Study: Yogyakarta Earthquake 2006," 2013.
[30] B. Kusumasari and Q. Alam, "Bridging the Gaps: the Role of Local Government Capability and the Management of a Natural Disaster in Bantul, Indonesia," *Nat. hazards*, vol. 60, no. 2, pp. 761–779, 2012.
[31] R. Bolin, "Disaster Characteristics and Psychosocial Impacts," *Disasters Ment. Heal. Sel. Contemp. Perspect.*, no. 3–28, 1985.
[32] E. T. Gerrity and B. W. Flynn, "Mental Health Consequences of Disasters," *public Heal. consequences disasters*, pp. 101–121, 1997.
[33] P. S. Houts, P. Cleary, and T. Hu, "The Three Mile Island Crisis: Psychological," *Soc. Econ. Impacts Surround. Popul. (University Park. PA, Pennsylvania State Univ.*, 1988.
[34] R. W. Perry and M. K. Lindell, "Psychological Consequences of Natural Disaster-Review of Research on American Communities," *Mass Emergencies*, vol. 3, no. 2–3, pp. 105–115, 1978.
[35] J. Twigg *et al.*, "Self-Recovery from Disasters," 2017.
[36] R. B. Olshansky, "How do Communities Recover from Disaster? A Review of Current Knowledge and an Agenda for Future Research," in *46th annual conference of the association of collegiate schools of planning*, 2005, vol. 27, pp. 1–19.
[37] Y. Okuyama, M. Sonis, and G. J. D. Hewings, "Economic Impacts of an Unscheduled, Disruptive Event: a Miyazawa Multiplier Analysis," in *Understanding and interpreting economic structure*, Springer, 1999, pp. 113–143.
[38] G. Horwich, "Economic Lessons of the Kobe Earthquake," *Econ. Dev. Cult. Change*, vol. 48, no. 3, pp. 521–542, 2000.
[39] K. S. Pribadi, D. Kusumastuti, S. A. H. Sagala, and R. Wimbardana, "Post-Disaster Housing Reconstruction in Indonesia: Review and Lessons from Aceh, Yogyakarta, West Java and West Sumatera Earthquakes," in *Disaster Recovery*, Springer, 2014, pp. 197–223.
[40] G. P. Smith and D. Wenger, "Sustainable Disaster Recovery: Operationalizing an Existing Agenda," in *Handbook of disaster research*, Springer, 2007, pp. 234–257.
[41] H. Srinivas and Y. Nakagawa, "Environmental Implications for Disaster Preparedness: Lessons Learnt from the Indian Ocean Tsunami," *J. Environ. Manage.*, vol. 89, no. 1, pp. 4–13, 2008.
[42] U. UNEP, "Making Tourism More Sustainable: a Guide for Policy Makers," *United Nations Environ. Program.*, 2005.

IOP Conf. Series: Materials Science and Engineering **674** (2019) 012044 doi:10.1088/1757-899X/674/1/012044

Using DEMATEL and ANP for development of post-mining area at sub-district Pongkor gold mining, West Java

E B Leksono[1*] and R Agustian[2]

[1] Department of Industrial Engineering, Universitas Muhammadiyah Gresik. Gresik. Indonesia

[2] PT. SUCOFINDO (Persero), Denpasar Branch. Indonesia

* Email: eko_budileksono@umg.ac.id

Abstract. The objective of this paper proposes strategy for development of post-mining area at sub-district Pongkor gold mining, Bogor, West Java. The strategy of development designed using DEMATEL and ANP. The DEMATEL used to determine of important and influence level, then ANP used to determine potency weighted for development strategy. From the DEMATEL, 5 (five) potency from 8 (eight) is feasible to develop while post-mining of Pongkor gold mining, namely: training and education area, tourism of underground mine, natural & culture tourism, conservation area, and sports facilities. The development priority can determine based on weight of potency by ANP. Based on rank of weight by ANP, potencies for development are training and education area, tourism of underground mine, natural & culture tourism and conservation area.

1. Introduction

Area of sub-district Pongkor gold mining was managed by Business Unit of Gold Mining, PT. Aneka Tambang (Persero) Tbk (PT. ANTAM Tbk). This area located in District of Nanggung, Bogor, West Java, Indonesia. Area of sub-district Pongkor gold mining is part of the Taman Nasional Gunung Halimun & Salak (TNGHS). The exploration licenses of Pongkor gold mining will expire March 9, 2021[1]. Based on regulations, PT. ANTAM Tbk must have programmed as post-mining responsibility and reclamation. PT. ANTAM Tbk must recover from mining area with sustainable development program. The sustainable development programs aim to sustain the economic, environmental and social.

Roadmap of the sustainable development of the post-mining at the sub-district Pongkor gold mining has been introduced by the name AGROGEOEDUTOURISM concept. The AGROGEOEDUTOURISM concept will try integration of agronomy, geology, education, and tourism in the area post-mining. The AGROGOEEDUTOURISM designed for 2 (two) zone, namely mining tourism and natural (environmental) tourism [2].

This article proposes of development strategy of the post-mining area at district Pongkor gold mining based on environmental and social potency and society perception. The methodology in this article uses survey with in depth interviews. The output of the survey processed by the decision-making trial and evaluation laboratory (DEMATEL) and the analytical network process (ANP). DEMATEL used to determine important, influence level and relations between variable or attribute [3] then ANP used to determine of potency weighted for development strategy.

IOP Conf. Series: Materials Science and Engineering **674** (2019) 012044 doi:10.1088/1757-899X/674/1/012044

2. Literature review

2.1. The decision-making trial and evaluation laboratory (DEMATEL)

The DEMATEL is a multi-attribute decision making (MADM) method that is used as a tool to help in decision making [4]. The DEMATEL is used to analyze the component structure from decision variables. The DEMATEL can be used to view direct or indirect relationships between variables or attributes [5]. Furthermore, the DEMATEL can describe the important level and level of influence of an attribute or a variable on system and then the DEMATEL uses matrix system to get all causal relationships of an attribute or a variable with the others [3], [5].

2.2. The analytical network process (ANP)

ANP is a general theory of relative measurement used to derive a composite priority ratio from individual ratio that represents relative measurement of the influence of elements that interact with control criteria [6]. The ANP provides a way to input judgment and measurements to derive ratio of priorities scale for the distribution of influence among the criteria and groups of criteria in the decision making process [5]. ANP is a qualitative multi-attribute decision-making approach that provides structured communication to address business model [7]. The weight of indicators on performance measurement can be determined by ANP. The ANP can help decision-makers to determine strategy to improve performance based on relationship between indicators and weight of indicator. On the performance measurement, ANP usually combined with the DEMATEL [4], where the DEMATEL is used to relate indicators and perspective and then the relationship created is used to make the ANP model. Finally, the ANP model can determine weight of attributes or criteria.

3. Methodology

This study has applied DEMATEL and ANP. The DEMATEL used to determine the relationship between the potency of development, and then the weight of potency of development calculated by ANP. The following steps were utilized in this study:

a. Survey to identify of the potency of development
b. Survey to identify important level of potency with Likert Scale (1 = very unimportant, 2 = unimportant, 3 = neutral, 4 = important, 5 = very important).
c. Validation of pre survey results by calculating the weighted average (WA) and level of consensus (LC) using the criteria cut off: WA ≥ 4.2 & LC ≥ 0.5 (economic) and WA ≥ 4.5 & LC ≥ 0.7 (environmental and social aspects) [8].

$$WA_n = \frac{\sum (R_i x LS)}{\sum R} \quad (1)$$

WA_n = criteria n, R_i = respondent i, LS = Likert score, R = total of respondent

$$LC_n = \frac{\sum LS}{\sum R} \quad (2)$$

Where, LC_n = LC of criteria n, LS = Likert score for agree, R = total of respondents.

d. Survey to identify level of influence between potency with 0 – 4 scale (0 = no influence, 1 = low, 2 = normal, 3 = strong, 4 = very strong). Survey result were processed to the DEMATEL steps [5].
e. Building direct relation matrix (A) based on average of influence value from a_i to a_j.

$$A = [a_{ij}]_{nxn}$$

$$A = \begin{bmatrix} 0 & a_{12} & \cdots & a_{1n} \\ a_{21} & 0 & \cdots & a_{2n} \\ \vdots & \vdots & \cdots & \vdots \\ a_{n1} & a_{n2} & \cdots & 0 \end{bmatrix} \quad (3)$$

f. Building normalization matrix (X)

$$X = A \times s \quad (4)$$

$$s = \min\left[\frac{1}{\max \sum_{j=1}^{n} a_{ij}}, \frac{1}{\max \sum_{i=1}^{n} a_{ij}}\right] \quad (5)$$

g. Building total relationship matrix (T),

$T = X(I - X)^{-1}$; I as identity matrix. (6)

h. Calculation of the level of importance and influence between potency.

Element of $T = \left[t_{ij}\right]_{nxn}, i, j = 1, 2, \ldots, n$; where, i = rows and j = column (7)

D and R represent direct and indirect relationships from row and column.

$$D = \left[\sum_{j=1}^{n} t_{ij}\right]_{nx1}, (i = 1, 2, \ldots, n) \quad (8)$$

$$R = \left[\sum_{i=1}^{n} t_{ij}\right]_{nx1}, (j = 1, 2, \ldots, n) \quad (9)$$

Where, (D + R) indicates level of importance and (D – R) indicates influence level.

i. Building significant matrix to describe the relationship between potency. Steps to prepare for significant matrix:
   - Calculation of average of T :

$\overline{X} = \frac{\sum T}{\sum ij}$; Where, i = sum of row and j = sum of column (10)

   - Reduction of all T with $\overline{X}$ or (T - $\overline{X}$), the value of significant matrix indicate the level of relationship

j. Development of the ANP structure model based on the significant matrix. This step uses super decision software.

$$W = \begin{bmatrix} W_{11} & \cdots & W_{1j} & W_{1n} \\ \vdots & \vdots & \vdots & \vdots \\ W_{i1} & \cdots & W_{ij} & W_{in} \\ W_{ni} & \cdots & W_{nj} & W_{nn} \end{bmatrix} \quad (11)$$

k. Survey to perform pair-wise comparison between the potency of development with scale 1 – 9. The survey result was processed to ANP steps [7]. The validity of the pair-wise comparison could be considering the inconsistency value. If value of inconsistency < 1 so the pair-wise comparison is valid [6].

l. Calculation of potency weight.

4. Results and discussions

4.1. Identification of potency

Identification potency for development of post-mining area at district Pongkor gold mining based on survey with location mapping and in depth interview. Besides that, identification of potency based on AGROGEOEDUTOURISM concept and sustainability aspects. The relations of potency with AGROGEOEDUTOURISM concept and sustainability aspects were descript in table 1.

IOP Conf. Series: Materials Science and Engineering **674** (2019) 012044 doi:10.1088/1757-899X/674/1/012044

Table 1. The potency relations with AGROGEOEDUTOURISM concept and sustainability aspects

Potency of development	Concept-based	*Sustainability aspects*
Agriculture	AGRO – EDU	Economic
Breeding ruminants and aquaculture	AGRO – EDU	
Centre of a small and medium enterprise (SME)	AGRO – GEO –EDU – TOURISM	
Conservation area	(ECO) – EDU – TOURISM	Environmental
Tourism of underground mine	GEO – EDU – TOURISM	Social
Natural & culture tourism	(ECO) – EDU – TOURISM	
Training and education area	AGRO – GEO – EDU	
Sport facilities	EDU – TOURISM	

4.2. Potency validation

Validation of potency on development of post-mining area at sub-district Pongkor gold mining has been done with processed survey output. The survey was conducted through focus group discussion (FGD) followed by 109 respondents. Respondents consisted of village officials, community leaders and youth leaders around the mining area of the sub-district Pongkor. Validation of output survey based on value of weighted average (WA) and level of consensus (LC). Then, valid potency on table 2 can be processed to determine the importance and influence level.

Table 2. Criteria validation by WA and LC

Potency	Disagree	Agree	Respondents	WA	LC	Validation
ECONOMIC ASPECTS						
Agriculture	3	106	109	4.890	0.972	Valid
Breeding ruminants and aquaculture	6	103	109	4.780	0.945	Valid
Centre of small and medium enterprise (SME)	6	103	109	4.780	0.945	Valid
ENVIRONMENTAL ASPECTS						
Conservation area	13	96	109	4.523	0.881	Valid
SOCIAL ASPECTS						
Tourism of underground mine	1	108	109	4.963	0.991	Valid
Natural & culture tourism	1	108	109	4.963	0.991	Valid
Training and education area	4	105	109	4.853	0.963	Valid
Sport facilities	13	97	109	4.569	0.890	Valid

4.3. Level of importance and influence

The DEMATEL stage was used to make relationships started from building direct relationship matrix based on average values from respondents [5] and finished at establishing relationships between criteria or attributes [9]. Table 3 shows initial matrix as direct relationship matrix and called matrix A. Table 4 shows normalization matrix, and Table 5 shows the inverse matrix. Finally, table 6 shows total relation matrix. The total relation matrix can be processed to determine level of importance and influence of potency.

Table 3. Initial matrix (A)

	Tourism of underground mine	Natural & culture tourism	Training and education area	Agriculture	Breeding ruminants and aquaculture	Conservation area	Sport facilities	Centre of SME	Sum
Tourism of underground mine		3	3	3	2	3	2	3	19
Natural & culture tourism	2		3	3	3	3	3	3	20
Training and education area	3	3		3	3	3	3	3	21
Agriculture	2	3	3		3	3	2	3	19
Breeding ruminants and aquaculture	2	2	3	3		2	2	2	15
Conservation area	3	3	3	3	3		2	2	19
Sport facilities	2	2	3	2	2	2		3	16
Centre of SME	3	3	3	3	3	2	2		18
Sum	17	19	21	20	18	18	16	19	

Table 4. Normalization matrix (X)

	Tourism of underground mine	Natural & culture tourism	Training and education area	Agriculture	Breeding ruminants and aquaculture	Conservation area	Sport facilities	Centre of SME
Tourism of underground mine		0.146	0.146	0.146	0.097	0.146	0.097	0.146
Natural & culture tourism	0.097		0.146	0.146	0.146	0.146	0.146	0.146
Training and education area	0.146	0.146		0.146	0.122	0.146	0.146	0.146
Agriculture	0.097	0.146	0.146		0.146	0.122	0.097	0.146
Breeding ruminants and aquaculture	0.097	0.097	0.146	0.122		0.097	0.073	0.097
Conservation area	0.146	0.146	0.146	0.146	0.146		0.097	0.097
Sport facilities	0.097	0.097	0.146	0.097	0.097	0.097		0.146
Centre of SME	0.146	0.146	0.122	0.146	0.122	0.097	0.097	

IOP Conf. Series: Materials Science and Engineering **674** (2019) 012044 doi:10.1088/1757-899X/674/1/012044

Table 5. Inverse matrix

	Tourism of underground mine	Natural & culture tourism	Training and education area	Agriculture	Breeding ruminants and aquaculture	Conservation area	Sport facilities	Centre of SME
Tourism of underground mine	1.93085	1.15804	1.21944	1.18083	1.07099	1.08388	0.94803	1.15226
Natural & culture tourism	1.04964	2.06133	1.25547	1.21326	1.14095	1.11276	1.01379	1.18477
Training and education area	1.11393	1.21869	2.15823	1.24313	1.14840	1.14057	1.03792	1.21432
Agriculture	0.98972	1.12343	1.18466	2.01871	1.07819	1.03288	0.92007	1.11863
Breeding ruminants and aquaculture	0.83997	0.91988	1.00907	0.95800	1.79177	0.86053	0.76209	0.91569
Conservation area	1.05137	1.15026	1.21373	1.17373	1.10308	1.95051	0.94207	1.10735
Sport facilities	0.88238	0.96537	1.05705	0.98502	0.92333	0.90183	1.73132	1.00067
Centre of SME	1.00795	1.10274	1.14334	1.12482	1.03774	0.99493	0.90207	1.97112

Table 6. Total relation matrix (T)

	Tourism of underground mine	Natural & culture tourism	Training and education area	Agriculture	Breeding ruminants and aquaculture	Conservation area	Sport facilities	Centre of SME	Sum (D)	D-R	D+R
Tourism of underground mine	0.93	1.158	1.219	1.181	1.071	1.084	0.948	1.152	8.744	0.878	16.61
Natural & culture tourism	1.049	1.061	1.255	1.213	1.141	1.112	1.013	1.184	9.032	0.332	17.731
Training and education area	1.114	1.218	1.158	1.243	1.148	1.14	1.038	1.214	9.275	0.034	18.516
Agriculture	0.989	1.123	1.184	1.018	1.078	1.033	0.92	1.118	8.466	-0.431	17.363
Breeding ruminants and aquaculture	0.839	0.92	1.01	0.958	0.79	0.86	0.762	0.915	7.057	-1.237	15.351
Conservation area	1.051	1.15	1.213	1.173	1.103	0.95	0.942	1.107	8.692	0.614	16.77
Sport facilities	0.882	0.96	1.057	0.985	0.923	0.902	0.731	1.00	7.447	0.189	14.704
Centre of SME	1.008	1.1	1.14	1.125	1.03	0.99	0.902	0.97	1.008	-0.38	16.95
Sum (R)	7.86	8.699	9.241	8.89	8.295	8.078	7.257	8.665	7.86		

From (D+R) on table 4 indicates that the existence of all potency is important. Besides, from (D – R) indicates that tourism of underground mine, conservation area, natural & culture tourism, sports facilities, and training & education area have value positive (+). It means that its potency has influence on development post-mining at sub-district Pongkor gold mining. Furthermore, the five of potency can be application for development area post-mining. Finally, the five potencies need to calculate to determine development priority. The development priority can be determined based on weight of potency. The ANP method can determine the potency weight because the ANP method accommodates the relation between potency for calculating the potency weight.

4.4. Relation between potency

The relation between potency can determine based on the significant matrix of DEMATEL. The significant matrix can build based on the total relation matrix. Table 7 shows the significant matrix.

Table 7. The significant matrix

	Tourism of underground mine	Natural & culture tourism	Training and education area	Agriculture	Breeding ruminants and aquaculture	Conservation area	Sport facilities	Centre of SME
Tourism of underground mine		0.111	0.173	0.134	0.024	0.037	-0.099	0.105
Natural & culture tourism	0.003		0.209	0.166	0.094	0.066	-0.033	0.138
Training and education area	0.067	0.172		0.196	0.102	0.094	-0.009	0.167
Agriculture	-0.057	0.077	0.138		0.031	-0.014	-0.127	0.072
Breeding ruminants and aquaculture	-0.207	-0.127	-0.038	-0.089		-0.186	-0.285	-0.131
Conservation area	0.005	0.103	0.167	0.127	0.056		-0.105	0.060
Sport facilities	-0.164	-0.081	0.010	-0.062	-0.124	-0.145		-0.046
Centre of SME	-0.039	0.056	0.096	0.078	-0.009	-0.052	-0.145	

From table 7, a minus (-) value on a significant matrix indicates that there is no relationship of influence between perspectives. Equally, a positive (+) value indicates that a row perspective influences a column perspective.

4.5. Weight of potency

Weight potency can calculate based on the pair-wise comparison between potency using ANP method. The ANP method can process by super decision software. Pair-wise comparison between potency was based on the average values of respondents. Before entering the data, the structure model had been created. The structure model of ANP by super decision software described in figure 1. Then, average values from respondents by super decision software described in figure 2. Inconsistency index from pair-wise comparison between potency is 0.088 (0.088 < 1), it means the perception of respondents is valid.

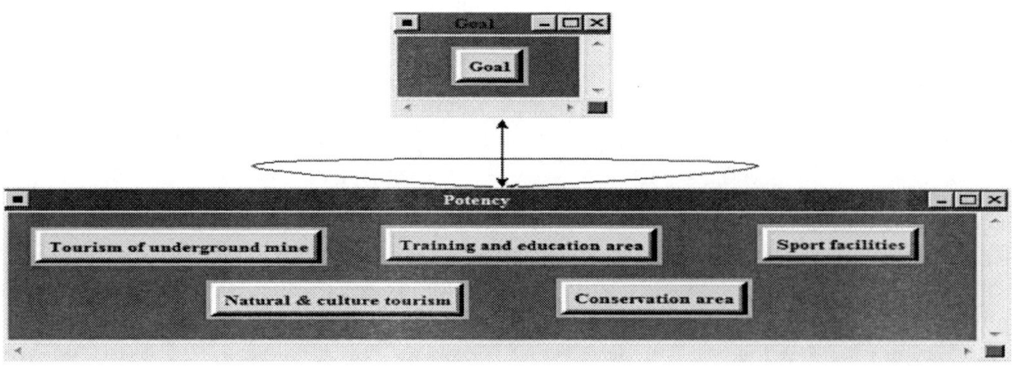

Figure 1. Structure model ANP by super decision software

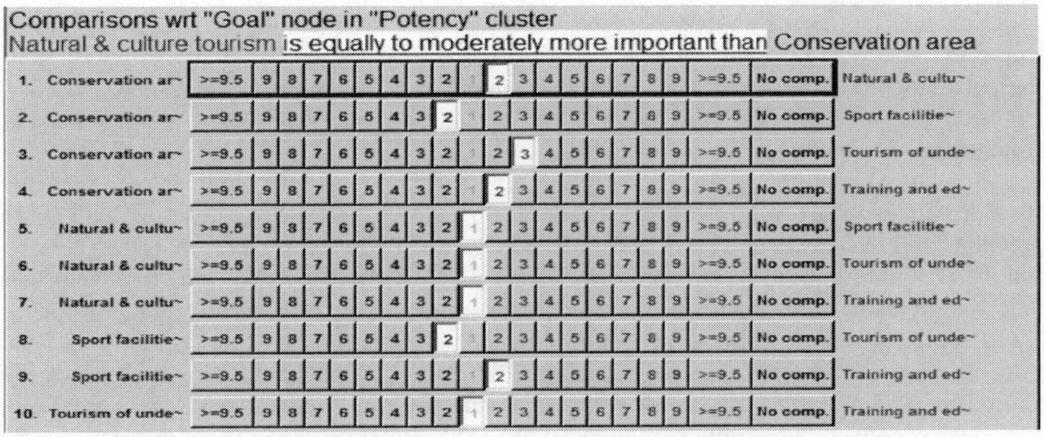

Figure 2. Pair-wise comparison between criteria's

After the pair-wise comparison, super decision software can process the weight of potency. Weight of potency described on priorities icon of the computation menu. Figure 3 shows the potency weight on the development of post-mining area at sub-district Pongkor gold mining.

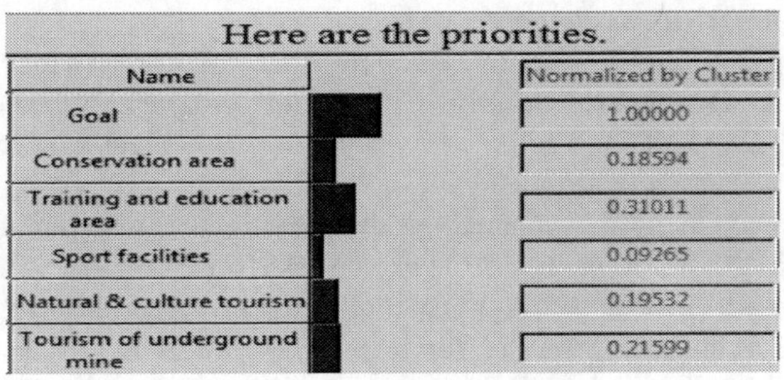

Figure 3. Weight of potency

Figure 3 shows that the training and education area (0.31011), tourism of underground mine (0.21599) natural & culture tourism (0.19532) and conservation area (0.18594) can be alternatives on development of post-mining area at sub-district Pongkor. Based on weight of criteria, training and education area has highest weight more than other criteria. It means, the district Pongkor gold mining very relevant for training and education centers, likes formal education and or non-formal education.

4.6. Discussion

Based on survey with location mapping and in depth interviews with some stakeholders, we found 8 (eight) potency of development for post-mining at sub-district Pongkor gold mining. The potencies need confirm to respondents before processed by the DEMATEL. From confirmation to 109 respondents, we need validation with WA and LC and we found that all of potency is valid. It is mean, the potencies are suitable for development in post-mining at sub-district Pongkor. The valid potencies were processed by the DEMATEL to determine of importance and influence level.

From the DEMATEL, all of the potency is important. Besides, only 5 (five) potency has influence on other potencies. Furthermore, 5 (five) potency from 8 (eight) valid potency were selected to develop on the post-mining at sub-district Pongkor. Finally, the 5 (five) potency need priority for developed. The development priority can be determined based on weight of potency. The ANP method can be used to determine the potency weight because the ANP accommodates the relation between potency in the set of the potency weight. The relation between potency was determined based on significant matrix.

Based on rank potency by ANP, potencies of training and education area (0.31011), tourism of underground mine (0.21599), natural & culture tourism (0.19532) and conservation area (0.18594) can be alternatives on development of post-mining area at sub-district Pongkor gold mining.

5. Conclusions

We found 3 (three) phenomena in this study. First, all of potency of development for post-mining at sub-district Pongkor ware valid potency. Its mean, 8 (eight) potency will relevance with community requirements and relevance with AGROGEOEDUTOURISM concept by PT. ANTAM (Persero) Tbk. Second, based on influence level, only 5 (five) from 8 (eight) potency relevance to develop, namely: training and education area, tourism of underground mine, natural & culture tourism, conservation area, and sport facilities. Third, based on weight of potency, only 4 (four) potency is feasible to develop on post-mining at sub-district Pongkor gold mining, namely: training and education area, tourism of underground mine, natural & culture tourism and conservation area. The limitation of this paper is every potency need alternative for developed. Finally, exploration alternatives for each potency that feasible to develop need for future research. For example, natural & culture tourism has development alternatives: free fall, forestry, tourism village, etc.

References

[1] "Tempo.co, Dec 27," *Tempo.co*, 2016.
[2] "Koran SINDO, March 13," *Sindo.com*, 2018.
[3] E. B. Leksono, Suparno, and I. Vanany, "Development of Performance Indicators Relationships on Sustainable Healthcare Supply Chain Performance Measurement Using Balanced Scorecard and DEMATEL," *Int. J. Adv. Sci. Eng. Inf. Technol.*, vol. 8, no. 1, pp. 115–122, 2018.
[4] I. Gölcük and A. Baykasoglu, "An analysis of DEMATEL approaches for criteria interaction handling within ANP," *Expert Syst. Appl.*, vol. 46, pp. 346–366, 2016.
[5] F. Chen, T. Hsu, and G. Tzeng, "A balanced scorecard approach to establish a performance evaluation and relationship model for hot spring hotels based on a hybrid MCDM model combining DEMATEL and ANP," *Int. J. Hosp. Manag.*, vol. 30, no. 4, pp. 908–932, 2011.
[6] T. L. Saaty, "Fundamentals of the analytic network process," in *ISAHP 1999, Kobe, Japan*, 1999, pp. 1–14.
[7] A. Bhattacharya, P. Mohapatra, V. Kumar, and P. K. Dey, "Green supply chain performance measurement using fuzzy ANP-based balanced scorecard : a collaborative decision-making approach," *Prod. Plan. Control*, vol. 25, no. 8, pp. 698–714, 2014.
[8] A. A. Feil, D. M. De Quevedo, and D. Schreiber, "Selection and identification of the indicators for quickly measuring sustainability in micro and small furniture industries," *Sustain. Prod. Consum.*, vol. 3, pp. 34–44, 2015.
[9] E. B. Leksono, Suparno, and I. Vanany, "Using DEMATEL approach to develop relationships of performance indicators on sustainable service only supply chain performance measurement," *IOP Conf. Ser. Mater. Sci. Eng.*, vol. 337, no. 12023, pp. 1–8, 2018.

Food processing industry and ergonomics handicraft creation from shellfish

E Nurmianto[1,*] and A Pamungkas[2]

[1] Industrial Engineering Department, Faculty of Industrial Technology, Sepuluh Nopember Institute of Technology, Indonesia
[2] Urban And Regional Planning Department, Faculty of Architecture, Design and Planning, Sepuluh Nopember Institute of Technology, Indonesia

[*] E-mail: nurmieko@gmail.com

Abstract. Ujung Pangkah Sub-district, Gresik has great potential of seafood in the form of shellfish but low selling price. This is because the shellfish are sold without being processed first but sold in the form of fresh shells. The problem can be solved when there are methods to process shellfishes into various types of foods and handicrafts with higher production capacity and socialization to the community about shell processing correctly and adequately. Therefore, efforts are made to increase the industry and household scale in food processing and handicraft production from shellfishes in Ujung Pangkah Sub-district. These efforts are carried out in the form of various training and food processing assistance into various foods (such as shell nuggets, and crispy shells), handicraft making (table clock, photo frame, pencil box), financial management, packaging techniques, and marketing management to improve their welfare. This innovation is shell processing into a variety of products that combine several functions of food processing, but safe, comfortable, and healthy for the user and useful for its products. Through this program is expected many entrepreneurs who use the potential shells.

Keyword: shellfish processing, food from shellfish, handicraft from shellfish

1. Introduction

1.1. The Potential of Shellfish

Most of the coastal areas possess potential fish and shells [1]. Most of the fishermen in collecting fish and shell not in good manner, it will increase musculoskeletal disorder [2]–[4]. Aspects of ergonomics are crucial to note not only in industry but also by coastal communities who work as fishermen. This prevents the possibility of back injury and musculoskeletal disorders [5]. Subdistrict Ujung Pangkah is a subdistrict in Gresik regency that has great potential in developing shrimp catching business. In Ujung Pangkah, there is much fish caught by fishermen, including shellfish, shrimp, mangrove crab, and grouper fish, snapper fish. Caught Shellfish has not been developed by fisherman whereas its natural potential is enough to support their life. Generally fisherman only sells shellfish that are small, so it has more economic values [6].

Following the government program that is the empowerment of rural communities and populist economy, it is necessary to explore all potential that exists in the countryside, such as excavation and utilization of natural resources and human resources [7] while academics help everything related to the needs and can benefit the rural communities. In Ujung Pangkah subdistrict, there are 70% of the productive age population work as a fisherman, fish farmers, and farmers [8]. Ujung Pangkah community mostly get an education through boarding school. Knowledge of shellfish processing is obtained from field experience. People in Ujung Pangkah sub-district still need further guidance to increase knowledge, ability, and transfer of shellfish processing technology.

Shellfish caught by a fisherman in the Ujung Pangkah sub-district is quite big, but it is underutilized. In fact, shellfish can be a source of food. Still available fishing grounds that have not been utilized by the community, such as channels of water entry or water irrigation in the area of aquaculture and estuaries. Shelter from the processing of shellfish is not a problem, because the results of this processing can be marketed in Ujung Pangkah or in markets and seafood restaurants.

The shellfish processing industry can be conducted by anyone who is no exception to a certain level of education [9]. Considering that most of the community members are able to do shell processing and cooperate with various parties and related institutions, besides optimizing from the utilization of shellfish from Ujung Pangkah community that is very high, it should also get attention to be optimized its existence as the main ingredient of shell fogging. In addition, shell waste can also be used for handicraft products.

New trends in food processing today is including new developments in techniques and methodologies for the manufacture, preservation, and monitoring of food [10]. Consumers' desire is oriented towards new foods that are comfortable and easily conserved. Consumers are also more aware of the need for safer products, increasing the demand for fresh food or minimally processed preservatives of synthetic chemicals. Another important trend is the consumer demand for healthy food. Processed shellfish by household-scale industries without preservatives and healthy to eat, other than that shellfish obtained fresh from the environment itself. Shellfish also contain nutrients that are good for health. Shellfish can be processed into various variations of food such as burgers, sticks, clams crispy, crackers, and others. This food is consistent with current trends with different ingredients that simultaneously empower coastal community economies.

Management of production and service activities is very important to emphasize due to the importance of competition in the business environment, maintaining market share, and customer satisfaction aspects. The active industry in the food sector is faced with great risks [11]. Therefore, marketing training should be undertaken to reduce undesirable risks for growth and survival.

Shellfish are mostly found in the tropics area, especially around the coastal area of the Gresik Regency. Since ancient clams, there are almost on the beaches in Gresik regency. All this time the utilization of shellfish waste is limited to complementary toys of children. So the price of the shell becomes very cheap, due to not yet optimal society in using shellfish.

Ujung Pangkah sub-district, Gresik regency has big potential in producing shellfish. During this time, shellfish produced by fishing communities are still sold in raw materials and not yet processed into finished products. Shells have many benefits such as can be made into handicrafts such as Box Tissue, Photo Frame, Wall Clock, Pencil Place, etc. It can improve the welfare of the local community [8].

1.2. The Benefit of Processing Shellfish

Shellfish could be processed into various foods (shell crackers, shell nuggets, shell burgers, shell sticks, and crispy shells) and shell handicraft products (table clock, photo frame, pencil case). Utilization of these shells can improve welfare and foster the creativity of the community, especially in coastal areas. Various processed foods from shellfish could also be utilized to attract consumers of coastal tourism. The shellfish itself has a high protein content, low calorie, low fat/cholesterol with a lower proportion of saturated fat, vitamin B12, and minerals such as copper, iron, and zinc [12].

1.3. Target and Outcomes

The target is to improve the welfare of the community that has been analyzed previously. The outcomes of this community service are the publication of food processing and handicraft from shellfish, method of processing shellfish to variety food (shellfish crackers, shell sticks, crispy shells, and shellfish burger), and the making of handicraft products from shellfish (Table 1).

Table 1. Predicted results of the training activities

Before the community partnership program	After the community partnership program
1. Yields of shellfish were purchased by middlemen at relatively low prices	1. The yield of shellfish is processed independently, resulting in high-value products
2. The community had not been able to process the shellfish into various processed based on the shell.	2. The community can process the shellfish into various processed products such as (shell crackers, shell nuggets, shell burgers, shell sticks, and crispy shellfish)
3. The community did not know how to process shellfish into handicraft products	3. The community can process shellfish into handicraft product, so resulting in sales at high prices
4. The majority of the community only work as cultivators of shellfish, income only from shellfish cultivation	4. The community has additional work by cultivating high-income shellfish

2. Method

The implementation method of training program activities of the household industry for food and handicraft from the shell in Ujung Pangkah sub-district, Gresik Regency, East Java Province described in the following diagram in the form of solutions offered to overcome the existing problems (Figure 1).

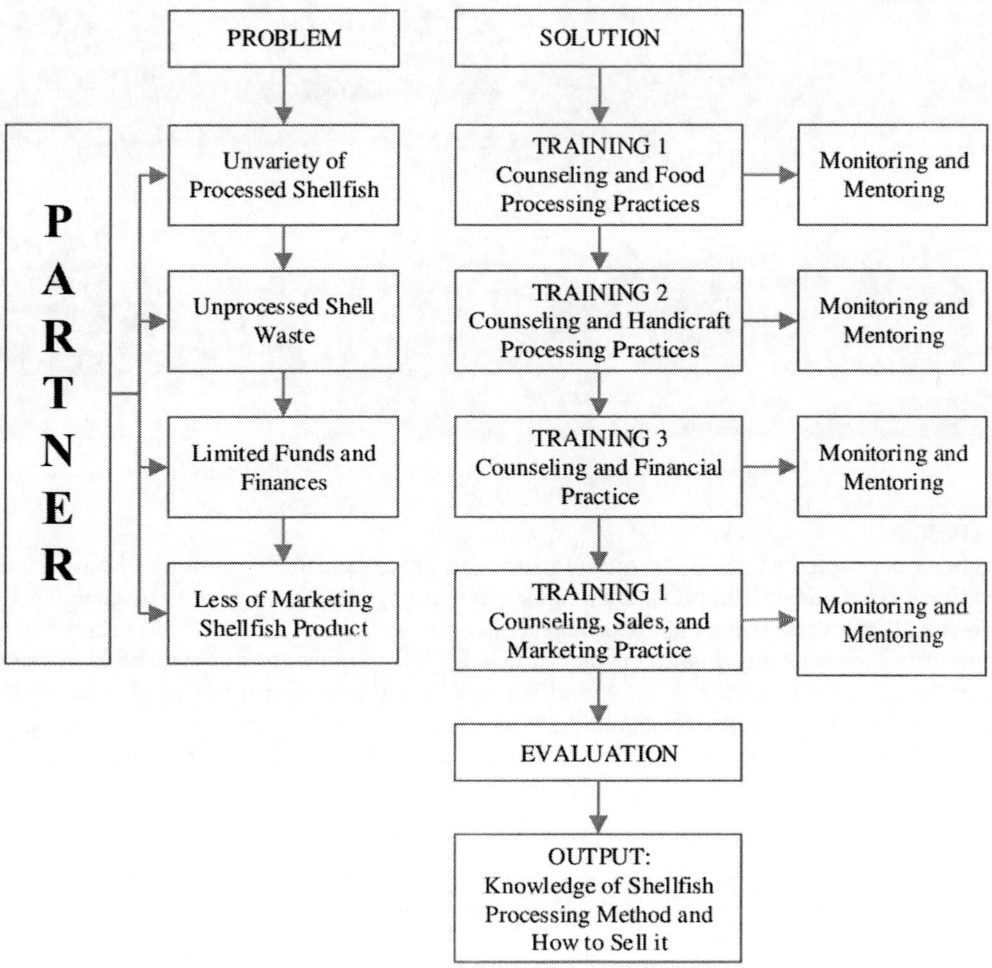

Figure 1. Method of implementation

3. Results and Discussion

Activities and training (Figure 2) that have been implemented are as follows:

1) Training of food processing from shellfish, to support this training it has been given shell processing tools. This training is taught how to make a burger, sticks, crackers, and crispy made from shell.
2) Training of making handicrafts. Participants know how to make handicraft products from shellfish become handicrafts as well as can develop their creativity.
3) Financial management training, participants know how to manage financially of shell business and can do simple financial bookkeeping
4) Training of basic packaging techniques. Participants know how to create attractive packaging as well as know-how to package the processed products.
5) Training of marketing techniques. Participants know and understand how to sell their products

Figure 2. Training and Empowering People Program

4. Conclusions

Its have been conducted some of the training for food processing for various shell-based food (shell crackers, shell nuggets, shell burgers, shell sticks, and crispy shellfish). The waste from shellfish has been processed into various handicraft products such as table clock, photo frames, and pencil cases. The training has successfully improved the knowledge of participants in terms of knowledge about processing shellfish into various culinary and various handicrafts, and it has been improved skills in processing shellfish into various foods and handicrafts along with entrepreneurial activities. As short, This program can assist the partners in improving economic welfare independently and business development in coastal areas.

References

[1] T. Thorvaldsen, "The importance of common sense : How Norwegian coastal fishermen deal with occupational risk," *Mar. Policy*, vol. 42, pp. 85–90, 2013.

[2] H. J. Lipscomb, D. Loomis, M. Anne, K. Kucera, S. Marshall, and L. Li, "Musculoskeletal Symptoms Among Commercial Fishers in North Carolina," vol. 35, pp. 417–426, 2004.

[3] M. Torner, G. Blide, H. Eriksson, R. Kadefors, R. Karlsson, and I. Petersen, "Musculo-Skeletal Symptoms as Related to Working Conditions Among Swedish Professional Fishermen," no. September, pp. 191–201, 1988.

[4] Q. Sholihah, A. Satria, A. Alim, and R. Fauzia, "Ergonomics Awareness as Efforts to Increase Knowledge and Prevention of Musculoskeletal Disorders on Fishermen," *Aquat. Procedia*, vol. 7, pp. 187–194, 2016.

[5] E. Nurmianto, *Ergonomi : Konsep Dasar dan Aplikasinya*, 2nd ed. Surabaya: Guna Widya, 2004.

[6] D. P. dan K. P. J. Timur, "Rekam Jejak Pembangunan Perikanan dan Kelautan di Jawa Timur."

[7] D. P. dan K. P. J. Timur, "Kementerian Kelautan dan Perikanan Akan Kembangkan Potensi Kelautan dan Perikanan di Jawa Timur," *Majalah Pro Ikan*, 2012.

[8] E. Nurmianto and N. Priyo, "Peran Ergonomi Makro Dan Desain Produk Kerajinan Dalam Mengolah Limbah Pantai Menjadi Produk Kreatif," 2010.

[9] E. Nurmianto, A. Pamungkas, and A. Sulisetyono, "Pengembangan Ekonomi Lokal Berbasis Perikanan Untuk Kemandirian Pulau Poteran, Kabupaten Sumenep.," 2016.

[10] F. J. Señorans, E. Ibáñez, and A. Cifuentes, "New Trends in Food Processing," *Crit. Rev. Food Sci. Nutr.*, vol. 43, no. 5, pp. 507–526, 2003.

[11] M. J. Rezaee, S. Yousefi, M. Valipour, and M. M. Dehdar, "Risk Analysis of Sequential Processes in Food Industry Integrating Multi-Stage Fuzzy Cognitive Map and Process Failure Mode and Effects Analysis," *Comput. Ind. Eng.*, 2018.

[12] Nurjanah, A. M. Jacoeb, R. N. Ulma, S. Puspitasari, and T. Hidayat, "Komposisi Kimia Kupang Merang (Musculista Senhausia) Segar dan Rebus," 2014.

Acknowledgment
This research of Food Processing and Ergonomics Handicraft Creation From Shellfish was financially supported by The Ministry of research, technology and higher education of the Republic of Indonesia through contract number 827/PKS/ITS/2018.

Design of mobile, portable and ergonomic fish smoke machine on a motorcycle

E Nurmianto[1], A Anzip[2] and Sampurno[3]

[1] Industrial Engineering Department, Faculty of Industrial Technology, ITS, Indonesia
[2] Mechanical Industry Engineering Department, Faculty of Vocational, ITS, Indonesia
[3] Mechanical Engineering Department, Faculty of Industrial Technology, ITS, Indonesia

* Email: nurmieko@gmail.com

Abstract. Built and Used On Motorcycle The Mobile, Portable, and Ergonomic Fishmaker are the design of a fish-shaped fog-shaped device transformed into a tool that is on a motorcycle. During this fuming, results are still not optimal because it still interferes with the eyes, nose, and lungs. Smoked fish is only natural without a touch of technology and innovation. The shape of this cupboard is considered appropriate because it has an optimal smoke capacity, a good level of strength so stable when the process of curing, has a practical design because it is easy to carry, dismantled and stored. This fumigation equipment uses motorcycles, with higher production capacity and socialization to the public about the use of proper and correct fish sauce. The innovation of this tool is the reduction of smoke that combines some smoke-reducing functions to be harmless but safe, comfortable and healthy for the users and effective for the products of processed seafood. Its innovative advantages are Practically shaped like cupboard, sturdy and sleek; Ergonomic can be designed and redesigned according to user size; Mobile can be taken anywhere and light; Portable can be off pairs making it easier and lighter to carry and can be used inside the house if the outside is raining; Smoke free because smoke is trapped in the cupboard and after the smoke is complete the valve can be opened then the smoke will be wasted through the chimney after that the fish can be taken; Hygienic and leave a delicious aroma on fumigation results.

Keywords: fumigation, motorcycle, portable, easy to move, ergonomic

1. Introduction

Indonesia as an archipelagic country has a vast maritime zone of 5.8 million km2 consisting of 2.3 million km2 of territorial waters, a territorial sea of 0.8 million km2 and an exclusive economic zone waters of 2.7 million km2. Indonesia has a capture fishery potential of 6.4 million tons per year. The abundant resources must be accompanied by sufficient fish processing process in order to become a leading commodity of Indonesia, especially East Java. So far the distribution of fish sales in the form of raw sales that can only survive in a matter of days only. Thus, the process of preserving fish proved capable of prolonging the rot of fish. There are various kinds of fish preservation, among others by salting, drying, filtering, fumigation, fermenting and cooling fish. The benefits of eating fish are known to many people, as in Japan and Taiwan are the main foods in everyday dishes that provide the

IOP Conf. Series: Materials Science and Engineering **674** (2019) 012047 doi:10.1088/1757-899X/674/1/012047

effect of youth and life expectancy is higher than in other countries. Fish processing as a preservation function causes people to consume more fish, one of which is the process of curing fish.

Therefore it is necessary to fish and portable fish fumes that can accommodate more fish in one curing and safe for the health of the fishermen and users even able to be taken anywhere with a motor vehicle. The solution offered to overcome this problem is by Designing Soymaker Fish On Motorcycle The Mobile, Portable And Ergonomic is a portable fish and portable fish that can accommodate more fish combined with an ergonomic motorcycle. Considering the deficiencies of prior curing tools, these fumigators must be made of durable material in order to survive longer than previous tools and through QFD and Anthropometry testing [1]–[3] to analyze the feasibility of products in terms of consumers and also ergonomic tool users.

Output targets from this research are the result of appropriate technology in the form of the Design of soymaker fish on a motorcycle which mobile, portable, ergonomic and environmentally friendly.

2. Objective and Research Benefits

2.1. Research Objective

The general objective of this research is to support the government in the program of empowering the commodities of coastal resources with easy and ergonomic methods. While the specific purpose of this research are to created a prototype design of fish motorcycle on motorcycle the mobile, portable and ergonomic with the use of motor vehicles as an effort to create a fish fog device that is mobile, portable, and environmentally friendly and apply the design of mobile motorcycle, portable and ergonomic motorcycle modulator that is mobile, portable and environmentally friendly in coastal area of East Java.

2.2. Research Benefits

The benefits of this research are:

1) Fishermen community can more quickly perform fish processing from the sea.
2) Sucker Fish On Motorcycle has advantages in terms of mobility when compared with a fish-shaped smoker.
3) Suckers On Motorcycle can sustain the economic improvement of coastal communities.
4) Sucker Equipment On Motorcycle has a large fish storage capacity in one fog process, so the output of the product is more than the conventional tool.
5) Implementation of Soymaker Fish On Motorcycle does not cause health and environmental problems (environmentally friendly), because the combustion system is done in a closed with a chimney on the top.

3. Literature Review

3.1. Potential of Coastal Resources of Indonesia

Indonesia as a state of the archipelago has a vast maritime zone of 5.8 million km2 consisting of 2.3 million km2 of territorial waters, the territorial sea of 0.8 million km2, and 2.7 million km2 of exclusive economic zone waters. Indonesia has a capture fishery potential of 6.4 million tons per year. Newly exploited at 63.5% or 4.1 million tons per year. The level of exploitation rate is still far from its sustainable potential (Ministry of Marine Affairs and Fisheries, 2009).

East Java is one of the provinces that have the potential of marine fishery resources consisting of pelagic fish and demersal fish. The area of marine fisheries management in south-eastern Java has enormous potential because it is directly opposite to the Indian Ocean and has fish potency, especially big pelagic groups like tuna (Thunnus sp) and skipjack (Katsuwonus Pelamis). Including three major coastal areas in East Java, namely Lamongan, Probolinggo, and Mojokerto. These three coastal areas produce a wide range of marine products. The abundant resources must be accompanied by sufficient fish processing process in order to become a leading commodity of Indonesia, especially East Java. So

IOP Conf. Series: Materials Science and Engineering **674** (2019) 012047 doi:10.1088/1757-899X/674/1/012047

far the distribution of fish sales in the form of raw sales that can only survive in a matter of days only. Thus, other processes such as fumigation, proven to prolong the rot of fish.

3.2. Fish processing

3.2.1. Conventional Fish Fogging

Smoking defined as the process of penetrating volatile compounds in fish produced from wood-burning [2], [4] [4], which can produce products with specific flavors and flavors, long shelf life due to antibacterial activity, inhibit enzymatic activity in fish that may affect the quality of smoked fish. The chemical compounds of wood smoke are generally phenols (which act as antioxidants), organic acids, alcohols, carbonyls, hydrocarbons and nitrogen compounds such as nitrous oxide, aldehydes, ketones, esters, ether, and further penetrates fish meat

3.2.2. Health Effects of Conventional Cooking Equipment

Smoking (Figure 1) is one of the livelihoods of Indonesians in coastal areas besides fishers. Fumigation is one of the traditional processes of fish preservation and anti-microbial compounds contained in the smoke from burning fuels such as aldehydes, alcohols, ketones can inhibit the growth of bacterial decay and the occurrence of protein coagulation in fish so that fish can survive from decaying bacteria. Coconut shell fuels are included in the category of hardwood species consisting of lignin, cellulose, hemicellulose, and ash with moisture content of 6% -9%. Smoke generated by the burning of coconut shells in the fogging environment is a problem that exists in the curing room fish.

Figure 1. Conventional Fish Smoke

3.3. Previous Research on Motorcycles

The background of this research is that the two-wheeled motor vehicle in Indonesia, called the Motor (Figure 2), is a vehicle that is in great demand by the public as a motor vehicle whose price is affordable and can be used for various transportation purposes. Starting from self-transport (1 person), along with other people (2 people), along with 2 others (3 people), husband and wife with 1 or 2 children (4 adults and 2 children), carrying gallons of water, computer, merchandise such as household groceries, bakery cakes and others [5].

The use of motors with excess capacity is not allowed in the Traffic Act and is dangerous to both themselves and other users. Some motorcycle users have tried to add parts of the motor to increase capacity, such as adding a rear-mounted box, pulling a two-wheeled train behind it, and some installing a side train permanently.

Figure 2. Motorcycle with passengers and dangerous goods on the highway

To get the right anthropometric size of Indonesian children, we can compare the size of British children with Indonesian children of the same age. From the ergonomic technique and Figure 2, we can create a scheme that shows how minimal and maximally this sidecar can be used by the child based on anthropometry that has been processed.

Figure 3 shows that the children aged 5 and 8 years can easily board a sidecar in the position of 1 person who was transported. Also, in a position where the feet can be straightened easily in a sidecar. While in children aged 12 years, the legs began to bend when on the sidecar.

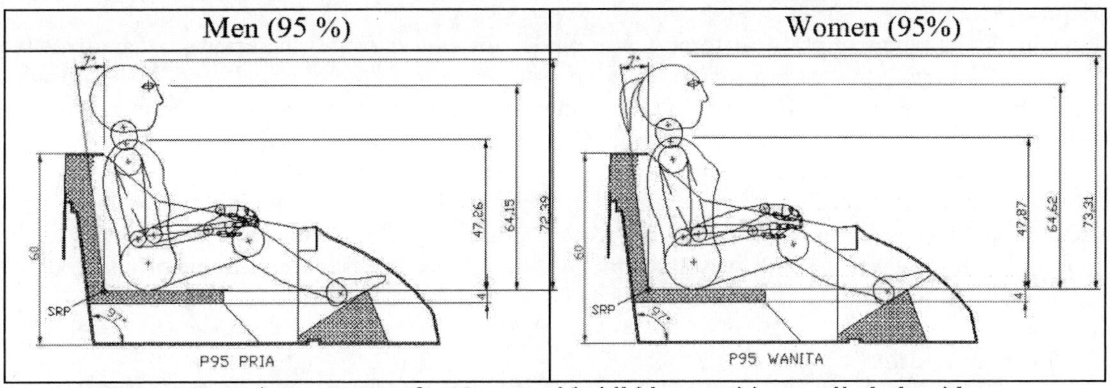

Figure 3. Anthropometry of a 12-year-old child in a position to climb the sidecar

3.4. Design of Curing Equipment of Motorcycle (on Motorcycle)

Musculoskeletal symptoms were reported by 215 fishermen, followed by 6 month interval for 18 months. Exposure information is collected through field observation and in-depth ethnographic interviews that allow the potential for ergonomic stress to be identified and cataloged with tasks and work stages. Symptoms that caused work disruption in the last 12 months were reported at 38.5% of the initial cohort. Lower back symptoms are the most common cause of work disorder (17.7%), followed by pain in the hands or wrists and shoulders (7%). Symptoms in the body region are more likely to be reported among individuals who are not full-time fish, and those who work other parts of the work or throughout the year have a much lower prevalence of symptoms; both may reflect the effect of healthy workers. A number of ergonomic stressors are identified at all stages of fishing with exposure variability dictated by some unpredictable factors such as weather; but also with the type of boat, equipment, crew size, and level of experience. Reduce associated ergonomic exposures [6], [7], [8]. Working among these traditional workers is important, regardless of whether they directly cause or contribute to their musculoskeletal symptoms, or aggravate the existing pathology.

To make it easier for fishermen to smoke fish from one coast to another. So in the design of the fish smoke tool located above the motorcycle. Based on the above conditions, then comes the idea of the Ergo-Design of Fish Cultivation Tool. This tool is the stuff to process fish smoke with capacity 120 in one process. This tool has the privilege that most of the cigarette smoke does not disturb the public (environmentally friendly) is marked with a chimney that is 3 meters high so that the smoke can be wasted into the free air and does not disturb the community around the SMEs. In addition, this tool is equipped with doors that can be opened at the front and rear so easy in the placement of fish before and after the fogging process.

East Java Province has become a region capable of producing high commodity sea products, but its process is still not well developed. Lack of capital makes them unable to increase production capacity and limited marketing. They still use traditional technology.

Because of such circumstances, there are still opportunities to improve efficiency and effectiveness through Small Business Development Programs in accordance with the situation and conditions of the small industry. The design of management, human resources, production, finance, and design will, in turn, enhance the competitiveness of small industries. The existing labor force in SME Fish Production has gained the skills of making fish from years of work experience and transmitted informally.

The conventional tool used for fumigation is a tool for baking goat satay or chicken satay. This tool is not sufficient because it can only burn about 20 to 30 fish in a process of fish exfoliation. That is once the process of curing fish takes about an hour, then the condition has not been able to meet the number of orders from consumers. In addition, this tool has a shortage during the processing of smoked fish, most of the smoke disturbs neighbors or the surrounding community or less environmentally friendly.

3.5. Biomechanics

Biomechanics is a human physical force that includes the power or physical power of humans while working and studying the workings and design of equipment to fit the physical capabilities of humans when doing work activities. Biomechanics encompasses the human body as a system consisting of interconnected elements connected, through joints and muscle tissue [9], [10].

Occupational biomechanics is a part of applied mechanics that studies the physical interaction between workers with machines, materials, and equipment aimed at minimizing complaints on the muscular-skeletal system in order to increase work productivity. Biomechanics is also used to evaluate a job, so it can produce a better way to work by minimizing the force and moment charged to the operator in order to avoid work accidents.

3.6. Nordic Body Map

Nordic Body Map is one of the most commonly used forms of ergonomic checklist questionnaires to find out one's discomfort with a product or object [11]. Nordic Body Map also aims to map the level of complaints of a person as material analysis and a particular action. The questionnaire uses a human body image that is divided into 9 main parts, namely, neck, shoulder, upper back, elbow, lower back, wrist, waist, knee, and heels/feet.

4. Method

4.1. Testing Method

4.1.1. Quality Function Deployment (QFD)

Quality Function Deployment (QFD) is a methodology in the process of designing and developing products or services capable of integrating 'consumer sounds' into the design process [12], [13]. QFD is a way for companies to identify and meet the needs and desires of consumers of products or services it produces. QFD is a methodology for translating consumer wants and needs into a product design that has specific technical requirements and quality characteristics.

4.1.2. Motorcycle Anthropometry Testing
Two-wheeled motor vehicles in Indonesia id called motorcycle lasim, is a vehicle that is in great demand by the public as a motor vehicle that is affordable and can be used for various transportation purposes. The physical dimensions of motorcyclists related to motorcycle design and other detail parts such as fairing or other protective devices whose effectiveness depends on the relative position of the rider's body part.

Motorcycle design provides a problem for ergonomics experts because it has narrow constraints (constrain) so difficult for them (ergonomics experts) to design a motorcycle by the size of the user's body in general [5]. There are two types of problems found in designing an ergonomic motorcycle, namely: 1. Variations in the size of the motorist's body size; and 2. Relationship attitude of the rider body with the effectiveness of work when motorists use the motor.

Information or data on the anthropometry of human sitting attitude while riding very few motorcycles in the United Kingdom (UK). Previous studies have found the fact that motorcyclists in the UK generally have a height that is not compatible with the motorcycle they use [14]. In the design of this tool, anthropometric tests will be performed [15] to identify the motorist's ergonomics and comfort level for motor designs and fish fumes installed on the motor.

4.1.3. Nordic Body Map *Operator* / User Testing
In knowing the user's complaints related to the smokers on a motorcycle, researchers use Nordic Body Map. *The Nordic Body Map* [11], [16], (Nurmianto et al, 2015) is an ergonomic checklist questionnaire aimed at mapping out a person's complaint level in terms of 9 major body parts, namely a) neck, b) shoulders, c) top, d) elbows, e) lower back, f) wrist, g) waist, h) knee, and i) heel/leg.

4.2. Designing Curing Equipment on Motorcycle
Figure 4 shows a fish cure tool that is mobile, portable, and environmentally friendly.

Figure 4. Appropriate technology in the form of proposed fish and portable fish fogging tool

5. Results

5.1. Design of Fish Cultivation Tool that is Mobile and Portable
The proposer has made the design of a mobile and portable fish fogging apparatus for immediate application. This tool will be applied to various types of motorcycles with modifications that have been determined by the proposer. In the implementation, the tool will be installed on the motorcycle seat that had previously been given a wooden plinth. The wooden board serves to protect the motorcycle seat so as not to be damaged, so that motorcycle owners can use the motor with dual function (the use of personal motorcycle and selling smoked fish).

The Curing Tool Fish on Motorcycle has a length of 81 cm and a height of 97 cm. At the bottom of the tool there is two fish storage that has been done fumigation. Each of these components has a length of 31.2 cm and a height of 36 cm. At the center has a length of 21 cm, where this size can be flexibly adjusted to the width of the seat of the motor. The design of the curing device on motorcycle is shown in Figure 5.

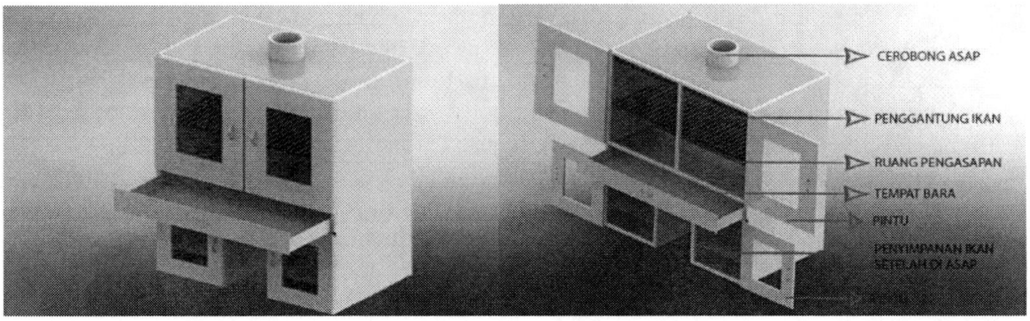

Figure 5. The design of a fish smoke appliance on an ergonomic motorcycle

In Figure 5, the design of Fish Curing Tool on Motorcycle is in closed condition and looks in (open). It has 7 components that include chimneys, fish bags, fogging chambers, coals, upper doors, fish storage after the smoke, and the bottom door. Chimney works to remove smoke generated from the coal, so the temperature and the process of curing fish can be regular. The next component is a fish-made iron, which is used as a place to lay the fish when the curing process takes place. Fishing contained in the fogging room, space where the curing of fish takes place. Below the fogging chamber there is a place of coals, the place where charcoal or briquettes as a coal-fired laid. Coal place is equipped with sliding doors to facilitate the user in entering the charcoal or briquettes in the process of fogging. When the process of fumigation takes place, then the top door will be closed so that the smoke and heat during the process can be evenly and stable. In addition, this tool also has fish storage that has been through the process of fogging. Thus, users can store fish that have been done fumigation and do the process of fumigation with other fish.

In the implementation, the use of Fish Curing Tools on Motorcycle to conduct the process of curing fish does not take place on a motorcycle, but the top tool must be lowered first. This tool is equipped with a sliding wheel which is useful to facilitate the top tool component that can be lowered and automatically 4wheel buffer will come out. The designs related to the separation mechanism of the Curing Tool of lower and upper fish are as in Figure 6.

Figure 6. Separation of Upper and Lower Catching of Fish Curing Devices

Figure 6 shows the design of the separation mechanism of the bottom and upper fish fogging. In practice, the bottom tool will remain attached to the motorcycle seat, while the top tool will be lowered by shifting to the process of curing the fish. At the bottom of the tool there are 4 wheel buffers that will come out automatically when this component is lowered. When the process of curing fish has been completed, the device is re-installed on the seat of the motor by bending the four-wheel buffers and restore the starting position.

6. The Next Stage Plan

6.1. Toolmaking and Testing Tools

After the finished product is made, then testing the Fish Sucker on Motorcycle to the fisherman community in Coastal Area (Lamongan, Bangkalan, Gresik, Surabaya, and Pasuruan). Testing is done directly to the fishermen community by riding a motorcycle that has been installed a fish sneakers and fish curing test using the tool. Furthermore, the researchers conducted interviews with samples of fishers who have used a fish smoker on motorcycle to fill the Nordic Body Map questionnaire so that it can be known the complaints experienced by the sample.

6.2. Evaluating Tools

A motorcycle that has been tested then evaluated to improve the design of the tool to be more ergonomic in its application. In addition, complaints felt by user samples with the Nordic Body Map method will be analyzed for consideration of design improvements.

References

[1] R. S. Bridger, *Introduction to Ergonomics*. Netherlands: Taylor & Francis, 2003.
[2] E. Nurmianto, *Ergonomi : Konsep Dasar dan Aplikasinya*, 2nd ed. Surabaya: Guna Widya, 2004.
[3] William J. Wiehagen and Fred C.Turin, "Ergonomic Assessment of Musculoskeletal Risk Factors at Four Mine Sites : Underground Coal , Surface Copper , Surface Phosphate , and Underground Limestone," Pittsburgh, PA, Information Circular 9475, 2004.
[4] E. Nurmianto, "Ergonomics smoke machine for indigenous people in Indonesia," *MATEC*, 2015.
[5] Z. Djunaidi and R. Arnur, "Risiko Ergonomi Ketidaksesuaian Desain dan Ukuran Tempat Duduk Sepeda Motor terhadap Antropometri pada Mahasiswa Ergonomic Risks of Incompatibility of Design and Size of Motorcycle Seat againts Anthropometry among College Students," *J. Kesehat. Masy. Nas.*, vol. Vol. 9, No, pp. 243–248, 2015.
[6] G. a Mirka, G. Shin, K. Kucera, and D. Loomis, "Use of the CABS methodology to assess biomechanical stress in commercial crab fishermen.," *Appl. Ergon.*, vol. 36, no. 1, pp. 61–70, Jan. 2005.
[7] A.-C. Falck and M. Rosenqvist, "What are the obstacles and needs of proactive ergonomics measures at early product development stages? – An interview study in five Swedish companies," *Int. J. Ind. Ergon.*, vol. 42, no. 5, pp. 406–415, Sep. 2012.
[8] J. Jonkman, J. M. Bloemhof, J. G. A. J. van der Vorst, and A. van der Padt, "Selecting food process designs from a supply chain perspective," *J. Food Eng.*, vol. 195, pp. 52–60, 2017.
[9] Patrick G Dempsey, "A critical review of biomechanical , epidemiological , physiological and psychophysical criteria for designing manual materials handling tasks," *Ergonomics*, vol. 41, no. October 2013, pp. 37–41, 1998.
[10] M. F. Antwi-Afari, H. Li, D. J. Edwards, E. A. Pärn, J. Seo, and A. Y. L. Wong, "Biomechanical analysis of risk factors for work-related musculoskeletal disorders during repetitive lifting task in construction workers," *Autom. Constr.*, vol. 83, no. April, pp. 41–47, 2017.
[11] P. K. Hembecker, D. C. Reis, A. C. Konrath, L. A. Gontijo, and E. A. Eugenio, "Investigation of musculoskeletal symptoms in a manufacturing company in Brazil: a cross-sectional study," *Brazilian J. Phys. Ther.*, vol. 21, no. 3, pp. 175–183, 2017.
[12] L. Cohen, *Quality Function Deployment : How to make QFD Wok for you*. New York: Addison-

Wesley Publishing Company, 1995.

[13] J. Marsot, "QFD : a methodological tool for integration of ergonomics at the design stage," vol. 36, pp. 185–192, 2005.

[14] P. Ward, "Centre for Motorcycle Ergonomics & Rider Human Factors Advanced Training & Rider Performance (Final report) Prepared for the Institute of Advanced Motorists Rossukorn Saikayasit School of Psychology Centre for Motorcycle Ergonomics & Rider Human Factors Advanced Training and Rider Performance," no. December, 2010.

[15] A. S. Amrutkar and N. R. Rajhans, "Ergonomic Posture for Motorcycle Riding .," no. April, 2017.

[16] E. R. Vieira, M. V. G. Buckeridge Serra, L. Brentini de Almeida, W. Vieira Villela, J. Domingos Scalon, and P. R. Veiga Quemelo, "Symptoms and risks for musculoskeletal disorders among male and female footwear industry workers," *Int. J. Ind. Ergon.*, vol. 48, pp. 110–116, 2015.

[17] E. Nurmianto, U. Ciptomulyono, Suparno, and S. Kromodihardjo, "Manual Handling Problem Identification in Mining Industry: An Ergonomic Perspective," *Procedia Manuf.*, vol. 4, no. December, pp. 89–97, 2015.

Acknowledgments

Authors acknowledge funding support from Directorate General of Higher Education, Ministry of Technology Research and Higher Education, Republic of Indonesia through contract number 1420/PKS/ITS/2018.

Reducing the natural disaster impacts by using an appropriate water testing technology

D Rahmawati [1,3*], J S Adiansyah [1], Alpiana [1,3] and Isfanari [2]

[1] Mining Technology Departement, Engineering Faculty, Universitas Muhammadiyah Mataram, Indonesia
[2] Civil Engineering Departement, Engineering Faculty, Universitas Muhammadiyah Mataram, Indonesia
[3] Muhammadiyah Disaster Management Centre, West Nusa Tenggara (MDMC NTB), Indonesia

[*] Email: dee.rahmawati@gmail.com

Abstract. The natural disaster might cause various impacts to the human including water supply. Polluted water sources should be treated prior consumed. The water treatment process refers to the water testing results. This paper was aimed to determine an effective water testing technology for emergency situations associated with natural disasters. The methods used were literature review and experiment. Various water testing methods during emergency response were explored by a terature review. In addition, the accuracy of manual and digital water testing kit, including pH and turbidity, were conducted through an experiment. The research resulted that the effectiveness of turbidity tubes ranged from 80 to100%. While the pH kit effectiveness ranged from 80 to 94%. There is no rule to classify the best technology during emergency response. The chosen technology depends on each humanitarian agency's condition including financial and human resources.

1. Introduction

A natural disaster is an incident resulting from the natural process which caused a negative impact on human life. The damage level depends on the community and environment affected [1] [2]. Shortly after the disaster, the condition of the community is unstable and vulnerable, and they require immediate aid from various stakeholders including government, company, and NGOs. The government and private sectors often collaborated, and they usually shared responsibility among them to respond to the disaster [3]. In addition, the humanitarian agencies oblige to follow three protection principles which stipulated the Humanitarian Carter as follows: a right to receive humanitarian assistance, a right to life with dignity, and a right of protection and security. Therefore, the humanitarian agencies have four emergency response standards including 1) Water Supply, Sanitation, and Hygiene Promotion (WASH), 2) Food Security and Nutrition, 3) Shelter, Settlements, and Non Food Items, 4) Health Services [4] [5].

A natural disaster including floods and earthquake could affect the freshwater supply. For instance, the flood affects water turbidity and the earthquake could damage freshwater distribution networks. In addition, the tsunami is a potential trigger for mixing the seawater and freshwater that might disturb the freshwater supply [6]. The availability of freshwater supply is very crucial within the first 3 days of natural disasters. Lack of freshwater supply during the natural disaster increases the diseases including diarrhea that transmitted by feces to mouth [5]. The common problem associated with freshwater supply during the natural disaster is the water quality that usually below the standards, so the water must be treated further. The Ontario storm in 2000 was one of the examples of freshwater pollution caused by bacteria. The pollution has caused 7 people died and more than 2300 people sick due to E.Coli and Campylobacter [6]. Two technical root causes were found, namely inadequate water quality measurement and treatment. Another lesson learned from Bima flood in December 2016 also indicated the gap in knowledge of water supply technology. In this case, The Muhammadiyah Disaster Management Center of West Nusa Tenggara (MDMC NTB) was unable to supply the fresh water for bathing, washing, and other personal use, because no applicable water treatment technology in the field. Therefore MDMC NTB provides package water for drinking and cooking (consumed water). In general, this paper aimed to determine an effective water testing technology in the emergency response associated with natural disasters.

2. Method

The methods of this paper are literature review and experiment. Various water testing methods during emergency response were explored by a literature review. The effectiveness of manual and digital water testing kit, including pH and turbidity, was observed by the experiment. The digital kit used pH meter and turbidity meter. The manual kit used turbidity tube and pH and Chlorine kit for pool and spa. The data analysis was used descriptive design which is shown in the table.

3. Result and Discussion

3.1. Water testing for emergency response

The sources of water should be assumed infeasible to be consumed during the emergency condition, so the humanitarian agencies must treat the water from various sources to fulfill needs. Currently, the most effective water treatment system for an emergency condition is Point of Use Water Treatment Technologies (PoUWT), for instance, coagulant-flocculants, filtration, and disinfectant [7][8]. In order to determine the water treatment phases, the water quality must be measured. The water quality measurement kit can be categorized into two types, namely, manual portable kit (e.g. indicator test for chlorine residual in water, turbidity tube [9], pH indicator); and digital portable kit (pH meter, Turbidity meter, Chlorine meter). Some humanitarian organizations have developed the technologies to measure the water quality and treat the polluted water in the emergency situation as displayed in Figure 1.

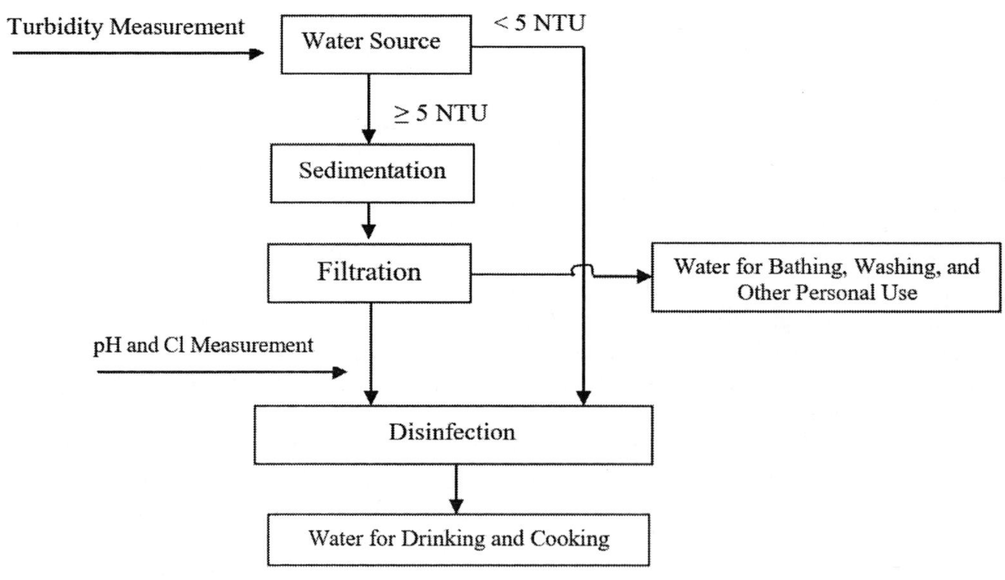

Figure 1. Water Treatment Flow Chart

The water source can be surface water (River, Lake, Spring) or groundwater (Well, Boreholes). The first step is conducting the turbidity measurement. If the water turbidity is ≥ 5NTU, the water must be flow into sedimentation process, and the coagulant is added. Secondly, the water is filtered using cotton cloth for low capacity treatment and sand filter for higher capacity treatment. The water from filtration process is ready to be used for domestic purposes including bathing, and washing, but it must be treated further by disinfectant prior consumed (drinking and cooking). There are several methods of disinfecting processes, such as, chlorination, boiled the water, and leave the water under the sunlight which called SODIS method [4] [8]. Escherichia Coli (E.Coli) bacteria will inactive when exposed by heat 70°C or higher for a minimum 3.5 seconds [10].

The standard of water quantity used in emergency response is 7,5 – 15 l/day/person. Meanwhile, the standard of water quality is shown in Table 1 [10].

Table 1. Water Quality Standards in Emergency Response

Parameter	Threshold
Coli bacteria-contaminated with feces/100 ml of water	0
Turbidity	< 5 NTU
Residual Chlorine	Max. 5 mg/l
pH	6 - 8

3.2. Turbidity and pH measurement

3.2.1. Turbidity Measurement.

The turbidity measurement in this paper was done by using a turbidity tube and turbidity meter. The turbidity tube is a transparency tube with a black cross/viewing disc at the bottom. The observer keeps pouring the water sample until they can no longer see the black cross/viewing disc, then the observer read out the scale on the outside of the tube in NTU (Nephelometric Turbidity Unit). Meanwhile, turbidity meter is a digitalize kit in which the observer pour the water sample into a small clean container, put the water sample on the turbidity meter then read the turbidity. Table 2 provides the result of turbidity measurement in some water samples.

Table 2. Result of Turbidity Measurement

Water Samples	Turbidity meter (NTU)	Turbidity tube (NTU)	Turbidity tube effectiveness
1	11,5	9,7	84%
2	7,8	6,9	88%
3	8,6	7,3	85%
4	15	12	80%
5	0,47	<5	100%
6	61,3	57,9	94%
7	104	100	96%

Table 2 indicated that the turbidity measurement by tube and turbidity meter generated different results. A turbidity meter is more accurate to measure less than 5 NTU; then turbidity tube can be measured only a maximum of 5 NTU. The results of measurement as shown in table 1 indicated that the effectiveness turbidity tube ranged from 80 to 100%. However, the observation result using turbidity tube varies for each person. In order to reduce the high difference, the observation must be conducted by at least two observers

3.2.2. pH Measurement

pH measurement in this paper was done by using pH meter and manual kit pH and chlorine for the pool. pH measurement by pH meter follows the steps: pour the water sample into a clean container, then put the probe into the water and read the pH. Meanwhile, to measure pH in the manual, the observer fills the vial to the mark with water sample and adds 4 drops of red phenol solution, then place cap on vial and invert several time to mix. Finally, compare the observer's reading with the pH color standards. Table 3 provides the result of pH measurement in some water samples:

Table 3. Result of pH Measurement

Water Samples	pH meter	pH kit	Effectiveness of pH Kit
1	6,04	6,8	88%
2	6,01	6,8	88%
3	6,06	6,8	89%
4	5,99	6,8	88%
5	7,113	7,6	94%
6	7,119	7,6	94%
7	6,99	7,6	92%

Table 3 indicated that the pH measurement using pH meter and pH kit resulted in slightly different values. The data showed that the effectiveness of the pH kit was around 88 - 92%. The measurement scale of pH kit can be read with the lowest limit of pH kit is around 6,8, so the acidity less than 6,8 was approached as 6.8. In general, the difference measurement result of using manual kit and the digital kit was found. Table 4 presented the advantages and disadvantages of those measurement technologies.

Tabel 4. Differences between Digital and Manual Portable Kit

Technology	Advantage	Disadvantage
Digital Portable Kit	Very accurate Easy to use	High cost Need Power Supply (e.g. battery) Need calibration
Manual Portable Kit	Low cost Easy to use No need power supply No need calibration	Less accurate

The precision of digital portable kit is very high, so the result of measurement very accurate. In contrast, the measurement using a manual portable kit depends on the observer, so the result could be less accurate. Both technologies are easy to use. However, the digital portable kit needs a power supply to operate, and it must be calibrated in order to maintain accuracy. The price of the digital kit is very high, and it becomes a barrier for many projects in the developing country. The lowest price of measurement kit and self-made measurement kit using local raw materials were chosen, it was more likely as well [9] [11] [12].

4. Conclusion

Water supply technologies consist of a water testing kit and water treatment is required to assist emergency responses situation. Some of those technologies are less effective in the emergency situation associated with natural disasters. Most water supply technology developed in normal condition which is different from actual disaster conditions. There is no rule to classify the best technology in emergency response. The chosen technology depends on each humanitarian agency's financial and human resources support. Based on natural disaster risks in West Nusa Tenggara, the appropriate technology for water testing should be low cost, easy to use, and simple in design. Although the manual portable kit is less effective in measurement than the digital one but the cost is more effective.

References

[1] M. Ozcelik, "Alternative Model for Electricity and Water Supply After Disaster," *J. Taibah Univ. Sci.*, vol. 11, no. 6, pp. 966–974, 2017.
[2] A. Pratibha and A. Archna, "At the Whim of Nature 'Natural Disasters': Causes and Prevention," *Soc. Issues Environ. Probl.*, vol. 3, no. 9, 2015.
[3] T. Simon, A. Goldberg, and B. Adini, "Socializing in Emergencies—A Review of the Use of Social Media in Emergency Situations," *Int. J. Inf. Manage.*, vol. 35, no. 5, pp. 609–619, 2015.
[4] S. I. Ali and K. Kadir, "Water Treatment," *WASH Emergencies| HIF*, 2016.
[5] S. C. for Humanitarian Response, "The Sphere Project: Humanitarian Charter and Minimum Standards Disaster Response." Mc Connan (ed) Geneva, 1998.
[6] S. E. Hrudey, E. J. Hrudey, and S. J. T. Pollard, "Risk Management for Assuring Safe Drinking Water," *Environ. Int.*, vol. 32, no. 8, pp. 948–957, 2006.
[7] E. Butler *et al.*, "Point of Use Water Treatment with Forward Osmosis for Emergency Relief," *Desalination*, vol. 312, pp. 23–30, 2013.
[8] D. Lantagne and T. Clasen, "Point-of-Use Water Treatment in Emergency Response," *Waterlines*, vol. 31, no. 1–2, pp. 30–52, 2012.
[9] E. Myre and R. Shaw, "The Turbidity Tube: Simple and Accurate Measurement of Turbidity in the Field," *Michigan Technol. Univ.*, 2006.
[10] S. Saimah, M. B. Sudarwanto, and H. Latif, "Dekontaminasi Bakteri Escherichia Coli dan Staphylococcus Aureus Pada Sarang Burung Walet dengan Perlakuan Pemanasan (Decontamination of Escherichia coli and Staphylococcus aureus in Edible Bird's Nest Using Heat Treatment)," *J. Kedokt. Hewan-Indonesian J. Vet. Sci.*, vol. 10, no. 2, pp. 143–147, 2016.
[11] H. Rush and N. Marshall, "Case Study: Innovation in Water, Sanitation and Hygiene," *Cent. Res. Innov. Manag. Univ. Bright. Bright.*, 2015.
[12] H. Jones and B. Reed, *Water and Sanitation for Disabled People and Other Vulnerable Groups: Designing Services to Improve Accessibility*. WEDC, Loughborough University, 2005.

Operator assignment model for minimizing total operator cost in a flexible flow shop environment

A Maruf *, **S Zahra and M M Isnaini**

Department of Industrial Engineering, Bandung Institute of Technology, Indonesia

* Email: maruf@itb.ac.id

Abstract. This research proposes a mix-integer linear programming model for operator assignment in a flexible flow shop environment. The objective is to minimize the total cost of operator transportation and operator exchange between machine. The model was developed considering practical constrain whereas machine are operated semi automatically, allowing an operator to be assigned to more than one machine in a certain period of time. Therefore, number of operator could be less than the number of machine. The proposed model result in a reduction up to 60% for operator transportation and operator exchange compared to current practice. Current output of proposed model will assign operator based on the number of available operator. Further research will be conducted by relaxing the number of operator to increase the operator utilization

1. Introduction

This paper arises from a practical case study of a textile industry. One of the textile process is creating a yarn by twisting together of fibers, also known as spinning process. The spinning process is categorized as a flexible flow shop system, which different type of yarn will go through the same stage of machine sequences. In each machine stage there are one or identical parallel machines. Different types of yarn are produced depending on the textile job order and due date. Thus, sequencing of yarn jobs is necessary to fulfill a certain objective. There are studies for job sequencing in a flexible flow shop [1]. In this case study, each machine has different level of automation, thus requiring less of operator time than the process time of the machine. In other words, it is possible that the operator shifts from one machine to another and conduct different jobs. Consequently, the number of operator could be less than the number of machines. Most research on operator assignment consider the number of operator is fixed during a certain time bucket in a planning horizon [2].

This research focus on the operator assignment problem. The objective of the research is to minimize the total cost of traveling and number of operator exchange between assignment, as seen in Figure 1.

The paper will be divided into four sections. The following section will discuss the model development. Section 3 will describe the case study and merit of the proposed model. The last section will point out remarks and further research activity.

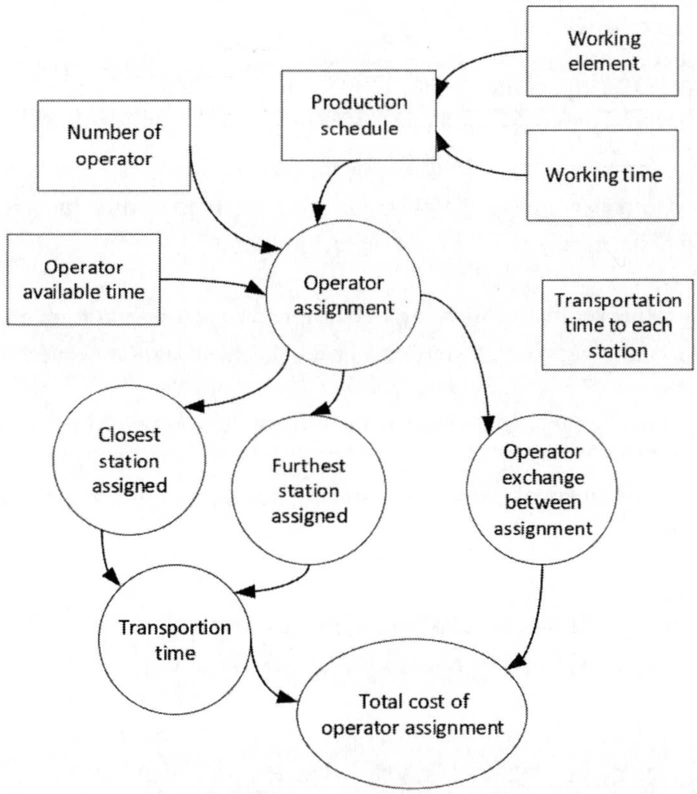

Figure 1. Influence diagram of interest

2. Model development

The case study has similar characteristics of the operator assignment problem in an assembly line system [3]. Each operator is assigned certain task within a certain planning horizon. Differences is that operator must move between stations [4]. Gebennini [5] has developed a model for assigning operators into a flexible manufacturing system. The objective function is to minimize operators' walking time. Gebennini also assumes that an operator may not shift to other machine until the job is finished. Considering the practical fact that machine has certain level of automation, the proposed model will relax this assumption enabling operator to move to other machine within the same time bucket if it fulfill the operator capacity within the time bucket. The proposed model will also consider a practical constrain for minimizing operator exchange. The objective function of case study is to minimize operator transportation time and operator switching between assignments.

The proposed model assumes that that the distance between station is linier in a flow shop layout [6]. The operator origin point will be set at the medium point between the closest and furthest assigned station. The operator skill level and walking speed are assumed the same.

2.1. Notation

The following notation of the proposed model.

Indexes

- $j = 1, \ldots\ldots, J$: station index;
- $k = 1, \ldots, K$): operator index;
- $t = 1, \ldots\ldots, T$): time buckets in the planning horizons;

Parameters

- C_I: traveling cost

IOP Conf. Series: Materials Science and Engineering **674** (2019) 012049 doi:10.1088/1757-899X/674/1/012049

- C_2: operator exchange cost
- τ: time bucket length
- T^{man}: operator's available time
- δ_{jt}: amount of time for each operator required at station *j* during time *t*
- f_{jt}: frequency of operator on station *j* during time bucket *t*.
- β_{jt}: binary parameter set to 1 if station *j* is working a job in time bucket *t*; 0 otherwise
- d_j: walking distance of station *j* from the origin point

Decision variables

- x_{jkt}: binary value set to 1 if operator *k* is assigned to station j in time bucket *t*, 0 other wise
- y_{jkt}: binary value set to 1 if station *j* is the furthest station served by operator *k* in time bucket *t*
- z_{jkt}: binary value set to 1 if station *j* is the closest station served by operator *k* in time bucket *t*
- d_{jkt}: walking distance of station *j* assigned to operator *k* from the medium point of the operator *k*

2.2. Model formulation

The proposed mixed-integer linear model is formulated as follows:

$$\min C_1 \sum_{j=1}^{J}\sum_{k=1}^{K}\sum_{t=1}^{T} \bar{d}_{jkt} f_{jt} + C_2 \sum_{j=1}^{J}\sum_{k=1}^{K}\sum_{t=1}^{T} \bar{b}_{jkt} \quad (1)$$

Subject to

$$\sum_{k=1}^{K} x_{jkt} = \beta_{jt}, \quad \forall j, \forall t \quad (2)$$

$$y_{jkt} \geq x_{jkt} - \sum_{h=j+1}^{J} x_{hkt}, \quad \forall j < J, \forall k, \forall t \quad (3)$$

$$y_{Jkt} = x_{Jkt}, \quad \forall k, \forall t \quad (4)$$

$$\sum_{j=1}^{J} y_{jkt} \leq 1, \quad \forall k, \forall t \quad (5)$$

$$z_{1kt} = x_{1kt}, \quad \forall k, \forall t \quad (6)$$

$$z_{jkt} \geq x_{jkt} - \sum_{h=1}^{j-1} x_{hkt}, \quad \forall j > 1, \forall k, \forall t \quad (7)$$

$$\sum_{j=1}^{J} z_{jkt} \leq 1, \quad \forall k, \forall t \quad (8)$$

$$\bar{d}_{jkt} \geq \frac{1}{2}\left(\sum_{i=1}^{J} d_i y_{ikt} + \sum_{i=1}^{J} d_i z_{ikt}\right) - d_j x_{jkt} - M(1 - x_{jkt}), \quad \forall j, \forall k, \forall t \quad (9)$$

$$\bar{d}_{jkt} \geq d_j x_{jkt} - \frac{1}{2}\left(\sum_{i=1}^{J} d_i y_{ikt} + \sum_{i=1}^{J} d_i z_{ikt}\right), \quad \forall j, \forall k, \forall t \quad (10)$$

$$\bar{b}_{jkt} \geq \frac{1}{2}\left(\sum_{i=1}^{J} d_i y_{ikt-1} + \sum_{i=1}^{J} d_i z_{ikt-1}\right) - d_j x_{jkt} - M(1 - x_{jkt}), \quad \forall j, \forall k, \forall t \quad (11)$$

$$\bar{b}_{jkt} \geq d_j x_{jkt} - \frac{1}{2}\left(\sum_{i=1}^{J} d_i y_{ikt-1} + \sum_{i=1}^{J} d_i z_{ikt-1}\right), \quad \forall j, \forall k, \forall t \quad (12)$$

$$\sum_{i=1}^{J} 2\bar{d}_{jkt} f_{jt} + \sum_{i=1}^{J} f_{jt}\delta_{jt} x_{jkt} \leq T^{man}, \quad \forall k, \forall t \quad (13)$$

$$\bar{d}_{jkt} \geq 0, \quad \forall j, \forall k, \forall t \quad (14)$$

$$\bar{b}_{jkt} \geq 0, \quad \forall j, \forall k, \forall t \quad (15)$$

$$x_{jkt}, y_{jkt}, z_{jkt} \in [0,1], \quad \forall j, \forall k, \forall t \quad (16)$$

Where

- Constraint (2) assure that an operator could be assigned at one machine in a certain time;
- Constraints (3)-(5) identifies the last station assign to each operator in each time bucket;
- Constraints (7)-(8) identifies the first station assign to each operator in each time bucket;
- Constraints (9)-(10) calculate the of the operator medium point between station *j* assigned to operator *k* in time bucket *t*;
- Constraint (11)-(12) is the binary setting to identify in number of operator exchange;
- Constraint (13) is the capacity for each operator for traveling time and assignment time at station *j* in time bucket *t*;

IOP Conf. Series: Materials Science and Engineering **674** (2019) 012049 doi:10.1088/1757-899X/674/1/012049

- Constrain (14)-(16) define non-negative and binary variables.

3. Case study

The case study is undertaken in a textile company, specifically in the spinning department. The spinning department has six stages process: open bale, blowing, carding, drawing breaker, drawing finisher and open-end. There are two set of drawing breaker, three set of drawing finisher and three set of open-end. The current distance between station is shown in Table 1. The traveling time given are the standard traveling time measured from different operator.

Table 1. Traveling time

From	To Station	Time (minutes)
	1, Open bale	0,19
	2, Blowing	1,60
	3, Drawing breaker 1	2,20
	4, Drawing breaker 2	2,44
	5, Drawing finisher 1	2,80
Origin Point	6, Drawing finisher 2	3,07
	7, Drawing finisher 3	3,16
	8, Open end 1	3,42
	9, Open end 2	3,74
	10, Open end 3	3,86
	11, Open end 4	3,97

The company is currently practicing a job sequencing algorithm to minimize lateness. The job sequence is then transformed into the operator load to conduct the job sequencing as shown in Table 2. Note that the Table 2 does not directly inform the job sequence but depicts the necessary workload to operate the job. The workload given in are standard time for operating the machine induced by the job sequencing. Currently the time bucket is set to 30 minutes and operator time break is set at 8 and 0, where no job are assigned.

Table 2. Production schedule

Stasiun	*Time Bucket*															
	δjt															
	1	**2**	**3**	**4**	**5**	**6**	**7**	**8**	**9**	**10**	**11**	**12**	**13**	**14**	**15**	**16**
1	30,00	30,00	9,99	10,48	0,00	0,00	0,00	0,00	0,00	0,00	0,00	0,00	0,00	0,00	0,00	0,00
2	0,00	10,71	0,00	0,00	0,00	10,71	0,00	0,00	0,00	10,71	0,00	0,00	0,00	0,00	10,71	0,00
3	29,87	17,20	15,91	1,81	15,91	1,81	15,91	0,00	0,00	15,91	17,20	15,91	29,87	15,91	1,81	15,91
4	15,74	17,70	7,90	11,28	7,90	17,70	7,80	0,00	0,00	14,72	17,80	7,80	12,56	16,72	7,80	9,90
5	9,90	5,54	13,74	17,60	7,90	10,46	18,72	0,00	0,00	7,80	7,80	21,48	7,80	7,80	17,80	13,58
6	25,96	14,30	16,10	11,96	27,32	15,74	11,10	0,00	0,00	15,52	27,32	15,52	13,76	11,86	22,54	19,74
7	17,08	26,64	11,42	14,42	18,42	21,86	17,42	0,00	0,00	15,52	12,54	25,52	17,32	15,52	13,76	25,86
8	18,30	15,52	23,76	14,52	9,88	13,30	28,32	0,00	0,00	18,74	12,54	10,64	23,76	15,52	17,32	28,74
9	8,57	6,54	2,08	6,56	2,50	6,95	2,50	0,00	0,00	5,73	2,50	6,95	3,33	4,48	2,91	6,95
10	6,90	7,37	2,50	5,31	2,91	7,79	2,50	0,00	0,00	5,73	7,37	2,50	2,50	5,73	7,79	1,25
11	8,55	2,50	6,56	2,08	6,54	4,89	2,91	0,00	0,00	2,91	6,14	5,29	3,33	5,73	2,50	7,37

The current operator assignment of the company is depicted in Figure 2. The operator assignment is based on trial and error conducted by the supervisor and being refined from time to time.

Station	*Time Bucket*															
	1	**2**	**3**	**4**	**5**	**6**	**7**	**8**	**9**	**10**	**11**	**12**	**13**	**14**	**15**	**16**
1	1	1	1	1												
2		2				2				5					6	
3	2	2	2	2	2	2	2			2	2	1	1	1	2	2
4	3	3	3	3	3	3	3			3	3	3	3	3	3	3
5	4	4	4	4	3	3	3			4	4	4	4	4	4	4
6	3	3	3	3	4	4	4	Time break		3	3	3	4	4	4	4
7	3	3	3	4	4	4	4			4	4	4	3	3	3	3
8	4	4	4	4	4	5	4			4	4	4	4	4	4	5
9	5	5	5	5	5	5	5			5	5	5	5	5	5	5
10	6	6	6	6	6	6	6			6	6	6	6	6	6	6
11	7	7	7	7	7	7	7			7	7	7	7	7	7	7

Figure 2. Current operator assignment

The proposed model has been solved using Ilog Cplex^{TM}. The optimal solution is depicted in Figure 3. Seven operators have been assigned. The total transportation is reduced from 636 meters to 204 meters and there is less operator exchange between station. Allocation of operator assignment is currently proportional to the workload, where only 1 operator is need for station 9, 10 and 11. The supervisor verify the feasibility of only allocating 1 operator for machine 9,10 and 11. In addition, station 4, 5, 6 dan 7, which previously was relocated an additional of two operators. The above facts were confirmed at site where as many work-in-processes was queueing for operator to setup.

Station	*Time Bucket*															
	1	**2**	**3**	**4**	**5**	**6**	**7**	**8**	**9**	**10**	**11**	**12**	**13**	**14**	**15**	**16**
1	3	3	3	3												
2		1				3				1					1	
3	1	1	1	1	1	1	1			1	1	1	1	1	1	1
4	6	6	6	6	6	6	6			6	6	6	6	6	6	6
5	6	6	6	6	6	6	6			3	3	3	3	3	3	3
6	5	5	5	5	5	5	5	Time break		5	5	5	5	5	5	5
7	4	4	4	4	4	4	4			4	4	4	4	4	4	4
8	2	2	2	2	2	2	2			2	2	2	2	2	2	2
9	7	7	7	7	7	7	7			7	7	7	7	7	7	7
10	7	7	7	7	7	7	7			7	7	7	7	7	7	7
11	7	7	7	7	7	7	7			7	7	7	7	7	7	7

Figure 3. Proposed operator assignment

4. Conclusions and further research

This paper proposes a mixed integer linier model for operator assignment in a flexible flow shop environment. The model minimizes the total operator assignment cost consisting of traveling cost and operator exchange cost. Practical constrain has been considered and a case study has been presented. Further development of the model will relax the number of operator available to increase operator utilization.

References

[1] Silvia C, Reis V, Morais A, Brilenkov I, Vaza J, Pinheiro T, Neves M, Henriques M, Varela M L, Pereira G, Dias L, Pernandes N O and Carmo-Silva S 2017 A comparison of production

control systems in a flexible flow shop *Procedia manufacturing* **13** 1090-95.

[2] Ernst A, Jiang H, Krishanamoorthy M and Sier D 2004 Staff scheduling and rostering; a review of application, methods and models *Europen journal of operational research* **153** 3-27.
Van den Berg J, Belien J, De Bruecker P, Demeulemeester E and De Boeck L 2013 Personal scheduling: a literature review *Europen journal of operational research* **226(1)** 367-85

[3] Sungur B and Yavuz Y 2006 Assembly line balancing with hierarchical worker assignment *Journal of Manufacturing Systems*. **37** 290-98.

[4] Eiselt H A and Maianov V 2018 Empolyee positioning and workload allocation *Computer & operations research* **35(2)**, 513-24

[5] Geneninni E, Zeppetella L, Grassi A and Rimini B 2016 Minimizing operators' walking times into a linear system layout *IFAC* **49-12** 1709-14.

[6] Heragu S S 2016 *Facilities design* CRC Press p 128

IOP Conf. Series: Materials Science and Engineering **674** (2019) 012050 doi:10.1088/1757-899X/674/1/012050

Determination of lot size orders of furniture raw materials using dynamic lot sizing method

M Djunaidi[1,2,*], B A R Devy[1], E Setiawan[1,2] and R Fitriadi[1,2]

[1] Industrial Engineering Department, Universitas Muhammadiyah Surakarta, Indonesia

[2] Center of Logistics and Industrial Optimization Study, Universitas Muhammadiyah Surakarta, Indonesia

[*] Email: much.djunaidi@ums.ac.id

Abstract. Wisanka is an indoor and outdoor furniture manufacture and export company in Indonesia. In the production activities, inventory control of raw materials is very important and needs to be handled properly so that the production process can run smoothly and orders from consumers can be fulfilled in a timely manner. Problems faced in the company that there is no optimal planning in determining the size of ordering of raw materials. Reservations made during this only based on the quantity of raw materials ordered can meet the needs during a certain period without considering the total cost incurred. The main purpose of this research is to determine the policy of controlling raw material requirement for the coming period by paying attention to the inventory cost incurred. In this research used lot sizing technique based on Silver Meal algorithm and Wagner Within algorithm. The use of lot sizing technique based on Wagner Within algorithm resulted in a cost efficiency of 3.6% for klabang material while Silver Meal of 2.9% for semi-polished rattan and 3.4% for leles compared to the lot sizing technique applied by the company.

1. Introduction

Inventories are managed by a business or company to anticipate some future demand [1]. Supply is made to deal with uncertainty in business. This uncertainty can be either out of stock, demand that exceeds estimates or overstock. Timing and quantity of supply is one way to deal with out of stock or overstock. Refueling is the company's attempt to place orders to suppliers for the purpose of storing inventory. Companies often get into trouble because of unavailability of materials (inventory), or too much material stacked in warehouses [2]. Overstock can cause a risk of damage to higher material inventories, higher storage costs, and the emergence of idle funds. Out of stock may cause delays in the production process or demand fulfillment, resulting in loss of customers due to delayed order fulfillment [3].

The use of the Lot Sizing technique is appropriate for use in determining the quantity of inventory orders in which in addition to minimizing the number of orders, it can also minimize the cost of direct inventory and inverse cost of inventory orders [4]. The optimal solution for single-sizing economic lot-sizing (ELS) problems has been known since Wagner and Whitin published their dynamic programming analysis [5]. The use of the Wagner-Within method continues to be developed, such as the efficient implementation of the non-concave cost (WWA) formula [6], with the application of branch and bone techniques to achieve faster solutions [7], and integration of the WWA algorithm with the horizon-planning theorem [8], and developed a near-optimal lot-sizing model for material requirements planning systems [9], as well as development based on theoretical planning and economic concepts, which is Economic Part Period (EPP).

Silver and Meal developed a heuristic method for selecting case lot size from deterministic time-varying demand rate and discrete opportunity to recharge, later known as the Silver-Meal method [10]. The Silver - Meal method can be used to determine the calculation of the value of inventory, especially the safety stock, reorder point, and lot sizing, then determine the best method [11], to calculate inventory cost and determine order quantity and frequency [12].

Wisanka is one of furniture manufacturer and exporter company in Indonesia. Wisanka manufactures various types of furniture, including outdoor furniture, indoor furniture, rattan furniture, and lighting furniture. The raw materials used are rattan, mahogany frame, and woven material. The furniture production process is based on order from consumer (make to order). Therefore the production strategy of the company is made to order causes the demand for furniture needs become very volatile. The supply of raw materials should be sufficient to anticipate the fluctuating demand.

Optimal planning in determining the size of raw material ordering becomes a problem in Wisanka. Reservations made so far only consider the quantity of raw materials that can meet the needs during a certain period without considering the inventory cost to be spent by the company, the quantity of ordering of raw materials is done based on the intuition of the number of products ordered. In addition, the company has also never conducted an evaluation and comparing existing inventory systems with other methods.

Proper handling of raw materials into a requirement for the production process to run smoothly and orders from consumers can be met in a timely manner. This research proposes the use of two methods, namely the Silver – Meal algorithm and Wagner – Whitin algorithm. The consideration of the selection of the silver-meal algorithm method is to provide an optimal local cost value on every raw material repurchase and over a long period of time has a probability that it can provide a lower cost. The selection of the Wagner – Whitin algorithm method is to provide the optimal solution used to determine the minimum cost [13].

2. Literature Review

2.1. Silver Meal Algorithm

The Silver-Meal algorithm is one of the methods for planning and controlling the supply of raw materials based on the cost period stating that the purchase of materials is only done at the beginning of the period, while the cost of storing is only charged to material stored over a period by considering the capacity of the warehouse [14]. The steps in calculating the silver-meal algorithm are as follows [13]:

1) Calculate the first lot size starting from period T. Lot size = requirement (net requirement) in period T, then calculate message cost and store cost in that period (if there is save fee). After calculating the total cost multiply by 1 divided by a sequence of period (if lot of first-order then multiplied 1/1 if lot second order then multiplied 1/2).
2) The second lot size, the need for the next period plus the first lot size, calculate the total cost per period. After counting multiply by 1/2 because it is the second lot order. The formulas that can be used are:

$$\mathrm{TRC(T)} = \frac{\mathrm{C+P_h \sum_{k=1}^{T}(k-1)\, R_k}}{\mathrm{T}} \tag{1}$$

Where *TRC (T)* = relevant total cost in period T; *T* = period (T = 1, 2, 3, ..., T); *C* = order cost per period; *h* = percentage of shelf cost per period; *P* = purchase cost per unit; P_h = storage cost per period; and R_k = average demand in period k

3) Compare total cost per current period with before, if *TRC(L)* = *TRC(L-1)* as well as lot size = warehouse capacity back to step two and *TRC(L)* > *TRC (L-1)* go to step 4 and if not then go back to the first step.

4) Lot size in the period

$$\sum_{t=T}^{L}(t-T)dt \quad (2)$$

5) If $T = L$, and the time frame of the planning horizon has been reached, then the algorithm can be stopped. If it does not, return to the first step.

2.2. *Wagner Whitin algorithm*

The Wagner-Whitin algorithm is one heuristic method that uses a dynamic programming approach. This method can provide optimal value for dynamic demand or lot sizing problems to find the optimal solution [15]. The steps of the Wagner-Within algorithm are as follows:

1) Calculating the total cost of the inventory variable for each possible alternative ordering in a given period with N periods, which falls into the total cost of the variable is the cost of the message and the save cost by the following formula:

$$Z_{ce} = C + hP\sum_{i=c}^{e}(Q_{ce} - Q_{ci});\ 1 \le e \le c \le N \quad (3)$$

Note:
C = ordering cost (IDR/order)
h = cost saving per unit per period (IDR/ unit/period)

$$Q_{ci} = \sum_{i=c}^{e} D_i \quad (4)$$

D_i = demand in period i.
C = initial limit of the period covered by order Q_{ci}
e = maximum limit of the period covered by order Q_{ci}

2) Calculate the minimum cost (f_n) in period c to e with an inventory assumption at the end of period 0. This algorithm will start with $f_0 = 0$ and calculate the f_n value. The formula for calculating the value of $f_e = min(Z_{ce} + f_{c-1})$ for c = 1, 2, 3, ..., e.
3) Find the optimal solution f_N by counting the number of orders backward by looking for the minimal value of each column that meets the period on the same line. The f_N value is the total cost value and the optimal lot order is calculated by the formula shown as below:
   a. The last order is made in the period c to meet demand from period c to period N. The formula as shown below:

$$f_N = Z_{ce} + f_{c-1} \quad (5)$$

   b. Order prior to the last order must be made in the period v to period v - 1. The formula is shown below:

$$f_{c-1} = Z_{vc-1} + f_{v-1} \quad (6)$$

   c. The first order is made in period 1 to meet the demand from period 1 to period u - 1. The formula is shown as below:

$$f_{u-1} = Z_{u-1} + f_0 \quad (7)$$

2.3. *Research Methods*

This research was conducted to evaluate raw material procurement in the past and to plan procurement for the future in Wisanka, one of the UKM furniture in Indonesia. The procurement data of raw materials during the period of June 2017 - May 2018 is used as the basis for forecasting raw material demand in the period of June 2018 - December 2018. Forecasting methods used to follow the pattern of raw material demand in the past, taking into account the minimal error rate. The analysis is done by calculating the total cost of raw material procurement that will come with the company's applied policy, then compared with the total cost of procurement if applying Silver Meal method and the use of Wagner Whitin algorithm.

3. Results and Discussion

Data on the amount of raw material consumption used in the period of June 2017-May 2018. This data is used as a basis in the process of forecasting demand for the period June 2018 - December 2018, to illustrate the forecast of raw material needs in the period. Furniture raw materials are calculated based

IOP Conf. Series: Materials Science and Engineering **674** (2019) 012050 doi:10.1088/1757-899X/674/1/012050

on the identification that includes classification A (which has a big handling priority). The selection of these raw materials is based on classification calculations using the ABC method, where this method focuses inventory control on high-value inventory types rather than low-value ones. The ABC analysis divides the inventory into three classes based on the value (or volume) of inventory [16].

Based on the calculation of the ABC method of phase 1, out of 23 raw materials calculated, it is included in category A of 7 types of raw materials, category B of 5 types and category C of 11 species. Grouping stage 2 uses the ABC method, for 7 types of raw materials in category A in grouping phase 1, obtained 3 types of raw materials belonging to category A. The three raw materials are klabang, semi-polished rattan, and leles, respectively.

Figure 1. Klabang.

Figure 2. Semi-polished rattan.

Figure 3. Leles.

Based on the pattern of past data in the period of June 2017 - May 2018 (see Table 1) that is seasonal, forecasting is done using 3 methods: double exponential smoothing (DES), linear trend (LT), and winters method additive (WMA). While the calculation of forecasting accuracy is done by 3 methods: mean absolute deviation (MAD), mean squared error (MSE) and mean absolute percentage error (MAPE). Forecasting results to be selected based on the smallest forecasting error. Forecasting results for July 2018 - December 2018, are shown in Table 2.

Table 1. Procurement data for the raw materials demand (on kgs)

Periods	Klabang	Semi-polished rattan	Leles
June 2017	1,274.00	136.00	359.85
July 2017	2,762.50	690.25	318.00
August 2017	2,133.25	455.50	652.50
September 2017	4,879.00	520.00	524.70
October 2017	4,879.00	750.75	824.70
November 2017	2,917.00	550.75	937.00
December 2017	2,337.00	739.25	517.85
January 2018	3,872.50	1,070.00	860.50
February 2018	1,100.75	423.50	456.25
March 2018	2,593.50	606.60	562.10
April 2018	3,845.50	436.50	744.25
June 2018	2,468.50	799.00	451.10

Table 2. Forecast result for the raw materials demand (on kgs)

Periods	Klabang	Semi-polished rattan	Leles
June 2018	3,004.09	751.15	681.49
July 2018	3,016.74	774.68	693.91
August 2018	3,029.39	798.21	706.34
September 2018	3,042.04	821.75	718.76
October 2018	3,054.69	845.28	731.18
November 2018	3,067.33	868.82	743.61
December 2018	3,079.98	892.35	756.03

3.1. Silver Meal methods

Calculation of lot size with Silver Meal method needs to consider the capacity of the warehouse. Based on the steps described in Eq. (1) and Eq. (2), calculations can be made for raw material demand and order times. After calculating the combined cost of a period and the combination with the following period, a comparison of the total cost per period for each combination is performed. In addition, also pay attention to the accumulation of orders with warehouse capacity. The calculation will be repeated for another combination of periods to find the optimal result or average cost per smallest period exceeding the warehouse capacity. Silver Meal method processing results can be shown in Table 3.

Table 3. Result of lot sizing for raw material using Silver Meal methods

# Order	Klabang		Semi-polished rattan		Leles	
	Periods (week-)	Amount (kgs)	Periods (week-)	Amount (kgs)	Periods (week-)	Amount (kgs)
I	1	5,266.65	1	1,924.48	1	1,905.15
II	8	5,304.59	11	1,854.81	12	1,998.33
III	15	5,342.52	20	1,972.48	23	1,127.83
IV	22	5,380.48	-	-	-	-

3.2. Wagner-Whitin methods

The wagner-whitin algorithm can provide optimal value for the problem of dynamic sizing request in accordance with certain period horizon. Based on forecasting results and known inventory cost, the Wagner-Within algorithm steps are performed using Eq. (3) to Eq. (7), to determine the minimum cost. The minimum cost can be calculated assuming that the condition of the inventory amount at the end of the period is zero. The results of Wagner-Whitin algorithm processing can be shown in Table 4.

Table 4. Result of lot sizing for raw material using Wagner-Whitin algorithms

# Order	Klabang		Semi-polished rattan		Leles	
	Periods (week-)	Amount (kgs)	Periods (week-)	Amount (kgs)	Periods (week-)	Amount (kgs)
I	0	4,512.46	0	1,138.49	0	1,028.44
II	6	2,262.55	6	387.34	6	346.96
III	9	4,553.57	8	598.66	8	529.75
IV	15	5,348.86	11	404.99	11	356.27
V	22	2,303.66	13	410.87	13	359.38
VI	25	2,309.99	15	1,485.13	15	1,282.68
VII	-	-	22	434.41	22	371.80
VIII	-	-	24	446.17	24	378.02
IX	-	-	26	446.17	26	378.02

3.3. Company Policy
The company has its own policy in lot sizing. The determination of lot sizing is not based on cost, but uses needs over the next several periods adjusted to the lead time of arrival of raw materials. The company has set raw material policy to meet production requirement for 2 weeks. So orders are made in periods 1, 3, 5, 7, 9, 11, 13, 15, 17, 19, 21, 23, 25, and 27.

3.4. Cost Comparison
Total inventory cost is calculated by summing up the ordering cost, storage cost, and purchase cost. The ordering cost is obtained from the ordering frequency multiplied by the cost of order. The cost of storage is obtained from the stored cost multiplied by the volume of stored material and the length of storage. The purchase cost is obtained from the amount of raw materials multiplied by the price of raw materials. Table 5 shows the value of ordering costs, storage costs and raw material prices observed.

Table 5. Result of lot sizing for raw material using Silver Meal methods

Raw Material	Ordering cost (IDR)	Storage cost (IDR)	Material price (IDR)
Klabang	1,168.75	408.00	8,000.00
Semi-polished rattan	376.45	713.00	13,500.00
Leles	360.44	637.00	12,000.00

Comparison of method is done to find method which give optimum cost so that can be used to do planning of optimal raw material inventory. The result of cost comparison can be seen in Table 6.

Table 6. Comparison of inventory cost between company policy, Silver Meal method and Wagner Whitin algorithm

Raw materiaks	Company policy (IDR)	Total inventory cost			
		Silver Meal		Wagner Within	
		Cost (IDR)	Saving	Cost (IDR)	Saving
Klabang	191,057,396	185,951,619	2.7%	**184,225,871**	**3.6%**
Rotan Semi Poles	85,454,341	**82,960,217**	**2.9%**	84,342,360	1.3%
Leles	67,177,269	**64,898,619**	**3.4%**	65,941,121	1.8%

The result of raw material requirement planning can provide the minimum cost solution that company issued in terms of inventory cost by using Silver Meal algorithm method. Therefore the results of this research can be used to help make company policy in June-December 2018. The savings for each raw material that is klabang using Silver Meal method with total cost of IDR 184,225,871,- and result in savings of 3.6%. As for raw materials of semi-polished rattan and leles with total cost of IDR 82,715,424,- and IDR 64,402,867,-, respectively. The application of the Silver Meal method is expected to provide savings of 2.9% for semi-polished rattan and leles by 3.4% compared to company policies.

4. Conclusion
Analysis of the plan calculation of optimal raw material needs based on forecasting method, by looking at the accuracy of the smallest forecast errors, in the case observed is based on linear trend method. The result of raw material requirement analysis done by Wagner Whitin and Silver Meal method can minimize the cost of the company in terms of inventory cost when compared to the fulfillment of raw materials using the company's policy. In the case observed, for planning for the period of June - December 2018, optimal conditions can be achieved for the raw materials of klabang using Wagner Within method can save by 3.6%. As for the raw materials of semi-polished rattan and leles using Silver Meal method, can save by 2.9% and 3.4%, respectively.

References

[1] Adeyemi S L and Salami A O 2010 *Journal of Social Science* **23**(2) 135-142
[2] Hasian D P 2012 *Jurnal Optimisasi Sistem Industri* **11**(1) 203-207
[3] Saputra P H and Zeplin J H T 2014 *AGORA* **2**(2) 1-9
[4] Yamit Z 2005 *Inventory Management* (Yogyakarta: Ekonisia)
[5] Wagner H M and Whitin T M *Management Science* **9** 89-96
[6] Evans J R 1985 *Journal of Operations Management* **5** 229-235
[7] Jacobs F R and Khumawala B M 1987 *Production and Inventory Management* **28** 39-43
[8] Saydam C and McKnew M 1987 *Production and Inventoty Management* **28**(2) 15-19
[9] Gaither N 1981 *Production and Inventory Management* **22**(4) 75-89
[10] Silver E A and Meal H C 1973 *Production and Inventory Management* **14** 64-75
[11] Nadyatama D, Aini Q and Utami M C 2016 *Proc of 4th International Conference on Cyber and IT Service Management* Bandung
[12] Rohmah D U M, Argo B D and Deoranto P 2016 *International Journal of Research in Advent Technology* **4**(12) 28-34
[13] Mbota H K W, Tantrika C F M and Eunike A 2015 *Jurnal Rekayasa dan Manajemen Sistem Industri* **3**(1) 178-188
[14] Wohos I P, Mandagi R J M and Walangitan D R O 2014 *Tekno Sipil* **12**(61) 25-34
[15] Basuki 2016 *Industrial Engineering Journal* **5**(1) 29–34
[16] Hudori M 2017 *Jurnal Citra Widya Edukasi* **9**(2) 1-10.

Human error analysis using sherpa and heart method in Batik Cap production process

I Pratiwi*, **M Masita, H Munawir and R Fitriadi**

Department of Industrial Engineering, Universitas Muhammadiyah Surakarta, Jalan Jendral A Yani Tromol Pos 1 Pabelan – Surakarta, Indonesia

* Email: Indah.Pratiwi@ums.ac.id

Abstract. The batik industry is one of the creative industries which development keeps increasing every year. Almost all of Batik Cap production process activities are done manually. Therefore, production operators play an important role in influencing the quality of Batik Cap. This research was conducted to identify the human error factor that happened in the Batik Cap production process using SHERPA and HEART method. The potential for human error that occurs is described with SHERPA tabulation. The human error assessment is calculated using the HEART method of human error probability value that occurs in the production process operator. The result of this research is 26 potential errors of overall activity in 6 work stations divided into 19 error in action error, 5 error on checking error and 2 error in selecting error. The highest probability of human error value in stamping the work station on the fabric stamping activity.

Keywords: Batik Cap, Human Error, SHERPA, HEART

1. Introduction

Human Reliability is the probability of human success in carrying out his work elements. The goal is to identify high-risk areas, identify the factors that lead to human error and how to repair an existing system so as to reduce adverse human error and pose a hazard. The most commonly used method is Systematic Human Error Reduction and Prediction (SHERPA) because it is suitable for errors related to human skills and habits, more detailed and consistent in error identification. The next method is Human Error Assessment and Reduction Technique (HEART) is a method designed as a method of quantifying human error risk quickly, simply and easily understood by engineers and human factors specialists [1]. Small businesses Batik Supiyarso in the District Laweyan Surakarta, where the process of making Batik done with the stamping known as Batik Cap. Stages of the production process are the process of cutting the fabric, the process of stamping, the coloring process, the "*penglorodan*" process, the process of drying, and the process of arrangement of batik fabric [2]. Produced batik quality is determined by the quality of the process being undertaken and also the quality of raw materials. The most influential production process of batik quality in outline occurs in the stamping and coloring. In the coloring section, the most common defects are in the form of uneven color gradations (5%), curved fabric (2%), and uneven dyeing (3%) [3]. Research in the textile industry with the highest Human Error probability results in Dyeing Department at occupation number 6, human error probability value 54.4% [4].

Humans have limitations in their work to make mistakes. Analysis of the occurrence of human error associated with skills and habits more detailed and consistent in the identification of errors using

IOP Conf. Series: Materials Science and Engineering **674** (2019) 012051 doi:10.1088/1757-899X/674/1/012051

the input hierarchy of the basic level task is the method of SHERPA [5][6]. While the HEART method is used in determining the chances of error in every job activity [7]. The HEART method is based on the principle that every time a task is performed there is a possibility of failure and this possibility is affected by one or more EPC (Error Producing Condition), the factors that have a significant effect on performance are indicated by the largest HEP value. This condition can then be applied to the best-case scenario of the estimated probability of failure under ideal conditions to then get a chance of final error [8].

This study attempted to find an alternative quality improvement of batik produced by identifying the possibility of human error that occurs in the process of Batik Cap so that it can be searched way to reduce the existence of the error.

2. Method

This study was conducted to identify the type of error operator/human error that contribute to the occurrence of defective products. Also to know the magnitude of operator error probability in every product defect event.

The data was collected through interviews directly to the person in charge of production and 9 operators by giving a questionnaire to determine the frequency of factors causing the error to occur, ie 0 to 1 value for each of the *driving factors*. The assessment range is the range of the probability value of the contribution to the occurrence of human error in order to assess the value of APOA (Assessed Proportion of Affect) by determining the probability value of error for each factor that has been identified.

After the data successfully collected proceed to the stage of data processing, namely:

1) Hierarchical Task Analysis (HTA). Error identification data is then performed breakdown task into sub-task using HTA which is used as reference determination identification of the work failure operator.
2) Data Processing SHERPA
   a) Human Error Identification (HEI), HTA Data is used to identify errors related to human error using SHERPA Error Mode table reference.
   b) Consequence Analysis, which occurs due to human error that may occur.
   c) Ordinal Analysis Probability, indicated by low, medium, and high.
   d) Analysis of Strategy, explains remedial solutions to overcome human error which may occur.
3) HEART Data Processing
   a) Classify the task type to one of 8 GTT categories. HTA data used for the identification of errors related to human error using the reference table owned by GTT HEART.
   b) Determine EPCs and APOA. The result of error identification is translated into EPCs language to determine EPC value and used as a reference for a questionnaire to determine APOA value.
   c) Calculates the HEP value. To determine the highest error probability level of each work station.

3. Results and Discussion

Making HTA for 6 work stations making Batik Cap. Mori Cutting Working Station with 2 female workers, 35-40 years old. Stamping Working Station with 5 male workers, age 48-60 years old. Coloring Working Station with 2 male workers, 35-37 years of age. *Penglorodan* Working Station, with 1 male worker, age 55 years. Drying Working Station, with 1 male worker, 37 years old. Batik Stacking Working Station with 1 female worker, age 56 years.

Table 1. Hierarchical Task Analysis In "Batik Cap" making Process

Working Station	*Hierarchical Task Analysis*
1. Mori Fabric Cutting	1.1. Measuring Fabrics 1.2. Cutting Fabrics 1.3. Stacking Fabrics on Hangers
2. Stamping	2.1. Cutting the Wax 2.2. Setting up the work table 2.3. Installing Fabrics on the Work Desk 2.4. Taste the Fabrics 2.5. Folding Fabrics that have been stamped
3. Coloring	3.1. Color mixing 3.2. Pouring colors into Peder Machine 3.3. Installing a fabric 3.4. Coloring and Color Locking Process
4. *Penglorodan*	4.1. Boiling Fabrics 4.2. Washing Fabrics
5. Drying	5.1. Lifting Fabrics to the Dryer 5.2. Pulling Fabrics from the Dryer
6. Stacking Batik Fabrics	6.1. Sorting Fabrics 6.2. Packing Fabrics

The data of work failure identification was obtained from an interview and brainstorming with Batik business owners and also with a total of 9 employees (Table 2).

Table 2. Identification of Operator's Working Failure

Activity Description		Failure Description	Cause of Failure	Failure Effect	Corrective action
No	*Task*				
1.1	Measuring Fabrics	Incorrect size of fabric	The operator forgot to count the size The operator is not focused due to the intervention of the other party	The wrong size of fabric / missed	Repeat again
1.2	Cutting Fabrics	Fabric is hard to cut Pieces of fabric are not symmetrical	The cutter/scissors are not sharp The operator is not used to cutting The operator is not focused due to the intervention of the other party	Cain defects	Cutting the defect fabric
1.3	Stacking Fabrics on Hangers	There are pieces of fabrics that are still folded / not tidy	The operator did not inspect the fabric after it was laid out The operator hurried when setting the fabric	Fabric is not tidy, hard to taste	Repeat again
2.1	Cutting the Wax	The operator did not cut the wax	Cutlery can not be used	It will not fit in the frying pan	Sharpen the cutting tool
2.2	Preparing the Work Desk	Plastic cover of wet foam torned	The operator did not check the desk when it's going to be used Operator is less aware because of their age	Wet fabric can not be stamped	Drain the fabric
2.3	Stacking Fabrics on the Work Desk	There are pieces of fabrics that are still folded	The operator missed to check the fabric The operator hurried when setting up the fabric	Fabric is not tidy, hard to taste	Repeat again

Activity Description		Failure Description	Cause of Failure	Failure Effect	Corrective action
No	***Task***				
2.4	Stamping the Fabrics	Wrong stamping	The stamping tool can not be used The operator hurriedly stamping the fabric Operators are less aware because of their age	The tasting does not fit the motive	Can not be fixed
2.5	Folding the stamped fabrics	The operator folds the fabric imperfectly	The operator hurried while folding the fabric	Damage the wax on the fabric	Repeat again
3.1	Color mixing	Incorrect color	The operator did not check the results Less experienced operators Operators are disturbed due to intervention from others	Cain defects	Repeat
3.2	Pouring Color into Peder Machine	The operator pours the color with a less precise dose	The operator hurries when pouring colors	Spilled color solution	Can not be fixed
3.3	Laying the Fabrics	The operator puts the fabric in the wrong position	The operator is not accustomed to installing the fabric to the peder The operator hurried when putting the fabric	Fabric wrinkled	Can not be fixed
3.4	Coloring and Color Locking Process	Uneven or dirty staining results	Did not do color testing on fabric Inexperienced operator Peder machine not cleaned	Fabric color does not match	Can not be fixed
4.1	Boiling Fabrics	The color of the fabric is faded	Boiling too long Too much starch solution	The color of the fabric is faded	Can not be fixed
	Washing Fabrics	Waxes are still attached	Boiling is not stirred continuously Too little starch solution	Waxes are still attached	Repeat again
5.1	Lifting Fabrics into Dryer	Uneven drying	The operator does not level the fabric in the dryer	Inhibits work	Repeat again
5.2	Pulling Fabrics from the Dryer	Uneven drying	The operator did not pull the fabric on the dryer	Inhibits work	Repeat again
6.1	Sorting Fabrics	False sorting of good/bad quality fabrics	The operator did not check thoroughly	Incorrect placement of fabric sort	Complete check
6.2	Packing Fabrics	The packaging is not strong	The operator hurriedly packs batik fabric	The packaging results are easily removed	Re-checking

After making HTA and collecting work failure identification data, it is then used as a reference for making a tabulation of SHERPA (Table 3).

IOP Conf. Series: Materials Science and Engineering **674** (2019) 012051 doi:10.1088/1757-899X/674/1/012051

Table 3. Tabulation of Systematic Human Error Reduction and Prediction

Task number	Task	Error Mode	Error Description	Consequences	Correction	Error Probability
1.1	Measuring Fabrics	A8	The operator did not measure the size of the fabric	Inhibits work	Repeat again	*Low*
		A2	The operator wrongly calculates the size of the fabric	Inhibits work	Repeat again	*Medium*
1.2	Cutting Fabrics	A7	The operator cuts the fabric not straight	Fabric defects	Cut out a defective fabric	*Medium*
1.3	Stacking Fabric on Hangers	C2	The operator did not check the crumpled fabric	Difficult to be stamped	Repeat again	*Low*
2.1	Cut out the candle	A8	The operator Did not cut the wax	It is not fit in the frying pan	Cut out the candle	*Low*
2.2	Prepare the Work Desk	A8	The operator did not wet the sponge with water	Interfere the work	Wetting the sponge	*Low*
		A8	The operator Did not give the fabric on the wet sponge	Interfere the work	Do it again	*Low*
		C2	The fabric is hollow	Wet fabric can not be stamped	Replace the glass fabric	*Medium*
		C2	The fabric is not tightened with the desk	Wet fabric can not be stamped	Repeat again	*HIgh*
2.3	Stacking Fabrics at the Work Desk	A9	The operator organizes the fabric not straight	Inhibits work	Repeat again	*High*
2.4	Taste the Fabrics	A7	The Operator wrongly stamped the motive	wrong fabric motive	Can not be fixed	*Medium*
2.5	Folding Stamped Fabrics	A9	The operator folds the fabric not thoroughly	Damage the wax on the fabric	Can not be fixed	*Medium*
3.1	Color mix	S2	The operator chooses a wrong mix	Fabric color is wrong	Do it again	*Medium*
3.2	Pouring Color to Water Glass	A6	The operator pours the colors not on the color box	Color does not appear	Do it again	*Low*
3.3	Laying Fabric	A2	The operator did not tie the fabric to the clamp	Inhibits work	Do it again	*Low*
3.4	Coloring and Color Locking Process	C1	The operator did not clean the peder	Color does not match	Can not be fixed	*Medium*
4.1	Boiling Fabrics	A8	The operator did not pour the starch solution	Waxes stuck together again	Repeat again	*Low*
		A4	The dose of starch solution is less	Waxes are still attached	Repeat again	*Medium*
		A4	The dose of starch solution solution is excess	The color of the fabric is faded	Can not be fixed	*Low*
		A8	Not stirring at the time of *penglorodan*	Waxes are still attached	Repeat again	*Low*
4.2	Washing Fabrics	C2	Checking of fabric is incomplete	Waxes are still attached	Repeat again	*Low*
5.1	Lifting Fabric to Dryer	A9	Not leveling the fabric in the fabrices	Uneven drying	Repeat again	*Low*
5.2	Pulling Fabric from the Dryer	A2	Not pulling fabric from the dryer	Uneven drying	Repeat again	*Low*
6.1	Sorting Batik Fabric	S2	Wrong choosing good/bad quality fabric	Incorrect packing location	Recheck	*Low*
6.2	Packing Batik Fabric	A4	Not folding in the same size	Not tidy	Repeat again	*Medium*
		A3	Packaging is not strong	The packaging results are easy to open	Repeat again	*Low*

After making HTA, the lowest level of HTA is used as a reference in HEART data processing (Table 4).

Table 4. Calculation of Human Error Assessment and Reduction Technique

Task	Generic Task	Human Unreliability Value	Calculation				HEP
1.1	E	0.02	EPC	16	36		0.032768
			Total HEART Effect	3	1.06		
			Proportion	0.3	0.4		
			Assessed Effect	1.6	1.024		
1.2	E	0.02	EPC	23	31	36	0.0270766
			Total HEART Effect	1.6	1.2	1.06	
			Proportion	0.4	0.3	0.5	
			Assessed Effect	1.24	1.06	1.03	
1.3	E	0.02	EPC	31	36		0.0211744
			Total HEART Effect	1.2	1.06		
			Proportion	0.2	0.3		
			Assessed Effect	1.04	1.018		
2.1	E	0.02	EPC	23			0.0236
			Total HEART Effect	1.6			
			Proportion	0.3			
			Assessed Effect	1.18			
2.2	E	0.02	EPC	17	38		0.04032
			Total HEART Effect	3	1.02		
			Proportion	0.5	0.4		
			Assessed Effect	2	1.008		
2.3	E	0.02	EPC	17	36		0.044792
			Total HEART Effect	3	1.06		
			Proportion	0.6	0.3		
			Assessed Effect	2.2	1.018		
2.4	E	0.02	EPC	15	23	31	0.0513
			Total HEART Effect	3	1.6	1.2	
			Proportion	0.4	0.4	0.7	
			Assessed Effect	1.8	1.24	1.14	
2.5	E	0.02	EPC	31	36		0.0228
			Total HEART Effect	1.2	1.06		
			Proportion	0.6	0.3		
			Assessed Effect	1.12	1.018		
3.1	E	0.02	EPC	31	17	36	0.0354
			Total HEART Effect	1.2	3	1.06	
			Proportion	0.4	0.3	0.4	
			Assessed Effect	1.08	1.6	1.024	
3.2	E	0.02	EPC	31			0.02036
			Total HEART Effect	1.06			
			Proportion	0.3			
			Assessed Effect	1.018			
3.3	E	0.02	EPC	31	36		0.0228
			Total HEART Effect	1.2	1.06		
			Proportion	0.6	0.3		
			Assessed Effect	1.12	1.018		
3.4	E	0.02	EPC	17	31	36	0.0354
			Total HEART Effect	3	1.2	1.06	
			Proportion	0.3	0.4	0.4	
			Assessed Effect	1.6	1.08	1.024	
4.1	E	0.02	EPC	31	31	36	0.02442
			Total HEART Effect	1.2	1.2	1.06	
			Proportion	0.4	0.5	0.3	
			Assessed Effect	1.08	1.1	1.012	

Task	Generic Task	Human Unreliability Value	Calculation				HEP
4.2	E	0.02	EPC	31			0.0212
			Total HEART Effect	1.2			
			Proportion	0.3			
			Assessed Effect	1.06			
5.1	E	0.02	EPC	31	38	5.1	0.02218
			Total HEART Effect	1.2	1.02		
			Proportion	0.5	0.4		
			Assessed Effect	1.1	1.01		
5.2	E	0.02	EPC	31	38	5.2	0.02182
			Total HEART Effect	1.2	1.02		
			Proportion	0.4	0.5		
			Assessed Effect	1.08	1.01		
6.1	E	0.02	EPC		31	36	0.02253
			Total HEART Effect		1.2	1.06	
			Proportion		0.5	0.4	
			Assessed Effect		1.1	1.02	
6.2	E	0.02	EPC		31	36	0.02213
			Total HEART Effect		1.2	1.02	
			Proportion		0.5	0.3	
			Assessed Effect		1.1	1.01	

Based on observations and interviews that is tabulated in the SHERPA tabulation (Table 5).

Table 5. Classification of Error Types

Number	Error Types	Total
1	Action Error	19
2	Checking Error	5
3	Selecting Error	2

Based on Table 5, the total of 18 tasks generated 26 error descriptions which are included in 3 types of errors classified according to SHERPA error mode ie error in action error, checking error and selecting error. The entire error description generated the consequences that have been described so that with the explanation there is a potential error that need to be wary of.

In table 6 HEP values are sorted from the largest to the smallest values to find out which task has the highest probability of operator failure.

Table 6. The sequence of Human Error Probability On Operator Production of "Batik Cap"

No	HEP	*Task*
1	0.0513	2.4 Stamping the Fabrics
2	0.0448	2.3 Organize the Fabric on the Work Desk
3	0.0403	2.2 Preparing the Work Desk
4	0.0354	3.1 Mixing Color
5	0.0348	3.4 Coloring and Color Locking Process
6	0.0328	1.1 Measuring Fabrics
7	0.0271	1.2 Cutting Fabrics
8	0.0238	4.1 Boil the Fabrics

No	HEP	*Task*
9	0.0236	2.1 Cutting Waxes
10	0.0228	3.3 Installing Fabric in Peder Machine
11	0.0228	2.5 Folding Fabrics which already Stamped
12	0.0225	6.1 Sorting Batik Fabric
13	0.0222	5.1 Lifting Fabric to Dryer
14	0.0220	6.2 Packing Batik Fabric
15	0.0218	5.2 Pulling Fabric from the Dryer
16	0.0212	1.3 Setting the Fabric on Hangers
17	0.0212	4.2 Washing the Fabrics
18	0.0204	3.2 Pouring Color to Peder Machine

Based on Table 6, the result of the HEP calculation shows the probability of human error in the production process of Batik Cap, which is 0.0204-0.0513. The biggest potential error occurs in the task of tasting the fabric with HEP of 0.0513. This Task has the highest number of EPCs that is 4 EPC, meaning it has the most number of error types where each EPC shows errors contained in this task. Description of the error in the task of tasting the fabric is one of the tasting motives, because the canting cap tool is less reliable due to some eroded angle, the operator in a hurry to do the tasting, the operator is exhausted due to standing position and exposed to heat radiation from the stove for 7 hours. operators are subject to limitations due to the influence of age and less experienced operators. The errors found did not fully show that the operator is wrong with the stamping motive, it is only the tasting results that are irregular and messy. There is also the effect of stamping sticking less visible so that the waxless attached to the fabric resulting in leakage color during the coloring process.

When compared to the value of HEP obtained, the size of HEP through measurement using the HEART method is determined not only by how much the proportion weighting which is the result of the questionnaire, also determined by the Generic task that is the Human unreliability selection which determines the general characteristics of each job. There are a number of 9 Generic tasks that indicate the level of operator unreliability. and EPC selection that indicates the conditions in the task that may cause errors. The larger the selected EPC number the smaller the EPC value is obtained. EPC is adjusted to the conditions in each job. The type of EPC varies depending on the situation that can lead to human error.

Processes that occurred on the production floor have a risk of accident because when the operator uses the machine, it must be equipped with personal protective equipment used to reduce the risk of accidents [9]. The results of the evaluation and analysis of the work system can provide suggestions for improvement to prevent the occurrence of occupational diseases or decreased productivity [10].

4. Conclusion

There are 26 human error potential descriptions of all Batik Cap production activities at 6 work stations divided into 19 errors during action error, 5 error at checking error and 2 error at the time of selecting error. The result of the probability assessment of human error is indicated by the HEP value of 0.0204-0.0513. The biggest potential error occurs in the task of stamping the fabric with the HEP of 0.0513. This is because the canting stamping tool is less reliable due to several corners that have been eroded, operators in a hurry to stamp, operators who are exhausted due to standing position and exposed to radiation heat from the stove for 7 hours, operators are subject to limitations due to the influence of age and less experienced operators.

Reference

[1] Bell J, Holroyd J 2009 *Heal Saf Exec Res Rep* ;RR679
[2] Pratiwi I, Kartikasari I 2018 *AIP Conf Proc 1977*. 2018;20051:20051-1-0020051–6.

[3] Agus Widaryanto, Bariyah C 2015 *Pros Simp Nas Teknol Terap* ;I14.
[4] Abbas FA, Ariwibowo B 1994 *Proceedding 8th Int Semin Ind Eng Manag* ;108–10.
[5] Stanton N 2005 *Handbook of Human Factors and Ergonomics Methods* ©
[6] Harahap FA 2012 *Skripsi Fak Tek Univ Indones Depok.*
[7] Rahmania T, Ginting E, Buchar *2013 e-Jurnal Tek Ind FT USU* ;2(1):58–65.
[8] Safitri DM, Astriaty AR, Rizani NC 2015 *Human Reliability Assessment dengan Metode Human Error Assessment and Reduction Technique pada Operator Stasiun Shroud PT . X. J Rekayasa Sist Ind* ;4(1):1–7.
[9] Putro FC, Helianty Y, Desrianty A 2015 *J Reka Integr Jur Tekik Ind Itenas* ;3(2):173–84.
[10] Pratiwi I, Nurkhasanah 2017 *Pros Semin Nas Teknol Terap* ;147–55.

IOP Conf. Series: Materials Science and Engineering **674** (2019) 012052 doi:10.1088/1757-899X/674/1/012052

Environmental benefit of public fleet fueled by LPG/Vi-Gas in Magelang City, Indonesia: A Simulation Study

E M Widodo[1], M Setiyo[2], M I Rosyidi[1], T A Purnomo[1], B Waluyo[2], B C Purnomo[2] and R Murniningsih[3]

[1]Department of Industrial Engineering, Universitas Muhammadiyah Magelang, Magelang, Indonesia.
[2]Department of Automotive Engineering, Universitas Muhammadiyah Magelang, Magelang, Indonesia.
[3]Faculty of Economics and Business, Universitas Muhammadiyah Magelang, Magelang, Indonesia.

E-mail: setiyo.muji@ummgl.ac.id

Abstract. This paper presents a simulation study on the utilization of LPG/Vi-Gas for public transport in Magelang City. Estimates of CO_2 reduction per year are calculated theoretically and based on data from NEDC experiences. Both of these methods show positive results, reducing CO_2 by theoretical calculations and comparing with NEDC experiences are 13.6 and 11 tons per year, respectively. In conclusion, the utilization of LPG/Vi-Gas for public fleets in Magelang city is feasible to be implemented as the Government's effort to reduce CO_2 emission from transportation sector.

1. Introduction

In recent decades, air pollution in big cities has caused anxiety for analysts and environmental stakeholders [1]. A decrease in air quality appears to be a logical compensation for industrial, household, and transport activity. Implementation of low technological standards, fuel quality, and less strict emissions thresholds cause pollution from evaporation and tailpipe. In Indonesia, air quality levels in several major cities such as Jakarta, Bandung, and Surabaya have proven far above the WHO standards [2].

Currently, Magelang is the city ranked 85th in Indonesia, with a population of more than 124 thousand people. Compared to Surabaya or Bandung with a population of more than 2 million, Magelang may be too small. However, with an area of 18.12 km^2, the population of Magelang is 6,864 and is ranked 20th nationwide (landscape of Magelang city is presented in Figure 1). The addition of population, expansion, and construction of houses and shopping centers, road widening, and tree felling are other reasons for the decline in air quality as the emission absorber becomes reduced. If the problem of pollution due to transportation is not planned from now, it is not impossible Magelang city will also experience problems like other big cities that have become the attention of stakeholders [3]–[7]. The presence of Transportation on Demand (ToD) such as Gojek and Grab also increase traffic density. Although ToD provides advantages for its users in terms of comfort and speed, in macro is the burden of energy. The need for fuel in the city with ToD services becomes larger which means it also contributes more exhaust emissions to the environment.

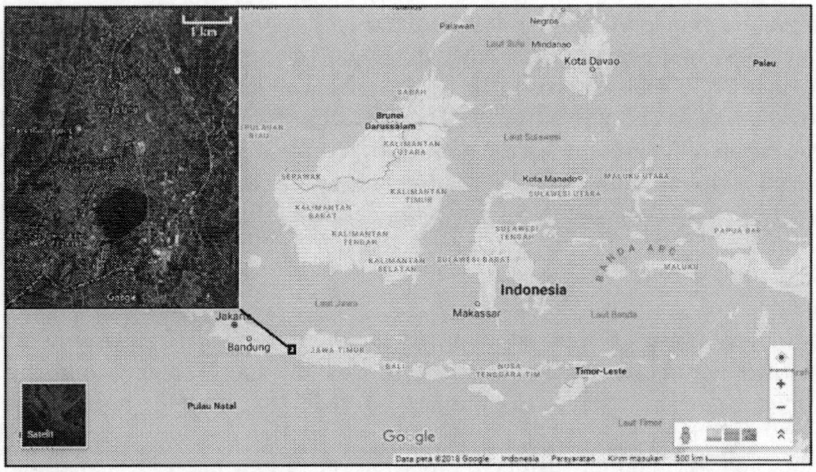

Figure 1. Location and landscape of Magelang city taken from google map

1.1. Current Status of Public Fleet

Around the year 2013, the Government of Magelang City implements city branding "A Million Flower City" to support tourism Borobudur Temple. Improved view of the city is expected to be able to improve visitor comfort. However, this city branding program has not been continued with the modernization of public fleet. Currently, connectivity in Magelang (residential, market, school, tourist spot, etc.) is served by 12 lines with more than 300 units of fleets. Unfortunately, the existing fleet is dominated by cars produced in the 2000s and earlier as presented in Figure 2. With vehicles powered by carburetor technology and poor maintenance, the trend is low fuel efficiency and high exhaust emissions.

Figure 2. Photographic views of the existing condition of the public fleet in Magelang

1.2. Opportunity to Switch to Cleaner Fuel

In 2017, Pertamnia as a state-owned company has completed a fuel station in Magelang with LPG for a vehicle called Vi-Gas. Previously, Pertamina has also implemented similar programs in Jakarta, Bogor, Surabaya, Denpasar, Yogyakarta, and several other big cities. The installation of the Vi-Gas dispenser in Magelang is expected to attract the interest of car owners in Magelang and surrounding areas to use

Vi-Gas, whether implemented in a full dedicated or bi-fuel system, where Vi-Gas and Gasoline can be used interchangeably [8]. If car owners have not been interested to convert their cars to Vi-Gas independently because of conversion costs, the Government can implement a free converter kits program to public fleet owners, as has been practiced in several cities such as Jakarta, Bogor and Palembang [9]. Vi-Gas is chosen because the pressure in the tank is low (0.8-1.2 MPa) compared to CNG (20 MPa), so the required infrastructure investment is also relatively low. Although Vi-Gas is a mixed fuel of propane and butane with different densities, it forms a homogeneous mixture [10]. The carbon content of Vi-Gas lower than gasoline is also an opportunity to produce lower exhaust emissions.

1.3. Objective

Currently, environmental issues and reduction of Greenhouse Gas (GHG) emission may be more important than other automotive issues, as an effort to maintain the environmental sustainability. Therefore, the objective of this study is to estimate CO_2 reduction from the implementation of LPG/Vi-Gas for public fleet in Magelang City. The simulation was carried out with two models based on carbon equilibrium theory and based on the results of emissions testing from NEDC experiences.

2. Method

Learning from WLPGA's annual report, there are two models of LPG applications for vehicles that convert existing vehicles by adding LPG kits and purchasing new vehicles already equipped with LPG kits [11]–[13]. The first model (convert existing vehicles by adding LPG kits) can be implemented at low cost, relatively. Converter and Mixer (CM) of LPG kits can be directly utilized even on carburetor engines. However, because it is applied to old vehicle technology, the resulting emissions are not as good as OEM products. In the second model, it generally uses Vapor Phase Injection (VPI) or Liquid Phase Injection (LPI) of LPG kits, which results in good engine performance as well as lower emissions. As a consequence, a much higher initial investment is needed. Therefore, the environmental benefits in this study are modeled in the first scenario, with consideration of much lower initial investment and can be implemented in the short-term.

3. Result and Discussion

3.1. Fleets Operational Data

Every year, the Department of Transportation of Magelang City conducts inspection and data collection on public transport. Data at the beginning of 2018 is presented in Figure 3. It is known that of the 335 units of the public fleet, only 306 units operate daily (about 91%). Then, of the 12 existing lines, each unit covered a distance average of 137 km per day.

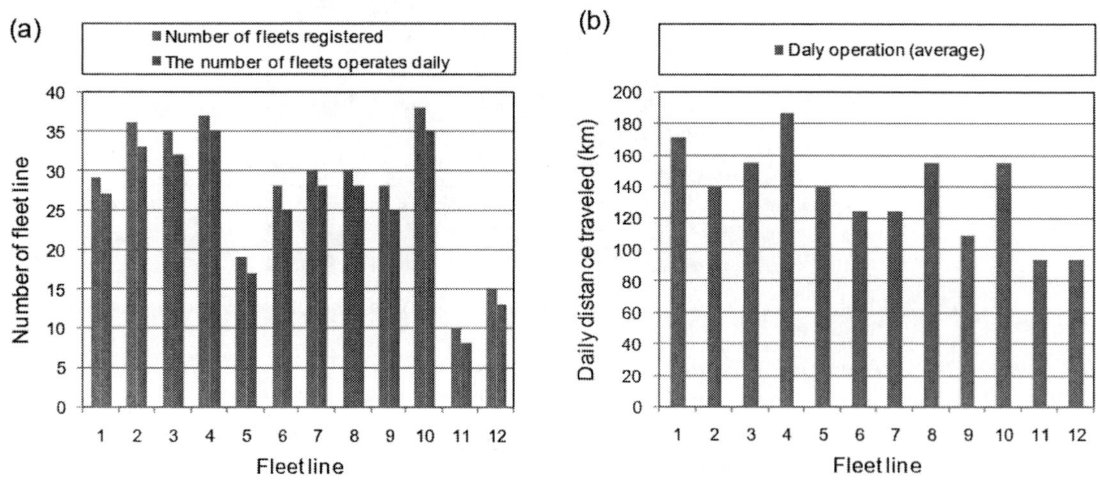

Figure 3. Public fleet operational data

3.2. CO_2 Simulation by Theoretically

LPG is a fuel with a lower carbon content than gasoline. If the combustion process is assumed by complete combustion, where LPG and gasoline do not produce CO and HC emissions, then the amount of CO_2 produced per km can be known. Gasoline has a density of 0.75 kg/liter so that 1 liter of gasoline weighs 750 grams. Gasoline consists of 87% carbon, which means one liter consists of 652 grams of carbon. To burn all the carbon content into CO_2 takes 1740 grams of oxygen. The amount is 652 + 1740 = 2392 grams of CO_2/liter of gasoline. Meanwhile, the density of LPG is 0.55 which means 1 liter of LPG weighs 550 grams. LPG consists of 82.5% carbon or 454 grams of carbon per liter of LPG. To burn all the carbon content into CO_2 takes 1211 grams of oxygen. The amount is 454 + 1211 = 1665 gram CO_2/liter of LPG [14]. Tailpipe CO_2 with energy consumption of 13,311 MJ/km (≈1 liter gasoline) can be presented in Figure 4.

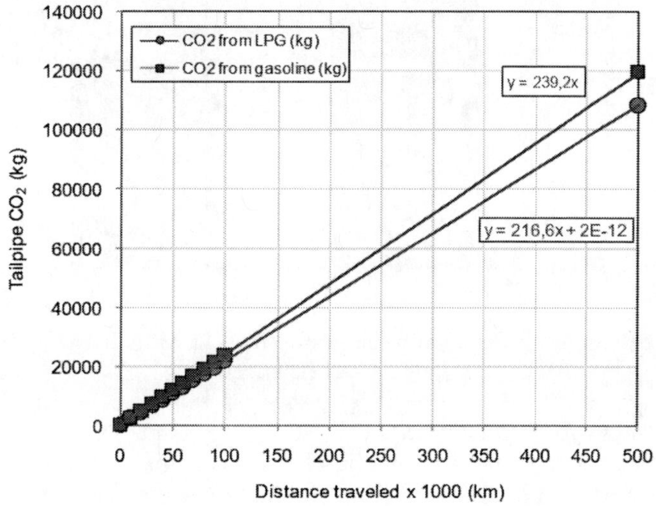

Figure 4. Tailpipe CO_2 emission from gasoline and LPG

With the energy content approach, Gasoline and LPG have a slightly different energy content (LHV) of 44.15 and 46.28 MJ/kg, respectively [15]. From the data available, 306 unit fleets serving 12 lines in Magelang City will require gasoline of 59.984 liters/year (44,449 kg/year) or equivalent to 1,986,237 MJ and then produce CO_2 of 143,626 kg (143.5 tons). The energy of 1,986,237 MJ is equivalent to 42,918 kg of LPG (78,032 liters) and then produce CO_2 of 129,924 kg (129.9 tons). From these calculations, there is a potential for CO_2 reduction of 9% or about 13.56 tons per year.

3.3. CO_2 Simulation from NEDC Experiences

From the literature study, there are many reports on the emission comparison of vehicles fueled by LPG and gasoline [16]–[20]. However, in this study, we will use data reported by Institut Francais du Petrole [21] because it uses the New European Driving Cycle (NEDC) with a variety of samples. Table 1 presents the emissions comparison of gasoline and LPG vehicle using NEDC.

Table 1. Emissions comparison of gasoline and LPG vehicle using NEDC [21]

Emission	Gasoline (g/km)	LPG (g/km)
CO	1.03	1.094
HC	0.075	0.061
NO_x	0.05	0.016
CO_2	171.2	152.9

From the data presented in Table 1, it appears that differences in CO, HC, and NOx emissions are not significant, while CO_2 shows a significant difference. With existing vehicle data, the use of gasoline and LPG will produce CO_2 of 102.69 and 91.72 tons/year, respectively. From the experience of testing with NEDC, there is a chance of decreasing CO_2 by 11% or equivalent to 10.98 tons/year. A comparison of theoretical simulations and NEDC experiences is presented in Figure 5.

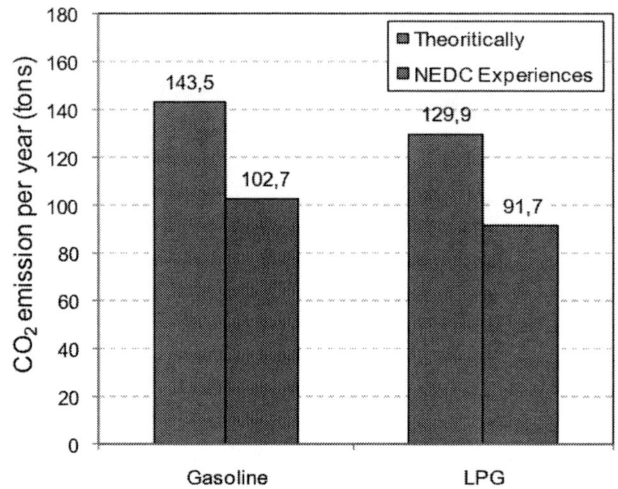

Figure 5. Estimation of CO_2 emissions from the public fleet in Magelang City by theoretical simulations and NEDC experiences

3.4. An alternative strategy for implementation

In recent years, owners and operators of public fleets have anxiety due to the presence of online transportation or Transportation on Demand (ToD) so that their income decreases. Therefore, the most rational conversion program from gasoline to LPG/Vi-Gas is implemented by the local government through a grant scheme or cooperation with the company through a CSR scheme. Since the fuel conversion program involves energy, environmental, supply and technological readiness, the local government can form a consortium with the Department of Transportation, the Environment Agency, Pertamina as a State-Owned Enterprise, a converter kits provider, and a university to conduct continuous research. In Magelang, efforts to reduce CO_2 from the transportation sector are an important issue to support city branding as a clean city and also to support tourism area connectivity.

4. Conclusion

Through this simulation study, we found that applying LPG/Vi-Gas to public fleets in Magelang is worth considering by Local Government in order to reduce CO2 emissions. From theoretical calculations and comparing with NEDC experiences, there is a potential reduction of CO2 by 13.6 and 11.0 tons per year, respectively. To realize the conversion program, a consortium financing scheme involving the Local Government, Department of Transportation, the Environment Agency, Pertamina as a State-Owned Enterprise, a converter kits provider, and a university to conduct continuous research is the most appropriate for the vehicle owner to receive benefits as soon as possible.

References

[1] R. . Colvile, E. . Hutchinson, J. . Mindell, and R. . Warren, "The transport sector as a source of air pollution," *Atmospheric Environment*, vol. 35, no. 9, pp. 1537–1565, Mar. 2001.
[2] B. P. Resosudarmo, "Indonesia's clean air program," *Bulletin of Indonesian Economic Studies*, vol. 38, no. 3, pp. 343–365, 2002.
[3] Satya Widya Yudha, "Air Pollution and Its Implementations for Indonesia: Challenges and

Imperatives for Change," *AirQualityAsia High-Level Session*, no. April. 2017.
[4] K. Hang Leung, "Indonesia's Summary Transport Assessment," *ADB Papers on Indonesia*, no. 15, 2016.
[5] Ministry of Transportation Indonesia, "Indonesian Country Report on Environmentally Sustainable Transport Implementation [Presentation Slides]," *The 4th Regional Environmentally Sustainable Transport (EST) Forum in Asia*, 2011.
[6] Asian Development Bank, "Indonesia: Country Environment Analysis," 2005.
[7] M. Steckdaub and R. Sekartini, "Environmental Policy & Vehicles Inspection in Indonesia," no. December. pp. 1–9, 2001.
[8] M. Setiyo, B. Waluyo, M. Husni, and D. W. Karmiadji, "Characteristics of 1500 CC LPG fueled engine at various of mixer venturi area applied on Tesla A-100 LPG vaporizer," *Jurnal Teknologi*, 2016.
[9] M. Setiyo, S. Soeparman, N. Hamidi, and S. Wahyudi, "Techno-economic analysis of liquid petroleum gas fueled vehicles as public transportation in Indonesia," *International Journal of Energy Economics and Policy*, vol. 6, no. 3, pp. 495–500, 2016.
[10] M. Setiyo, S. Soeparman, N. Hamidi, and S. Wahyudi, "Characteristic of LPG compositions in the fuel line during discharging process," *International Journal of Technology*, vol. 8, no. 1, 2017.
[11] World LPG Association, "Autogas Incentive Policies, 2015 Update," Neuilly-sur-Seine, 2015.
[12] World LPG Association, "Autogas Incentive Policies," Neuilly-sur-Seine, 2016.
[13] World LPG Association, "Autogas Incentive Policies, 2017 Edition," Neuilly-sur-Seine, 2017.
[14] Ecoscore, "How to calculate the CO2 emission from the fuel consumption?," 2018. [Online]. Available: http://ecoscore.be/en/info/ecoscore/co2. [Accessed: 20-Jul-2018].
[15] I. Staffel, "The Energy and Fuel Data Sheet," *Claverton Energy Research Goup*, 2011. [Online]. Available: http://www.claverton-energy.com/the-energy-and-fuel-data-sheet.html. [Accessed: 21-Jan-2018].
[16] J. Lau, W. T. Hung, and C. S. Cheung, "On-board gaseous emissions of LPG taxis and estimation of taxi fl eet emissions," *Science of the Total Environment*, vol. 409, no. 24, pp. 5292–5300, 2011.
[17] C. H. Lai, C. C. Chang, C. H. Wang, M. Shao, Y. Zhang, and J. L. Wang, "Emissions of liquefied petroleum gas (LPG) from motor vehicles," *Atmospheric Environment*, vol. 43, no. 7, pp. 1456–1463, 2009.
[18] S. Alam, B. Hyde, P. Duffy, and A. Mcnabola, "Assessment of pathways to reduce CO2 emissions from passenger car fleets : Case study in Ireland," *Applied Energy*, vol. 189, pp. 283–300, 2017.
[19] T. Tasic, P. Pogorevc, and T. Brajlih, "Gasoline and Exhaust Emission Comparison," *Advances in Production Engineering & Management*, vol. 6, no. 2, pp. 87–94, 2011.
[20] P. Bielaczyc, A. Szczotka, and J. Woodburn, "A comparison of exhaust emissions from vehicles fuelled with petrol, LPG and CNG," *IOP Conference Series: Materials Science and Engineering*, vol. 148, no. 1, pp. 0–10, 2016.
[21] Institut Francais du Petrole, "EETP: European Emission Test Programme," Rueil-Malmaison, 2004.

Acknowledgment
This paper is part of the PDUPT funded by the Ministry of Research Technology and Higher Education, Republic of Indonesia. The researchers also thanked the Head and Staff of the Department of Transportation of Magelang City.

Design tools for underground pipe welding on weldingtie-in parts using value engineering method

R A M Puteri[*] and D Romansyah

Industrial Engineering Department, Engineering Faculty, Universitas Muhammadiyah Jakarta, Cempaka Putih Tengah 27, Central Jakarta, Indonesia

[*] Email: renty.puteri@gmail.com

Abstract. In the industrial world, tools are needed to carry out all work activities, one of the aids on pipe connection. The research entitled the design of underground pipe welding tools in the welding tie-in section with value engineering methods. As for the background of this title, because the company does not achieve the target on the welding tie-in, it is caused by the process of fit-up is too long, narrow space and the tools used too much. Has a problem formation how to design a tool to facilitate fit up underground pipe (tie-in) and how the final welding process before and after using tools. And the purpose of the research to design the underground pipe welding aids with value engineering methods so that faster job time and more efficient workforce can also know the performance and result of implementation. This research uses field observation method directly with the following stages, functional analysis phase, creative stage, evaluation phase, percentage stage, and implementation. In data collection and data processing using Nordic Body Map questionnaire to find out the types of complaints experienced by workers during the preparation process of the tool up to the welding process on the welding tie-in. From this research is generated tool welding process in the underground pipe welding in the form of clamp aids made of iron plate with total cost IDR 2,713,000.00. Based on the results of the Nordic Body Map questionnaire, after using the tool cannot find a complaint on the worker, and the amount of labor from the initial process to the end is enough with six people on each team/group.

1. Introduction

PT. XYZ is one of the companies engaged in construction services, including oil and gas since 2003 to date has contributed to the distribution of gas pipelines owned by government. One of the BUMN-owned companies that use services from PT. XYZ is PT. PGN (Persero). In the gas pipeline welding process, especially pipeline and piping are divided into 2 parts, namely pipe-up welding (line-up) and underground pipe welding (tie-in) (Figure 1).

A Pipeline is a network that is used as a distributor of water pipes, sewage, gas, or fluid hydrocarbons from sources (for example reservoirs, steam plants, oil and gas wells, refineries) by pipes which are channeled to distributors. The pipe can be stored on the ground (above ground), or in the ground (underground) or even in the sea. The pipe diameter used in the pipeline is quite large, between 6-48 inches apart from that the distance is very far, and tends to be straight.

A Piping is system that connects between equipment in one facility using a pipe system, with a short distance with a pipe diameter of 1-4 inch. The Piping is used to distribute gas from the metering to industry or households. Welding of pipes on the ground (line-up) is the process of connecting pipes by bringing together the two ends of the pipe with parallel pipe position, using a buffer or commonly called shapod and threepods. In the process of welding the pipe on the ground (line-up) almost no obstacles were found both in terms of work time and progress targeted by the company unless there are natural disruptions such as rain.

Figure 1. Line Up Welding

Underground welding (tie-in) is the process of connecting pipes with limited space for underground pipe connection work (tie-in) requiring a very long time compared to the line-up welding work. This is due to space limited motion, the position of the pipe to be connected is not aligned, so that workers must adjust / fit-up first so that the two parts of the pipe can be straight and can then be connected. The adjustment / fit-up process is done on both parts of the pipe by channeled from both sides by using a beam as a base, but if the position of the pipe is lower and one is higher then use blockchain and three pods to lift one section of the pipe to be parallel. During this process, welders cannot produce productivity.

To work on pipe settings (fit-up) requires a workforce of at least 7 people including welders, of which the task is 2 people at the bottom to install and operate channeled, 3 people to install and operate the chain block, 1 person to lowering work tools such as beams and at the same time keeping the water pumping machine. If the work position is channeled by water/drainage, the average work time required to set a pipe connection of 4 - 6 inch diameter requires at least 4 hours of welding work.
From the aforementioned problems, it is necessary to have a tool that facilitates the fit-up process so that the target of the company can be achieved, the tools that are planned to be made will be used for the tie-in connection in the form of pipe clamps using hydraulic double as a tool for push / unify the two parts of the pipe in the welding process, as for the tools used at present include jacks, wood beams, iron plates, blockchains, three pods and jack clamps.

2. Related Works

Designing is a fundamental proposition that transforms something that already exists into something better, through three processes: identifying problems, identifying methods and implementing a problem-solving. In other words, programming, drafting, and implementation of the design (John Wade, 1997).
Aids are a tool used to assist workers in doing a job so that the work is easier to do, does not endanger workers and the work becomes more effective. Welding is one of the techniques for joining metal by melting some of the parent metal and filler metal.
Value Engineering is an array of methods to reduce production costs or use of goods and services without reducing the quality and performance that exists (Chaidir, 2007).
This method is a step that must be done by conducting research or design. The steps in value engineering are as follows:

1) Information stage, which is the stage of seeking as much information as possible about the existing problems, the factors that influence the problem and the plan for improvement.

2) Function Analysis Phase, which is the stage of analyzing each of the main functions and supporting functions.
3) Creative Stage, that is at this stage the identification of alternatives that can be selected in product planning is identified and identifying costs that can be reduced without reducing the value of the product's function.
4) The evaluation phase is, namely evaluating alternative alternatives in product planning and determining the best alternative used for product planning reference.
5) Orientation Stage, which is the stage of finding problems and preparing a research plan.
6) Development phase, namely the stage of determining the best alternative and developing the alternatives chosen in product planning.
7) Presentation Stage, which is the stage of presenting from an alternative chosen to the management team to get approval for the alternative.
8) The implementation phase, which is the implementation stage and implementing of selected alternatives to product planning that has been approved by the team management

3. Method
Figure 2 shown the research flow, which is including the research background, preliminary studies, research objective, and data processing.

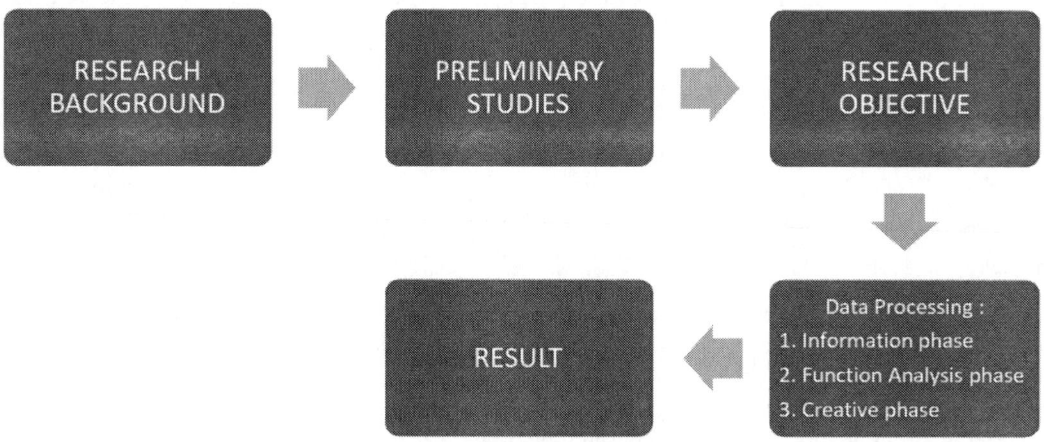

Figure 2. Research flow

4. Results and Discussion
Based on the comparison of the analysis of functions that exist in each design alternative, it can be seen that alternative designs 1, 2 and 3 are able to fulfill the function of each component, both primary and supporting functions, so that it can facilitate the process of fit-up work on underground pipe welding section in the tie-in section of the gas pipe installation work at PT. XYZ.

In terms of aesthetics, design alternative 1 is almost entirely made of iron plate base material so it requires painting to have a good appearance but the locking system still uses nuts and bolts with the composition all of 2 bolts on the left side and on the right side, so that it takes longer time when using it. In alternative designs 2 is almost entirely made of stainless steel so it does not require painting the appearance is good. But it has not been categorized as perfect because it uses hydraulic/one-sided jack so that the fit-up process of both parts of the pipe still has no precision. In the alternative design 3 is made entirely of iron and steel with double hydraulic / jack with the main function as ring clamping locking. This makes it easier for workers in the fit-up process because it can be set from both sides of the jack to get the desired fit-up results (Figure 3). In addition, the number of workers in the pipe fit-up process in the tie-in hole is done by 2 workers. Looking at the price function, alternative design 1 is IDR 2,566,000.00, alternative design 2 is IDR 5,977,500.00 and alternative design 3 of IDR. 2,713,000.00. In terms of ease of use and time of work and the difference in manufacturing costs that are not too far between alternatives 1 and 3 that is equal to IDR 147,000.00. So it can be decided that the selection of alternative designs falls to the alternative design 3 of making underground pipe welding aids in the tie-in section of PT. XYZ.

Figure 3. The Tools

5. Conclusion

Based on the analysis of the selection of alternative designs 1, 2 and 3 using the value engineering method, the alternative design is chosen 3, the consideration is easier in operation, simple, practical and fit-up results as expected with a manufacturing cost of IDR 2,713,000.00. Based on the position and how to use the tool during the fit-up process, the workers are easier to load work tools and do not need as many people. Based on data calculations, the ignition of the target in the welding tie-in job experienced a very significant increase of 25.04%, this after using underground pipe welding aids in the welding tie-in. Based on the Nordic Body Map questionnaire, there were no pain complaints from the workers when the pipe fit-up was 100% at the upper neck, back, waist and hips to 0%, from 80% of the left upper arm, the right upper arm became 0%, from complaints of 60% complaints on the left shoulder, right shoulder, left palm and right palm to 0%, and from 40% complaints on the left knee, right knee becomes 0%.

References

[1] ASME B31.8, 2012, Gas Transmission and Distributor Piping System.
[2] Brawi Mohammed Ali, 2014 Value Engineering pada Industri Kontruksi Bangunan, UI Press, Jakarta.
[3] BSN, 2009, Sistem Penyaluran Pipa dan Distribusi Pipa Gas, SNI.
[4] Danny L, 2006, Jay Mandelbaum, Value Engineering Handbook, IDA.
[5] Dyah Ika Rinawati, Gregorius BudhiWisnu S, 2011, PerancanganAlat Bantu GunaMereduksi Beban Otot Yang Diterima Oleh Pekerja Fine Fokus Adjustment Di PT Arisa Mandiri Pratama,Universitas Diponegoro, Semarang.
[6] Ginting Rosnani, 2014, PerancanganProduk, GrahaIlmu, Jakarta.
[7] Hardianto Iridiastadi, Yassierli, ErgonomiSuatuPengantar, PT RemajaRosdakarya, 2016.
[8] IK9, 2016, Pekerjaan Pengelasan, Instruksi Kerja Konstruksi Pia Baja, Core Competencial Project Division PMO Infrastructure, CCP-D003-0000-QWV-WI-001 PT. Perusahaan Gas Negara.

[9] Ir. Chadir Anwar Makarim, MSCE, Ph,D, 2007, BukuPeganganInstrukturTerjemaahBebas, Value Engineering e-learning Module, GDN, Glad Batch 3
[10] K. Minolta, 2006, Komunikasi Warna Presisi, CV Dainan Tech.
[11] Nurmianto Eko, 1996, Ergonomi Konsep Dasar dan Aplikasi, EdisiPertama, GrahaWidya, Jakarta.
[12] P. Gupito, Skripsi Perancangan Alat Bantu Judgement ColorMathching Pada BagianRnD PT. Kyz Dengan Metode Value Engineering,Fakultas Teknik Universitas Muhammadiyah Jakarta, 2016.
[13] Prosedur, Panduan Konstruksi Pemasangan Jaringan Pipa Gas. PT. Perusahaan Gas Negara, SBU I Jakarta.
[14] PQR 04 ASME IX, Welding Process SMAW, No. Reg. 959/TS/WPS/2016 PT. Kinarya Gemilang Adhitama, Jakarta.
[15] R. Ginting, *Perancangan Produk,*GrahaIlmu, 2009.
[16] Tarwaka , PGDip.Sc., M. Erg, *Ergonomi Industri Dasar-dasar Pengetahuan Ergonomi dan Aplikasi di Tempat Kerja,*Edisi II, Surakarta – Indonesia, 2015.
[17] WPS 01, *Main Procedure For Mainline and Tie-in*, Welding Procedure Specification, WPS/KGA/API/MT/01. PT. Kinarya Gemilang Adhitama, Jakarta. API 1104 Edition 2013.

Hyperspectral image segmentation using discriminant independent component analysis and swarm optimization approaches

Murinto * and D P Ismi

Department of Informatics Engineering Universitas Ahmad Dahlan Yogyakarta

* Email : murintokusno@tif.uad.ac.id, dewi.ismi@tif.uad.ac.id

Abstract. Hyperspectral data consists of hundreds of bands. The high dimension of hyperspectral data is a challenge for the researcher to design efficiency and accuracy image segmentation algorithm. In this paper, a new approach to extraction features and independence of hyperspectral image proposes using Discriminant independent component analysis (DICA) and multilevel thresholding techniques based on Otsu problems for image segmentation are introduced. Image segmentation is initial process of image analysis and recognition. Otsu's problem with multilevel thresholding is solved in each band from hyperspectral data that has been reduced dimensionality using DICA. The main purpose using multilevel thresholding is to get an optimal threshold that maximizes variance between classes. The swarm optimization approach used in this study is Darwinian particle swarm optimization (DPSO). The result of experiment showed that DPSO was better compared to other swarm optimization approaches. DPSO shows a statistically significant increase, both from CPU time processing and fitness value. DPSO technique is able to find optimal threshold with greater variance between classes and smaller search times compared to particle swarm optimization (PSO).

Keywords: Discriminant Independent Component Analysis, Darwinian Particle Swarm Optimization, Image Segmentation, Particle Swarm Optimization

1. Introduction

Multispectral and hyperspectral are two groups in spectral imaging. Multispectral and Hyperspectral are two groups in spectral imaging. Hyperspectral imagery (HSI) has more bands, usually more than one hundred bands and its spectral resolution is higher when compared to multispectral image. A hyperspectral image is often obtained through imaging spectrometers, such as HYDICE (Hyperspectral Digital Imagery Collection Experiment) and AVIRIS (Airborne Visible InfraRed Imaging Spectrometer). Each spectral vector corresponds to reflectance value of certain pixels in all spectral wavelengths. The hyperspectral image consists of hundreds data channels. The high dimensions of hyperspectral data make it difficult for researcher to design accurate and efficient image segmentation algorithms on such types of images. A large number of spectral bands show high-dimensional data and present challenges to segmentation, classification and image analysis. Most of the methods used are designed for gray-level analysis, color image or multispectral image not for hyperspectral images.

Applying this method to HSI will be a big challenge because of the increasing dimensions of data. Feature reduction is searching for a vector set that represents reduction dimensions of the observation

IOP Conf. Series: Materials Science and Engineering **674** (2019) 012054 doi:10.1088/1757-899X/674/1/012054

data, a process that produces a smaller number of features through a combination of existing channels. Each pixel of a hyperspectral image is a vector containing all spectral information provided through different spectral channels at different wavelengths. The dimensions of vector depend on number of spectral channels obtained through sensor and, in the case of hyperspectral data often number of spectral channels reaches hundreds.

In fact, feature selection or feature extraction techniques are often displayed as pre-processing on hyperspectral data analysis [1]. Such processing is often carried out in multispectral imagery to improve class to separate or to eliminate certain types of noise. One technique for reducing data dimensions is to use principal component analysis (PCA) [2], which estimates eigenvalues data to display a projection into a new feature space where maximum variance of data is maintained. PCA only involves statistics up to the second-order; this is probably not effective when used for hyperspectral data [3]. To overcome problems associated with dimensional reduction of data using PCA, several algorithms has been proposed such as kernel principal component analysis (KPCA) [4]. Other data dimension reduction techniques that have been widely used in recent years are Independent Component Analysis (ICA) [5]. This technique projects data into a new space where components are not independent (independent), estimated from a statistical point of view. Multivariate data with lower dimensions and independent features in discriminant independent component analysis (DICA) were obtained using entropy maximization [6] [7].

Segmentation is an important part of image processing that attempts to identify pixel intensity corresponding to a predefined class. Thresholding is the simplest segmentation method because it works by taking a certain threshold value (T). Pixels with higher intensity values than T are labeled as first-class while the rest are labeled second class. When an image is segmented into two classes, this is called the bi-level thresholding (BT) and requires only one T value. On the other hand, when pixels are separated into more than two classes, this task is called multilevel thresholding (MT) and requires more than one T value [8].

Image segmentation plays an important role in the field of remote sensing hyperspectral analysis. As an example to improve the classification results, the integration of the segmentation process and classification in the process of hyperspectral image processing has been carried out [9]. In certain cases to decide whether a pixel is inserted into a particular class simultaneously, it will be based on the feature vector and some other additional information obtained from the segmentation step. An accuracy of mind segmentation is needed in this approach to be more effective. A number of multispectral and hyperspectral image segmentation methods have been introduced in some literature. Some of these methods are mostly based on region-merging techniques, where the nearest image segments are combined with each other based on their homogeneity. Segmentation can be classified into four types: methods based on histogram thresholding, methods based on texture analysis, methods based on clustering and methods based on split-merging regions. The thresholding technique can be divided into two different groups, first, the optimal thresholding method [10] [11], where threshold optimal search makes the threshold classes on the histogram achieve the desired characteristics. Usually the threshold is chosen by optimizing the objective function (fitness function). Second, the thresholding method is based on the characteristics, detecting the threshold limit through some characteristic measurements of the histogram. Although this method is fast so that it is suitable for multilevel thresholding, the threshold number for this approach is difficult to determine and requires further special things.

Several algorithms have been proposed in several literatures relating to optimal thresholding. Some of these studies are related to bi-level thresholding, while others relate to multilevel thresholding problems. Bi-level thresholding reduces an optimization problem to determine the threshold T that maximizes the between class-variance and minimizes class variance (within-class variance) [12]. At bi-level thresholding, the problem is solved through the T value search. This approach is easy to implement but has drawbacks that are generally the old computation. The search for n-1 optimal threshold involves fitness evaluation $n(L-n+1)^{n-1}$ threshold combinations. This means that this method is not suitable from the point of view of execution time calculation. The task of determining n

- 1 optimal threshold on n-level imagery can be formulated as a multidimensional problem. Bio-inspired algorithms have been used in situations where conventional optimization techniques cannot find satisfactory solutions or the time needed to solve them too much, for example, when a function that is optimized is discontinuous and cannot be derived differentiated and / or presents too many nonlinear parameters.

One of the known bio-inspired algorithms is particle swarm optimization (PSO) [13]. The PSO consists of a number of particles that collectively move in the search space (for example the pixels of the image) to find global optimality (e.g. maximizing inter-class variance from the distribution of intensity levels in a given image). However, a common problem with PSO and similar optimization algorithms are that the algorithm can get stuck at a locally optimal point, and the algorithm might work with some existing problems but it might also fail on other problems. Another swarm optimization approach is the Darwinian Particle Swarm Optimization (DPSO) introduced by Tillet [14] to overcome the vulnerability in local optima.

This research proposed a method for discriminant analysis based on the extraction of independent features obtained by entropy maximization namely discriminant independent component analysis (DICA). Fisher's criteria and the sum of marginal negentropy of extracted independent features are maximized simultaneously. While the segmentation method used is swarm optimization approach namely particle swarm optimization (PSO) and Darwinian particle swarm optimization (DPSO).

2. Multilevel Thresholding for Optimization Problem

Multilevel segmentation techniques provide an efficient way to image analysis. The problem faced is usually the automatic selection of an optimal n-level threshold. A threshold can be found using Otsu's method (Otsu, 1979) [12], which can maximize variance between classes. Pixels of an image is represented in the L gray level. Number of pixels at level i as n_i , the total number of pixels denoted through $N = n_1 + n_2 + + \cdots + n_L$. The gray level histogram is normalized and is considered probability distribution equation is written in the equation (1).

$$p_i = \frac{n_i}{N}$$
$$p_i \geq 0\ ,\ \sum_{i=1}^{L} p_i = 1 \qquad (1)$$

Images are divided into classes K_0 and K_1 by a threshold at level k in bi-level thresholding problem. K_0 is pixels with levels $[1,2,..,L]$ and K_1 is pixels with levels $[k+1,..,L]$. Probabilities of class occurrences and mean class levels can be written as equations (2) and (3) respectively. While the average total level of the image μ_T can be written as equation (4).

$$\omega_0 = \sum_{i=1}^{k} p_i \quad \omega_1 = \sum_{i=k+1}^{L} p_i \qquad (2)$$
$$\mu_0 = \sum_{i=1}^{k} ip_i\,/\omega_0 \quad \omega_1 = \sum_{i=k+1}^{L} ip_i\,/\omega_1 \qquad (3)$$
$$\mu_T = \sum_{i=1}^{L} ip_i \qquad (4)$$

Two relations can be written as equation (5) for several choices of k.

$$\omega_0\mu_0 + \omega_1\mu_1 = \mu_T,\ \omega_0 + \omega_1 = 1 \qquad (5)$$

Furthermore, the objective function of the Otsu method can be defined as equation (6).

$$\text{Maximizes } \sigma_B^2 = \omega_0(\mu_0 - \mu_T)^2 + \omega_1(\mu_1 - \mu_T)^2 \qquad (6)$$

Otsu's method can be extended to multilevel thresholding problems. It is assumed that there is a threshold, which divides the image into m + 1 class : K_0 for $[1,2,..,k]$, K_1 for $[k_1+1,...,k_2]$ and K_m for $[k_m + 1, ..., L$. Then from the Otsu's function, it can be written as (7).

$$\text{Maximizes } \sigma_B^2 = \omega_0(\mu_0 - \mu_T)^2 + \omega_1(\mu_1 - \mu_T)^2 + \cdots + \omega_m(\mu_m - \mu_T)^2 \qquad (7)$$

One method used to perform segmentation processes is to use the thresholding method. Threshold image thresholding T partitioning an image I into parts of a particular region based on the value of T itself. Suppose there is an L level of intensity in each RGB component of an image, and these levels are within the range of values (0,1,2, ... L-1), then it can be defined in equation (8).

$$P_i^K = \frac{h_i^K}{N}\ ,\ \sum_{i=1}^{N} P_i^K,\ K = \{R, G, B\} \qquad (8)$$

IOP Conf. Series: Materials Science and Engineering **674** (2019) 012054 doi:10.1088/1757-899X/674/1/012054

Where is i intensity, $0 \leq i \leq L-1$. K is image components, $K = \{R, G, B\}$. N is the total numbers pixels in image. h_i^K is number of pixels for intensity level i correspondence with component K. While total average (mean) is mean that combined from every image component as an equation (9).

$$\mu_T^K = \sum_{i=1}^{L} i\, P_i^K,\ K = \{R, G, B\} \tag{9}$$

from 2-*level thresholding* can be extended to find n-level *thresholding* where is *n-1* level *threshold* $m_j^k, j = 1,2,..,n-1$. The image can be written as an equation function (10).

$$F^k(x,y) = \begin{cases} 0, f^k(x,y) \leq t_1^k \\ \frac{1}{2}(t_1^k + t_2^k),\ t_1^k < f^k(x,y) \leq t_2^k \\ \vdots \qquad \vdots \\ \frac{1}{2}(t_{n-2}^k + t_{n-1}^k),\ t_{n-2}^k < f^k(x,y) \leq t_{n-1}^k \\ L, \qquad f^k(x,y) > t_{n-1}^k \end{cases} \tag{10}$$

Where si x and y are widht (W) and length (H) image pixels write as $f^k(x,y)$ dwith intensity level Lin each RGB component. In this situation, pixels of the image will be divided into n classes $D_1^k, \ldots, D_n^k$ which represents multiple objects or certain features on a particular object. The method to get the optimal threshold is by maximizing the variance between classes through equation (11).

$$\sigma_B^{k^2} = \sum_{j=1}^{n} w_j^k (\mu_j^k - \mu_T^k)^2,\ k = \{R, G, B\} \tag{11}$$

Where is j represented specific class so that w_j^k and μ_j^k are the probability of occurrence (event) and mean from class j. Probability of occurrence w_j^k and classes $D_1^k, \ldots, D_n^k$ given as equations (12).

$$w_j^k = \begin{cases} \sum_{\substack{i=1 \\ C=\{R,G,B\}}}^{t_j^k} P_i^C, & j = 1 \\ \sum_{\substack{=t_{j-1}^k+1 \\ C=\{R,G,B\}}}^{t_j^k} P_i^C, & 1 < j < n \\ \sum_{i=t_{j-1}^k+1}^{L} P_i^k, & j = n \end{cases} \tag{12}$$

While the mean of each class μ_j^k calculated using equation (13).

$$\mu_j^k = \begin{cases} \sum_{\substack{i=1 \\ C=\{R,G,B\}}}^{t_j^k} \frac{P_i^k}{w_j^k q}, & j = 1 \\ \sum_{\substack{=t_{j-1}^k+1 \\ C=\{R,G,B\}}}^{t_j^k} \frac{P_i^k}{w_i^k}, & 1 < j < n \\ \sum_{i=t_{j-1}^k+1}^{L} \frac{P_i^k}{w_i^k}, & j = n \end{cases} \tag{13}$$

This n-level thresholding optimization problem is reduced to a search optimization problem for the threshold t_j^k which maximizes 3 objective functions (i.e. fitness function) of each RGB component written in the form of equation (14).

$$\varphi^k = \max_{\substack{1<t_j^k<\cdots<t_{n-1}^k<L \\ C=\{R,G,B\}}} \sigma_B^{k^2}(t_j^k) \tag{14}$$

IOP Conf. Series: Materials Science and Engineering **674** (2019) 012054 doi:10.1088/1757-899X/674/1/012054

3. Swarm Optimization Approach

Particle swarm optimization (PSO) technique first introduced by Kennedy and Eberhart (1995), is a stochastic optimization technique similar to the behavior of a flock of birds or sociological behavior of group humans. The basic idea of PSO is to involve a scenario in which a flock of birds searches for food sources in an area. All birds do not know exactly where the food, but with each iteration they will know how far the food will be found. The best strategy will be followed by birds that are close to food and also from the best previous position achieved. PSO is built with the concept of optimization through a swarm particle. The original PSO algorithm is written in the form of velocity updated equations and position updates as shown in equation (15) and equation (16) sequentially.

$$v_{ij}^{t+1} = v_{ij}^{t} + c_1 r_{1j}^{t} * \left(P_{best,i}^{t} - x_{ij}^{t}\right) + c_1 r_{2j}^{t} * \left(G_{best} - x_{ij}^{t}\right) \quad (15)$$

$$x_{ij}^{t+1} = x_{ij}^{t} + v_{ij}^{t+1} \quad (16)$$

The algorithm is controlled through individual experience (pbest), and overall experience (gbest) and the current movement of the particles to determine their next position in search space. Experiences are accelerated through two acceleration constant factors c_1 and c_2, two random numbers r_1, r_2 which is generated with a range value between 0 and 1. N-size population and the D dimension are denoted as $x = [x_1, x_2, .., x_N]^T$. Every particle $x_i (i = 1,2,3, \dots, N)$ given as $x_i = [x_{i1}, x_{i2}, \dots, x_{iD}]$, initial velocity was written as $v = [v_1, v_2, .., v_N]^T$. Then, next velocity $x_i (i = 1,2,3, \dots, N)$ given as $v_i = [v_{i1}, v_{i2}, \dots, v_{iD}]$. Whereas index i varies from 1 to N and index j varies from 1 to D.

The basic modification of PSO algorithm which is usually carried out includes how to increase convergence speed, control exploration, and exploitation trade-offs, overcome convergence or premature stagnation problems, clamping techniques, problem of boundary value, initial and final conditions. In PSO, particle speed is very important. At each step of the PSO process, all particles are processed through speed adjustments for each particle movement in each dimension of the search space. There are two characteristics in PSO namely exploration and exploitation. Exploration is the ability to explore different areas of the search space to get optimal good, while exploitation is the ability to focus search in the search area to improve the expected solution. When the speed increases, the position of the particles will be updated quickly. And as a result, particles will leave the search space boundaries and there is a possibility that they will drift away from the search space. Therefore, to control this difference, the particle speed will be reduced to remain within the search space boundaries. Several techniques were developed to increase the speed of convergence, in order to balance the exploration and exploitation trade-offs and find a good quality of completion for PSO, namely the speed of flanking techniques (velocity clamping), inertial weight (w), constriction coefficient (s). A common problem with optimization algorithms is being trapped into optimal locality. Certain algorithms can work well on one problem but may fail in another. If an algorithm can be designed to adapt to the fitness function, adjust itself to the fitness landscape, a stronger algorithm with wider application, without the need for special engineering problems will result.

In specific implementation of PSO, a group of test settlers was utilized. For applying natural selection with a single flock, an algorithm must detect when stagnation has occurred. In finding a better natural selection model using the PSO algorithm, Darwinian Particle Swarm Optimization (DPSO) was introduced by Tillet et al [14]. To analyze the general state of each flock, the suitability of all particles is evaluated and the environment and the best individual position of each particle are updated. If a new global solution is found, new particles will appear. The particles are removed if the flock fails to find a more suitable state in the specified number of steps (Algorithm 1). Some simple rules are followed to remove flocks, remove particles, and spawn new flocks and new particles: i) when the herd population is below the minimum limit, the herd is removed, and ii) the worst performing particles in the herd are removed when the number of steps is the maximum limit (SC_c^{max}) without increasing the fitness function achieved. After deletion of particles, instead of being set to zero, the counter is reset to the value close to the threshold number, according to the equation (17).

$$SC_c(N_{kill}) = SC_c^{max}[1 - \frac{1}{N_{kill}+1}] \quad (17)$$

With N_{kill} is number of particles removed from swarm during periods where there is no improvement in fitness. To spawn a new flock, a swarm of particles must never be removed and the maximum number of flocks must not be exceeded. However, the new flock is only made with probability p = f / NS, where f is a random number in [0, 1] and NS is the number of flocks. The algorithm of DPSO is shown in Algorithm 1.

Algorithm 1: DPSO [14]

Main Program Loop (1 Step)

For each swarm in the collection
Evolve the swarm (Evolve Swarm Algorithm: Right)
For each swarm in the collection
Allow the swarm to spawn
Delete 'failed' swarms

Evolve Swarm Algorithm
For each particle in the swarm
Update Particles' Fitness
Update Particles' Best
Move Particle
If swarm gets better
Reward swarm: spawn particle: extend swarm life
If the swarm has not improved
Punish swarm: possibly delete particle: reduce swarm life

4. Experiment Results

Data reduction and image segmentation based on the swarm optimization approach in this paper was implemented using Matlab R.2016a in a computer Intel® Core ™ i5-4210U CPU@1.70 GHz. The data used in this study are AVIRIS Indian Pines hyperspectral image data obtained from AVIRIS sensors above the Northwestern Indiana region in 1992. Images size is 145 x 145 pixels. The number of image bands is 220 spectral bands (some bands contain noise and water absorption) [15]. Figure 1 shows the image of Aviris Indian Pines and its histogram.

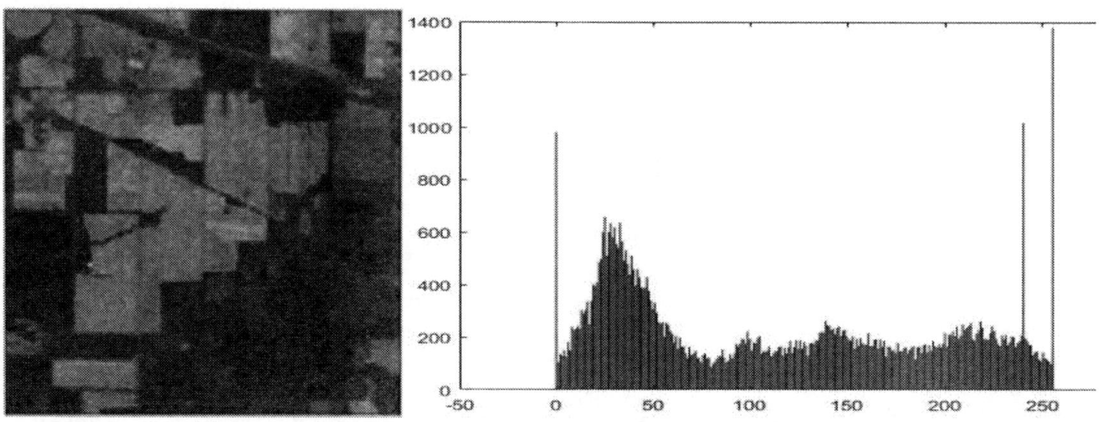

(a) (b)

Figure 1. Aviris Indian Pines data set: (a) Aviris Indian Pines 3 band (b) Image Histogram

The data dimension reduction method used in this paper is discriminant independent component analysis (DICA). The number of independent components (IC) is the first, second and third independent components. Aviris Indian Pines Image which consists of 3 independent components is then subject to a segmentation process. Figure 2 shows the original image as well as the first, second and third independent components (IC's) of Aviris Indian Pines image.

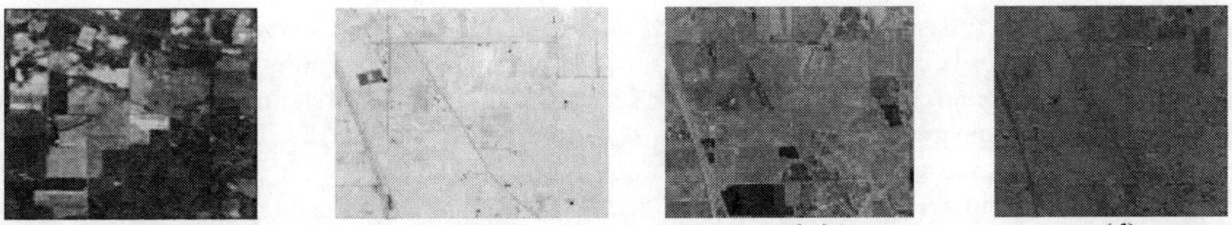

(a) (b) (c) (d)

Figure 2. Aviris Indian Pines data set: (a) RGB Image (b) - (d) the first, second and third independent components

The image segmentation technique used in this research is based on swarm optimization approach. Two swarm optimization approaches are used namely particle swarm optimization (PSO) and Darwinian Particle Swarm Optimization (DPSO). While the quality of segmentation images produced from these two techniques is measured by using a measure of segmentation quality through the PSNR, SSIM and PSE values produced [16] [17]. Initializing PSO and DPSO parameters is shown in Table 1.

Table 1. Parameter Initialization of DPSO and PSO

Parameter	DPSO	PSO
Iteration numbers	200	200
Population number	100	100
C1	0.5	0.5
C2	0.5	0.5
Vmax	4	4
Vmin	-4	-4
Xmax	255	255
Xmin	0	0
Min Population	-	15
Max Population	-	60
Numbers of Swarm	-	4
Min Swarm	-	2
Max Swarm	-	6
Stagnancy	-	8

CPU processing time of the data set used in this paper was tested on the PSO and DPSO algorithms respectively for thresholding levels of 8, 10 and 12. The average CPU processing time was obtained from 20 runs. The results are shown in Table 2.

Table 2. Average CPU Processing Time

Level	DPSO	PSO
10	5.5383	10.7495
11	5.8246	11.7717
12	5.9267	14.1271

Table 2 shows the optimal threshold value, fitness, and the quality of segmentation results from the swarm optimization approach using Particle Swarm Optimization (PSO) and Darwinian Particle Swarm Optimization (DPSO). The quality parameters of the segmentation results used are peak signal to noise ratio (PSNR), mean square error (MSE) and structural similarity index of SSIM). Fitness value (between-class variance) at each level of image segmentation using multilevel thresholding is sought at 10.11 and 12. Levels of optimal threshold and fitness showed in Table3 are calculated from Aviris Indian Pines images which have been reduced to 3 bands. Table III shows that DPSO is better than PSO in terms of fitness value, PSNR, SSIM, and MSE. The fitness value generated by using the DPSO method is better because PSO may be trapped at optimal locality while the DPSO uses natural selection to avoid stagnation. From this it can be seen that DPSO can find the optimal threshold value with a faster time compared to PSO. From this it can be concluded that the DPSO method is more recommended to be used as an image segmentation method, especially in high-dimensional images such as hyperspectral images.

Table 3. The threshold value, fitness, PSNR, SSIM and MSE Segmentation of Aviris Indian Pines

Methods (level)	Threshold Value	Fitness	PSNR	SSIM	MSE
DPSO(10)	24,49,77,103,134,165,195,219,239	7260.8	29.7830	0.9338	203.82
DPSO(11)	19,40,63,87,112,139,167,193,218,237	7270.2	30.6581	0.9392	187.44
DPSO(12)	18,38,61,84,108,131,156,180,204,225,242	7277.1	31.4600	0.9391	180.99
PSO(10)	18,38,62,87,113,143,175,202,232	7258.6	29.6198	0.8828	354.19
PSO(11)	16,36,60,84,109,136,165,194,217,237	7269.9	30.5942	0.9126	251.62
PSO(12)	14,30,50,73,95,121,148,173,198,220,229	7276.1	31.3749	0.9287	218.98

The results of Aviris Indian Pines image segmentation using the Darwinian Particle Swarm Optimization (DPSO) and Particle Swarm Optimization (PSO) approaches are shown in Table IV. The segmentation results along with the histogram are shown at the threshold level T = 10.11,12.

Table 4. Results after Application of DPSO and PSO Approach to Aviris Indian Pines

Method	T=10	T=11	T=12
DPSO	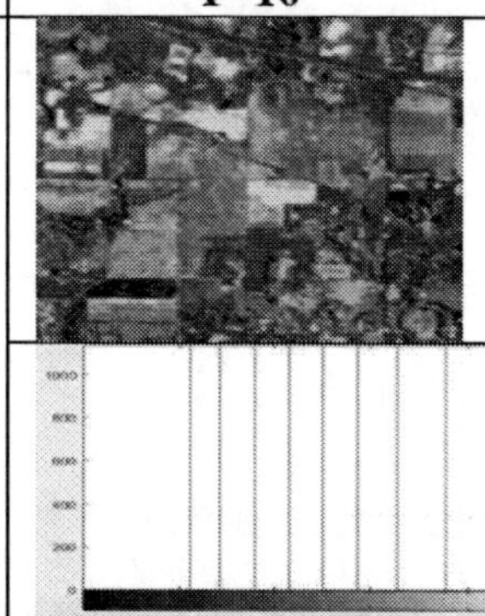	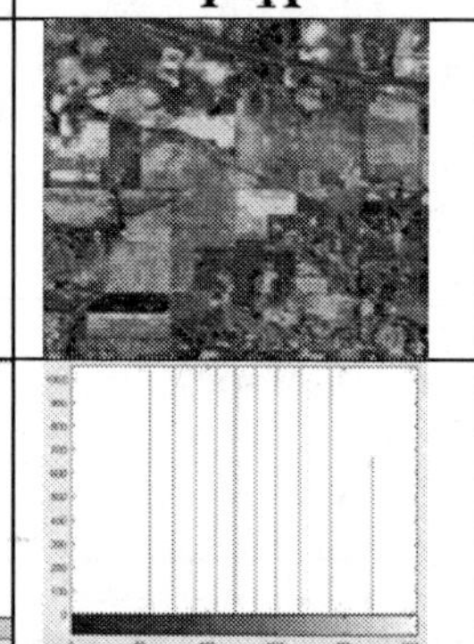	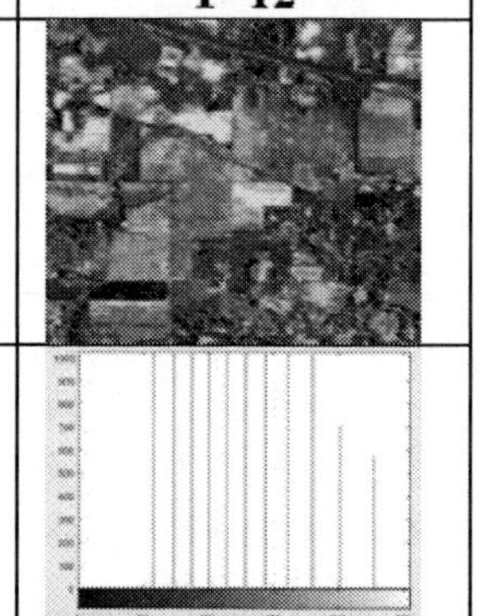
PSO	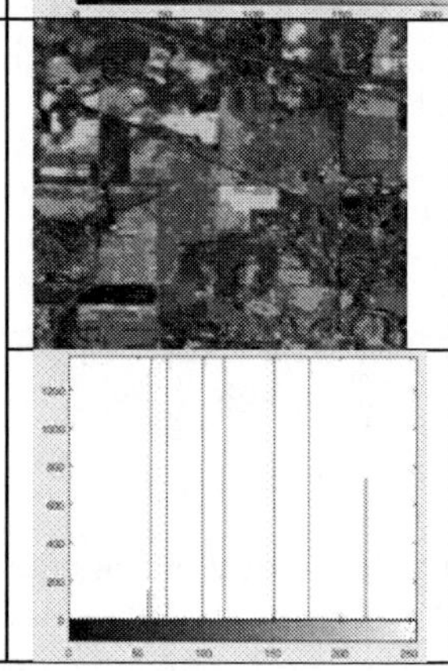	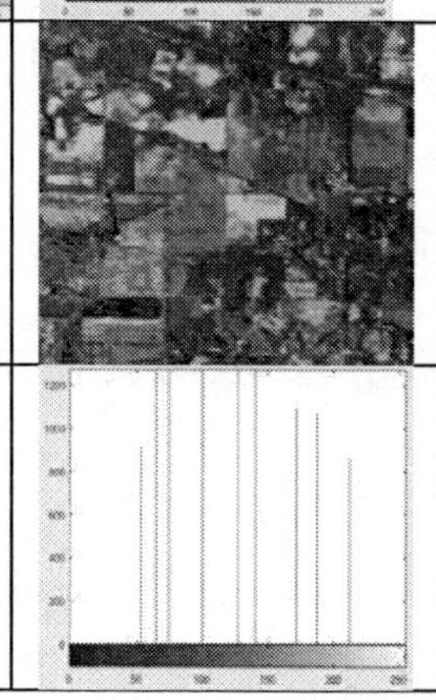	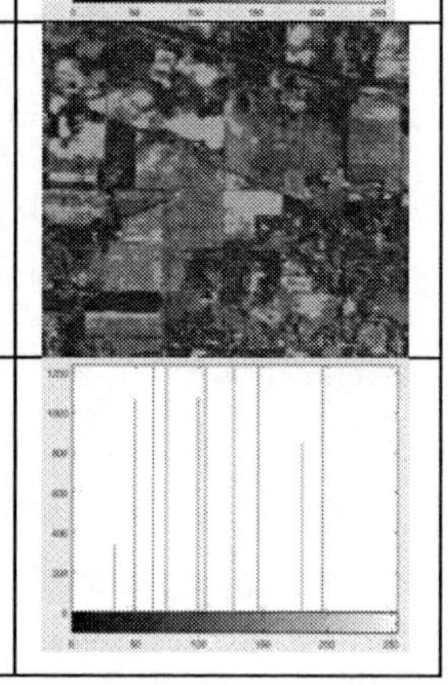

5. Conclusion

In this study an approach has been proposed for the segmentation of hyperspectral images using data dimension reduction Discriminant Independent Component Analysis (DICA). The method used to overcome the Otsu optimization problem is done by finding the threshold value using multilevel thresholding based on Darwinian particle swarm optimization (DPSO). The experimental results show that the DPSO method is better compared to the Particle Swarm Optimization (PSO) method in the context of CPU processing time and the fitness value generated.

Acknowledgment
This research is part of the research funded with the contract number: PSNI-112 / SKPP / III / 2018 contract number. The author said thank you to DP2M-RISTEKDIKTI for the funds provided so that this research can be realized.

References

[1] Landgrebe, D.A. 2003 Hyperspectral image data analysis as a high dimensional signal processing problems. *IEEE Signal Processing Magazine*, 19:17–28.
[2] Jolliffe, I. T. 1986. Principal Component Analysis and Factor Analysis. In *Principal component analysis* (pp. 115-128). Springer New York.
[3] Richards, J. A.(2005 Analysis of remotely sensed data: The formative decades and the future. *IEEE Transactions on Geoscience and Remote Sensing*, *43*(3), 422-432.
[4] Fauvel, M., Chanussot, J., and Benediktsson, J. A. 2012 A spatial-spectral kernel-based approach for the classification of remote-sensing images. *Pattern Recognition*, *45*(1), 381-392.
[5] Hyvärinen, A., Karhunen, J., & Oja, E. 2001 *Independent component analysis* (Vol. 46). John Wiley & Sons.
[6] Dhir, Chandra Shekhar, Jahyung Lee,and Soo-Young Lee.2012 Extraction of independent discriminant features for data with asymmetric distribution, *Knowledge and information systems* 30.2. 359-375
[7] Murinto and Nur Rochmah Dyah PA. 2017. Dimensionality Reduction using Hybrid Support Vector Machine and Discriminant Independent Component Analysis for Hyperspectral Image. International Journal of Advanced Computer Science and Applications(ijacsa), 8(11).
[8] Oliva, D., Cuevas, E., Pajares, G., Zaldivar, D., & Perez-Cisneros, M. (2013). Multilevel thresholding segmentation based on harmony search optimization. *Journal of Applied Mathematics*, *2013*.
[9] Ghamisi, P., M.S. Couceiro, N.M.F. Ferreira and L. Kumar. 2012 Use of Darwinian particle swarm optimization technique for the segmentation of Remote Sensing images. *Proceedings of the IEEE International Geoscience and Remote Sensing Symposium*, Jul. 22-27, IEEE Xplore Press, Munich.
[10] Pun, T. (1980). A new method for grey-level picture thresholding using the entropy of the histogram. *Signal processing*, *2*(3), 223-237.
[11] Otsu, N. (1979). A threshold selection method from gray-level histograms. *IEEE transactions on systems, man, and cybernetics*, *9*(1), 62-66.
[12] Akay, B. (2013). A study on particle swarm optimization and artificial bee colony algorithms for multilevel thresholding. *Applied Soft Computing*, *13*(6), 3066-3091.
[13] Eberhart, R., & Kennedy, J. (1995, October). A new optimizer using particle swarm theory. In *Micro Machine and Human Science, 1995. MHS'95., Proceedings of the Sixth International Symposium on* (pp. 39-43). IEEE.
[14] Tillett, J., Rao, T., Sahin, F., & Rao, R. (2005). Darwinian particle swarm optimization.
[15] AVIRIS Dataset. (2002).[cited 2016 june 1].Available from ftp://ftp.ecn.purdue.edu/biehl/MultiSpec/92AV3C .tif.zip
[16] A.M. Eskicioglu, P.S. Fisher, "Image Quality Measures and Their Performance", *IEEE Transactions on Communication*, vol. 43, no. 12, pp. 2959-2965, December 1995.
[17] Murinto, Winiarti, S., Ismi, D. P., & Prahara, A. 2017 Image enhancement using piecewise linear contrast stretch methods based on unsharp masking algorithms for leather image processing. In *Science in Information Technology (ICSITech), 2017 3rd International Conference on* (pp. 669-673). IEEE.

Extraction of EEG signals using the discrete wavelet transforms

H hindarto[1,*], A muntasa[2]

[1] Informatics, Faculty Of Engineering, Universitas Muhammadiyah Sidoarjo, Indonesia
[2] Informatics, Faculty Of Engineering, Universitas Trunojoyo Madura, Indonesia

[*] E-mail: hindarto@umsida.ac.id

Abstract. This study focuses on feature extraction for Electro Encephalo Graph (EEG) signals using the Discrete Wavelet Transform method. The EEG signal is used to move up the cursor and the down cursor. In each subband of the EEG signal wave the average value is taken to characterize the EEG signal. Backpropagation Neural Network is used as an EEG signal classification to determine whether the up cursor or the down cursor. The data used in this study are EEG data derived from BCI competition 2003 (BCI Competition 2003). Decision-making is done in two stages. In the first stage, the mean value of each wavelet subband is used as a feature extraction of the EEG signal data. This feature is an input to the Backpropagation Neural Network. In the second stage of the classification process into two classes of class 0 (for the up cursor) and class 1 (for the down cursor), there are 260 training data file of EEG and 293 signals from EEG signal data testing file, so the whole becomes 553 data files of EEG signals. The result obtained for EEG signal classification is 77.2% of the tested signal data.

Keywords: EEG, Mean, Discrete Wavelet Transform, BackPropagation

1. Introduction

Electroencephalography (EEG) is a method of monitoring of Electrophysiology to record electrical activity in the brain. The electrodes are placed along the scalp electrodes are invasive, although is sometimes used as in electrocorticography. EEG measures the voltage fluctuations resulting from the flow of ions in the neurons of the brain. In the context of clinical, EEG refers to the spontaneous electrical activity of brain recordings during the time period, as recorded from multiple electrodes placed on the scalp [1]. Diagnostic applications generally focus on the potential associated with EEG spectral content.

Placement of electrodes on the scalp following the system specified already 10-20 System. The proper electrode placement and good are the main requirements for obtaining EEG recording are good and trustworthy. In addition, the cleanliness of the scalp, the condition of the electrodes, and EEG machine compliance subjects while recording is also very influential to get good results. Hans Berger stated that human brains have electrical activity that is continuous and can be recorded. Brain activity may be possible to transmit the order to the electronic equipment with the help of Brain-Computer Interface (BCI) [2]. Most of BCI using spontaneous mental activity (for example, imagine moving the fingers, hands, or the entire arm, etc.) To generate an electroencephalogram (EEG) signals that can be distinguished [3]. EEG signals that can be distinguished from later converted into external actions. During the last few years, a variety of evidence has been evaluating the possibility to recognize some mental task EEG signals from [4]. However, how to improve the performance of the EEG signal recognition in signal processing is still a major problem. The procedure of introduction mainly includes classification and feature extraction, where the extraction of features plays an important role in classification. This paper mainly focuses on the extraction of features.

Currently, the feature extraction method for image motor EEG mainly includes some methods include using the method of Fast Fourier transform (FFT) [5] [6], Fourier spectral features are calculated by the method of Fourier transformation using Welch windowed segment signal. The main disadvantage of this method is the method using only information and not to use the frequency domain information time. However, research shows that the combination of information frequency and time domain information can improve the performance of the classification of EEG signals [7]. Autoregressive (AR) is a method of the spectrum of AR, band power calculated in multiple frequency bands and the amount of power used as independent variables [8]. In addition, model multivariate autoregressive coefficient model used as a feature [9]. Time-frequency analysis by Wang et al, use of time-frequency analysis as a useful tool for the components of the EEG during motor imagery oscillating [10]. As we all know, the components of the EEG oscillating produced during motor imagery is a time and frequency related, therefore, this method of obtaining promising results. However, the components of the EEG oscillating simultaneously can cause a shift in the slow cortical potentials. A combination of two related signals may be used to enhance the information being extracted. Methods of time-frequency EEG components considering only oscillates. Utilizing wavelet transform coefficients, i.e., extracting coefficients of wavelet transform on the useful frequency bands according to the transcendent information [11]. However, the mechanism of production of EEG is rather complicated, so it is difficult to get accurate information on the transcendent and somewhat inflexible.

Based on the above research aft the background was designed as follows, section 2 describes the materials and methods used in the search feature and classification of EEG signal, part 3 describes the result of the extraction feature process and the classification EEG signals, and section 4 describe the conclusions of this research.

2. Materials and Method

2.1. Materials

The DataSet is taken from a healthy subject. The subjects were asked to move the cursor up and down on the computer screen, while the potential of cortical taken. During recording, the subject receives visual feedback from the potential the slow cortical (Cz-Mastoids). Cortical leads to the positive movement of the cursor to the bottom of the screen. Cursor movement causes cortical negative up. Each experiment 6s. During each experiment, Only the interval of 3.5 minutes per experiment is provided for training and testing. 256 Hz sampling rate and record length of 3.5 produces 896 point data [15].

2.2. Discrete Wavelet Transform

Comparison Discrete Wavelet Transforms (DWT) with continued Wavelet transform (CWT), DWT is considered a relatively easier implementation. How to obtain a representation of the time and scale of a signal using digital filtering techniques and operations, sub-sampling is the basic principle of DWT. Series high-pass filter and low-pass is a series that is used first to missed signals. The operations, subsampling is used to sample the output signals from an EEG taken half of each output signals EEG, the process is the process of decomposition of one level (Figure 1). Filter the output low-pass and high-pass used as an insert in the decomposition process of the next level. This process is repeated until the desired level of the decomposition process. Output filter high-pass and low-pass filters are merged into one, referred to as the coefficients of wavelet signal information, containing the results of the transformation that has been compressed [12].

The operations, sub-sampling that eliminates redundant signal information, wavelet transform has become one of the most data compression methods that are reliable. High-pass filter pair off and low-pass that is used must be a Quadrature Mirror Filter (QMF), i.e. a pair of filters that meet the following equation:

$$h[L - 1 - n] = (-1)^n \cdot g(n) \tag{1}$$

Where h [n] is the filter high-pass, g [n] is the low-pass filter and L is the length of each filter.

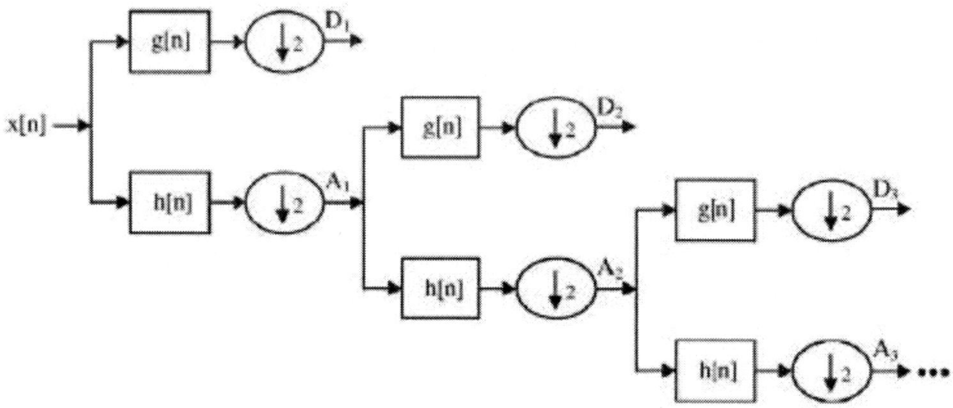

Figure 1. sub-band Decomposition of discrete wavelet transforms

Wavelet coefficients calculated using daubechies wavelet sequence 2 features more suitable for detecting changes in the EEG signals. In this study, the EEG signal is decomposed into the details of the D2.

To reduce the dimensions of the vector features, set up the statistical wavelet coefficient is used. The following statistics feature is used to represent the time, frequency distribution of EEG signals, i.e., the average value of the wavelet coefficients at each sub-band.

2.3. Backpropagation Neural Network

Neural network (ANN) is a system of information processing that has characteristics similar to a network of nerve Biology [13]. This means that the neural network is one of the information processing system designed by parroting the workings of the human brain in solving a problem by making the learning process through changes in the weighting of synopsis. JST is a method that can find a non-linear relationship between the load and the economic factors that vary as well as other factors that can make adjustments against the changes that occur. JST can be applied with good forecasting is the field [14]. To predict what will happen, we need a technique of divination to determine the process of planning and decision-making. JST backpropagation is a technique that can be used for forecasting. Mostly used in Backpropagation network many layers or can be called also multi-layer in hopes of minimizing the error on the result of the calculation technique conducted by the network. There are three main steps, i.e. enters data into the input network (feedforward), performs calculations and turning from error propagation (backpropagation) and performs renewal of weights and biases (adjustment). After discovering a pattern of network i.e. the values of the weights and biases, the network can be used to determine the output of any insert (testing).

In this study, the process of data classification is done by separating EEG signals into two parts, namely the data for the training process as many as 268 the data vector and the data for the test process data are used as much as 293 the data. This network has 4 input node (x_1, x_2, ..., x_4) comes from the feature DWT, 10 nodes in hidden layer 1 (z_1, z_2, ... the z_{10}), 15 nodes in hidden layer 2 (w_1, w_2, ... w_{15}), and outputs the binary type to identify conditions (y_1, y_2). Research on network architecture can be seen in Figure 2. The output pattern with 2 outputs in the form of binary target. The type of pattern can be seen in table 1.

Table 1. The Output Vector Patterns

No	Classification Of Data	The Output Pattern
1.	Up Cursor	0
2.	Down Cursor	1

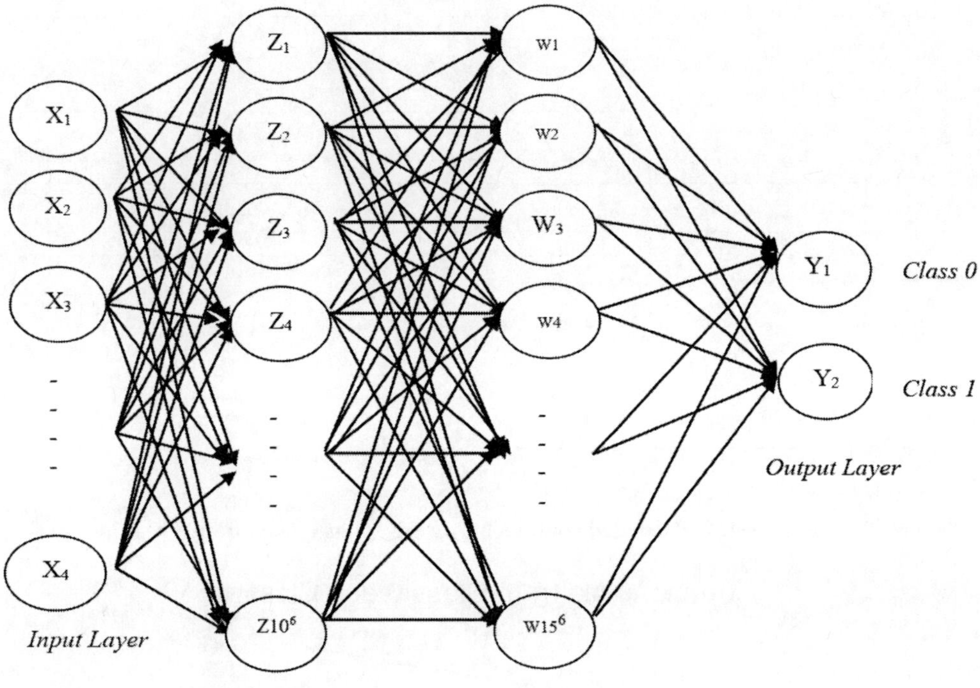

Figure 2. Architecture Backpropagation Neural Network with two hidden layers

3. Results and Discussion

The data used for research that is using data from the BCI Data set He 2003 competition. Data set It is composed of 6 channels (the electrode taped in the scalp amounted to 6 electrode sensor, resulting in a 6 channel signal EEG). The data set is that it consists of the training and data Testing data.

The amount of data a lot, will cause long computation process is caused by lots of data that are processed, so with a little feature that results in a fast computing process. At the research, one signals EEG has only taken the average value and the maximum value of each subband extraction process of DWTuntuk from the Foundation of the characteristics for the identification process.

Figure 3 is a recording of EEG is divided based on the frequency of the sub-band wavelet coefficient, resulting in A5, D5, D3, and D4 from the DWT. Frequency sub-band wavelet (0-4 Hz), (4-8 Hz), (8-16 Hz) and (16-32 Hz) was made into a set of feature extraction are taken from each sub-band EEG signals.

Figure 4 shows that for the average value in each subband for class 0 and class 1 has the value difference. The difference in value is not the same show that level of classification by taking the average value quite well. Classification using a Backpropagation Neural Network implemented using features of the average value of the DWT as input. In this study, a set of training amounted to 260 data samples and test data 293 sample data. On the training process used data as much as 260 sample data (from normal subjects) for channel 1. To test process data is used as many as 293 sample data (from normal subjects) for all channels. Table 2 is the distribution of the sample classes in the training data collection and validation. Data obtained from subjects in the training process is a way to improve the ability of Backpropagation. To train a backpropagation using a training data set, whereas To verify the accuracy and effectiveness of Backpropagation was using a test data have been trained to detect the movement of the cursor moves up and down.

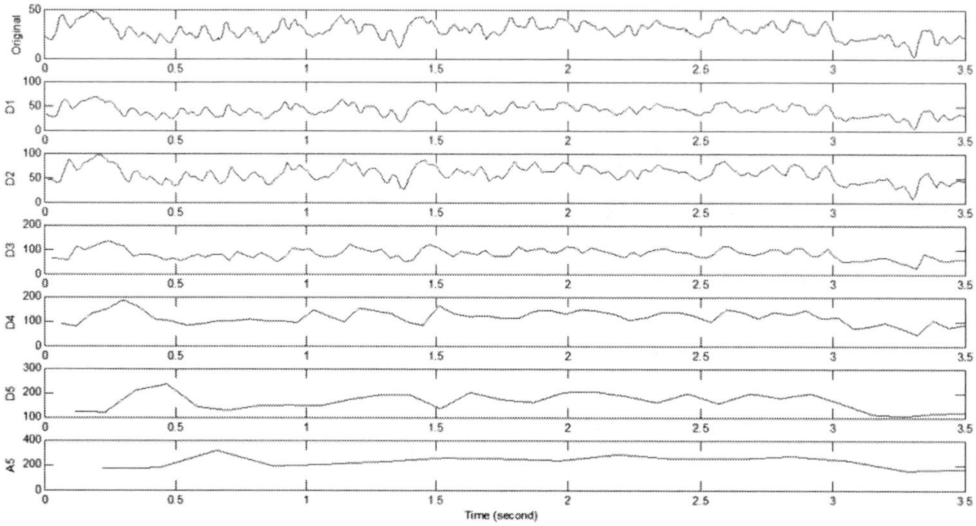

Figure. 3. Approximate and detailed coefficients of EEG signal taken from a healthy subject

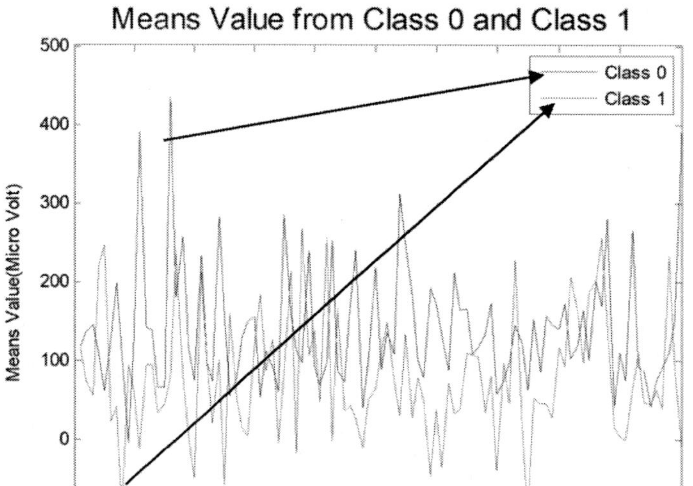

Figure 4. The mean value of the coefficient of the Sub Band Wavelet transforms

Table 2. The sampling distribution of the class at the training data and test data

Class	Training set	Test sets
Up Cursor (class 0)	130 x 6 Channel	293 x 6 channels (mix)
Down Cursor (class 1)	130 x 6 Channel	

The results of the extraction process characterized with DWT used a neural network, input for this study using the method of back Propagation (4-10-15-2) i.e. 4 input that comes from the characteristics of EEG signal and 2 the hidden layer (Figure 5). Two hidden layers are 10 nodes in the hidden layer node on 15 1dan hidden layer 2, and 2 targets (the movement of the cursor to the top of the class 0 and cursor movements down to grade 1). In addition to using the architecture network 4-10-15-2, to test the study also uses additional and subtraction hidden layers.

The process of training is the process of searching the best weight value with the smallest error value of the acquisition of the target output is desired, this is done in the initial process of identification. Process mapping is done after the process of training, namely the identification signals of the EEG up the cursor and down cursor movement based on the weights that already obtained in the process of training.

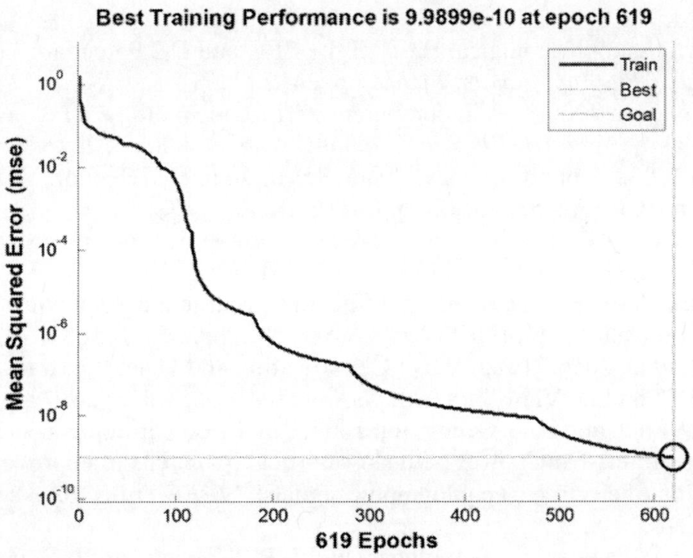

Figure 5. A training neural network of Performance using 2 hidden layers

260 data training from channel 1 in 619 the training period and the size of the steps for adaptation parameter has a value of 9, initial 998.10^{-10}. Performance Backpropagation using 2 hidden layers capable to perform the training process with minimal error limits, so 100% had the accuracy of the training process.

From table 3 to see that channel 6, which occupied a good degree of accuracy compared to the other channel with a value accuracy of 77.2%.

Table 3. Backpropagation accuracy results with 2 hidden layers for all channels

	Channel 1	Channel 2	Channel 3	Channel 4	Channel 5	Channel 6
Accuracy	60,0 %	63,1 %	67,1 %	68,0 %	73,4 %	77,2 %

From table 4 visible that by using 2 hidden layers on backpropagation can already achieve the accuracy value of 77.2% of the testing process.

Table 4. The influence of the number of Hidden Layer of different neural network

	MSE (1 Hidden Layer)	**MSE (2 Hidden Layer)**	**MSE (3 Hidden Layer)**
Time	13 second	15 second	34 second
Iteration	1000	619	916
MSE	$8,14.10^{-2}$	$9,99.10^{-10}$	$9,88.10^{-10}$
Accuracy	67,7 %	77,2 %	74,1 %

4. Conclusion

The study introduced the Discrete Wavelet to extract features by taking the average value in each subband signals EEG. The process of classification of EEG signals is divided into two classes, namely class 0 and class 1. This research, using EEG signals 553 file data for training and testing. The accuracy of the classification of Backpropagation reaches 77.2% of test data. To generate better results, the work of researchers who will come will examine the appropriate search techniques for the extraction of features and classification of EEG signals to command moves the cursor. The research results obtained are compared with the methods already researched.

References

[1] H. D. Lux, 1981. "Neurophysiological Basis of the EEG and DC Potentials," *Electroencephalogr. Basic Princ. Clin. Appl. Relate. Fields*, pp. 24–28.
[2] J. R. Wolpaw *et al.*, Jun 2000. "Brain-computer interface technology: a review of the first international meeting.," *IEEE Trans. Rehabil. Eng.*, vol. 8, no. 2, pp. 164–73.
[3] M. Planck *et al.*, 2003."Support Vector Channel Selection in BCI Support Vector Channel Selection in BCI," vol. 51, no. 120, pp. 1003–1010.
[4] J. D. R. Millán, 2003. "Adaptive brain interfaces for communication and control," *HCI Spec. Issue User Interfaces Age*, vol. 113, pp. 767–791.
[5] T. A. H. Fatehi and A.-B. R. Suleiman, 2011."Features Extraction Techniques of EEG Signals For BCI Application," *Naif Arab Univ. Secur. Sci.*, pp. 35–40.
[6] Shakshi and R. Jaswal, 2016. "Brain Wave Classification and Feature Extraction of EEG Signal by using FFT on Lab View," *Int. Res. J. Eng. Technol.*, vol. 3, no. 7, pp. 1208–1212.
[7] B. D. Mensh, J. Werfel, and H. S. Seung, jun 2004. "BCI Competition 2003--Data set Ia: combining gamma-band power with slow cortical potentials to improve single-trial classification of electroencephalographic signals.," *IEEE Trans. Biomed. Eng.*, vol. 51, no. 6, pp. 1052–6.
[8] D. J. McFarland, L. A. Miner, T. M. Vaughan, and J. R. Wolpaw, 2000. "Mu and beta rhythm topographies during motor imagery and actual movements," *Brain Topogr.*, vol. 12, no. 3, pp. 177–186.
[9] C. W. Anderson, E. A. Stolz, and S. Shamsunder, 1998. "Multivariate autoregressive models for classification of spontaneous electroencephalogram during mental tasks," *IEEE Trans. Biomed. Eng.*, vol. 45, no. 3, pp. 277–286.
[10] T. Wang, J. Deng, and B. He, 2004. "Classifying EEG-based motor imagery tasks by means of time-frequency synthesized spatial patterns," *Clin. Neurophysiol.*, vol. 115, no. 12, pp. 2744–2753.
[11] H. Adeli, Z. Zhou, and N. Dadmehr, 2003. "Analysis of EEG records in an epileptic patient using wavelet transform," *J. Neurosci. Methods*, vol. 123, no. 1, pp. 69–87.
[12] Y. S. Shin, P. Investigator, and D. Unlimited, *Monterey , 1993. California I1 l.*
[13] K. Hornik, M. Stinchcombe, and H. White, 1989. "Multilayer feedforward networks are universal approximators," *Neural Networks*, vol. 2, no. 5, pp. 359–366.
[14] T. Özel and A. Nadgir, 2002. "Prediction of flank wear by using back propagation neural network modeling when cutting hardened H-13 steel with chamfered and honed CBN tools," *Int. J. Mach. Tools Manuf.*, vol. 42, no. 2, pp. 287–297.
[15] http://bbci.de/competition/ii/

Acknowledgment

The authors are grateful to the Chairman of the Muhammadiyah University of Sidoarjo who gave time for the research that the researcher and the Directorate of Research and Community Service, Directorate General of Research, Research and Development of the Ministry of Research, Technology and Higher Education of the Republic of Indonesia supported the fund for this research.

Analysis of the security level of modified CBC algorithm cryptography using avalanche effect

N R D P Astuti, I Arfiani and E Aribowo

Informatics Department, Faculty of Industrial Technology, Universitas Ahmad Dahlan, Prof Soepomo Street, Janturan, Umbulharjo, Yogyakarta, Indonesia

Email: rochmahdyah@tif.uad.ac.id, ika.arfiani@tif.uad.ac.id, ekoab@tif.uad.ac.id

Abstract. There have been many modifications of cryptographic algrotima aimed at increasing the security of the algorithm. One cryptographic algorithm that has been modified is CBC (Cipher Block Chaining) cryptography. CBC is a block-based modern cryptographic algorithm. The disadvantages that exist in CBC is the encryption process in each block plainteks using the same key, resulting in if the cryptanalyst finds the key that is used it will be easy to decrypt the ciphertext. Modifications made to CBC aimed at increasing security is by combining the Vigenere algorithm in determining the key and block transposition on the final reading of the encryption process. Parameters for assessing the algorithm are good or not by looking at the value of avalanche effect. A cryptographic algorithm will be categorized as good if the value of avalanche effect ranges between 45% -60% and bernilia is very good if the value of avalanche effect is 50%. The results of this study indicate that the modified CBC cryptographic algorithm resulting from the encryption process of several plaintexts has Avalanche effect values in the range above 50% and below 60%, which means that the modified CBC algorithm is of good value.

Keywords: Modified CBC, modified performance algorithm, Avalanche effect.

1. Introduction

Cryptography is a mathematical technique related to information security aspects such as confidentiality, data integrity, user authentication, and data authenticity [1]. There have been many modifications made in cryptography aimed at increasing the security power of the algorithm. One of the cryptographic algorithms that have been modified is Cipher Block Chaining (CBC). In CBC mode, each block of plaintext is XORed with the previous ciphertext block before being encrypted. This way, each ciphertext block depends on all plaintext blocks processed up to that point. To make each message unique, the initialization vector must be used in the first block [2]. The CBC encryption process uses the same key on each block. This results in vulnerability, by finding the key and initial vector used, the kripanalis will very easily decrypt the encrypted ciphertext. To improve the security of the CBC, modification of the encryption process and the encryption result reading process is carried out. Modifications made to the CBC algorithm are by combining the Vigenere algorithm for determining keys and block transposition on the reading of the encryption process [3].

To determine the performance of the modified CBC algorithm, it is necessary to test the security level of the algorithm. Measurement of security level will be measured by the analysis of the avalanche effect. According to Webster, A. F .; Tavares, Stafford E. (1985), that a very well-known parameter used to analyze security (randomness) of encryption algorithms is with avalanche effect [4]. The test of avalanche effect measurement is done by changing the number of encrypted text bits (cipher) because a one-bit change in the key text or plaintext will produce significant changes to the ciphertext. The avalanche test is considered best if half of the ciphertext bits are changed according to the Strict Avalanche Criterion (SAC) [5].

Analysis with the value of the avalanche effect will prove the performance of the modified CBC algorithm. This paper will show how to measure the level of security of the modified CBC algorithm with the value of avalanche effect, the measurement is done by encrypting several plaintext forms and calculating the value of the avalanche effect from the encrypted ciphertext.

2. Method

2.1. CBC (Cipher Block Chaining)

Cipher Block Chaining Algorithm (CBC) is a cryptography method based on block. The series of plaintext bits are divided into bits of the same length, for example 64 bits. The same plaintext block will be encrypted into the same ciphertext block when the same key is used. This operation is applied to modern cryptography algorithms that are already operating at the bit level (0 or 1) or groups/block bits and not characters.

Suppose that block plaintext (*P*) which is m-bit in size is expressed as a vector $P = (p1, p2, ..., pm)$ which in this case pi is 0 or 1 for $i = 1, 2, ..., m$, and block ciphertext (C) is $C = (c1, c2, ..., cm)$ which in this case ci is 0 or 1 for $i = 1, 2, ..., m$. If the plaintext is divided into *n* blocks, the plaintext blocks are expressed as $(P1, P2, ..., Pn)$. For each plaintext *Pi* block, the constituent bits can be expressed as $Pi = (pi1, pi2, ..., pim)$ vectors. Encryption and decryption with key *K* are expressed in succession with the equation.

$$E_K (P) = C \tag{1}$$

for decryption, and

$$D_K (C) = P \tag{2}$$

The *E* function must be a one-to-one correspondent function, so

$$E^{-1} = D \tag{3}$$

The CBC mode scheme is shown in Figure 1 and Figure 2.

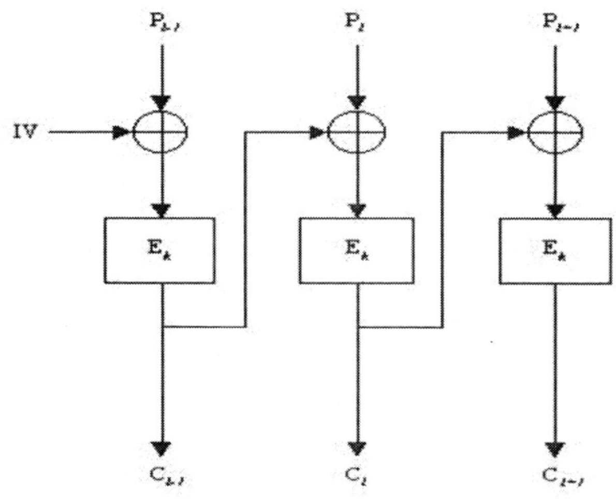

Figure 1. Scheme of the CBC method encryption process (Bruce 2013)

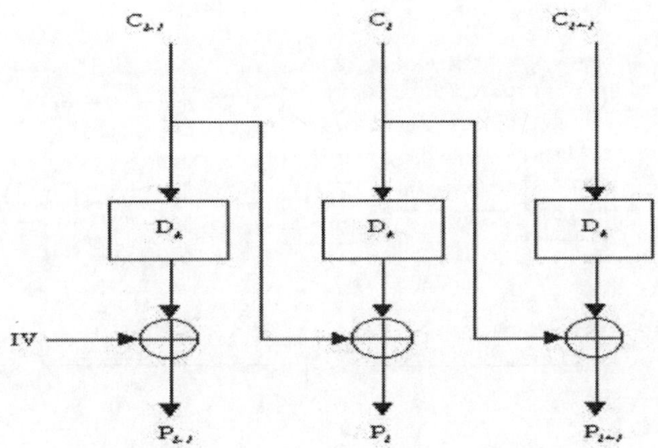

Figure 2. Scheme of the CBC method decryption process (Bruce 2013)

Mathematically, encryption with CBC mode is declared as

$$C_i = E_K\,(P_i \oplus C_{i-1}) \tag{4}$$

and decryption as

$$P_i = D_K\,(C_i \oplus C_{i-1}) \tag{5}$$

For the first plaintext block encryption process, use C0 as the initial vector (initialization vector or IV). Vector Initialization (IV) does not need to be kept secret. Initialization Vector (IV) which is a random data as the first block. Initialization Vector (IV) has no meaning, it is only used to create each unique ciphertext block. Identical plaintext blocks encrypted into ciphertext blocks differ only if the previous plaintext block is different. If the previous plaintext block is the same, it is possible that the ciphertext is the same, to prevent this, Initialization Vector (IV) is used.

2.2. Modified CBC

The result of improvisation in the modified CBC method is that there are two encryption processes in the modified CBC technique. The first modification encryption *(E1)* is in the key placement process. The key to the CBC method which is initially fixed in value for each block, by combining the Vigenere cipher method the key will always change following the key length in each block. Changing each key on each block will make it difficult for the cryptanalysis to estimate the connectedness of each plaintext block which ultimately leads to the entire message.

The second modification encryption (*E2*) is a continuation of the first process, the modification is found in the first encryption ciphertext reading process (*E1*) done by the block transposition method, this modification is intended to reduce the weakness of CBC, namely the ciphertext reading using the monotonous block concept so that it is easily kripanalis to find out the message sent, by changing the reading position of the encryption result by using block transposition, the kripanalis will find it difficult to estimate.

The modified CBC method scheme for the encryption process can be seen in Figure 3. Description of the modified CBC scheme as follows, on encryption 1 (*E1*) will be used Vigenere cipher. . The first plaintext block will be XORed with the value IV (Initialization Vector), the XOR result will be encrypted using the key from the vinegere cipher and generate an intermediate ciphertext. The encrypted ciphertext will be the initial vector on the next plaintext block. And the encryption process will end until the last plaintext block.

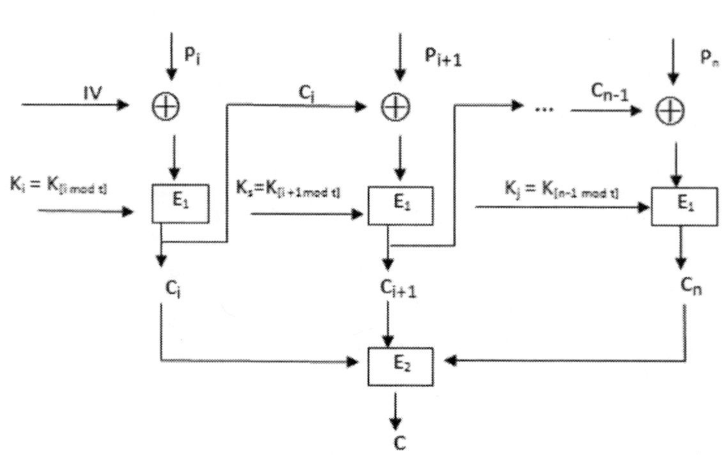

Figure 3. The modified CBC algorithm scheme for the encryption process (Astuti 2012)

Encryption process 2 (*E2*), the final ciphertext reading is done in Block Transposition. The intermediate output from C_i to C_n will be read in the Block Transposition or transposition column, the output of the process is the final Ciperteks (*C*) which will be converted into hexadecimal. In CBC modification with block transposition the ciphertext reading process will be sequential according to the column. Whereas for the modified CBC decryption process scheme is shown in Figure 4.

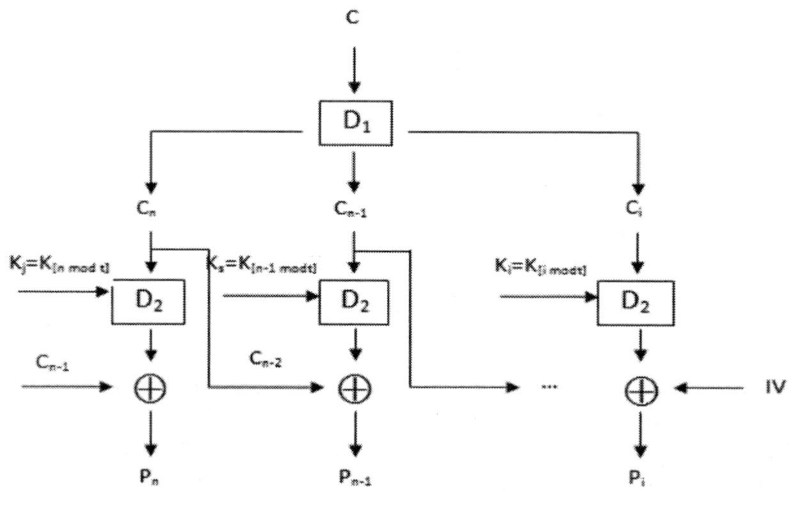

Figure 4. Schematic of the modified CBC algorithm decryption process. (Astuti 2012)

The decryption process is the opposite of the encryption process, Decryption Process 1 (*D1*), is the process of reading ciphertext done in Block Transposition. The ciphertext will be made per block according to the block key from C_i to C_n. Result *D1* is an intermediate plaintext.In decryption 2 (*D2*) intermediate plaintext, Vigenere cipher will be used. The current ciphertext block is XORed with the plaintext block that was decrypted before, the result will be XORed with the key. Determination of key index K In the modified CBC technique, it is stated successively by the equation:

$$K_i = K_{[i \, mod \, t]} \tag{6}$$

the *i* key is obtained from the value of *i* modulus with *t*, where *t* is the key length and *i* is the index with values from 1 to *n*. For the second plaintext and so on the key index becomes *Ks* where s is the plaintext index with a value of 1 to *n*, the *Ks* equation becomes

IOP Conf. Series: Materials Science and Engineering **674** (2019) 012056 doi:10.1088/1757-899X/674/1/012056

$$K_s = K_{[i+1 mod\ t]} \tag{7}$$

while for the *n-th* plaintext, key *Kj* determination using the equation

$$K_j = K_{[n-1\ mod\ t]} \tag{8}$$

j is the key index with values from *1* to *t*, the key index will be *t* if the modulus is 0. So that the encryption and decryption equation of the CBC modification are respectively

$$C_i = E_{Ki}(P_i \oplus C_{i-1})$$
$$C_{i+1} = E_{Ks}(P_{i+1} \oplus C_{i-1})$$

....

...

...

$$C_n = E_{Kj}(P_n \oplus C_{i-1})$$

and decryption is

$$P_i = D_{Ki}(C_i \oplus C_{i-1})$$
$$P_{i+1} = D_{Ks}(C_{i+1} \oplus C_{i-1})$$

....

....

....

$$P_n = D_{Kj}(Cn \oplus C_{i-1})$$

2.3. Avalanche Effect

One characteristic to determine whether or not a cryptographic algorithm is by looking at the avalanche effect. Changes in one bit from plaintext or one key bit must produce changes in many bits of the password text. This property is known as the Avalanche Effect [6]. The key will produce changes to many bits in the ciphertext. An avalanche effect is said to be good if the changes in the bits produced range from 45-60% (50% is a very good result). This is because these changes mean making a difference that is quite difficult for cryptanalysts to carry out attacks. The value of the avalanche effect is formulated with [7]:

$$Avalanche\ effect\ (AE) = \frac{\Sigma\ bit\_change}{\Sigma\ total\_bit} \times 100\ \%$$

2.4. Research Flow

The process of this research is planned with the following lines:

1) The course of the study begins with reviewing the theory of the process of measuring avalanche effects from existing literature, books, and journals.
2) The results of the theoretical study are made in the business process in the form of a flowchart avalanche effect analysis process.
3) Development of applications as tools for measurement using specific program languages.
4) Testing CBC algorithm performance Modified with plaintext with extension .txt and .doc.
5) The results of the application are data from the measurement of plaintext and ciphertext.
6) Measurement data are analyzed to produce conclusions on the performance of the algorithm.

The research flow is shown in Figure 5.

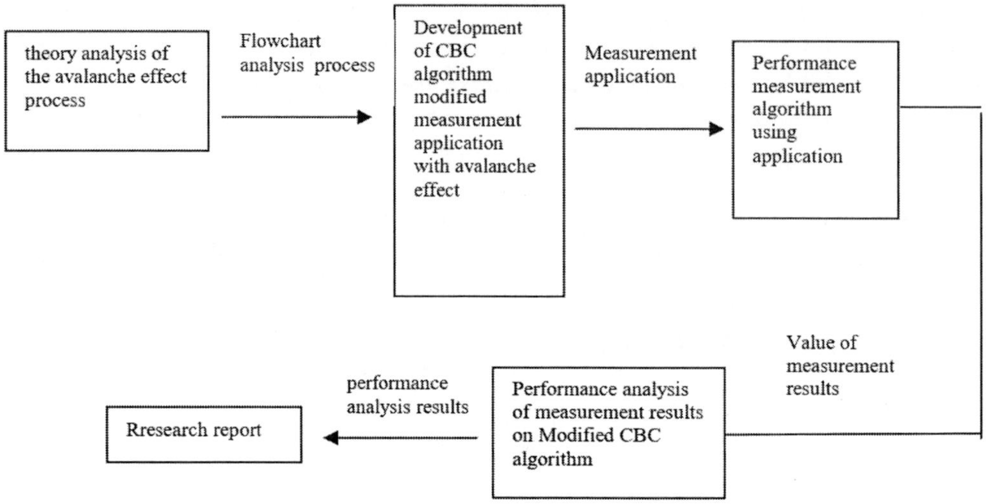

Figure 5. The research flow

3. Results and Discussion

One characteristic to determine whether or not a cryptographic algorithm is by looking at the avalanche effect. Changing bits in plaintext and key will cause significant changes to the resulting ciphertext [8]. Testing the modified CBC cryptographic algorithm performance with the value of the Avalanche effect is done by giving the same plaintext using capital letters and plaintext using lowercase letters. The key is given the same. The results of measuring the value of avalanche effect can be seen in table 1 and table 2.

Plaintext: "Ahmad Dahlan University informatics engineering"

Key: uad produces 8 bits in hexa.

Table 1. Test results of plaintext using lowercase letters

Number of Cipherteks Key bits	The original plaintext and in hex values	Chipertext	AE(%)	Analysis of AE
8	"infromatics" = 696E666F726D6174696373	32ACAF2B51AEFAB1D048CA	50%	very good
8	"informatics engineering" = 696E666F726D617469637320656E 67696E656572696E67	32A02ACAC0AF2B02BD1A99BEFA D74B1E3F904BB58CA80A	56%	good
8	"informatics engineering ahmad" = 696E666F726D617469637320656E 67696E656572696E672061686D61 64	32A02AFB2B02BFB2B02BFB46A66 DEEFAD752AC78FEB504BB5B432A 02AF	52%	good
8	"informatics engineering ahmad dahlan" = 696E666F726D617469637320656E 67696E656572696E672061686D61 64206461686C616E	32A02AFE1656057F42CAC0AFE85A 35336F42EFAD7529F58F1FD6AF412 ED6D5C1950157F0	51%	good
8	"informatics engineering ahmad dahlan university" = 696E666F726D6174696373656E67 696E656572696E6761686D616464 61686C616E756E69766572736974 79	32B0FC18400CAC3F0610032B0FC18 400D1AAF4F3027EFBDE0C168AB1E D07157E104DD7C50A00CAC3F0610 0	56%	good

Plaintext: "INFORMATICS ENGINEERING AHMAD DAHLAN UNIVERSITY"

Key: uad

Table 2. Results of testing plaintext using capital letters.

Number of Cipherteks Key bits	The original plaintext and in hex values	Chipertext	AE(%)	Analysis of AE
8	"INFORMATICS" = 494E464F524D4154494353	32995CAE8DEF96341065	54%	good
8	"INFORMATICS ENGINEERING" = 494E464F524D415449435320454E47494E454552494E47	32A822CAE08F2B023D1A9BBEFADF4B1E1F904B350CAA08	51%	good
8	"INFORMATICS ENGINEERING AHMAD" = 494E464F524D415449435320454E47494E454552494E472041484D4144	32A822732B823732B023F346A6ED6EFADF5A2C787E3D04B353C32A8227	52%	good
8	"INFORMATICS ENGINEERING AHMAD DAHLAN" = 494E464F524D415449435320454E47494E454552494E472041484D4144204441484C414E	32A822749657046E13CAC08FCA7A35376B07EFADF5A1558F0FC7BA412CD4F761954113A4	53%	good
8	"INFORMATICS ENGINEERING AHMAD DAHLAN UNIVERSITY" 494E464F524D415449435320454E47494E454552494E472041484D4144204441484C414E20554E4956455253495459	32A8227490F8CAE08DC263E72B023F29CF9D1A9BB583E74EFADF5A15FC4B1E1F8F749B104B353DDB348CAA089D243E	52%	good

From the results of testing two different types of paints, the results of the avalanche effect test for both plaintexts are between 50% - 56%, which means that the modified cryptographic algorithm has proven safety because it is in a good category from the avalanche effect.

4. Conclusion

Based on the results of this research, it can be concluded as 1) The modified CBC algorithm in terms of security is classified as good, so it can be used to improve the security of data from irresponsible people; 2) Modified CBC algorithm is said to be good according to the test results with the value of avalanche effect which is in the range of 50% is very good and a maximum of 56% in the good category; 3) The test uses the same key and plaintext whose contents are the same as letters, namely capital and lowercase; and 4) The testing process by breaking sentences into words and combining each word from the existing plaintext.

5. Sustainability of Research

Some suggestions that can be submitted and can be advanced research are: a) testing with other variables such as the complexity of the modification algorithm, and b) testing is done on data types other than text.

References

[1] Menezes, A.J., Oorschot, P.C. van & Vanstone, S.A., 1996. *Handbook of Applied Cryptography* 1st ed., New York: CRC Press Inc.

[2] William F. Ehrsam, Carl H. W. Meyer, John L. Smith, Walter L. Tuchman, "Message verification and transmission error detection by block chaining", US Patent 4074066, 1976

[3] Astuti, N.R.D.P., 2012. *Modifikasi Metode Kriptografi Modern Cipher Block Chaining (CBC) Dengan Algoritma Vigenere Cipher Dan Block Transposition.* Universitas Diponegoro,

Semarang.
[4] *Webster, A. F.; Tavares, Stafford E. (1985). "On the design of S-boxes". Advances in Cryptology - Crypto '85. Lecture Notes in Computer Science. **218**. New York, NY,: Springer-Verlag New York, Inc. pp. 523–534. ISBN 0-387-16463-4.*
[5] Ibrahim, M.S., Security Analysis of Secure Force Algorithm for Wireless Sensor Networks. , pp.1–7
[6] Ramanujam, S. & Karuppiah, M., 2011. Designing an algorithm with high Avalanche Effect. , 11(1), pp.106–111
[7] Sidorenko, A. V & Mulyarchik, K.S., The Analysis Of The Avalanche Effect In Discrete Chaotic Maps For Block Ciphers. , pp.32–35.
[8] Bruce, S., 2013. Applied cryptography. Journal of Chemical Information and Modeling, 53, pp.1689–1699

Acknowledgments

This research was supported by KEMENRISTEKDIKTI and LPMP of Ahmad Dahlan University in the research scheme of the beginner lecturer research number: PDP-094 / SKPP / III / 2018.

Design of 270 CC T2 gas capacity converter

F Azarul [1], M D Trisno [2] and D Dahlan [3]

[1] Mechanical Engineering Study Program, STT Muhamadyah Cileungsi-Bogor, Jalan Anggrek No. 25, Perum. PTSC, Cileungsi Village, Kec. Cileungsi, Bogor, Cileungsi, Bogor, West Java 16820, Indonesia
[2] Mechanical Engineering Study Program, Faculty of Industrial Technology - ISTN, Jl. Moh. Kahfi II, Jagakarsa, Jakarta 12640, Indonesia
[3] Mechanical Engineering Study Program Faculty of Engineering - Universita Pancasila, Srengseng Sawah, Jagakarsa, Jakarta 12640, Indonesia

Email: firmansyah.azarul@gmail.com [1], m.dwitris@yahoo.co.id [2], dddkamang@yahoo.com [3]

Abstract. Fuel motor is a machine that converts chemical energy into mechanical energy. In the design of this approach, we look for a gas converter model in accordance with its capacity according to the volume of the cylinder used, where the thing that needs to be considered is the performance of the gas converter itself, namely from the art form/model, fuel efficiency and large pressure settings. gas supply that enters the combustion chamber. The GX270 general motor is a fuel motor engine used as a test tool. The design was carried out at the STT Muhamadyah Cielungsi mechanical engineering energy conversion laboratory. To understand the variables that affect performance, in making this converter, things that need to be considered are variations in rotation and variations in load. From the design of the gas converter, the results of the design calculation are as follows: T torque = 19.12 (Nm), effective power Ne = 5.003066667 (kW), average effective pressure Pe = 889.4340741 (kPa), total usage hourly fuel mf = 0.0675 (kG / hour), specific fuel consumption Be = 0.013491725 (kg / kN.m), fuel ratio with air F / A = 0.13781, ideal air mass flow rate mv = 0,0,001138242 (m3 / s), volumetric efficiency 843,30666 (%) and thermal efficiency 15,78 (%), the greater the engine rotation used, the more fuel level will be consumed, as well with effective engine power.

Keywords: Kit converter, constant pressure, effective power, efficiency, General GX270T2 engine

1. Introduction

According to Law (UU) No. 45 of 2009 concerning amendments to Law No. 31 of 2004, stated that small fishermen are people whose livelihoods are fishing to fulfill their daily needs by using the largest fishing vessel 5 GT [1]. Most of the Small Fishermen use boats with gasoline engines. Fishing requires operational costs of around 70% for fuel oil (BBM). If the fuel price rises, the operational costs of fishing will be high so that fishermen's income decreases. The scarcity of BBM and its high price make it a separate problem for small fishermen. The use of cheap, convenient, safe and compatible alternative energy to drive small fishing boat engines is very necessary [2].

Based on the Presidential Regulation of the Republic of Indonesia Number 5 of 2006, concerning the National Energy Policy, the important agenda of the national energy policy is the development of alternative energy sources to reduce dependence on energy sources such as fuel oil (BBM). The government has called for fuel energy savings, given the availability of crude oil reserves that are increasingly lacking. There are many fuel resources that can be researched and utilized as alternative fuels, one of which is gas fuel in the form of Liquefied Petroleum Gas (LPG) which is one component of natural gas fuel with large enough reserves in Indonesia [3]-[6].

In general, boats/fishing vessels below 5 GT use a lot of gasoline engines. Therefore, there is a need for further research on modification of gas converters to produce design designs and create new gas converters for optimal boat/fish boat motors, in accordance with standards and prioritizing ergonomics, safety factors. Get it the right comparison of air and gas on a small fishing boat/fishing boat motor. Analyze and test machine performance validation, namely: Torque, effective power, fuel consumption. In optimizing fuel resources, one of the real steps in the development of energy conversion engine technology, through the study of engine modification.

This study is a continuation of previous research on converter kits. Some existing kit converters for the initial use of fuel oil after running are carried out by moving the gas fuel through the valve, and the membrane vacuum in the kit converter is not maximum. When the engine is off it does not automatically stop the gas flow.

This next research was carried out by modifying the design of the gas converter and the validation test. Modified gas converters designed to optimize the use of gas converters on gasoline motors. So that with the manufacture of gas converters it is expected to get the same engine performance, torque and effective power as the use of fuel oil with lower operating costs and easier operation compared to previous gas converters.

2. Literatur Review

2.1. Fuel Motor

Fuel motor is one type of heat engine, which is a machine that converts thermal energy to do mechanical work or convert chemical fuel energy into mechanical power. Before becoming mechanical power, the chemical energy of the fuel is converted first to thermal energy or heat by burning fuel with excess air. This combustion is carried out inside the heat engine itself and some are carried out outside the heat engine and the otto cycle of the gasoline engine is also called a constant volume cycle, where combustion occurs when the volume is constant [7] [8].

2.2. Fuel Motor Classification

The fuel motor can be classified into 2 (two) types. The classification of the combustion motor is the external combustion motor and the internal combustion motor in the following:

2.2.1. External combustion engine

The external combustion motor is where the combustion process takes place outside the engine itself. The heat from the fuel is converted into motion but exceeds the intermediate medium and then changes into mechanical power. In general, steam and turbine engines have characters that can only be used as large-scale starters for example, locomotives, ships, and power plants and are not good when used as multipurpose generator drives, motorbikes, and vehicles (cars) and can be seen as in Figure 1.

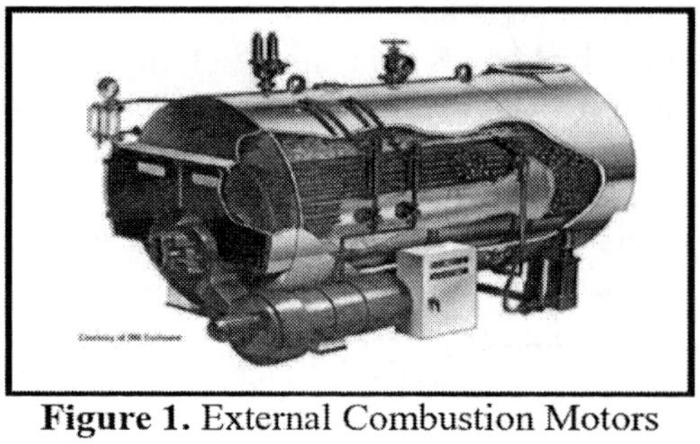

Figure 1. External Combustion Motors

2.2.2. Internal combustion motor

The internal combustion motor is the fuel combustion process that occurs inside the engine itself so that the heat from the direct combustion process can be converted into mechanical power construction and engine planning to be smaller and simpler, such as a diesel engine that can operate in high temperatures with repeated cycles and the use of this fuel motor has spread widely because it has a strong and reliable power besides that the use of fuel becomes more efficient and efficient.

The advantages of internal combustion motors are: Lightweight, small size, the power produced is large and very practical for vehicles. Another advantage can first be operated wherever there are air and fuel. Thus the area of operation is so large. Secondly, high thermal efficiency can produce considerable power with a relatively small amount of fuel.

The disadvantages of internal combustion motors are first valve opening and closing, air intake, exhaust combustion gas disposal and repeated bursts in each cycle causing vibration. The two can not be operated in a closed room for a long time because pollution will increase with the duration of use. The three fuels are limited, coal and wood cannot be used, because they cause ash. The internal combustion motor can be seen in Figure 2.

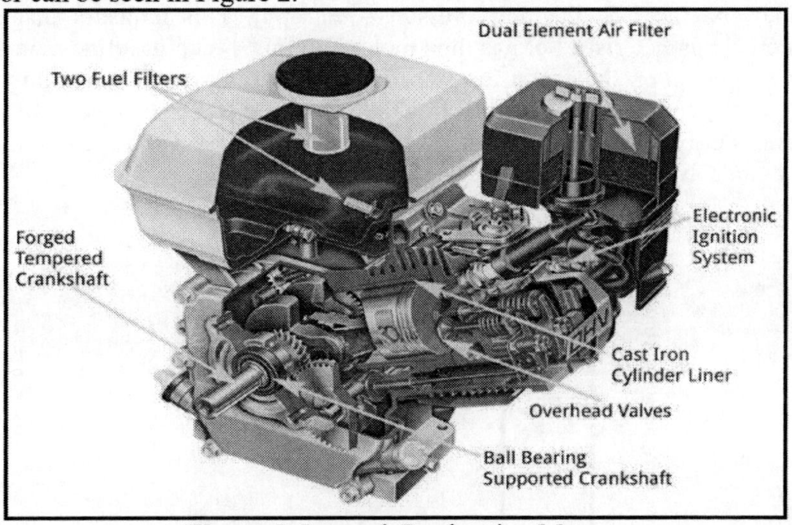

Figure 2. Internal Combustion Motor

2.2.3. Working Principle of four-step fuel motor (4-stroke)

4-step gasoline motor (4-stroke) undergoes one process at each step, namely: the Suction step begins with the movement of the piston from the upper die point (TMA) to the lower die point (TMB), the open suction valve and the exhaust valve are closed. Through the suction valve, the fuel mixture, air enters the combustion chamber.

Compression step The rotating crankshaft moves the piston to the TMA after reaching TMB. The intake valve and exhaust valve are closed. The fuel-air mixture is compressed, the pressure and temperature inside the cylinder increases, so that this mixture is flammable. This compression process is also called the pressure step, which is when the piston moves from TMB to the TMA and the two valves are closed. Work step where both valves are closed. At the time the piston reached the TMA, there arose a jump of electric sparks from the spark plug and burned a mixture of high-pressure and high-pressure air-fuel.

Exhaust step After reaching the TMB the crankshaft moves the piston to the TMA, the cylinder volume decreases. When removing the entrance valve is closed and the exhaust valve is open. The piston presses the combustion residual gas out of the cylinder. so that the cycle occurs repeatedly as in Figure 3 [9].

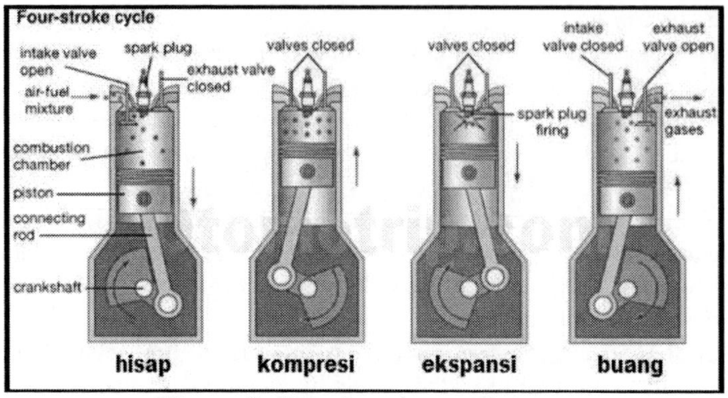

Figure 3. Step 4 engine work steps

The gasoline motor work cycle can be described in the indicator diagram, namely the P-V diagram (Pressure-Volume) and the T-S diagram (Pressure - entropy). This indicator diagram is useful for analyzing the internal characteristics of gasoline motors. In this 4-step gasoline motor, the combustion gas-only pushes the piston on the expansion step. Therefore, to allow the piston to move in the other three steps, some of the combustion energy during the expansion step is changed and stored in the form of flywheel kinetic energy. The way the engine works for the 4-step, in one cycle occurs in 4-steps. The steps that occur in the Otto 4-step engine are as shown in Figure 4.

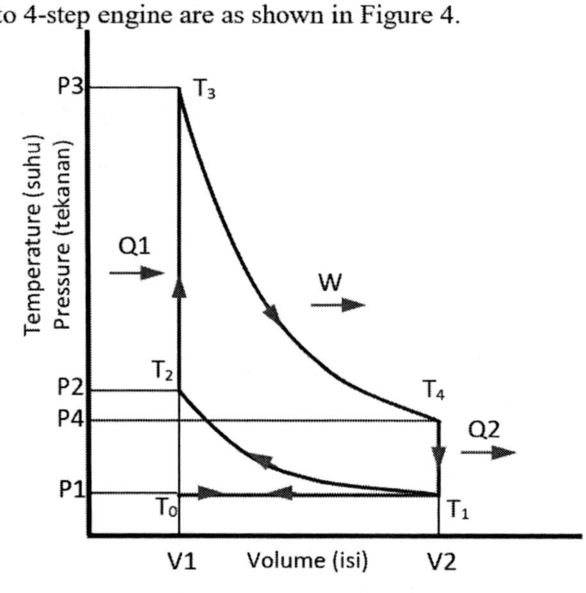

Figure 4. diagram of p-v motorcycle gasoline

2.3. Boat / Ship

Fishing vessels can be distinguished on non-motorized fishing vessels and motorized fishing vessels. This motorized fishing vessel is a vessel that uses a fuel motor as a driving source and propeller as a driving force. Based on the driving motor, fishing vessels can be distinguished on outboard engines and inboard engines. An outside motorized fishing vessel is a fishing boat that has a drive motor not located inside the hull but is mounted seated on the stern of the ship, on one side of the bulwark or on the stern deck of the ship. Deep motorized fishing vessels are fishing vessels that have a driving motor in the hull or below the deck in the engine room, mounted on the engine foundation so that the propeller shaft penetrates the stern wall or on the propeller height. [10] [4]

Outdoor motorized fishing vessels are often used by small-scale fishermen. In principle, this design is intended to facilitate maintenance. The motor used also has a lightweight construction, high engine speed, and its service life is only one to two years, so that efforts to extend the service life, more efficient fuel consumption and good durability during operation are problems that are often experienced by fishermen small scale [9], [11]-[19].

Judging from the combustion, the driving motor is divided into two types, namely otto and diesel fuel motor. The otto fuel motor or better known as the gasoline motor, the combustion process uses fire from a spark plug with gasoline, while the diesel engine combustion occurs due to an increase in the temperature of the air and fuel mixture due to piston compression reaching the flashpoint. Because of the principle of fuel ignition that occurs due to pressure, the diesel motor is called a compression ignition engine and the gasoline motor is called a spark-ignition engine [7]. The inboard motor is the engine/motor of the ship which is located inside the hull of the ship or bottom of the deck and is mounted seated on the engine foundation so that the propeller shaft penetrates the hull wall.

The outboard motor is the engine/motor of the ship that is not located inside the hull. The drive motor is usually mounted on the stern transom so that the propeller shaft does not penetrate the stern wall.

3. Gas Converter Design Method.

The concept of this research is the initial stage of making a product. The stage that will determine the end result of an analysis that will be carried out almost as a whole in this study, many things must be reviewed first. To simplify the steps of the study or in a more regular analysis process, a research flow chart is made as shown in Figure 5.

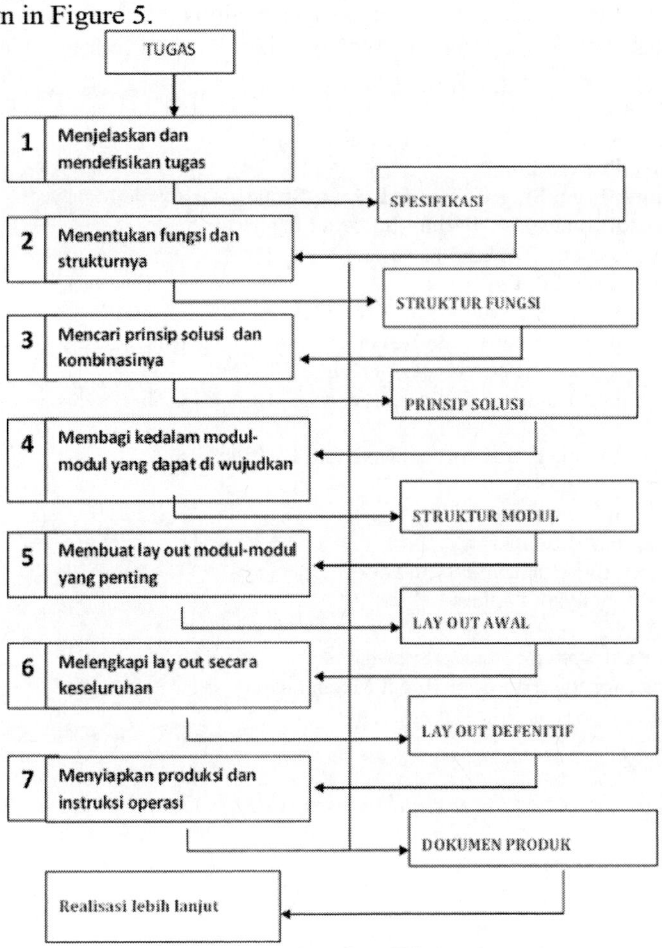

Figure 5. Design Diagram

From the design diagram above, the design results are obtained as shown in Figure 6.

IOP Conf. Series: Materials Science and Engineering **674** (2019) 012057 doi:10.1088/1757-899X/674/1/012057

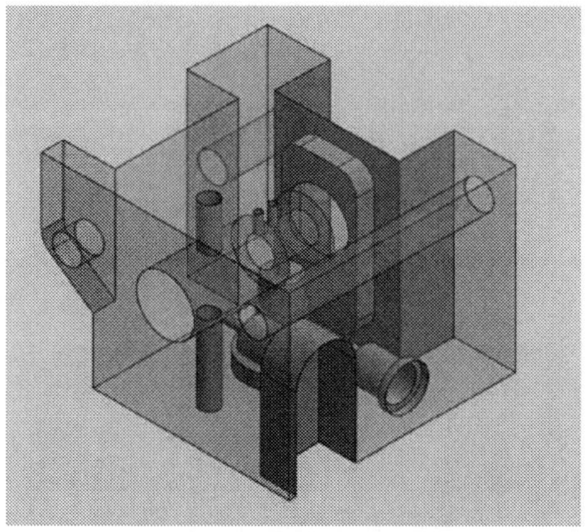

Figure 6. Converter Design

3.1. Selection of Driving Systems

The rotary drive system that is used as a rotary drive for this braking disk uses a 4-step engine because the fuel consumption is more efficient, the engine specifications that will be used in the analysis of the drive rotation of the disk system are machines with specifications as follows:

Specifications
Details of General GX270 T2 Technical Data :

- Horse Power: 8.0HP
- Engine Type: Single Cylinder, 25, 4-Stroke, Air-Cooled OHV
- Bore x Stroke: 77 x 58 mm (3.0 x 2.3 in)
- Displacement: 270 cm3 (16.5 cu in)
- Compression Ratio: 8.2: 1
- Net Horse Power Output: 6.6Kw (9 HP) / 3600 rpm
- Maximum Torque: 1.95kgf-m (19.12 Nm) / 2500rpm
- Net Torque: 17.7 Nm (13.1 lbs ft) at 2,500 rpm
- PTO Shaft Rotation: Counterclockwise (from PTO shaft side)

Ignition System: Transistorized Magneto Ignition
Starting System: Recoil

- Fuel Tank Capacity: 6 Liters
- Air Cleaner: Semi-Dry
- Carburetor: Horizontal type valve
- Oil Capacity: 1.1 Liters

Lubrication System: Forced Splash

- Governor System: Centrifugal Mechanical
- Dimension (L x W x H): 380 x 430 x 410 mm
- Weight: 25 kg

3.2. Determine the basic design of the main dimensions of the Gas Converter Based on the machine used.

The basic design of the main dimensions is adjusted to the data obtained by the machine and the calculation as follows:

- Horse Power: 8.0HP
- Engine Type: Single Cylinder, 25, 4-Stroke, Air-Cooled OHV
- Bore x Stroke: 77 x 58 mm (3.0 x 2.3 in)
- Displacement: 270 cm (16.5 cu in)
- Compression Ratio: 8.2: 1

- Net Horse Power Output: 6.6Kw (9 HP) / 3600 rpm
- Maximum Torque: 1.95kgf-m (19.12 Nm) / 2500rpm
- Net Torque: 17.7 Nm (13.1 lbs ft) at 2,500 rpm

3.3. Calculation of Gas Converter Dimensions.

Basic calculation of Gas Converter / Carburetor To design a Gas Converter / Carburetor in certain conditions, it is necessary and existing data, namely machine specification data:

- Cylinder diameter (D) = 77 mm
- Step length (L) = 58 mm
- Engine rotation (n) = 2500 rpm
- Cylinder capacity = 270 cc
- Power P = 6.6 kW

then the basic calculation starts from:

1) Calculate fuel volume Vbb

Vbb = phi x D x D x S / 4000 (1)

Where Vb is cc engine volume, Phi for constant whose value is 3.14, D is mm piston diameter, S for stroke step from tma to tmb mm. The calculation is Vb = 3,14x77x77x58 / 4000 = 269,94737 = 270 cc.

2) Calculate the diameter of the converter main hole/carburetor, (D).

D = K. √ (c.n) (mm). (2)

Where D is diamerter (mm), K is constant (0.6-0.9), C is cylinder volume (cc), and n is maximum rotation of the engine (rpm). The calculation is D is 0.65.√ 0.27.3600 = 20.26 => 20 mm. For gas used, the price of K is 0.25. D = 0.25.√0.27.3600 = 7.79934 => 8 mm. (https://www.facebook.com/spmmotor.pas/posts/129431253897523 and http://www.otosena.com/category/karburator/)

3) Relationship Between Torque and Engine Power - In-vehicle engine specifications, power is written in horsepower or kilowatt units at certain engine speed, and torque or torque with kgm units (Figure 7).

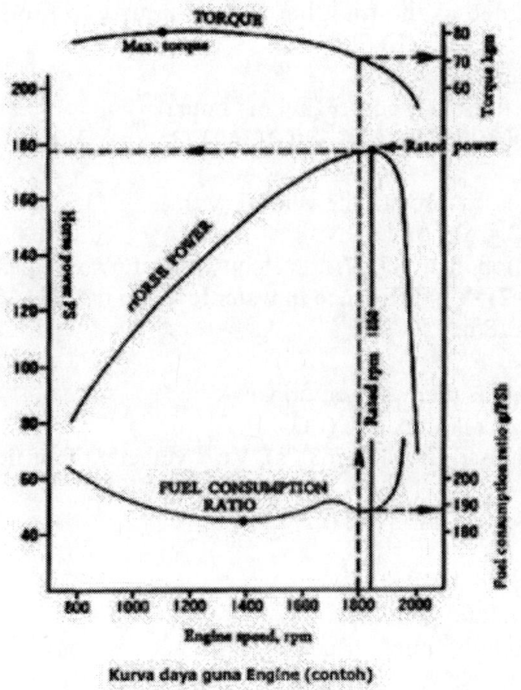

Figure 7. Engine performance graph

In the internal combustion engine, the combustion gas will press the piston connected to the crankshaft with the connecting rod. The compressive force of the gas produces a torque on the crankshaft and makes the crankshaft rotate. Power is torque multiplied by rotation (angular velocity): Moment of torque (Nm)

$$T = p.l \quad (3)$$

Where, p = force (kg); l = torque arm length (m); l = 0,81 m. From specifacations, maximum Torque is 1.95 kgf-m (19.12 Nm)/2500 rpm.

4) Effective shaft power (kW)

$$Ne = 2.\pi.n.T.10/60 \quad (4)$$

Where : Ne = Effective shaft power; T = Torque (Nm); n = Crank shaft rotation (rpm).
Ne = 2 . π . 2500 . 19,12 . 10/60; Ne = 5003066667 (kNm/hours); Ne = 5003066667 (kW)

5) Average effective pressure of Pe (kpa).

$$Pe = Ne / (V.z.n.a) .60.106 \quad (5)$$

Where, V = volume step 270 (cm3 / cc); Z = Number 1 cylinder; n = rotation 2500 (rpm); a = number of cycles of rotation = 0.5.
Pe = (50030.66667) / (270 .1.2500.0.5) .60.106 = 889.4340741kpa

6) The use of mf hourly fuel (kG / Hour).
The use of fuel is expressed in kg / h; the amount of fuel used as much as 10 grams in seconds is:

$$mf = 10 / t.\rho b.3600 / 1000 \ (kg / h) \quad (6)$$

Where:t = fuel consumption time of 10 grams; ρbb = fuel type = 0.01875 kg / m3.
mf = 10 /10 x ρbb x 3600 / 1000 = 0.0675 (kg / h)

7) Use of raw materials (Be).
The use of specific fuels is important for a motorbike. which is closely related to efficiency thermal motor. Specific fuel is defined as the fuel that is used hourly to produce

$$Be = mf / Ne. \ (kG / h) / kW \quad (7)$$

Where, mf = fuel consumption (kG / hour).

$$Ne = \text{effective shaft power} \ (kN.m / hour) \quad (8)$$

Be = 0.0675 (kg / hour) / 5.003066667 (kW) = 0.013491725 kg / kWh

8) The velocity of airflow passes through the venturi (vu).

$$vu = C\sqrt{(2.g.\Delta h)} \ (m / s) \quad (9)$$

Where, C = carbon composition in air 87; Vu = velocity of airflow rate passing through Venturi (m / s); G = earth gravity style (m / s2)=h = difference in water level on the manometer (mm).
Vu = 87.√(2.9,8.0,001)= 12.18 m / s.

9) The volumetic air flow rate is melwati venturi (mv).

$$mv = (\pi d \wedge 2) / 4 \ Vu \ x \ 106 \ (m3 / s) \quad (10)$$

where, mv = volumetric air mass flow rate (m3 / s); Vu = velocity of air flow; and D2 = diameter of the pipe venture (m).
mv = (π ⟦0.02⟧ ^ 2) /4.12.18= 0.001138242 (m3 / s).

10) Vu air flow rate (kg / h).

$$Vu = \rho u.mv.3600 \ (kG / h) \quad (11)$$

Where, Vu = air flow rate (kG / h); ρu = air density (kg / m3); and Mv = volumetric air flow rate (m3 / s).
Vu = 1,293.0,001138242 .3600= 5,29828921 (kG / h).

11) Comparison of fuel and air.

$$F / A = mf / Vu \tag{12}$$

Where, mf = fuel consumption (kg / h); and Vu = air mass flow rate (kg / h).
F / A = 0.0675 / 5,29828921 = 0.012739961

12) Ideal airflow rate (mid).

$$mid = V1.\ Z.\ n.a.\ \rho u.\ (kg / h) \tag{13}$$

Where, mid = ideal air flow rate (kg / h); V1 = step volume (m3); Z = number of cylinders; A = 0.5 for a 4 step motor; and ρu = period of air (kg / m3).
V1 = π / 4. D2. L = 3.14 / 4. 0.0772. 0.058 = 0,0002699 m3
Mid = 0,0002699. 1. 3600. 0.5. 1,293 = 0,628275509 (kg / h).

13) Volumetric efficiency.

$$nv = Vu / mid\ 100\% \tag{14}$$

where, nv = Volumetric efficiency (%); Vu = air flow rate (G / h); and mid = ideal air flow rate (g / h).
nv = (5,29828921) / (0,628275509) x100% = 843,30666%

14) Thermal efficiency ή t.

$$ήt = Ne / (mf.LHV)\ x\ 100\% \tag{15}$$

Where, =t = thermal efficiency; Ne = effective shaft power (kN.m / hour); MF = usage of fuel material (kg / hour); and LHV = Heating value of LPG = 11,220 cal / kg = 46,966, 92 (kJ / kg)
ήt = 50030.66667 / (0.0675.46966.92) x 100% = 15.78%.

4. Calculation results and design drawings

The calculation result data from the design can be seen in Table 1.

Table 1. Recapitulation of calculation result data design.

No.	Item	Unit	Results
1	Vbb	cc	269.94737
2	D	mm	10.91085653
3	T	Nm	19.12
4	Ne	kW	5.003066667
5	Pe	kPa	889.4340741
6	mf	Kg/hours	0.0675
7	Be	Kg/kNm	0.013491725
8	vu	m/s	12.18
9	mv	m^3/s	0.001138242
10	Vu	Kg/hours	5.29828921
11	F/A		0.012739961
12	mide	Kg/hours	0.628275509
13	ηv	%	843.30666
14	ηth	%	15.78121499

Based on Table 1, we can get the design in Figure 8 to Figure 10, and Figure 11.

IOP Conf. Series: Materials Science and Engineering **674** (2019) 012057 doi:10.1088/1757-899X/674/1/012057

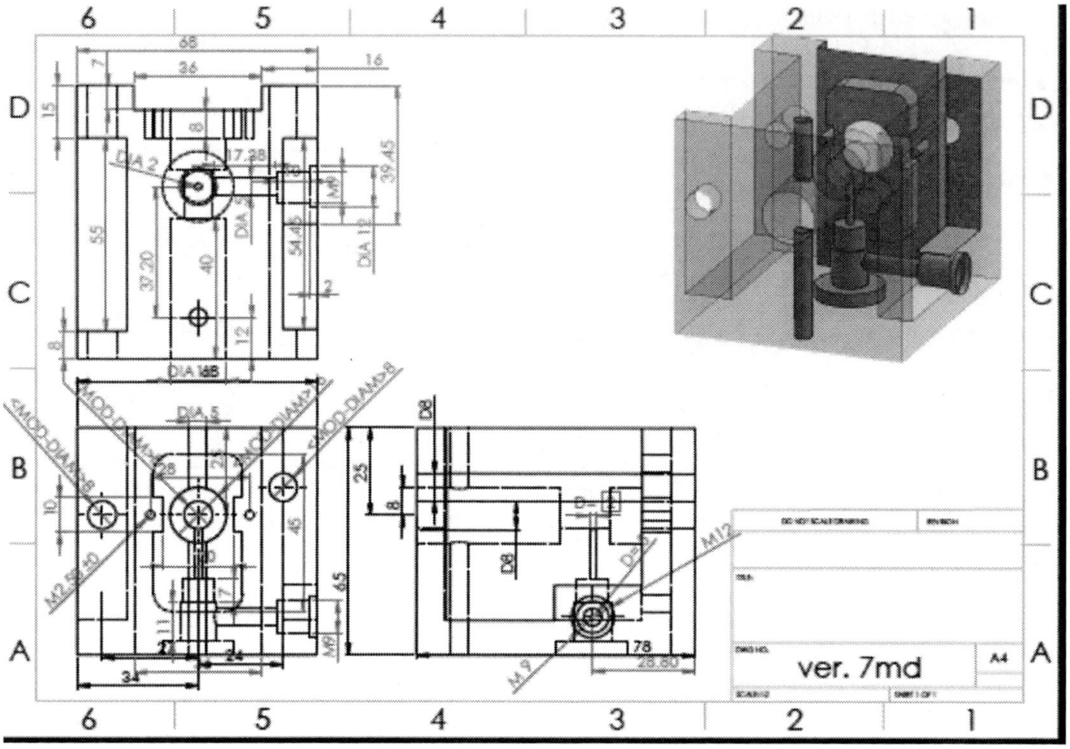

Figure 8. main case

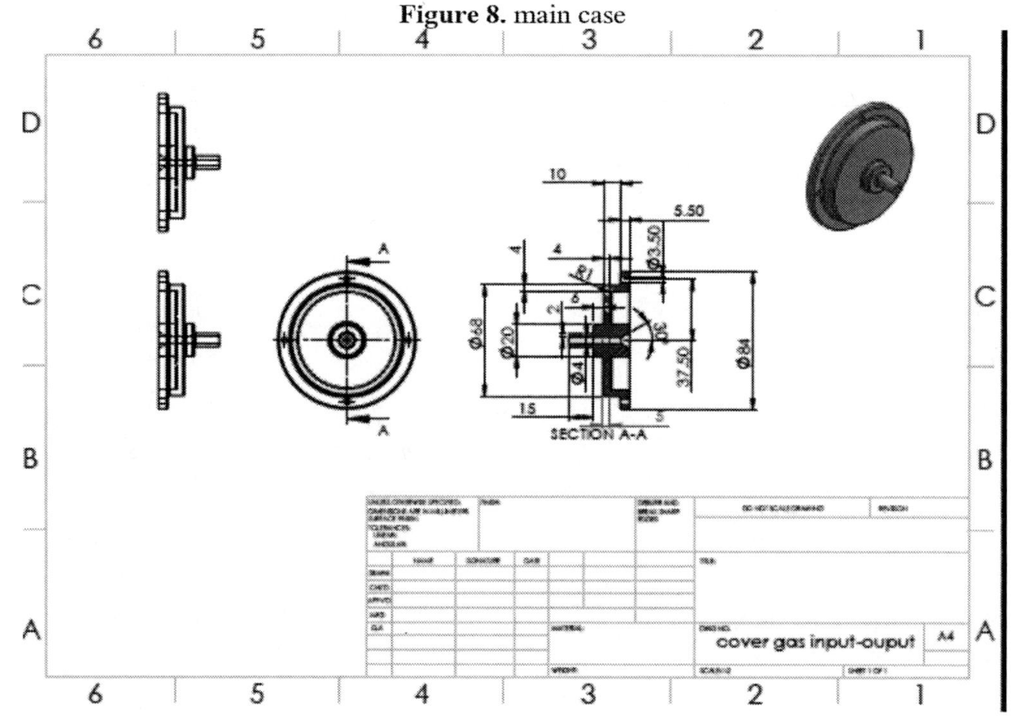

Figure 9. Cover for gas fire

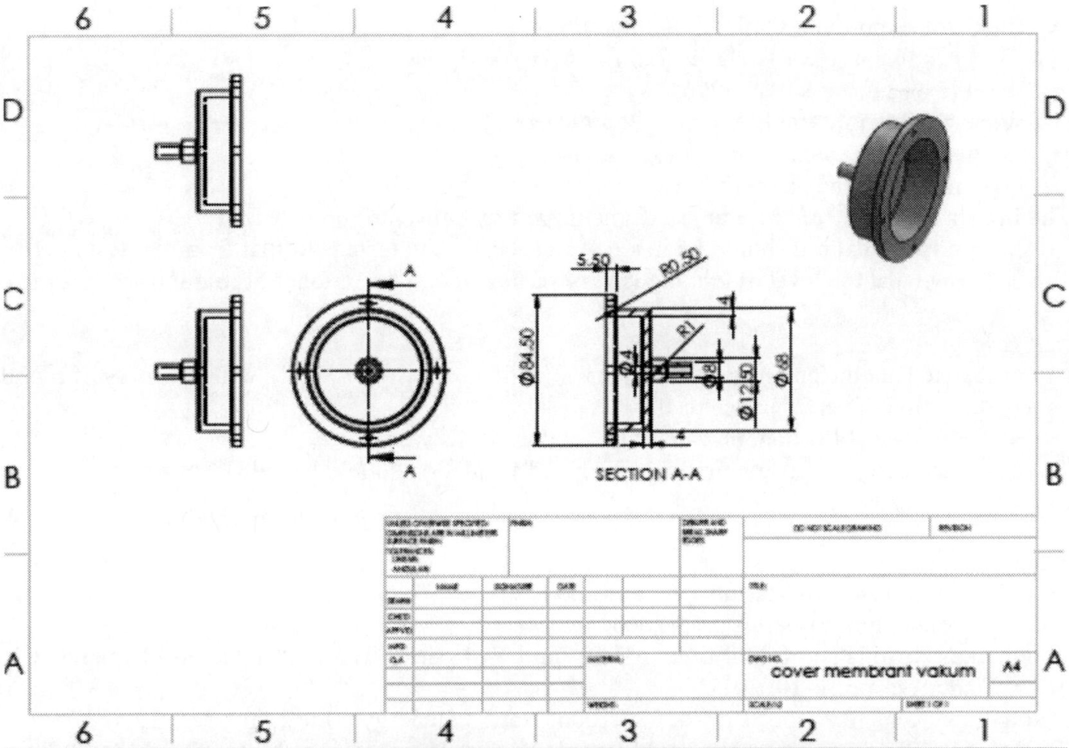

Figure 10. Cover vacuum

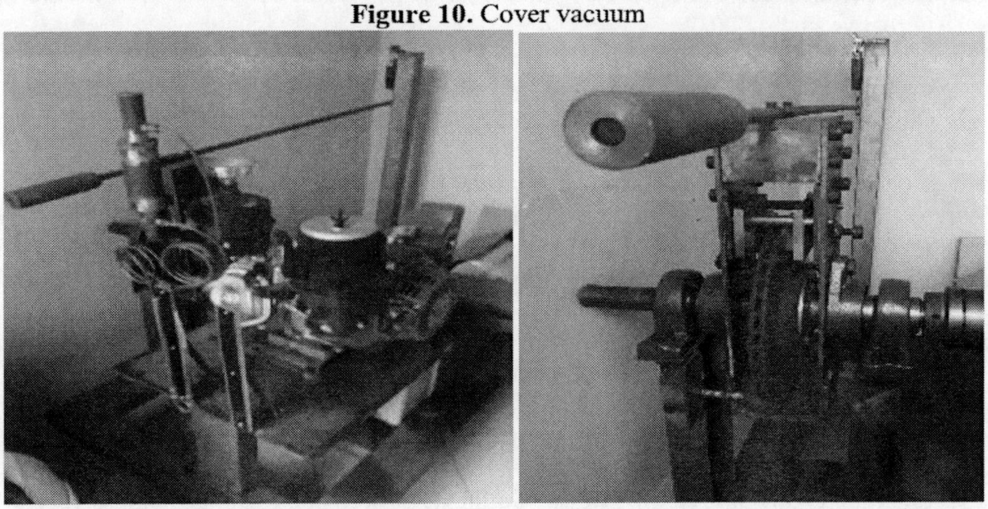

Figure 11. Visual Design

5. Conclusion and Suggestion

5.1. Conclusion

1) From the results of theoretical calculations the gas converter is obtained:
   - The diameter of 8 mm gas converter main hole.
   - 2500 rpm rotation.
   - Torque = 17.7 (Nm),
   - Effective power = 5.003066667 (kW).
   - Average effective pressure = 889,4340741 (kpa),

- Fuel consumption = 0.0675 (kg / hour),
- Specific fuel usage = 0.013491725 (gram.m / kN)
- Fuel to air ratio = 0.012739961
- Ideal airflow rate (mid). = 0.628275509 (kg / h)
- Volumetric efficiency. Nv = 843,30666%
- Thermal efficiency ἡt = 15.78%

2) The higher the torque on the crankshaft the more fuel consumption is used
3) To get validation data, it should be made a prototype, then tested, so that from the tests performed can be known that the level of braking is very influential on the amount of rotation on the crankshaft

5.2 Suggestion

1) Can be carried out the engine performance test with a capacity of 270 cc with a variety of rounds, so as to get maximum performance results
2) In testing for data collection, be careful to get the results we want
3) To obtain accurate results, maximum testing and validation should be held

References

[1] Anonymus_ Ministry of Maritime Affairs and Fisheries, "Law of the Republic of Indonesia number 31 concerning Fisheries p.4," 2004.
[2] Anonymus_ Ministry of Maritime Affairs and Fisheries, "Fleet Structure of National Capture Fisheries," pp. 1–10, 2011.
[3] Anonymus_Kemenkuham, "PP No. 5 of 2006: National energy policy, "pp. 1-5, 2006.
[4] B. Baruno, "Performance of Liquefied Petroleum Gas (Lpg) in Fuel Motor 6. 5 Mobile Phones as Alternative Fuels, "2014.
[5] A. Kuncoro and S. Wisnugroho, "The Design of a Dual Fuel Converter Kit (LPG - SOLAR)," no. November 2016, pp. 1–8, 2016.
[6] A. Amadhayanti, "Identification of Factors Affecting the Improvement of the Performance of CILINCING Fisheries UMKM Using a Converter Kit," pp. 1–6, 2017.
[7] Arismunandar W, Activator of Motorbike Burners, ITB Bandung. 1988
[8] T. Wiratno, S. Rahardjo, I. C. Engine, and M. Bensin, "Calculation of Power and Consumption of Fuel Gasoline YAMAHA YAMAHA 100 CC," vol. 12, no. 2, pp. 58–75, 2012.
[9] A. Kuncoro and S. Wisnugroho, "LPG as an Alternative Energy for Fuel Dual-Fuel Diesel Engine Fishing Vessels," November, pp. 1–12, 2016.
[10] Gulbrandsen O, Small-Scale Fisherfolk Communities in the Bay of Bengal. Food and Agriculture Organization (FAO) Fisheries Technical Paper. Madras. India. Gunarto, "1990.
[11] J. Fyson, "Design of Small Fishing Vessels. Fishing News (Books) Ltd. England., "1985.
[12] S. Bhuiyan, "Dual Fuel Performance of a Small Diesel Engine for Applications with Less Frequent Load Variations," no. 10, 2009.
[13] N. S. Rahardjo, Oktavian. Asikin, Zaenal. Budihardjo. Setiyobudi and Bangun. dwilarno., Use of LPG for Fishing Boat Motor ikanan 5 GT. Technical Guidelines for Capture Fisheries, Center for Fishing Development, Directorate General of Capture Fisheries (DJPT) Ministry of Maritime Affairs and Fisheries. 2013.
[14] M. Supono, S.Pd., ST., "Getting Closer to LPG (Liquified Petroleum Gas) As a Fuel for Stoves," Malang, Intermediate Widyaiswara at the Program. Stud. Plambing and Sanitation Dep. PPPPTK Bid building. Automotive and Electron., 2014.
[15] W. D. Yudisworo, "Alternative Study on the Use of ELPIJI GAS BBG for Conventional Gasoline Engine Fuels," 2014.
[16] A. Suyabodha, "Science Direct Comparison the Rate of Energy Consumption Between Gasoline 95 and LPG in Spark Driving Ignition Engine under Real Conditions Conditions the feasibility," Energy Procedia, vol. 118, pp. 164–171, 2017.

[17] P. Kristanto, D. Faculty, T. Industri, J. Engineering, M. University, and K. Petra, "Setting Idle Conditions and Acceleration in Gas-Fueled Motors," pp. 7–11, 2001.
[18] A. C. Arya, "Comparison of Car Performance Using Fuel Oil (BBM) and Gas Fuel (BBG)," vol. 01, no. 01, pp. 1–8, 2016.
[19] W. Pahl, G and Beitz, Engineering Design, The Design Council ,. London, 1984.

The effect of copper flame holder application on butane combustion characteristics in meso-scale combustor

Sudarman*, I Fauzi and A F H Soegiharto

Department of Mechanical Engineering, Faculty of Engineering, University of Muhammadiyah Malang, Jl. Raya Tlogomas No. 246 Malang 65144, Indonesia

Email: sudarman@umm.ac.id,

Abstract. This experimental research aims at determining the characteristics of butane combustion in Mesocombustor made of a quartz glass tube with a diameter of 3.5 mm and an outer diameter of 5 mm. The findings revealed that the flammability limit became wider with sharp curve narrowing. The narrowing of flammability limit did not occur in the other flame holder made of mesh stainless steel. Different from stainless steel conductivity, copper has declining thermal conductivity in each of temperature increase. The wider flame section was due to the increase of temperature in each flow velocity.

Keywords: mesocombustor; flame holder; thermal conductivity rate; flammability limit; visualization; temperature.

1. Introduction

Micropower generator with hydrocarbon fuel might become a potential alternative due to its greater fuel density than those contained in batteries [1][2]. One type of Micropower generator has two important parts, which are thermophotovoltaic (TPV) and micro or meso scale combustor. Thermophotovoltaic (TPV) is a material that is able to convert heat into ready-to-use electricity. Micro or meso scale combustor functions as a converter from chemical fuel into heat [3]. To simplify this discussion, micro/meso-scale combustor, hereinafter, is called mesocombustor.

Maintaining the flame stability in mesocombustor is difficult due to the short residence time of mesocombustor and high heat loss [4]. The main important issue to create flame stability in mesocombustor is to minimize thermal quenching due to heat loss [5].

An experiment of flame stabilizing in mesocombustor tube with stainless steel flame holder application had been conducted by Yuliati, et al. [6]. The experiment revealed the effect of thermal conductivity of mesocombustor partition on flame stability in mesocombustor. A similar experiment to stabilize the flame of liquid fuel was conducted by using quartz mesocombustor with an addition of steaming segment insertion of stainless steel flame holder [7]. The flame holder is a part of mesocombustor in which the flame is located nearby. Flame holder functions as heat recirculatory assistor from the heat into reactant. Heat recirculatory aims for the reactant's initial heating before entering the combusting zone, by minimizing the heat loss. [6-8].

One of the materials with the highest thermal conductivity is copper. This article seeks to examine the effect of copper flame holder application on butane combustion characteristics in quartz tube mesocombustor.

2. Method

The mesocombustor design can be seen in Figure 1 while the experiment installation is illustrated in image 2.2. The mesocombustor was made of quartz glass tube with an outer diameter of 5 mm and inner diameter of 3.5 mm. A copper flame holder with a thickness of 1 mm was placed between them. The flame holder material was copper with thermal conductivity in the room temperature of 386 W/m.K [9] and stainless steel with the thermal conductivity of 16.3 W/m.K [9]. The temperature was measured on the 2 mm partition of inlet (T1), the 2 mm of outlet (T2), the flame (T3), and the combustion gas in the outer layer of flame holder (T4). The scheme of experimental installation which is containing parts of 1. Compressor; 2. Butane Gas; 3. Air Flowmeter; 4. Gas Flowmeter; 5. Mesocombustor; 6. Camera; 7. Thermocouple; and 8. Mesocombustor holder are shown in Figure 2.

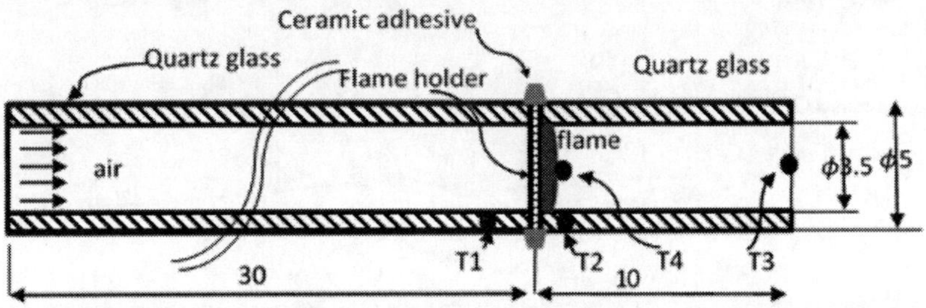

Figure 1. Mesocombustor and temperature measurement spot

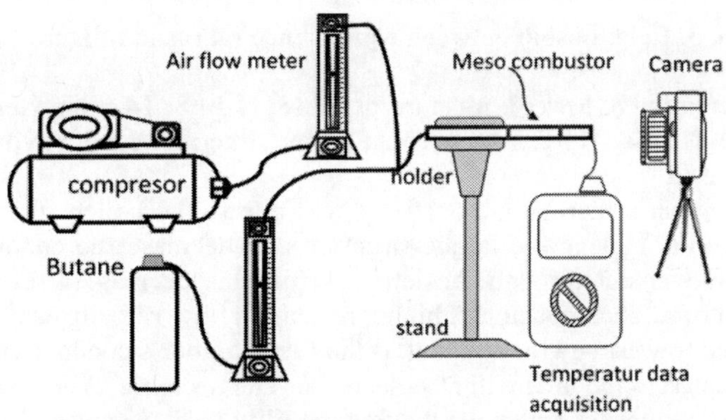

Figure 2. Experiment Installation Scheme

The flame visualization was captured by Nikon D5200 Camera. The fuel used in this experiment was butane gas (C4H10). The flowmeter was used to measure the flow velocity of air and gas into mesocombustor. The flow velocity of air and gas was made varied from the lowest to highest value, with a stable flame inside the mesocombustor.

The examined combustion characteristics involved flammability limit, flame visualization, and temperature. Meanwhile, the flame visualization was captured by a macro lens camera, and the temperature was measured by using thermocouple.

3. Result and Discussion

In Figure 3, a stable flame occurred in the equivalence ratio ϕ between 0.7 and 2.15, most of them occurred in equivalence ratio of $\phi>1$, or rich mixture zone. The copper's flammability limit was wider than the one made of stainless steel due to the higher thermal conductivity value (386 W/m.K for copper and 16.3 W/m.K for stainless steel). Higher thermal conductivity causes more heat energy distributed to the outlet, in order to heat the reactant. Copper has wider flammability limit than stainless steel, yet it has a deviation in the form of sharp narrowing and irregularity of flow velocity above 55 cm/s.

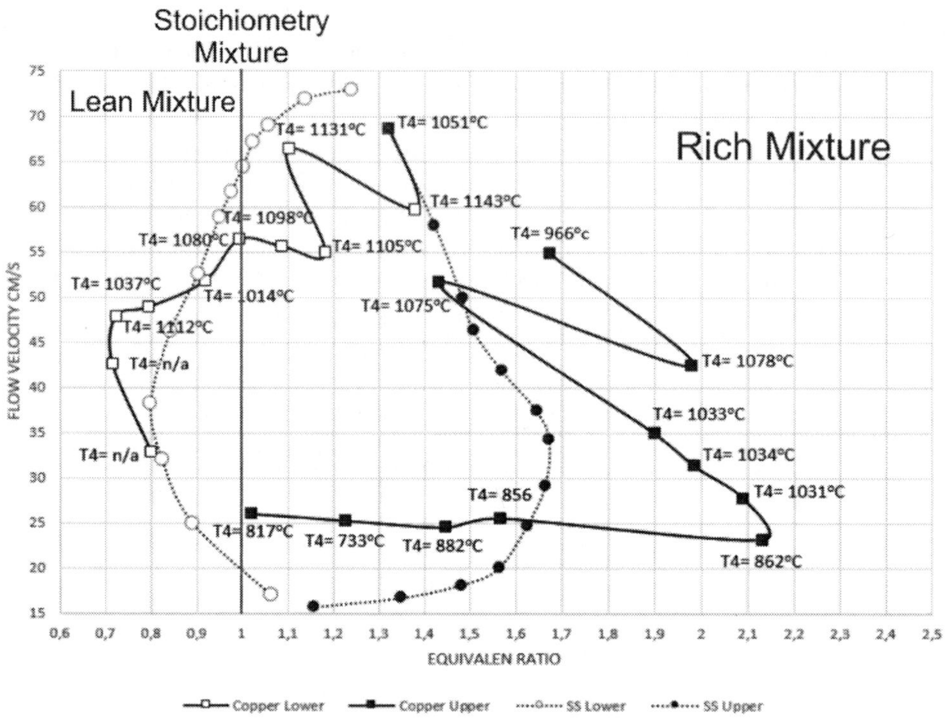

Figure 3. Relationship between equivalence ratio and mixture flow velocity

In Figure 3, the deviation of lower limit starts at φ1.18; U 54.8; T4 (flame temperature) 1033°C, and the upper limit starts at φ 1.4; U 51; T4 1131°C. Copper's thermal conductivity would decrease when the temperature increased.

The decrease of copper's thermal conductivity was due to the lattice structure's irregularity. The rising temperature would damage the lattice structure and decrease the conductivity rate due to the distributed electron movement through the lattice [10]. This decrease affected the amount of heat distributed to the inlet of mesocombustor. A higher reactant's flow velocity caused higher heat produced and higher heat needed to warm up the reactant. If the flame holder's conductivity decreases, the heated velocity to be recirculated to the inlet will also decrease. The decrease of heat velocity would affect the flame stability in mesocombustor and made the flammability limit narrow suddenly.

The stainless steel used in this research was grade 304 which had opposite characteristics with copper. Copper's thermal conductivity would decrease when the temperature increased while stainless steel's thermal conductivity would increase when the temperature increased. Based on the international standard table, the thermal conductivity of stainless steel grade 304 in the temperature of 100^{0}C was 16.3 W/m.K and temperature of 500^{0}C was 21.5 W/m.K. The increase of thermal conductivity caused a more regular flammability limit because higher temperature causes more heat needed to warm up the reactant. Thereby, by the increase of thermal conductivity, the needed heat energy to warm up the reactant as well as the flame stability could react well. This was proven by the stainless steel's flammability limit that reached the highest point in the mixture's flow velocity. The illustration can be seen in Figure 3.

Stainless steel has a lower thermal conductivity than copper; so, it has a narrower flammability limit than copper. On the other hand, copper has a higher conductivity yet there is a sudden narrowing in its flammability curve. The copper application for the flame holder in mesocombustor needs to be reconsidered because its thermal conductivity would decrease when the temperature increased. The flame and temperature visualizations are illustrated in Figure 4.

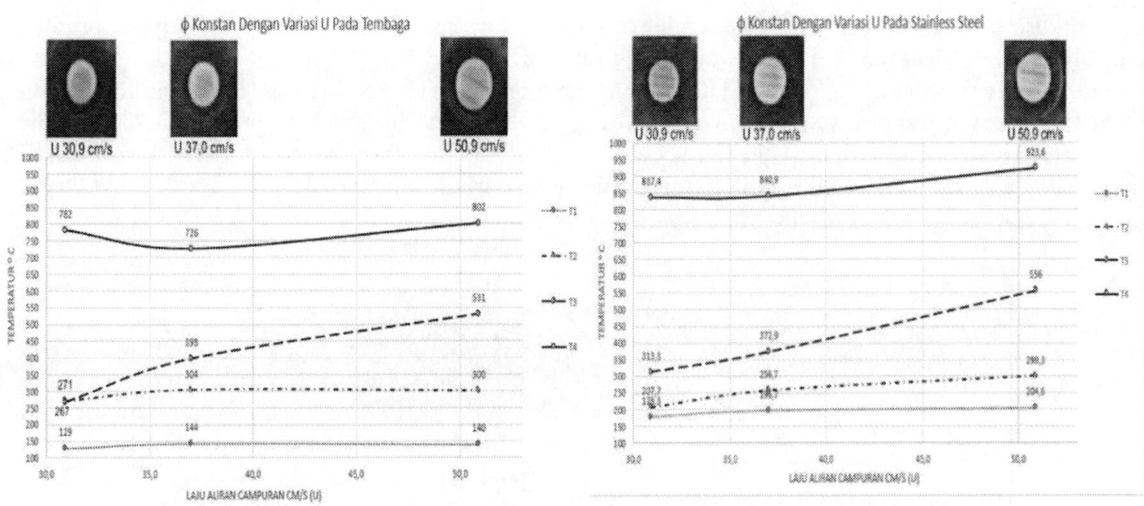

Figure 4. Flame visualization in φ 1,3 with varied mixture flow velocity (U)

Figure 4 illustrates the relationships between flow velocity, temperature, and flame visualization. The visualization above was taken in U 30,9 cm/s; 37 cm/s dan 50,9 cm/s and equivalence ratio (φ) was made constant at 1.3. The image above showed that the flammability limit got wider and the temperature increased as the mixture flow velocity increased.

In the flow velocity of lean mixture (U = 30,9 cm/s), the reactant supply for low mixture combustion was low, so the combustion would generate lower heat. The low heat was not sufficient to provide activation on reactant and produced gap without combusting reaction or quenching distance. The reactant in quenching distance cooled the combustion space, and the heat loss into environment causes the quenching distance got thicker. On the other hand, the flow velocity of high mixture (U = 50.9 cm/s) generated a higher heat level. The high heat level was able to transfer sufficient activation energy to reactant, which causes the quenching distance narrow. The increase in temperature was proportionate as the increase in flow velocity. The illustration was given in Figure 4.

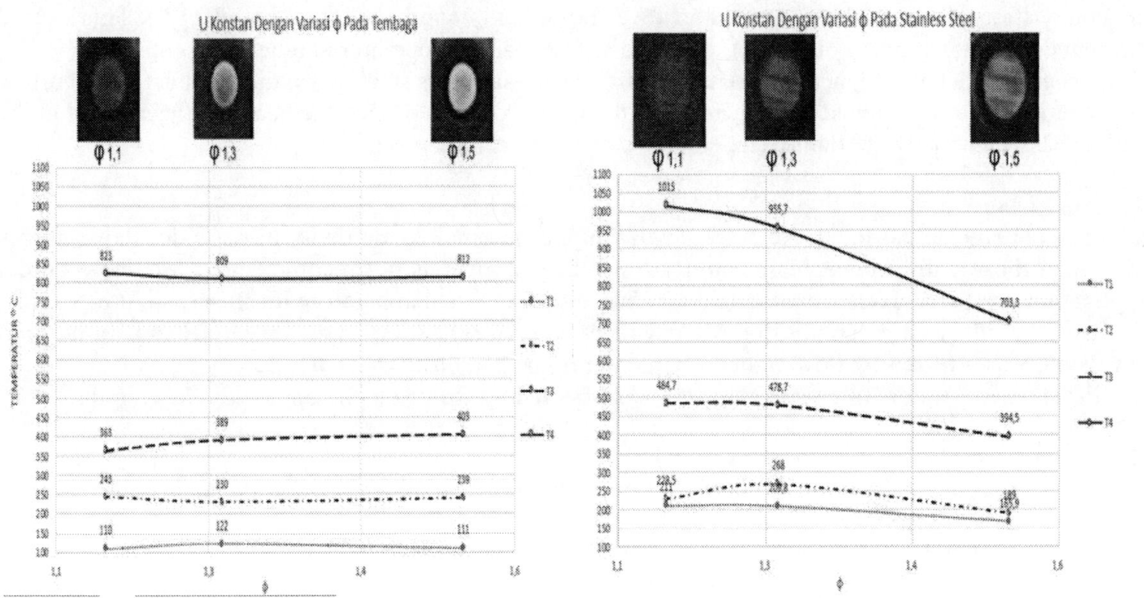

Figure 5. Flame visualization in U 38 cm/s with varied equivalence ratio (φ)

In Figure 5, it shows that in the equivalence ratio of 1.1, both stainless steel and copper produced dark blue flame. Meanwhile, in the equivalence ratio of 1.3 and 1.5, the color became bright blue. In lean mixture (equivalence ratio of 1.1), the flame became darker due to the lack of fuel in the mixture. Otherwise, in rich mixture (equivalence ratio of 1.3), the flame became brighter, and in the richest mixture (equivalence ratio of 1.5), the flame became much brighter. It can be concluded that rich mixture produced a brighter flame while lean mixture produced a darker flame color.

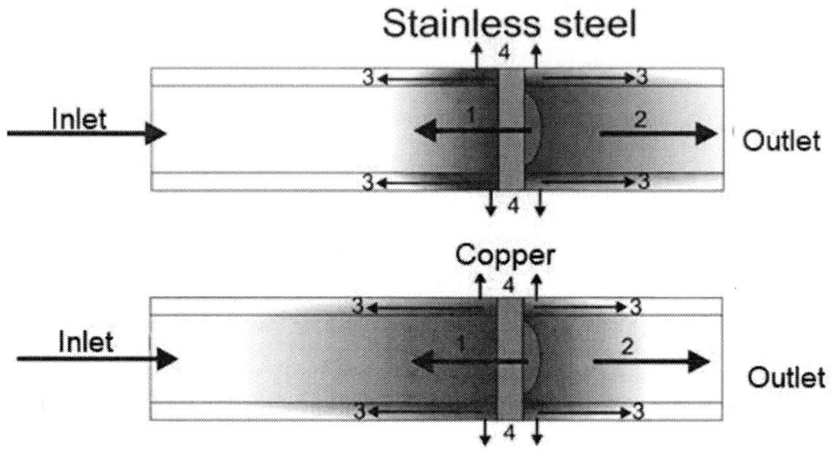

Note : 1: Heat convection into the inlet of mesocombustor (heat needed to warm up the reactant); 2: Heat convection to the outlet (heat emission with the gas); 3: Heat conduction through glass partition; 4: Heat conduction from the mesocombustor's partition to the environment.

Figure 6. Illustration of heat distribution in mesocombustor's inlet and outlet due to thermal conductivity difference

In Figure 6, it shows that stainless steel, which has a lower thermal conductivity than copper, when the combustion occurred, the generated heat was not quickly conducted into the inlet (A). This caused the heat concentrated in the outlet (B), which made the flame temperature higher than copper.

Copper, which had a higher thermal conductivity than stainless steel, when the combustion occurred, the generated heat was quickly conducted into the inlet (A). This caused the heat unconcentrated in the outlet (B), which made the flame temperature lower than stainless steel.

4. Conclusion

It can be concluded that the flame holder with a higher thermal conductivity has a wider flammability limit than those with a lower flammability limit. Pure copper is not recommended for flame holder application due to the decreasing thermal conductivity when the temperature increases; so, it influences the flame stability of mesocombustor. Meanwhile, flame visualization, in the form of flammability limit extension, was affected by flow velocity while the flame brightness was affected by equivalence ratio. Besides, the temperature of mesocombustor was affected by flow velocity and equivalence ratio.

References

[1] Maruta, K., Ju, Y. Microscale Combustion: Technology Development and Fundamental Research. Progress in Energy and Combustion Science. 2011; 37: 669-715.
[2] Dunn-Rankin, D., Leal, G.M., and Walther, D.C., Prog Energy Combust. Sci. 2005; 31: 422-465.
[3] Chou, S.K., Yang, W.M., Chua, K.J., Li, J., Zhang, K.L. Development of Micro Power Generators. ELSEVIER. 2011; 88: 1-16.
[4] Yuliati, L. Flame Stability of Gaseous Fuel Combustion Inside Meso-Scale Combustor with Double

Wire Mesh. Applied Mechanics and Materials. 20

[5] Hua, J., Wu, M., Shan, X,. Studies on Combustion Characteristics in Micro Combustor and it's Application. Chemical Engineering Research Trends. 2007; 1-70.

[6] Yuliati, L., sasongko, M.N., Wahyudi, S. Flammability limit and Flame Visualization of Gaseous Fuel Combustion Inside Meso-scale Combustor with Different Thermal Conductivity. Applied Mechanics and Materials. 2014; 493: 204-209.

[7] Hery Soegiharto, A. F., et al. (2017). "The Role of Liquid Fuels Channel Configuration on the Combustion inside Cylindrical Mesoscale Combustor." Journal of Combustion 2017: 1-9.

[8] Zhou, J., Wang, Y., Yang, J., Liu, Z., Wang, K., Cen. Applied Thermal Engineering. 2009; 29: 2373-2378.

[9] Standar Internasional . B-1. Nilai-nilai properti Logam.

[10] Koestoer, R.A. Perpindahan Kalor. Edisi I. Jakarta: Salemba Teknika. 2002: 135.

Reliability test prototype wind turbine savonius type helical as an alternative electricity generator

A M Siregar * and C A Siregar

University of Muhammadiyah Sumatera Utara, Medan 20238, Indonesia

* Email: ahmadmarabdi@umsu.ac.id

Abstract. The consumption of energy in the world today is still dominated by fossil fuels that will gradually run out. This period we must utilize renewable energy sources. One source of renewable energy that is environmentally friendly and we can use is wind energy. Wind turbines can be used to generate electrical energy so that it can solve one of the energy problems for people in need. A suitable wind turbine at low wind speed and medium wind speed is a type of Savonius turbine. The development of research is made prototype wind turbine. This test is carried out on a prototype of a vertical axis type wind turbine and aluminum flask. The rotor has a diameter of 350 mm and a height of 440 mm. The measured parameters were wind speed, rotation speed of rotor, voltage, and current of the generator. The performance of wind turbine savonius in this test at wind speed 4 m / s up to 6 m / s, maximum rotation of turbine shaft 35,40 rpm, electric power 6.94 watts at 6 m / s, at 6 m / s wind speeds produce wind power of 18.65 watts and produce electrical power of 6.94 watts.

1. Introduction

Wind power production has been under the main focus for the past decade in power production and tremendous amount of research work is going on renewable energy, specifically on wind power extraction [1]. With the rapid development of wind power industry, wind turbine reliability has become a hotspot in wind power[2]. Wind power is one of the most abundant sources of renewable energy in our country and environmentally friendly because it depresses CO2 emissions, therefore we can obtain unlimited electricity from wind energy[3]. Wind power is useful energy from wind[4]. Wind power plants convert wind power into electrical energy by using windmills or wind turbines. The way it works is simple enough that wind turbine spins are passed to the generator rotor where the generator has a copper winding that functions as a stator[5].

One of the wind turbines capable of starting to spin at low wind speeds because it has a high self-starting wind turbine is a vertical axis type savonius [6]. This wind turbine has a simple construction that can be applied on a small to medium scale. In this research will be seen and observed the reliability of prototype wind turbine savonius type. This type has been designed and manufactured by Siregar [7] which, the size of the rotor diameter 350 mm and height 440 mm, as well as 10 Watt and a diameter of 8 mm turbine shaft.

2. Wind Turbine

Based on the orientation of the rotor rotation, the wind turbine is divided into two types namely horizontal and vertical axis. The horizontal axis of rotation is parallel to the direction of the wind, while the vertical axis is opposite to the direction of the wind. Horizontal axis-based axis lift, slim blade, and high rotational speed. The rotor-based vertical axis drag force, wide blade and low rotational speed.

2.1. Savonius Wind Turbine

Type Savonius as shown in Figure 1, was created by a Finnish engineer SJ Savonius in 1929. This Vertical Axis Wind Turbine (TASV) is the simplest and most large version of an anemometer. Savonius turbine can rotate because of the force of the thrust from the wind so that the rotation of the rotor will not exceed the wind speed. This type of turbine is suitable for low power applications and is usually used at different wind speeds.

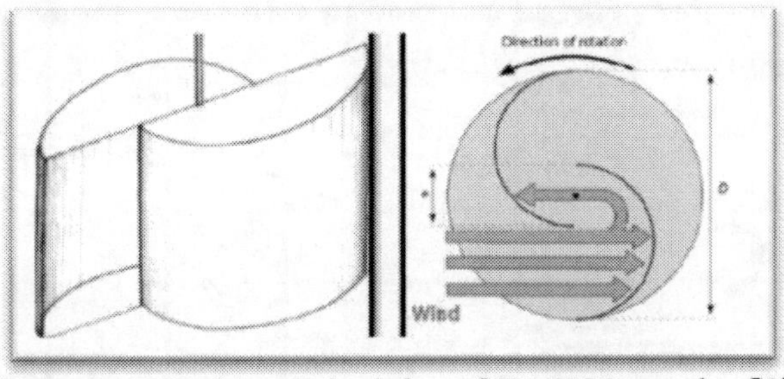

Figure 1. Working Principles of TASV Savonius [8]

One type of vertical axis wind turbine that can be used at low-speed winds is the Savonius wind turbine. The turbine construction is very simple, composed of two half-cylinder blades. On the development of this turbine Savonius undergo many changes in the shape of the rotor, as shown in Figure 2.

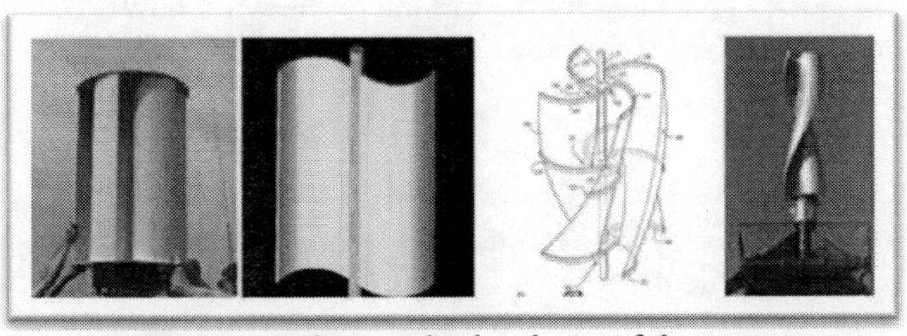

Figure 2. Changes in the shape of the rotor

2.2. Wind Power

Wind power, Pa defined as multiplication between the mass flow rate, $\rho\ Av$ with kinetic energy per unit mass $\frac{1}{2}V^2$ *wind* [7], can be expressed ;

$$P_a = \frac{1}{2}\rho\ Av^3 \tag{1}$$

Wind power is the input power carried by the wind to be converted by the turbine rotor into a shaft rotational mechanical power.

2.3. Electrical Power

Electrical power, P_g is the power out of the generator in the form of voltage (V) and current (I). By measuring the amount of voltage generated, the amount of power generator can be written:

IOP Conf. Series: Materials Science and Engineering **674** (2019) 012059 doi:10.1088/1757-899X/674/1/012059

$$P_g = V.I \tag{2}$$

2.4. Tip Speed Ratio

Tip speed ratio (TSR), λ is the ratio of the rotor tip to the free wind speed. For certain nominal wind speeds, the speed ratio tip will affect the rotor speed. The elevator type wind turbine will have a larger speed ratio tip compared to the drag type turbine. The magnitude of the speed ratio tip can be calculated by the equation;

$$\lambda = \frac{wr}{v} \tag{3}$$

Figure 3 shows the variation of speed ratio and power coefficient, Cp for various wind turbines.

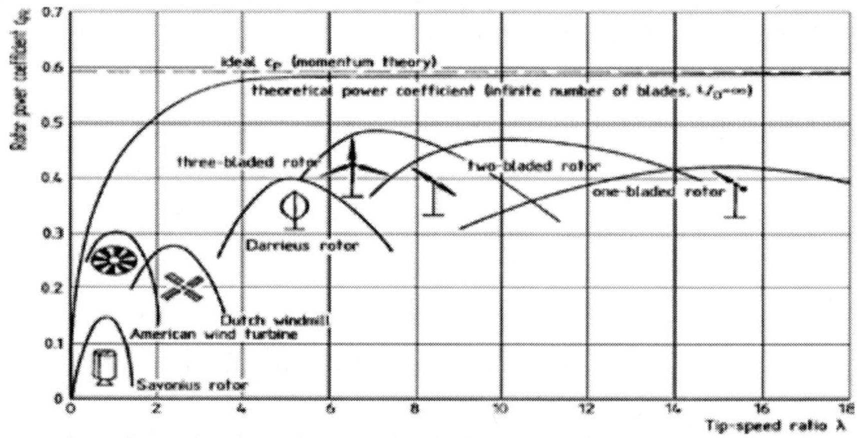

Figure 3. Value of speed ratio and power coefficient on different types of wind turbines [7]

2.5. Power Coefficient

To state the performance of a wind turbine is usually expressed in terms of power coefficients, Cp which is the ratio of the power generated by the turbine to the wind power obtained. Coefficient Wind turbine power can be calculated by the formula;

$$Cp = \frac{P_m}{P_a} \tag{4}$$

3. Experiment and Methods

The prototype of the wind turbine type savonius turbine has been made and placed on a wind tunnel for observation, as shown in Figure 4.

Figure 4. Prototype wind turbine type savonius turbine [5]

Wind turbine design savonius vertical axis helical type has the following dimensions and specifications;

- Rotor diameter = 350 mm = 0,35 m
- High rotor = 440 mm = 0,44 m
- Area of Turbine Cross section (rotor) A= 0,15 m^2
- Shaft diameter = 8 mm
- Rotor power = 10 Watt
- Maximum power Pm = 5,82 Watt

3.1. Measuring instrument

In testing wind turbine savonius helical type is used measuring tools, among others:

- Anemometer serves to measure wind speed.
- Tachometer serves to measure the rotation of the turbine shaft (rpm). Measurements are made on the wind turbine shaft.
- Multimeter serves to measure the quantity of electricity often known as VOM (Volt-Ohm meter) which can measure voltmeter, ohm-meter, and current (ampere meter).

3.2. Test Procedure

Test to determine the performance of wind turbine savonius helical type which have been made with the following test procedures:

- Setup of wind tunnel test equipment.
- Arrangement of measuring equipment research equipment
- Place the wind turbine savonius wind turbine in the wind tunnel suction test
- Prepare the measuring instrument (anemometer, tachometer, multimeter) and capture data simultaneously.
- Record the data.
- Data processing.

4. Results and Discussion

From the measurement and calculation results obtained as in Table 1

Table 1. Measurement and calculation results

wind speed *V*	round shaft *N*	wind power *Pa*	electrical power *Pg*	tip speed ratio λ	power coefficient *Cp*
m/s	*Rpm*	*Watt*	*Watt*		
4	20,40	5,53	1,68	0,093	1,052
4,5	23,20	7,87	2,28	0,094	0,740
5	28,60	10,79	3,52	0,104	0,539
5,5	32,80	14,37	5,66	0,109	0,405
6	35,40	18,65	6,94	0,108	0,312

4.1. Wind Speed Against Turbine Round

The effect of wind speed on the turbine rotor rotation is shown in Figure 5. In this case, the wind speed is directly proportional to the resulting rotation, meaning the greater the wind speed that touches the blade, the greater the turbine rotation produced, the greater the energy given by wind to the turbine, the turbine's energy can be turned into a bigger turn. With wind speed 6 m/s obtained shaft rotation 35,40 rpm, while wind speed 4 m/s obtained spindle shaft equal to 20,40 rpm.

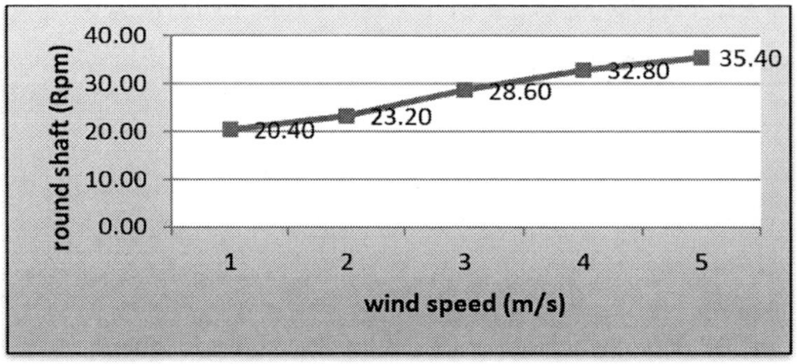

Figure 5. Effect of wind speed on the turbine spin

4.2. Axis Round Against Power

The graph is shown in Figure 6, that the increasing speed of rotation of the turbine shaft increased the the electrical power and wind power.

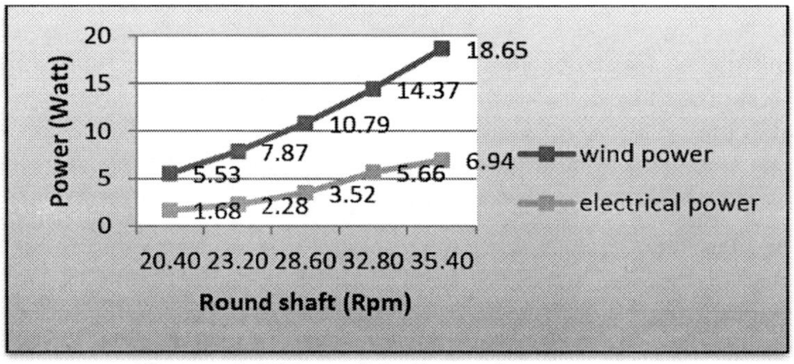

Figure 6. Effect of shaft rotation on power

4.3. Wind velocity with Power Coefficient and Wind Power

The power coefficient and wind power obtained are inversely proportional to wind speed, where the faster the wind, the smaller the power coefficient produced and the greater the winding power produced.

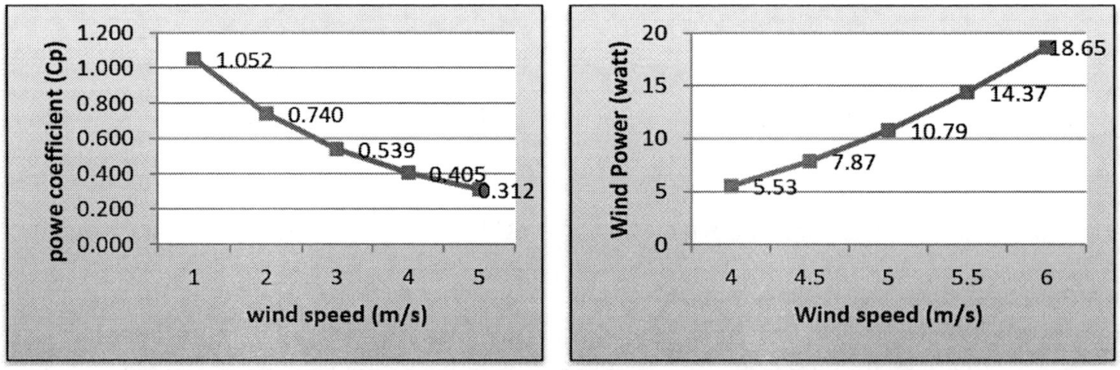

Figure 7. The Wind Speed Relation with the Power and Power Coefficients

5. Conclusions

The performance of Savonius prototype wind turbines can be taken as follows: (1) The resulting wind turbine rotation is directly proportional to the wind speed, where the wind speed of 6 m / s produces a wind turbine rotation of 35.40 rpm and a wind speed of 4 m / s produces a turbine rotation of 20.40 rpm; (2) The power coefficient at wind speeds of 6 m / s is 0.312; (3) At a wind speed of 6 m / s produces wind power of 18.65 watts and produces electrical power of 6.94 watts.

References

[1] N. Ramesh babu, P. Arulmozhivarman 2013, Wind Energy Coversion Systems A Technical Review, *Journal of Engineering Science and Technology, Vol. 8, No. 4 , 493 - 507*

[2] Yao Li, Caichao Zhu, Chaosheng Song, Jianjun Tan 2018. Research and Development of the Wind Turbine Reliability. *International Journal of Mechanical Engineering and Applications*. Vol. 6, No. 2, pp. 35-45. doi: 10.11648/j.ijmea.20180602.14

[3] Sumiati, R., Aidil Zambi. 2013. Bangun Miniatur Turbin Angin Pembangkit Listrik untuk Media Pembelajaran. *Jurnal Teknik Mesin Vol. 3, No. 2. 1-8.*

[4] Korpaasa M., Holena A. T. 2003. *Rancang* Operation and sizing of energy storage for wind power plants in a market system. Electrical Power and Energy Systems 25, 599–606. www.elsevier.com/locate/ijepes

[5] Karki R., Billnton R. 2001. Reliability/Cost Implications of PV and Wind EnergyUtilization in Small Isolated Power Systems IEEE TRANSACTIONS ON ENERGY CONVERSION, VOL. 16, NO. 4, DECEMBER.

[6] J. Sargolzay 2007. Prediction of The Power Ratio in Wind Turbine Savonius Rotors Using Artifical Neural Networks. Zahedan: Baluchestan University.

[7] Siregar, A. M. 2017. Design and manufacture of prototypes Dua tipe rotor turbin angin sumbu vertikal sebagai objek penelitian studi eksperimental. *Jurnal Teknovasi Volume 04, Nomor 02, 1–14 ISSN : 2540-8389*

[8] Kadir, A. 1995, Energi: Sumber Daya, Inovasi, Tenaga Listrik dan Potensial Ekonomi. Jakarta: Penerbit Universitas Indonesia.

[9] Hau, E. 2006. *Wind Turbines Fundamentals, Technologies, Application, Economics, Edisi Kedua*", Germany: Springer.

Application of catalytic converter copper catalyst with honeycomb surfaces to reduce emissions of flue gas in motorcycles

M A Siregar [*], C A Siregar, Muharnif, A M Siregar and I Maulana

University of Muhammadiyah Sumatera Utara, Medan 20238, Indonesia

[*] Email: munawaralfansury@umsu.ac.id

Abstract. The application of a catalytic converter is believed to be able to accelerate the process of chemical reaction in vehicle exhaust gas. This will have an impact on reducing the high rate of air pollution from motorcycles exhaust gases. This study will compare the exhaust emission value of the standard exhaust muffler with the exhaust that has been installed catalytic converter. Catalytic converter which is made from copper with honeycomb surfaces. The motorcycle used with an engine capacity of 125 CC. Testing is done based on three different engine rotation that is 2000 rpm, 3000 rpm, and 4000 rpm. The result, at 4000 rpm engine speed decrease CO emission level of 1.76%, HC of 73 ppm and CO_2 of 1.7%.

1. Introduction

The high number of motorcycle growth in Indonesia in 2016 reached 5,931,285 units [1]. This amount is certainly affecting the air quality in Indonesia caused by the exhaust emissions of gas, which affects the level of air pollution such as Carbon Monoxide (CO), Hydro Carbon (HC), Nitrogen Dioxide (NOx) and Sulfur Dioxide (SO_2) which has a devastating effect on the health of the human body and erodes the ozone layer present in the atmosphere.

There are many technologies developed to reduce the number of air pollution sourced from vehicle exhaust gases. So far the use of Catalytic Converter is the best way to control vehicle exhaust emissions [2]. This allows the reduction of CO, HC, NOx and SO_2 levels. Catalytic converters are believed to be able to meet future emissions requirements and emissions regulations [3, 4].

Catalytic converter serves to accelerate the oxidation of HC and CO emissions, as well as reducing NOx. In addition, the installation of a catalytic converter may alter harmful pollutants such as CO, HC, and NOx into harmless gases, such as Carbon dioxide (CO_2), water vapor (H_2O) and nitrogen (N_2) by chemical reactions. The conversion of these harmful pollutants is reflected in the following reaction:

$$CO \rightarrow CO_2 \tag{1}$$

$$HC \rightarrow H_2O + CO_2 \tag{2}$$

$$NOx \rightarrow N_2 + O_2 \tag{3}$$

At the equation of reaction 1 and 2 occurs oxidation reaction (addition of oxygen), while in reaction number 3 requires the release of oxygen (reduction). If the carbon in the fuel burns perfectly, there will be a reaction that produces:

$$C + O_2 \rightarrow CO_2 \tag{4}$$

IOP Conf. Series: Materials Science and Engineering **674** (2019) 012060 doi:10.1088/1757-899X/674/1/012060

If the air oxygen element is inadequate, combustion is not perfect so that the carbon in the fuel burns by the process as follows:

$$C + \tfrac{1}{2}O_2 \rightarrow CO \quad (5)$$

Materials used as catalytic converters are generally precious metals such as Palladium, Platinum, and Rhodium which are very expensive in the market. But the development of other types of materials has been done is copper coated with manganese. The result, capable of increasing the ability of carbon monoxide exhaust gas reduction [5]. In addition, Catalytic Converter made from aluminum oxide nanoparticles (Al_2O_3) as a catalyst has also been developed and found that the conversion efficiency of Al_2O_3-based catalytic converter is 99.5%, 92% for CO and HC emissions [6].

Purpose of Catalytic Converter made of a copper plate can also reduce CO_2 emissions of CO, HC, and CO_2 with a fairly good percentage of decrease [7]. Testing of catalytic converter made from copper oxide has also been done on diesel engines, resulting in the decay of HC value of 32%, NOx (61%) and CO by 21% [8].

The performance of the catalytic converter also depends heavily on the design and shape. This will affect the exhaust gas distribution. Simulation of conical pipe cone catalytic converter has been done. As a result, the incoming channel to the catalytic converter in the form of a cone pipe with a 300 angle is capable of significantly improving performance catalytic converter [9]. This paper will look at the performance of a copper-based catalytic converter with a honeycomb shape.

2. Experimental and Method

To conduct research, experimental equipment has been designed and developed to compare motorcycle exhaust emissions with the installation of catalytic converters made of copper with honeycomb form on vehicle exhaust and without catalytic converter (standard exhaust). The catalytic converter is installed in the exhaust and then connected to an SY-GA 401 gas analyzer to measure CO, HC, and CO2 levels.

Experiments conducted using a motorcycle with a cylinder capacity of 125 CC assembly year 2015 with a 4-stroke engine, data retrieval will be done on 3 revolutions of different machines i.e. at 2000 rpm, 3000 rpm, and 4000 rpm. The fuel used is the type of fuel gas with octane content (RON) 88. Test method is done by connecting between the exhaust and gas analyzer by using probe cable. Then start the engine with 2000 rpm rotation, 3000 rpm and 4000 rpm each for 2 minutes. Then performed data recording on the gas analyzer. The test equipment is shown in Figure 1.

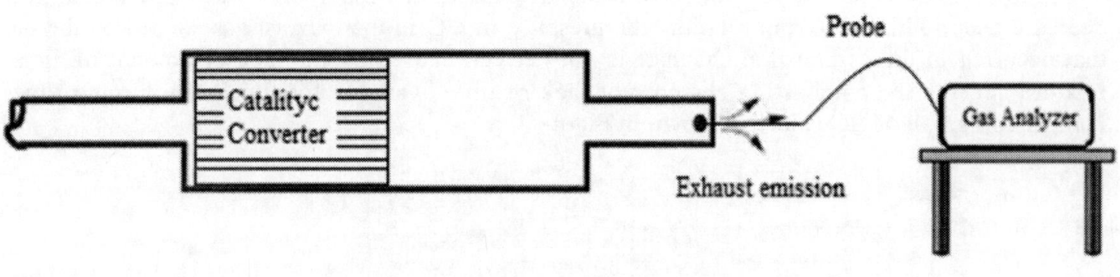

.**Figure 1.** Experimental Apparatus

Catalytic converter diameter outer circle 8 cm with a diameter of each hole 5mm and distance between hole 3 mm. Figure 2 shows the surface shape of the catalytic converter to be used in the test.

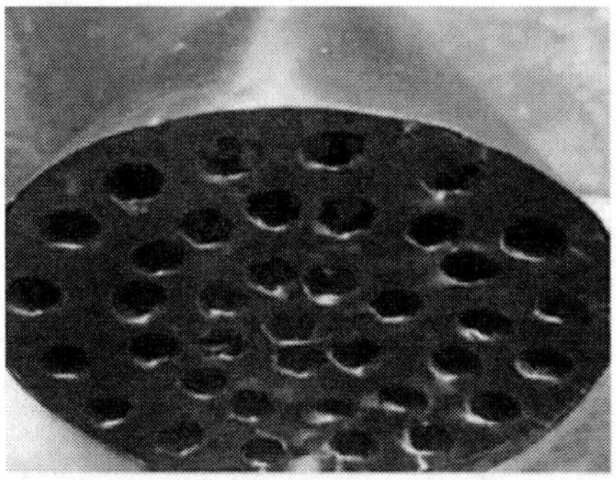

Figure 2. Catalytic converter surface honeycomb shape

3. Result and Discussions

Based on the results of testing at 2000 rpm engine rotation, 3000 rpm and 4000 rpm has been done, exhaust emission levels produced by the exhaust using catalytic converter is lower than the standard exhaust (without catalytic converter). At 2000 rpm engine speed there is a decrease in CO content of 0.18%, 3000 rpm engine speed of 1.19% and at 4000 rpm engine son of 1.76%. For the value of HC at 2000 rpm rotation, no impairment occurs, at 3000 rpm engine speed and 4000 rpm HC content has decrease respectively 4 ppm and 73 ppm. CO2 levels on exhaust emissions also decreased, at 2000 rpm engine speed decreased by 3.3%, decrease in CO_2 levels also occurred at engine speed 3000 rpm and 4000 rpm with a value of 1.6% and 1.7% respectively. This proves that the application of copper material catalytic converter can accelerate chemical reactions that occur in vehicle exhaust so as to reduce exhaust emission levels. The test results are shown in Table 1.

Table 1. Test Results

Engine Speed (rpm)	Value CO (%)		Value HC (ppm)		Value CO_2 (%)	
	Catalytic Converter	Standard	Catalytic Converter	Standard	Catalytic Converter	Standard
2000	3.59	3.77	165	165	5.1	8.4
3000	4.49	5.67	237	241	5.4	7.0
4000	4.01	5.77	187	260	4.6	6.3

Figure 3 shows that the overall CO value decreased in each round variation. the largest CO decrease occurred in 2000 rpm rotation, the presence of CO in the exhaust gas caused by the burning that occurred in the combustion chamber is not perfect, caused by the lack of amount of air in that mixture get into the combustion chamber or lack of time available for finish the burning time. The comparison graph of HC values is shown in Figure 4.

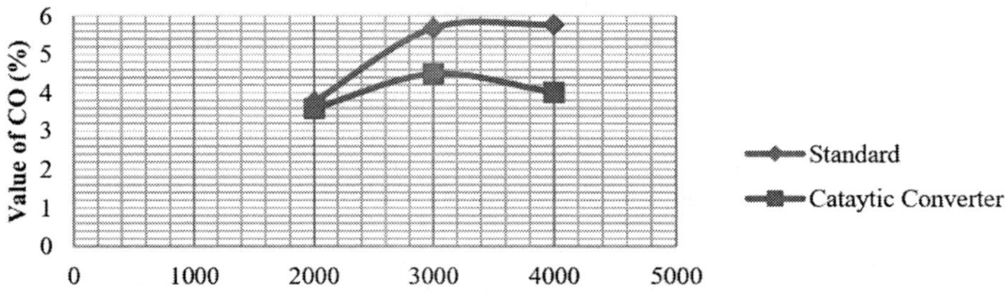

Figure 3. The CO of value comparison graph uses a catalytic converter and standard exhaust

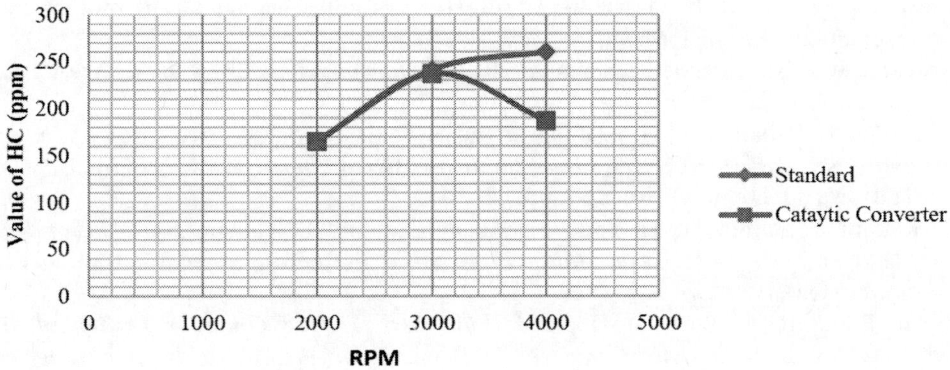

Figure 4. The HC of value comparison graph uses the catalytic converter and standard exhaust

From figure 4 above, the use of copper material as a catalyst can reduce the value of HC released from the exhaust. The test results show that overall HC emissions decrease in concentration in every variation of rotation. The presence of HC is caused by unburned fuel and exit into raw gas so that the fuel is split because the heat reaction turns into another HC cluster that comes out with the exhaust gas. The comparison graph of H_2O values is shown in Figure 5.

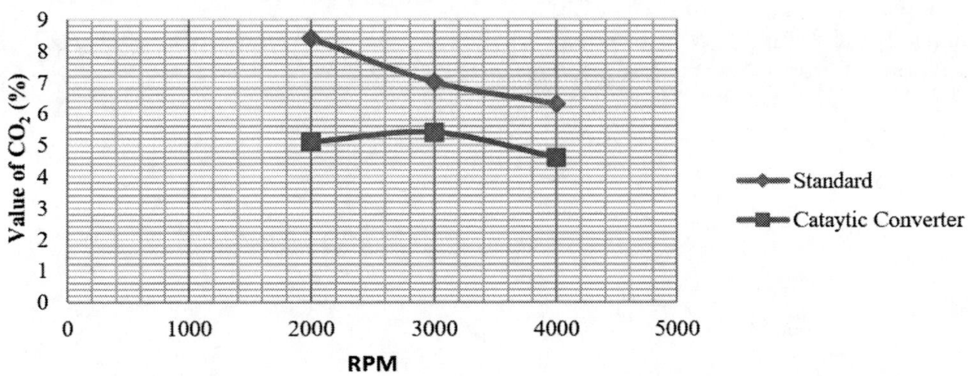

Figure 5. CO_2 of value comparison graph uses the catalytic converter and standard exhaust

Figure 5 shows that the influence of the Catalytic converter on CO2 emissions can oxidize well; this is due to copper increase the surface area of the catalyst. The results of the study showed a reduction in exhaust emission levels on the value of CO, HC, and CO2. The reduction of the value is caused by the installation of catalytic converter of copper material with the form of a honeycomb as a catalyst. Copper catalysts can increase the rate of reaction without permanently altering the chemical elements, and may work to form intermediates or adsorb the reacted substances.

4. Conclusion

Application of copper catalytic converter with honeycomb surface shape is able to reduce exhaust gas emission levels on motorcycles. The highest decrease in CO value was at 4000 rpm with a difference of 1.76%, for the highest decrease in HC value occurred at 4000 rpm with 73 ppm. While the decline for the highest CO2 value occurred at 2000 rpm engine speed of 3.3%. The decline in the value of the exhaust gases will help and reduce air pollution from motorcycle exhaust gases.

IOP Conf. Series: Materials Science and Engineering **674** (2019) 012060 doi:10.1088/1757-899X/674/1/012060

References

[1] https://www.bps.go.id/statictable/2017/11/23/1981/produksi-kenderaan-bermotor-dalam-negeri-unit-2000-2016.html accessed on May 24, 2018

[2] Pardiwala Julie M., Femina P., Sanjay P 2011 *Review paper on Catalytic Converter for Automotive Exhaust Emission* (Ahmedabad : Institute Of Technology, Nirma Unversity)

[3] R. Zhang., A. Villanueva., H. Alamdari., and S. Kaliaguine 2006 *Reduction of NO by CO over nanoscale LaCo1−xCuxO3 and LaMn1-xCuxO3 perovskites*, (Journal of Molecular Catalysis A: Chemical), vol. 258, pp. 22–34

[4] H. Iwakuni, Y. Shinmyou, H. Yano, H. Matsumoto, and T. Ishihara 2007 *Direct decompostion of NO into N2 and O2 on BaMnO3 – based perovskites oxides* (Applied Catalysis B: Environmental) vol. 74

[5] Irawan Bagus R.M., Purwanto P., and Hadiyanto H 2015 *Optimum Design of Manganese-Coated Copper Catalytic Converter to Reduce Carbon Monoxide Emissions on Gasoline Motor* (Elsevier) 23 p 86-92.

[6] Aalam C Syed., Saravanan Dr. C.G., and Samath C Mohamed 2015 *Reduction of Diesel Engine Emissions Using Catalytic Converter with Nano Aluminium Oxide Catalyst* (International Journal For Research In Emerging Science And Technology) Vol 2

[7] Maulana Imam 2018 *Analisis Catalytic Converter Dengan Bahan Tembaga Terhadap Emisi Gas Buang Pada Sepeda Motor 125cc* (Medan: University of Muhammadiyah Sumatera Utara)

[8] Venkatesan S.P., Udai Desai Shubham., and Heman Borana Karan., Goud Kagita Rajarshi Kushwanth., Kumar G Lakshmana., Kumar K Pavan 2017 *IOP Conference Series: Material Science and Engineering* **197** 012026

[9] Zainal Nurul Amina., Azmi Ezzatul Farhain., and Samad Mohd Arifin 2018 *IOP Conference Series: Material Science and Engineering* **328** 012029

IOP Conf. Series: Materials Science and Engineering **674** (2019) 012061 doi:10.1088/1757-899X/674/1/012061

Design and generating energy as a car alternator to be an alternative electricity

S Lubis[1,*] **and Cholish**[2]

[1] Department Mechanical Engineering University of Muhammadiyah Sumatera Utara Jl. Kapten Mukhtar Basri,BA No. 3 Postcode: 20238, Medan, Indonesia
[2] Departement Electrical Engineering University of Muhammadiyah Sumatera Utara Jl. Kapten Mukhtar Basri, BA No. 3 Postcode: 20238, Medan, Indonesia

* Email: sudirmanlubis@umsu.ac.id

Abstract. The alternator is a critical component. The car alternator issued AC electric power by utilizing wind power rotation is then converted into DC electrical power supplied to the accumulator before. Therefore, it is not required of operators to operate the working of the alternator. The voltage generated cells in a standard battery is 2V / cell. However, the working voltage is higher. As is known, in order to drain current through a battery, alternator voltage should be higher than the voltage of the batteries themselves. On the other hand must be maintained so that the voltage is not too high to prevent boiling of (gas) from the batteries. As the voltage regulator safe at 200 ° C should be maintained to 2.35 V / cell until 2.4 V / cell. For 12V battery it means that the working voltage of 14.1 V-14.

Keywords: Alternator, accumulator, Voltage

1. Introduction

The power plant can be Replaces like; power generation driven by solar energy, ocean wave energy and wind energy, which are still developed in limited companies. While the power plant can not be Replaces like; Hydroelectric Power Plant (HEPP), Diesel Power Plant (diesel), Gas Power Plant (power plant), Nuclear Power Plant (NPP), and so forth. In fear of this energy progressively reduced. Has done a lot of other possibilities utilization of natural resources, and everything possible could be used to generate electricity. Fulfilments of electric energy in remote areas, not reached by the grid.

Suitable electrical energy and the most efficient is the alternative power generation and solar power generation. This is supported by the location of the state of Indonesia located in the area equator enable the utilization of solar energy to be converted into electrical energy because solar rays shine all year round. From the description above, push research to try to exploit the car's alternator instead of a generator at power plant alternatives. Through this research will reveal how to use and performance of the car's alternator existing power plants. With the above-mentioned reasons, Researchers took the title of this research is "Design of Alternator Car for Alternative Electrical Energy Generation". Where electricity is produced from electrical energy conversion process that is commonly used is the engine AC generator, which is the main driving force can be manifold turbine engine, a diesel engine, or propeller engine. In the operation of power plants with a generator, because the reliability fluctuations, load size, then have two or more generators are operated with continuous duty, reserve, and turns to the generator.

2. Literature Review

2.1. Alternator

Car alternator (Figure 1) is a generator that functions as an electric energy supply for electric car needs such as lighting, indicator lights, ignition, fuel injection and electrical equipment other. The alternator has simple construction; the car alternator is some advantages when compared with another electric machine. The advantage is the alternator has not there are sparks between the brush and slip ring brush, as there is no commutator which is effect brushes become worn. More easy rotor and resistant to high rotation, and silicon diode.

Figure 1. Alternator

2.1.1. Rotor

Rotor (Figure 2) serves to generate a magnetic field. The rotor rotates along the shaft because the movement is called an alternator with a rotating magnetic field. The rotor consists of a core pole (pole cores), field coils, slip ring, shaft, and others. Shaped like a claw pole core and a field coil contained therein.

Figure 2. Rotor Alternator

Rotor serves to generate a magnetic field. The rotor rotates along the shaft because the movement is called an alternator with a rotating magnetic field. The rotor consists of: a core pole (pole cores), field coils, slip ring, shaft, and others. Shaped like a claw pole core and a field coil contained therein.

2.1.2. Slip ring or sliding ring

This tool from brass or copper which is mounted on the shaft by means of insulating material. This Slip ring rotates together with the shaft (as) and the rotor. Many slip ring No 2 and on each slip ring can shift borstal positive and negative, in order to strengthen Excitation Current to the winding magnet on the rotor.

2.1.3. Diodes (Rectifier)

In the alternator diode or rectifier (Figure 3), the diode has a function to convert the alternating current (AC) produced by the stator coil into direct current (Direct Current). Characteristics of a diode that can only be fed by a current in one direction. So the diode can be used as a current rectifier. In the conventional type alternator, there are six diodes, three diodes can enter so-called positive diode, and three other diodes is negative.

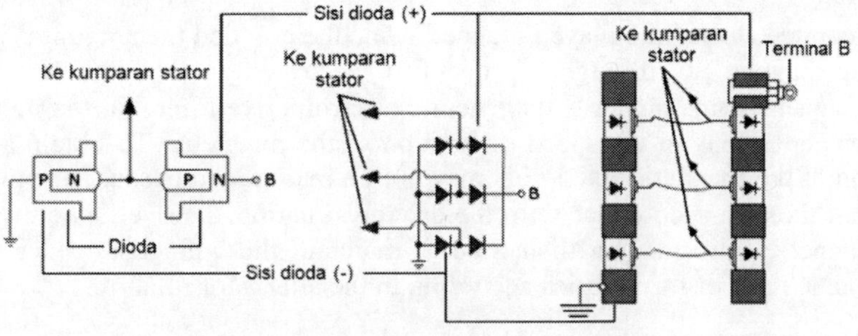

Figure 3. Diode (Rectifier) Alternator

As we can see in the Figure 3 on the diode circuit alternator, Then it appears that there are two diodes connected in series so that there are three pairs of diodes connected in series. One leg of the diode anode leg on the side of the negative diode coupled to each other and connected with the masses, while the cathode leg on the side of the positive diodes are connected together and also connected to the terminal B. The tip - stator coil is connected with the middle between the pair diode connected in series. It is the nature of the diode used for rectification. In fact function on the direction of P to N, when the voltage is less than a certain value, then the current can not flow. In silicon diodes, these prices usually range between 0.6 to 0.7 volt. When current is already flowing, it will continue to increase in size despite voltage changes almost nothing. The relationship between voltage and current vary, depending on the surrounding temperature. When the temperature rises, the current flows more easily.

2.2. *Working Principle Alternator*

At the time of the magnet (rotor) rotates in the stator coil will arise in the development of tension among both ends of the coil, will provide alternating current raise to. The relationship between the current generated in the coil with a magnet position is as shown in Figure 4. The highest flow will rise at the poles N and S reaches a distance that is closest to the coil. However every half rotation to flow in the opposite direction. Flow to shape sine wave called the "current phase frequently". Amendment 360 on the graph applies to one cycle and the number changes every second that happened called "frequency".

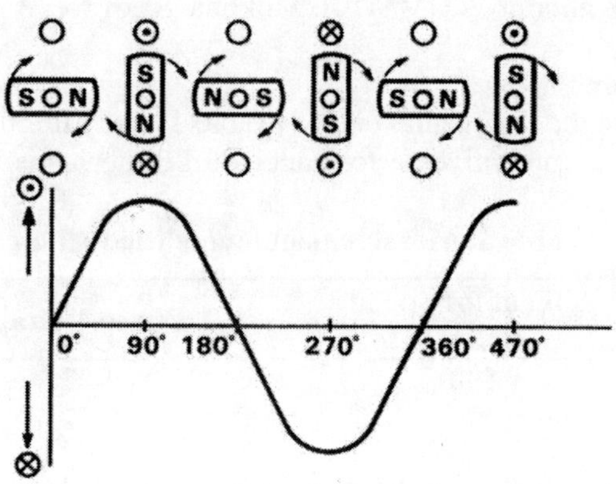

Figure 4. Sine Wave Generation of alternating current Phase One

Each coil A, B, and C is 120 °. At the time of the rotating magnetic among them, will rise alternating current at each coil. Figure 2.6 shows the relationship between the three alternating current with a magnet. The electricity has three alternating current is called "alternating current three-phase",

IOP Conf. Series: Materials Science and Engineering **674** (2019) 012061 doi:10.1088/1757-899X/674/1/012061

the car's alternator generate an alternating current three-phase. Typically, the components - automobile electrical components using electrical voltage 12 or 24 volt and the alternator's charging system should produce the voltage.

Electricity generated at the time played magnet inside the coil and the amount depends on the speed of rotation of the magnet. So, through a process of electro-magnetic induction, the faster the coil cutting lines of magnetic force bigger coils generate electromotive force. Furthermore, we can see that the voltage varies depending on the speed of rotation of the magnetic. To obtain a fixed voltage, the necessary rotation of the magnetic fixed, this may not be maintained because the machine will spin at speeds that are not fixed in accordance with the driving situation. To overcome this difficulty, in lieu of permanent magnet electro-magnet then used to maintain the voltage in order to remain. Electro magnet, the magnetic lines of force varies according to the alternator rotation.

3. Method

The method used in this research was started from collecting data, tool design and tool manufacture, and analyze. Measurements were made to get the output voltage and current that occurs in the equipment.

3.1. Research methods

The method to be performed in this study are:

1) The measurement results in the form of lists and charts of the gauges and that data can be stored in a computer.
2) The method used in the study with the data value of voltage and current produced equipment and the speed of the alternator.
3) From the measurement results will be seen the value of power measured mainly reactive power and power factor.
4) Analysis of the results of measurements and testing with the load.

3.2. Study Design The Forum

Draft measurement is made by measuring directly on the input and output current output instrument measuring instrument that has been prepared and measurement techniques that do are: 1) measuring voltage and current on tools Alternator; and 2) measuring at normal temperature Alternator tool.

3.3. Implementation of Research

Tests conducted on two conditions, namely cycles before and after the test load and before, housed in the Faculty of Engineering laboratory UMSU Jl. Mukhtar Basri No. 3 Medan.

4. Resultas and Discussion

The results of this study are the test results before (Table 1) and after the modified alternator (Table 2). From the data obtained the comparative performance of the alternator.

Table 1. Measurement unmodified Alternator

No.	Alternator speed (RPM)	1 Phase Voltage (Volt)
1	500	2.07
2	600	2,49
3	700	2.90
4	800	3.32
5	900	3.74
6	1000	4.15
7	1100	4.57

Table 1 shows that the measurement results with a wide variety of rotation speed of the alternator is followed by the results of different voltages. This indicates that the voltage is proportional to the rotational speed alternator.

Table 2. Measurement After Alternator modified

No.	Alternator speed (RPM)	1 Phase Voltage (Volt)
1	500	8.15
2	600	9.03
3	700	10.48
4	800	11.36
5	900	12.23
6	1000	13.69
7	1100	14.56

Based on the measurement results 1 phase output voltage of the alternator (Table 2), there are a wide variety followed by increasing the output voltage. On lap 500 rpm is the lowest voltage at the time before the modified, amounting to 2.07 V. The highest were at the time after modified, amounting to 8.11 V. The results of the measurement of the wind turbine and alternator after modifications in the field can be seen in Table 3.

Table 3. Measurement of the wind turbine and alternator field

Hour	Wind Speed (m / s)	DC voltage (Volt)
07:00	1.9	2.07
08:00	2.6	3.32
09:00	3.5	3.74
10:00	3.5	3.74
11:00	4.0	8.15
12:00	5.8	10.15
13:00	6.2	12.23
14:00	5.9	10.48
15:00	4.3	11.36
16:00	4.1	8.25
17:00	2.9	3.50
18:00	3.8	4.55

Based on the results of field measurements of the alternator output voltage depending on the speed of the wind that blows in the area Alue Naga wide variety followed by increasing the output voltage of the alternator. At 07:00 the wind speed of 1.9 m / s alternator voltage of 3.59 V. And high output voltage ranging from 11.00 in the amount of 14.11 V. The output voltage alternator achieved the highest time at 12.00 to 15.00 in the amount of 21, 20 V. this indicates that the speed increases as 11.00 till 16.00, so between 11.00 till 16.00 alternator power plants work efficiently. The alternator output voltage is still shaped AC voltage (AC) and should be rectified by the rectifier.

5. Conclusions

Horinzontal type multiblade wind turbine is suitable for areas that have average wind speed is low. The diameter of the copper wire in the stator with a smaller size will get a larger number of windings than before the alternator is modified, ie from 36 to 108 coil windings on the stator coil. Wind power generation by using horizontal type multiblade windmills that are placed on the shore Alue Naga

IOP Conf. Series: Materials Science and Engineering **674** (2019) 012061 doi:10.1088/1757-899X/674/1/012061

works efficiently at 11.00 till 16.00. For further research can be measured by using wind energy. Also recommended for further research to analyze alternative methods of generating wind energy.

References

[1] Ilhamd Fabillo, 2012, "Perancangan Pembangkit Listrik Tenaga Angin Skala Kecil (100va)" Proyek Akhir, Universitas Pendidikan Indonesia.

[2] Adityo Putranto. 2011, "Rancang Bangun Turbin Angin Vertikal Untuk Penerangan Rumah Tangga" Universitas Diponegoro.

[3] Situngkir, P. Putra S, 2011, "Rancang Bangun dan Uji Eksperimental Pengaruh Profil Dan Jumlah Sudu Pada Variasi Kecepatan Angin Terhadap Daya Dan Putaran Turbin Angin Savonius Menggunakan Sudut Pengarah Dengan Luas Sapuan Rotor 0,9 m2", Universitas Sumatra utara.

[4] Setiono,puji, 2006, "Pemanfaatan Alternator Mobil Sebagai Pembangkit Listrik Tenaga Angin", Universitas Negeri Malang.

[5] Habibie Ilham, 2012, "Perancangan Ulang Alternator Mobil Menjadi Generator Sinkron Kecepatan Rendah", Fakultas Teknik Universitas Syiah Kuala Darussalam, Banda.

An effect of different spark plug used and additional ethanol on engine performance and exhaust gas emission

Sarjito*, **I Prasetyo, T W B Riyadi, A D Anggono and M Effendy**

Department of Mechanical engineering, Faculty of Engineering, Universitas Muhammadiyah Surakarta, Jl. Ahmad Yani Tromol Pos I Pabelan Kartasura, 57162, Indonesia

* Email: sarjito@ums.ac.id

Abstract. The existence of fossil fuels is increasingly scarce nowadays. Therefore, finding alternative fuels to substitute petroleum fuels, especially gasoline that very huge used for a motor vehicle. One of the alternative fuels as a substitute for gasoline is bioethanol (BE). The aim of the research work was to analyze engine performance and exhaust emission level of vehicle using two different types of spark plug. Further investigation was addressed by varying additional bioethanol fuel made from raw material of cassava. The work was carried out experimentally both in the laboratory and engine test drive. The results show that by using platinum spark plug resulting better performance of engine. The additional bioethanol on fuel, the power, and torque generated from the engine increases with the increase of percent bioethanol in the fuel mixture. The fuel consumption is more efficient by using fuel of 30% BE, compared with engine operation using fuel of 100% pertalite. While the results of the exhaust emission on the vehicle show that by using fuel bioethanol, exhaust emissions decrease as the increase of bioethanol in fuel mixture than exhaust emissions generated when using fuel 100% pertalite. It was proved by statistical data analysis using a one-way ANOVA Varian (F test) for power, torque, and HC emission levels, found of F value less than F table. Therefore hypothesis null (Ho) was accepted. Further result in each variation of fuel there is no significant difference. However, for CO, found that counted F greater than F table, then Ho was rejected; therefore the result of each variation of fuel there is significantly different.

Keywords: Bioethanol, Cassava, Engine Performance, Emission.

1. Introduction

The scarce of fossil fuel is considerable interest to discuss recently. Its effect is really wide in every sector of life. The very close sector and really affected is in the transportation and industrial sector. The fluctuation supply and fuel price should make us aware that the stock of fuel on the earth is getting low day has consisted. Since fossil fuel is nonrenewable energy, then it encourages us to find that alternative. Actually, in Indonesia, there is a lot of renewable sources of energy, such as biodiesel from a plant of Ricinus *communis*, palm oil, and soy. Another alternative fuel is methanol and ethanol from biomass, sugar cane, corn, etc., which can be used as substitution of gasoline [1].

Other than that, its fossil fuel burning is giving a negative impact on the environment. Air quality that decreases day to day due to the burning smoke from fossil fuel is one of the effects that we can see clearly, the effect of greenhouse generated by gas of CO_2 resulted from the combustion of fossil

IOP Conf. Series: Materials Science and Engineering **674** (2019) 012062 doi:10.1088/1757-899X/674/1/012062

fuel. As we can see combustion of un-perfect fossil fuel will produce CO_2 that gradually increases in atmosphere. The radiation of sun spread to the earth should have been reflected back to space, but, the accumulation of its CO_2 will obstruct this reflection. As a result, radiation will be back absorbed by earth that finally increases the temperature of air on the earth [1].

Another CO_2 contributor is fossil combustion. The use of fossil fuel is getting increase significantly since industry revolution in 18th century. In that century, coal became a dominant source of energy than substituted by fossil fuel in the middle of 19th century. The two main sources of CO_2 emission globally were 2 types. Firstly, the electric generator coal-powered. Secondly is the combustion of motor vehicle engine. The emission of greenhouse effect should be reduced so that we should construct industrial systems and transportation that not depend on fossil fuel and coal. Therefore, to overcome this problem requires source of alternative energy which is reduce the use of fossil fuel, whilst it could reduce CO_2 emission. One of the energy sources in conjunction with reducing fossil fuel is fuel from vegetative namely [2].

Alcohol is produced from plants using microorganisms through fermentation process [2]. Introducing its alternative energy is also a struggle in reducing the use of fossil fuel in Indonesia and in the world typically. The form of energy source quite interesting to be developed since the availability in Indonesia and its properties are renewable. There are three categories of ethanol sources, which are liquid sugar cane, jell, and material matrix or lignocelluloses. All that material is relatively easy to be found and to be developed in Indonesia with very huge land area.

Therefore, probably become the most prospective alternative fuel and there are many studies addressed for developing such fuel. It is having some advantages compared with fossil fuel that could be directly mixture in the fuel tank, injected to combustion chamber and burned to reduce emission of exhaust gas. Such material gathered from renewable sources that unlimited in the form of tree that can be grew up perfectly, or biomass that contained sugar inside, jell or cellulose. By mixing and fossil fuel-based gasoline engines could help to enlarge supply of fuel, ensure safety deeply in supply of fuel widely, increase the efficiency of combustion and reduce hydrocarbon (HC) and emission carbon monoxide (CO). Other than that, the higher and gasoline mixing produce of good mixing, and the perfect combustion achieved. Therefore, reduce of environment problems, increase of agricultural economic growth and avoid of dependency to nations of producer of fossil fuel [3].

2. Literature Review

The work on utilization of as mixing of fuel was carried out by Mohsen Ghazikhani et.al. [4], on their research, an experimental study was done in conjunction with gas temperature effect and emission level on exhaust gas for two-stroke gasoline engine using gasoline and mixed with ethanol. The result showed that efficiency of combustion increase along with the increase of ethanol percentage on gasoline, since, properties of ethanol is easy evaporated hence overlapping in combustion chamber better and the dominant result in using ethanol was in reducing pollution significantly emitted from the engine which reduces CO by 35%, and that it the biggest reducing among other pollutants. Also, most of the emission increase with the increase of exhaust gas temperature, yet, hydrocarbon (HCs) on average decrease by 30% on increasing exhaust gas temperature.

Xiaochen Wang et.al. [5], on their work characterization of combustion, was done and exhaust gas emission were tested using fuel of E10W, E10 and pure gasoline (EO) with different load of engine of 20 Nm, 50 Nm, and 100 Nm, as well as in their experiment was done in constant engine revolution of 2000 rpm. The result of the experiment showed that compare with E0, E10W indicates greater than pick pressure on the cylinder at high load, also, the increase of pick released heat investigated for fuel of E10W at all operating conditions. The use of E10W increase of NO_x at about high load. However, at low load, E10W could reduce HC, CO, and CO_2 significantly. Compared with E10, E10W indicated pick of pressure greater than in the cylinder. Besides that, reducing emission of NOx revealed for E10W from 5 Nm to 100 Nm, meanwhile emission of HC, CO, and CO_2 slightly higher than small load and middle load.

Paolo Iodice et al. [6], carried out of a quite simple testing emission on a motorcycle at engine operation that has not reached work temperature by using gasoline fuel with ethanol. In their research, percentage of ethanol used was 10%, 20%, and 30% also, the emission measured was only CO and HC. The result shows that emission of CO and HC on the engine in cool conditions reduce with the use of ethanol-gasoline compared with the used of unleaded fuel, also, factor of emission level on cool engine that quantitatively as a function of ethanol percentage on fuel.

Bambang Sudarmanta et al. [7], undertaken of research experimentally to compare of compression ratio effect and time of ignition to ensure engine performance using fuel of E50. Result of their research showed that the increase of compression ratio increase engine performance by using fuel of E50, for an ignition time increase gradually from 18° before top death centre (BTDC) on engine revolution of 2000 rpm to 26° BTDC at engine revolution of 5000 revolution per minutes (rpm) by using fuel of E50. Then, by adding a compression ratio on the fuel of E50 reduce of specific fuel consumption by 13.42 % and increase thermal efficiency by 14.67 %.

Yanuandri Putrasari et.al. [8], carried out of investigation of mixing diesel fuel and ethanol of E2.5%, E5%, E7.5%, E10% and the experiment was done in vary of engine load of 0, 10, 20, 30, 40, 50, 60 Nm to ensure changing of parameter on diesel engine which was fuel consumption, thermal efficiency, exhaust gas temperature and lubricant temperature, meanwhile, characteristic of exhaust gas emission, the measured parameter was CO and HC. The result of the research that engine power indicated an increase in average effective pressure due to the increase of ethanol percentage, for specific fuel consumption reduced with the increase of ethanol percentage, then, reduce exhaust gas temperature and increase oil temperature that suitable with the increase of ethanol percentage on fuel. Also, the result of exhaust gas emission, characteristics of CO, and HC was decreased due to the increase of ethanol percentage.

This ethanol resulted from glucose fermentation followed by the distillation process. Ethanol stands for ethyl alcohol (C2H5OH); often called grain alcohol or alcohol. The feature of ethanol is un-colored fluids, easily evaporated and unique smelted. The density of ethanol is 0.7939 g/mL, and its boiling point of 78.32°C at 766 mmHg. Another characteristic of it is emulated in water and ether, as well as having heating value of 7093.72 kkal. Ethanol is used in vary of industry such as raw material in industry derivative from alcohol, mixing for a soft drink, mixing for sake or gin, pharmacy material and cosmetics, and mixing of vehicle fuel, increasing octane number, and gasoline alcohol (gasohol) [9].

The use of ethanol as an energy source in industry and vehicle will be reduced by CO_2 percentage that affected global warming. Quick or slow, fossil fuel sources (fossil fuel) will be running out as the limited deposits. Fossil fuel as an energy source that is nonrenewable, its limitation encourages industrial state glance ethanol (biofuel) as an energy alternative source. Other than that, continuously could be produced by microorganism, ethanol also environmentally friendly [10].

Some of the benefits of using ethanol as fuel are 1.) produced from renewable plants. 2.) Containing oxygen of about 35% therefore perfectly flammable. 3.) The use of gasohol reduces emission greenhouse. 4.) Combustion does not produce lead and benzene which having carcinogenic behavior (stimulate cancer). 5.) Reduce emission fine-particulates that dangerous for the health of humans. 6.) Easily solvable in water and not contaminant to water and soil.

The distillation process produced ethanol of 95%, for using as fuel requires much purer and reaching of 99.5% which is often to called Fuel Grade Ethanol (FGE). Considered the use of ethanol which varies, then, ethanol percentage which is utilizing should be different depending on the usage. Ethanol having 90-96.5% could be used in industry, where, ethanol with 96-99.5% can be used for much mixing soft drinks and pharmacy industry. Ethanol is used as fuel mixing for a vehicle that should be really dry and anhydrous in order to a corrosive, hence, ethanol should have percentage of 99.5-100%. The difference in percentage will be influencing to carbohydrate treatment process become glucose dissipated in water [11]. Some references the above were not clearly mentioned, from what the gasohol was made. In this work, investigation was focused in distinguish engine performance operate using two different spark plugs of standard and platinum. Moreover, the most unique of the

IOP Conf. Series: Materials Science and Engineering **674** (2019) 012062 doi:10.1088/1757-899X/674/1/012062

used in this work was explicitly made from cassava and also was made by our team, from the beginning till the final process. Further work carried out by forthcoming team as continuation research using biethanol made from durian seeds has been initiated in separate themes.

3. Methodology

The research methodology was constructed consecutively, start from gathering information, collecting tools and materials, followed by biethanol production, performance test of engine, and analysis of data are described as the following section.

3.1. Tools and materials

The tools prepared for this work was; 1.) Four stoke 110 cc gasoline engines 2.) Dynotest/Dynamometer. 3.) Emission Gas Analyzer, 4.) Tachometer 5.) Measurement glass.
Whilst, materials used in this research work are; 1.) Pertalite RON 90, 2.) FGE 99.7%, 3.) Standard spark plug and 4.) Platinum spark plug

3.2. Steps of production

The production was done through some steps of; (a) raw material preparation, cassava was destroyed using coconut solvent tools, then added of water 1.5 x weight of raw material to be boiled (b) liquidation process, where structure of powder/jell of cassava was broken chemically become a complex sugar with additional Enzyme Alfa Amylase (c) Next step was scarification process, where complex sugar was broken again to be a simple glucose with lower percentage of (10-12%) with additional Enzyme Gluco Amylase (d) Fermentation, in this step where flour has been changed to be simple glucose for differentiation in order to produce liquid of ethanol/alcohol with percentage of (2-5%) by adding ferment, this process occurred about 5-7 days in closed container (e) Distillation was done to separate alcohol from liquid result of fermentation. In the distillation process, at temperature of 78°C (equivalent with boiling point of alcohol) will be evaporating first than water having boiling point of 100°C. Then, the steam of at distillatory tool will be flowing to condenser then the steam to be condensate to be fluid of alcohol 70-95% (f) Dehydration, in purification ethanol 70-95% with some method, one of the method is chemically using limestone, that physically by absorption process using Zeolite Sintesis. The dehydration result is in the form of ethanol 99.6-99.8 % that is it so-called Fuel Grade Ethanol (FGE) or dry ethanol. Finally, its appropriate to be used as mixing fluid to the fuel.

3.3. Engine performance test

The testing of the engine is aimed to know optimal engine performance from a variety of spark plug both standard and platinum as well as by varying percentage of mixed to the fuel compared with the use of purely pertalite fuel. The parameter measured will cover power, torque, specific fuel consumption and emission of exhaust gas, when fuel was vary mixed with of BE 10%, BE 20% and BE 30% using the optimum spark plug.

3.4. Data analysis

The research work was analyzed using two methods; the first method was using analysis descriptive data. The second one uses statistical data analysis. Where, data logged was made in table, and presented graphically then, compared with statistic data analysis in vary of mixing levels of fuel 10%, 20%, and 30%. The statistical method consisted of t-test (double regression), One-Sample T-Test and Varian test of one way ANOVA (F-test) using SPSS software. Where, data found from the testing process inputted to SPSS program to be processed computerized to ensure the result whether the differences were significant or not for entire parameter investigated.

4. Result and Discussion

4.1. In the first step, the study was focused to investigate engine operation in difference type of spark plug.
In this work, two types of spark plug used were the type of standard and type of platinum. The typical parameter was power, torque, and specific fuel consumption operated in vary of engine speed to find the best performance as reference to the continuation investigation, as the following section.

4.1.1. Correlation between of power to engine speed using different spark plug and purely 100% pertalite.
It can be seen from Figure 1 that power indicated of increase in power using two different spark plugs when engine was operated using fuel 100% pertalite. The result shows that maximum power has resulted when the engine was operated using platinum spark plug of 7.14 HP at engine speed of 8900 rpm that relatively greater than that maximum power of engine when it is operated using standard spark plug of 6.84 HP at an engine speed of 9000 rpm. Using statistical data analysis proved by t-test (double regression), found that counted-T > table-T or (3,087 > 2,228). That way, Ho was refused. It's indicated that the platinum spark plug was significantly influenced by the power generated. It can be concluded that graphically, optimum power was on engine operated using platinum spark plug.

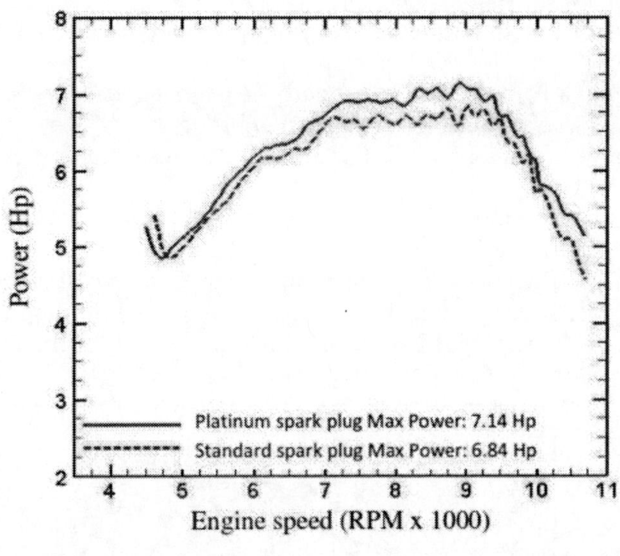

Figure 1. Correlation of power to an engine speed of two different types of spark plug

4.1.2. Correlation of torque to engine speed using different spark plug and purely 100% pertalite.
The comparison result of investigation torque to engine speed using different spark plug at engine speed of 4500 rpm resulted in the same torque of 6,20 Nm as shown in Figure 2, therefore, it can be underlined that graphically, torque was slightly increase using platinum spark plug compared with performance using standard spark plug. Further analysis was carried out using t-test (double regression), its resulting counted-T < table- T of (0,110 < 2,228) and counted-T<table-T, of (-0,932 < 2,228), therefore Ho was rejected. Hence, the result both of standard spark plug and platinum was not significantly influenced to the torque resulting. Overall, the result of using purely pertalite using different spark plugs proved that platinum spark plug more optimum than standard spark plug

particularly on power resulted. Therefore, platinum spark plug was kept for the forthcoming investigation of engine with different percentages of fuel mixing with ethanol.

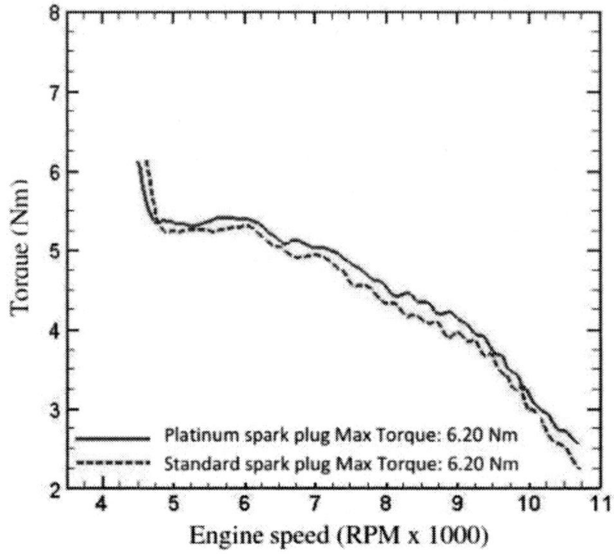

Figure 4.2. Correlation of torque to an engine speed of two different types of spark plug

4.2. The second step study was focused to investigate the engine operation in different percentages of Ethanol added and engine operated using platinum spark plug.
Result of investigation power and torque to engine speed are presented as the following section:

4.2.1. Correlation of power to engine speed using different percent of BE and platinum spark plug.
The investigation result, indicated in Figure 3 showed that maximum power resulted from an engine that is using fuel with 30% BE compared with resulted power using BE 20%, BE 10% and Pertalite 100%. Graphically, the increase of power due to the increase of BE to the fuel. The maximum power was resulted by engine at the use of BE of 30% which is 7,47 HP at the engine revolution of 8700 rpm.

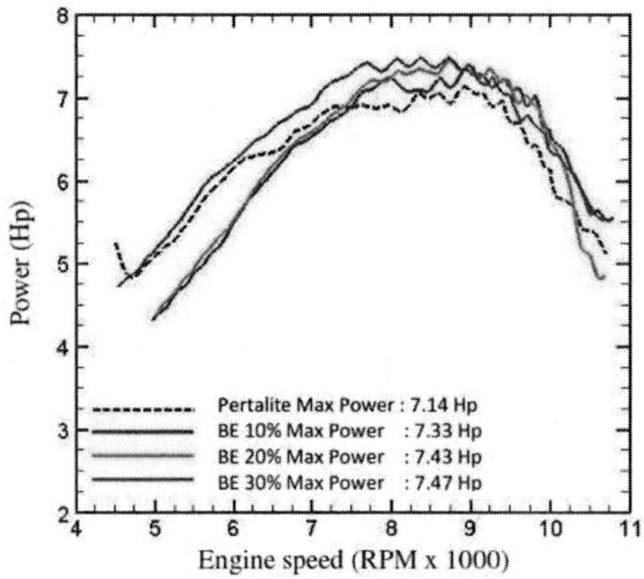

Figure 3. Correlation of power to an engine speed of 3 different % BE using platinum spark plug

From the result presented the above, it concluded that graphically, optimal power occurred at fuel with BE of 30% as that optimum power was increased. Investigation results of torque using vary of fuel with platinum spark plug can be seen in the following section.

4.2.2. Correlation of torque to engine speed using different percent of BE and platinum spark plug.
It was proved that from Figure 4 the resulting torque test by varying found that maximum torque resulted in fuel with BE of 30% showing from increasing torque compared with torque resulted by fuel with BE of 20% and BE 10%. From the above graph, the torque increase along with the percentage of mixing level. Maximum torque resulted in the use of BE of 30% was 5.51 Nm at engine revolution of 5700 rpm, that graphically, torque result slightly increase compared with the use of pertalite of 100%, yet, maximum torque resulted using fuel with BE of 30%, BE of 20% and BE of 10% slightly lower than maximum torque resulted using fuel pertalite 100% of 6,20 Nm.

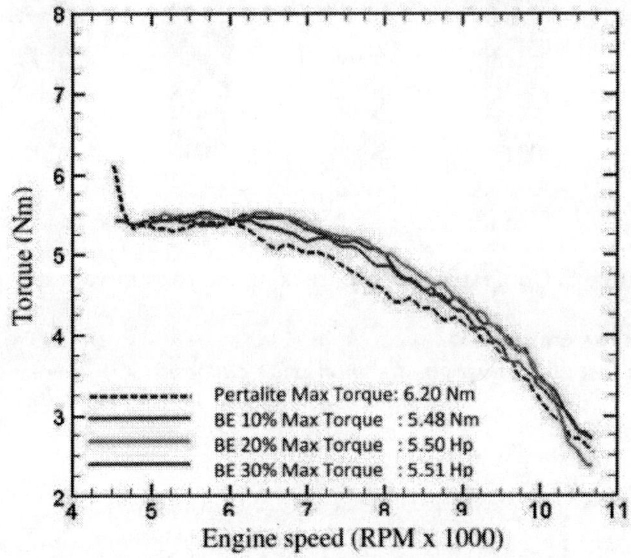

Figure 4. Correlation of torque to an engine speed of 3 different % BE using a platinum spark plug

The differences were not significant which is not point both of maximum power and maximum torque resulted from vary of mixing fuel with , and proved by statistical data analysis with one way ANOVA (F-test), resulted counted F-test < table-F or (0.119 < 2.82), which mean that Ho accepted, hence, there is no significant difference of power resulted by varying fuel mixing, meanwhile, for the torque resulting counted-F < table-F or (0,340 < 2,82), then Ho was accepted, and H_1 refused, it mean that there is not significantly different on torque resulted by varying fuel mixing. Thus, it can be concluded from statistical data analysis there is no significant difference both of power and torque resulted in varying fuel- mixing.

4.3. Result of specific fuel consumption
The result of the specific fuel consumption test, the investigation was carried out by operating a vehicle at the same condition, and the comparison of test distance to vary of mixing fuel is presented in Figure 5.

Based on a test result that fuel consumption by varying fuel mixture found that the use of fuel variation of BE 30% resulting much saving of fuel in test distance much longer. It is graphically with vary of fuel at difference testing distance (farer) along with compositional fuel mixing of pertalite much higher, yet, fuel consumption using BE 10% and BE 20% the testing distance getting lower (closer) compared with the use of 100% pertalite. Meanwhile, the use of fuel BE 30%, the testing distance was farer of 33,8 km compared with the use of 100% pertalite of 31,9 km with discrepancy of 1.9 km indicated that using BE 30% much saver by 5.9% compared with using 100% pertalite. However, using analysis data statistics using one-sample t-test, the value of T-test < table-T or (-413.211 < 3.182), therefore, Ho accepted; hence, average value testing distance using vary of fuel are not the same. It can be concluded that statistically, the average testing distance achieved by using vary of fuel is not the same. Thus, the use of BE 30% in testing uses distance farer compared with testing distance using 100% pertalite, BE 10% and BE 20% that indicated that the use of BE 30% are saver.

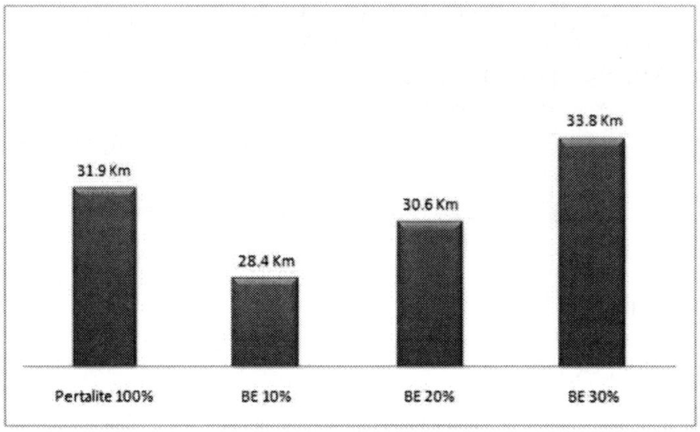

Figure 5. Comparison of the track on the road in vary of fuel mixing

4.4. Result of exhaust gas emission

The result of emission test et exhaust gas for monoxide carbon (CO) in vary of fuel mixing is shown in Figure 6,

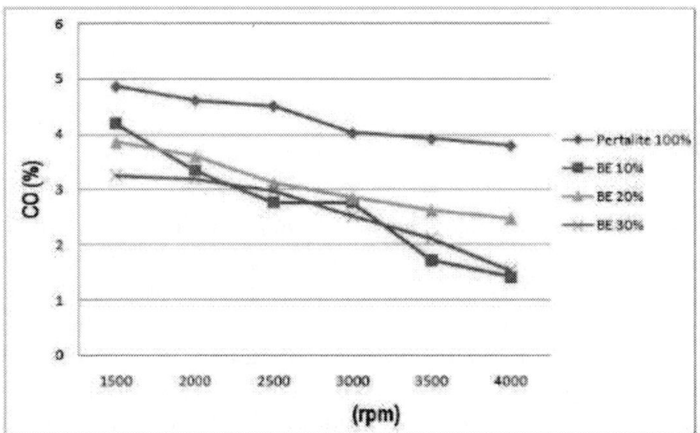

Figure 6. Comparison of CO emission to rpm using vary of fuel

The test of CO emission that CO found using fuel BE of 30% is lower compared with using 100% pertalite, and it can be seen on the engine revolution of 1500 rpm. The result using 100% Pertalite was 4,856% volume, and BE 10% of 4,181% volume, as well as BE 20% of 3,850% volume and BE 30% of 3,238% volume. It was occurred reducing CO of significantly, and proved statistically using one way ANOVA (F-test), of counted-F >table-F or (7,285 > 3,10), therefore, H1 was accepted, hence, there is significant difference emission of CO resulted by varying fuel mixing. It can be concluded that statistically CO resulted by using vary of fuels are not same. Thus, the use of fuel BE 30% the CO was less compared with the use of 100% pertalite, BE 10% and BE 20% that indicated that the use of BE 30% reduces CO significantly.

The testing result of the emission of exhaust gas of hydrocarbon (HC) is shown in Figure 7.

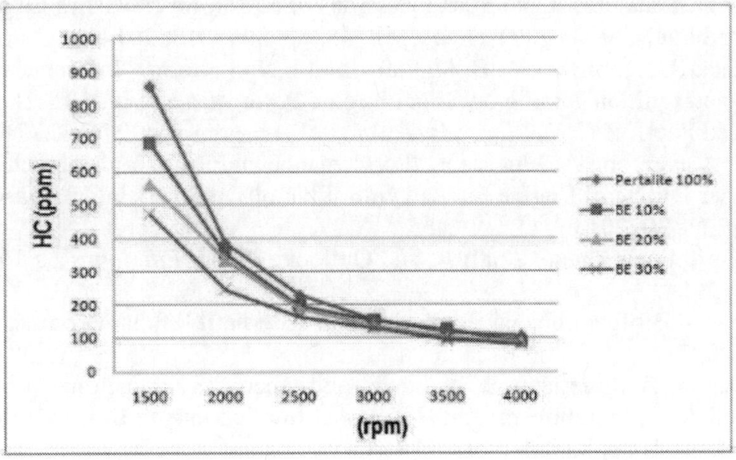

Figure 7. Comparison of HC emission to rpm using vary of fuel

The investigation result of HC found that the use of fuel BE 30%, resulting in emission of HC lower than the use of 100% pertalite at engine revolution of 1500 rpm. The result using100% Pertalite 100% is 858 ppm, BE 10% of 684 ppm, BE 20% of 563 ppm and BE 30% of 472 ppm, occurred at significantly reduce of HC, however, along with increasing rpm particularly at rpm of 2500 and more, the graph shows that changing not significantly, it almost say as same though using of fuel, such as engine revolution 3000 rpm resulting using 100% Pertalite of 153 ppm, BE 10% of 152 ppm, BE 20% of 136 ppm and BE 30% of 132 ppm consecutively, and there is no significantly different, and it proved that statistically using one way ANOVA (F-test), the counted-F < table F, (0,275 < 3,10), therefore, Ho was accepted, and there is no significantly different on HC emission resulted by varying fuel. Thus, it can be concluded that statistically, HC using vary of fuel was not giving changing on the graph significantly, especially at the engine revolution of above 2500 rpm.

5. Conclusions

The experimental test of gasoline engine was carried out in investigating the effect on different spark plugs used and the use of platinum spark plug type was found to be the optimum performance of the engine. Further testing was focused on distinguishing a vary of mixing levels of fuel. The investigation was followed by identifying exhaust gas emission of CO, CO_2, and HC. Moreover, the conclusion of the experimental result was done analytically and statistically.

References

[1] S. U. Handayani, "Pemanfaatan Bio Ethanol Sebagai Bahan Bakar Pengganti Bensin," *Gema Teknol.*, vol. 15, no. 2, pp. 99–102, 2006.

[2] Y. Indartono, "Bioethanol, Alternatif Energi Terbarukan: Kajian Prestasi Mesin dan Implementasi di Lapangan," *Fis. LIPI*, 2005.

[3] Y. Zhang, H. Zhao, M. Ojapah, and A. Cairns, "CAI Combustion of Gasoline and Its Mixture with Ethanol in a 2-Stroke Poppet Valve DI Gasoline Engine," *Fuel*, vol. 109, pp. 661–668, 2013.

[4] M. Ghazikhani, M. Hatami, B. Safari, and D. D. Ganji, "Experimental Investigation of Exhaust Temperature and Delivery Ratio Effect on Emissions and Performance of a Gasoline--Ethanol Two-Stroke Engine," *Case Stud. Therm. Eng.*, vol. 2, pp. 82–90, 2014.

[5] X. Wang, Z. Chen, J. Ni, S. Liu, and H. Zhou, "The Effects of Hhydrous Ethanol Gasoline on Combustion and Emission Characteristics of a Port Injection Gasoline Engine," *Case Stud. Therm. Eng.*, vol. 6, pp. 147–154, 2015.
[6] P. Iodice and A. Senatore, "Cold Start Emissions of a Motorcycle Using Ethanol-Gasoline Blended Fuels," *Energy Procedia*, vol. 45, pp. 809–818, 2014.
[7] B. Sudarmanta, B. Junipitoyo, A. B. K. Putra, and I. N. Sutantra, "Influence of the Compression Ratio and Ignition Timing on Sinjai Engine Performance with 50% Bioethanol-Gasoline Blended Fuel," *ARPN J. Eng. Appl. Sci.*, vol. 11, no. 4, pp. 2768–2774, 2016.
[8] Y. Putrasari, A. Nur, and A. Muharam, "Performance and Emission Characteristic on a Two Cylinder DI Diesel Engine Fuelled with Ethanol-Diesel Blends," *Energy Procedia*, vol. 32, pp. 21–30, 2013.
[9] C. Berg, "World Fuel Ethanol Analysis and Outlook; 2004," *Ediciones Le Monde Dipl. el Dipló*, 2004.
[10] I. Amri, "Dilema Biofuel Sebagai Sumber Energi Alternatif." Edisi Pertama. Kuala Lumpur. Hlm, 2007.
[11] P. J. Van Soest, J. B. Robertson, B. A. Lewis, and others, "Symposium: Carbohydrate Methodology, Metabolism, and Nutritional Implications in Dairy Cattle," *J. Dairy Sci.*, vol. 74, no. 10, pp. 3583–3597, 1991.

Acknowledgments
Authors would like to thank to all colleagues involved especially to co-authors for their valuable contributions to improve quality of this paper. The support obtained from Universitas Muhammadiyah Surakarta (UMS) for international publications is also gratefully acknowledged.

Convective heat transfer of nanofluids TiO_2/Thermo Oil XT 32 in concentric tube heat exchanger

B Suroso[1,*], S Kamal[2], B Kristiawan[3], H Irawansyah[4], B S Wibowo[5] and M Yani[1]

[1] Department of Mechanical Engineering, University of Muhammadiyah Sumatera Utara, Medan, Indonesia
[2] Department of Mechanical and Industrial Engineering, Gadjah Mada University, Yogyakarta, Indonesia
[3] Department of Mechanical Engineering, Sebelas Maret University, Surakarta, Indonesia
[4] Department of Mechanical Engineering, Lambung Mangkurat University, Banjar Baru, Indonesia
[5] Department of Mechanical Engineering, Bangka Belitung University, Bangka Indonesia

[*] Email: bektisuroso@umsu.ac.id

Abstract. This research was conducted to examine the heat transfer characteristic of nanofluids TiO_2/Thermo XT 32 oil on a concentric tube heat exchanger. TiO_2 nanoparticles with nominal diameter of 21 nm were dispersed in Thermo XT 32 oil as base fluid at three different particles concentration of 0.1%, 0.3%, and 0.5%, respectively. A straight horizontal stainless-steel tube of 1.2 m length, 0.012 m inner diameter, and 0.022 m outer diameter were used as a test section. Collecting data by varying the temperature of the heat transfer nanofluids and nanoparticle volume fraction. Nanofluids prepared using the method of two steps without the addition of a surfactant by using a magnetic stirrer for 1 hour then ultrasonic cleaner for 3 hours. The results of research conducted, the coefficient convection heat transfer of nanofluids TiO_2/ Thermo XT 32 oil increases along with the addition of fraction volumes; this is due to the increased thermal conductivity of the fluid as well with the rise of temperature against nanofluids TiO_2/Thermo oil XT 32. So it is very worthy of nanofluids applied in industries related to heat transfer processes.

1. Introduction

Heat transfer has an important role in many applications. Such is the Heat exchanger equipment using heat transfer principle. The development of science and technology is increasing rapidly, marked by the tendency of dimensions of a product that is getting smaller while increasing the ability of heat transfer. One way to increase heat displacement is fixing the properties of the fluid used. Generally the fluid used has low thermal conductivity. With added nanometer-sized solid particles are expected to improve the properties of the fluid.

Nanofluids are a mix of the cooling fluid with nano-sized solid particles or commonly called nanoparticles. Nanoparticle size ranging from 1 nm-100 nm. Choi and Eastman [1] introduced the term nanofluids in 1995. They investigated the addition of variations of carbon nanotube and metal particles of nano-sized who dispersed on several different fluids. The result is an increase in thermal conductivity value 10%-80% with the addition of nano-particles of 0.1% to 10%.

Based on the review conducted by the research Haddad, et al [2] more use of TiO_2 nano-particles with water base fluids, while the use of nano TiO_2 particles with fluid base oil is still a little bit. So the writers feel interested to conduct research using TiO_2 nanoparticles with Thermo oil base fluid XT 32. The selection of TiO_2 nanoparticles as fluid nanomaterials easily obtained due to the market and the prices are cheaper compared to relative nanoparticles. While the basic fluid used is Thermo XT 32 Oil, because the oil has the ability to work at high temperature.

Torii [3] research on convection heat transfer of nanodiamond particles suspension with water as the fluid essence. The results showed that the fluid viscosity nano is increasing along with the increased concentration of the nanoparticles. The value of the number concentration and large Reynolds nanoparticles strongly affect the magnitude of the value of the coefficient heat transfer, heat transfer coefficient increase with increasing concentrations of particle, and the value of the thermal conductivity of the fluid work.

Murshed et al. [4] conduct research on the characteristics of convection heat transfer of nanofluids TiO2/water under constant heat flux conditions. The results obtained that the fluid nano shows that there is an increase in convection heat transfer coefficient with increasing volumes of nanoparticles in aqueous solution. Duangthongsuk and Wongwises [5] measure a performance of heat transfer and pressure drop of nanofluids TiO2/air under turbulent flow conditions. The result value of thermal conductivity of nanofluids increases with increasing concentration of volume nanoparticles. The value of the thermal conductivity is higher than with the equation of validation, but still in the same trend. The value of the coefficient of heat transfer of nanofluid is higher compared to the fluid essence, As well as, increased Reynolds number and concentration of the used particle and ranges from 26% higher than the fluid essence.

Putra et al. [6] research on heat transfer of nanofluids convection TiO_2/ethylene glycol in a circular tube under constant heat flux conditions. The result is the value of the thermal conductivity of the fluid work experience increased from fluid base with the highest increase in variation of 1% (w/v) of 8.62%. The value of the heat transfer of nanofluids TiO2/EG as a working fluid in the heat exchanger can be seen from the Russell number value retrieved from the calculation. The value of the Reynolds numbers from 400 to 700 improved heat transfer. The highest increase was achieved at a concentration of 1% (w/v) of 30.55%.

Kristiawan et al. [7] measure the convective heat transfer of titanium (iv) oxide nanofluids under turbulent flow conditions. The results also revealed the drag reduction phenomenon in turbulent flow regime in which the friction factors of non-Newtonian nanofluids demonstrated lower trend than those of Newtonian fluid. In addition to the other mechanisms proposed by previous studies, this drag reduction phenomenon might be responsible for the heat transfer enhancement of nanofluids.

Irawansyah et al. [8] the results of the study concluded that the value of the thermal conductivity of nanofluids increases along with a rise in temperature and volume fraction. At the same temperature, increased concentration of the nanoparticles was able to increase the value of the thermal conductivity of the fluid. While at a fraction of the same volume, an increase in temperature also was able to increase the conductivity of nano-fluid. The increase in the value of the thermal conductivity of nanofluids high of approximately 12% compared with the substantially fluid obtained at a fraction of the volume of 1.5% with temperature plug 44.6 °C.

2. Experiment

2.1. Preparation of Nanofluids

The mixing of nanofluids in this research was done using a magnetic stirrer and ultrasonic vibration. Particles of nano TiO_2 is used to obtained from Sigma Aldrich Chemical Pvt Ltd. USA with the size of the diameter of 21 nm. The mixing of TiO_2 nanoparticles with Thermo XT 32 Oil without using surfactants, TiO_2 with concentrations of 0.1%, 0.3%, and 0.5% depressed against Thermo XT 32 Oil

IOP Conf. Series: Materials Science and Engineering **674** (2019) 012063 doi:10.1088/1757-899X/674/1/012063

with magnetic steering for 1 hour then continued by ultrasonic vibration for 3 hours, so there happen to clump (Figure 1).

(a) (b) (c) (d)

Figure 1. Preparation of Nanofluids Equipment; (a) Digital Scale, (b) Measuring Cup, (c) Magnetic Stirrer, (d) Ultrasonic Vibration

2.2. Experimental Setup

The experimental setup consists of a pump, cold water tank, hot nanofluids TiO_2 tank, water heater, concentric tube heat exchanger, manometer "U", flowmeter, thermocouples at the entry and the exit to measure the inlet and the outlet temperature. The flow of hot nanofluids goes into concentric tube heat exchanger is kept constant at the temperature of 60 °C, 80 °C, 100 °C, and 120 °C. In order to get accurate experimental results, first of all equipment is made stable by passing the water through the shell as well as the tube side area. To establish the laminar flow in the concentric tube, the flow rate of the hotter nanofluids TiO_2 is made constant as 2.0 LPM. Then the flow rate of the colder fluid is made constant as 2.0 LPM, by adjusting the flow meter. The schematic diagram of the experimental setup that we use is shown in Figure 2.

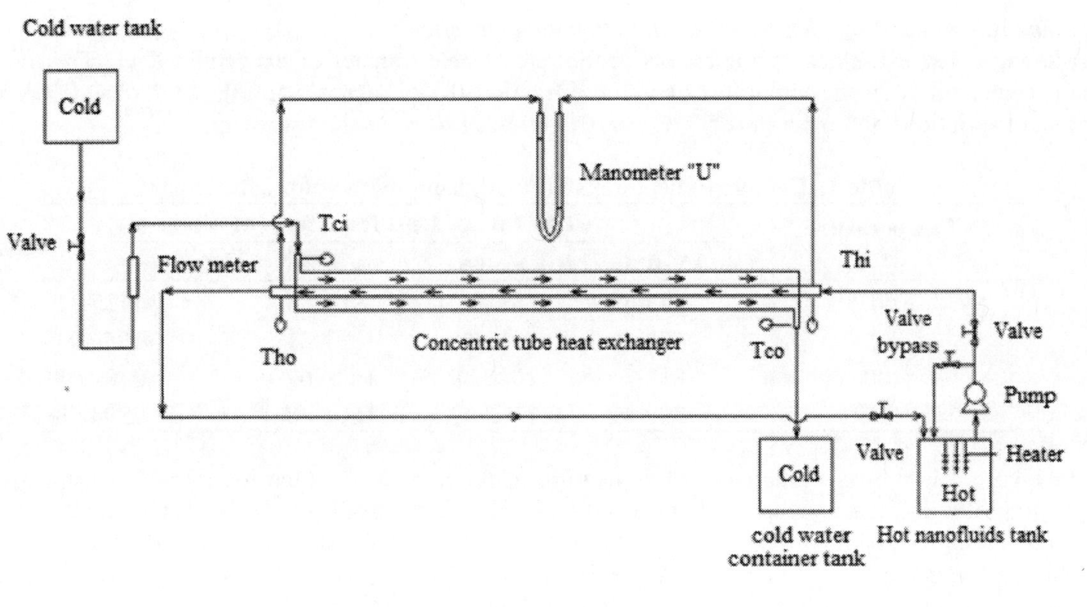

Figure 2. Schematic diagram of concentric tube heat exchanger

IOP Conf. Series: Materials Science and Engineering **674** (2019) 012063 doi:10.1088/1757-899X/674/1/012063

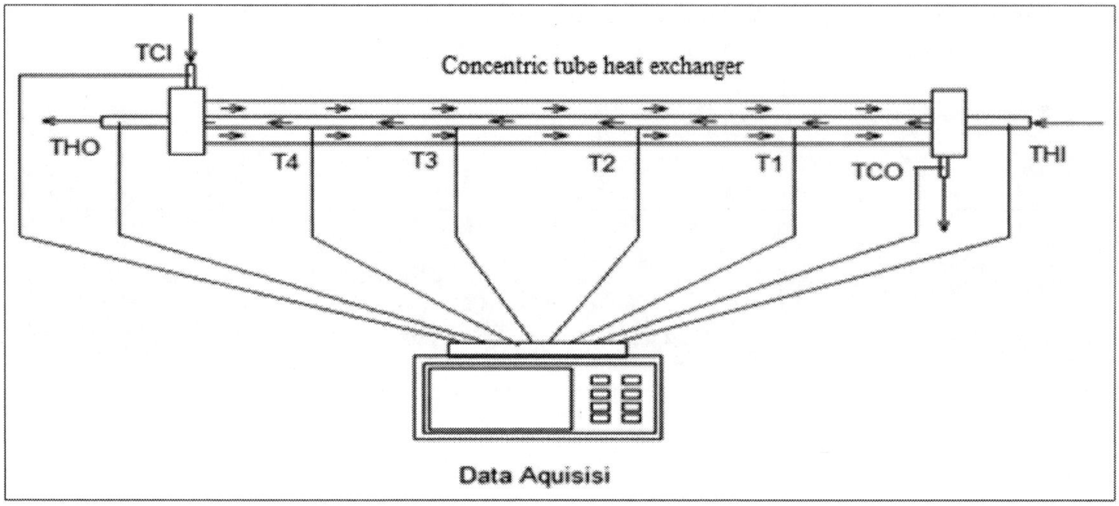

Figure 3. Schematic diagram data acquisition of temperature

2.3. *Experimental Procedure*

The equipment consists of 3 research systems, i.e. a system of measurement, the path system in pipe (inner tube), and the system of flow path on the annulus. The path is a path in the pipe is open. nanofluids heat in the tank, the pump is driven by the heat of nanofluids, nanofluids heat flows through concentric tube heat exchanger, and back to the tank hot nanofluids. Nanofluids heater controlled by thermos-controller to maintain a constant temperature in the tank hot nanofluids. The path of the flow in the annular path is open. The water in the tank is piped to the annulus heat exchanger, and it is coming out of annular immediately discarded.

3. Results And Discussion

3.1. *The rate of heat transfer relationship with volume fraction*

As shown in Table 1, showing the results of the rate of heat transfer of nanofluids TiO2/Thermo XT 32 oil (inner tube) on the variations of volume fraction 0 Vol% (basic fluid), 0.1 Vol%, 0.3 Vol%, 0.5% and variations of temperature 60 °C, 80 °C, 100 °C, 120 °C as the following.

Table 1. The rate of heat transfer relationship with volume fraction

Temperature (°C)	**The rate of heat transfer Q (W)**			
	0 Vol%	**0.1 Vol%**	**0.3 Vol%**	**0.5 Vol%**
60	443.69	635.15	906.83	1137.31
80	671.60	918.66	1381.81	1850.03
100	812.70	1080.39	1626.67	2184.05
120	999.49	1163.42	1801.48	2294.58

In Figure 4 is shown an increasing temperature of the nanofluids, then the rate of heat transfer is happening also increased. This is because the increased heat capacity base fluid Thermo XT 32 oil with the addition of nanoparticles. The highest rate of heat Transfer at a fraction of the volume of 0.5 Vol% and temperature of 120 ° C of 2294.58 W heat transfer rate and lowest at 0 volume Vol% fraction and temperature 120 ° C of 999.49 W or 129.57% fluid compared to his base.

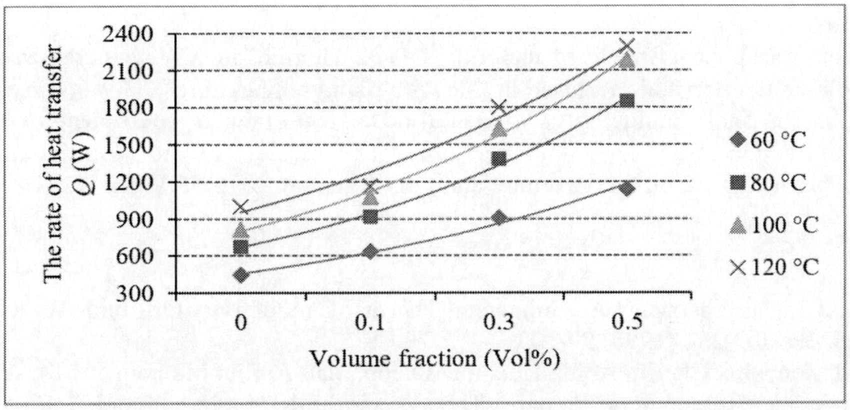

Figure 4. The rate of heat transfer relationship with volume fraction

3.2. *Convection heat transfer coefficient of relationship with volume fraction*

As shown in Table 1, showing the results of convection heat transfer coefficient of nanofluids TiO2/Thermo XT 32 oil (inner tube) on the variations of volume fraction 0 Vol% (basic fluid), 0.1 Vol%, 0.3 Vol%, 0.5% and variations of temperature 60 °C, 80 °C, 100 °C, 120 °C (Table 2).

Table 2. Convection heat transfer coefficient of relationship with volume fraction

Temperature (°C)	Convection heat transfer coefficient *h* (W/m².°C)			
	0 Vol%	0,1 Vol%	0,3 Vol%	0,5 Vol%
60	1315,55	1865,01	2972,22	4087,54
80	1347,17	1887,05	2995,47	4109,60
100	1367,98	1898,28	2976,52	4111,94
120	1398,12	1925,06	3001,32	4137,66

Based on Figure 4, the heat transfer coefficient convection in the pipeline within (inner tube) increases with the addition of nanoparticles against Thermo oil base fluid XT 32. So also with the influence of a rise in temperature of nanofluids. Heat transfer coefficient of convection on highest volume fraction of 0.5% and a temperature of 120 °C, i.e. 4137.66 W/m² (Figure 5). °C and the lowest at fraction 0% (Thermo XT 32 oil) and a temperature of 60 °C, i.e. 1315.55 W/m². °C. This is due to the increase in the rate of heat transfer of nanofluids.

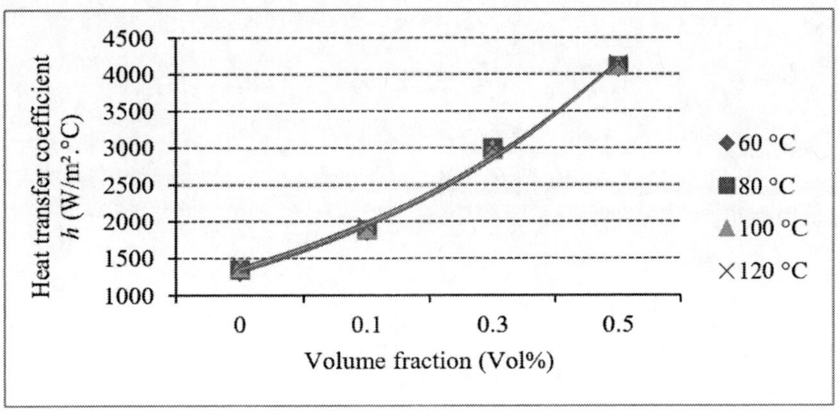

Figure 5. Convection heat transfer coefficient of relationship with volume fraction

IOP Conf. Series: Materials Science and Engineering **674** (2019) 012063 doi:10.1088/1757-899X/674/1/012063

4. Conclusion

Convection heat transfer coefficient of nanofluids TiO2/Thermo oil XT 32 increases along with the addition of volume fraction and increased in line with rising temperatures. This is caused by increased fluid properties (thermal conductivity) of nanofluids. Heat transfer coefficient of convection on highest volume fraction of 0.5% and a temperature of 120 °C, i.e. 4137.66 W/m².°C and the lowest at fraction 0% (Thermo XT 32 oil) and a temperature of 60 °C, i.e. 1315.55 W/m².°C.

References

[1] Choi S.U.S., dan Eastman J.A., Enhancing Thermal Conductivity Of Fluids With Nanoparticles, ASME FED 231 (1995) 99-105

[2] Haddad, Zoubida., Chérifa Abid., Hakan F. Oztop., dan Amina Mataoui., 2014, A Review on How the Researchers Prepare Their Nanofluids." International Journal of Thermal Sciences., 76, 168-189

[3] Torii, S., 2007, Experimental Study on Convective Heat Transfer of Aqueous Suspension of Nano-Diamond Particles, Proceedings of International Symposium on Ecotopia Science 2007, ISETS07

[4] Murshed, S. M. S., Leong, K. C., Yang, C. & Nguyen N. T., 2008, Convective Heat Transfer Characteristics of Aqueous TiO_2 Nanofluids Under Laminer Flow Condition, International Journal of Nanoscience, Vol, 7, No. 6. pp.325-331

[5] Daungthongsuk, W. & Wongwises, S., 2010, An Experimental Study on The Heat Transfer Performance and Pressure Drop of TiO_2-Water Nanofluids Flowing Under A Turbulent Flow Regime, International Journal Of Heat and Mass Transfer, Vol.53, pp. 334-334

[6] Putra, K. I. Kristiawan, B. Budiana, P. E, 2012, Heat transfer of nanofluids convection TiO2/ethylene glycol in circular tube under constant heat flux conditions, Mekanika, Vol. 10, No. 2, pp. 117-124

[7] Kristiawan, B. Kamal, S. Suhanan and Yanuar, 2015, Convective Heat Transfer Of Titanium (Iv) Oxide Nanofluids Under Turbulent Flow Condition, ARPN Journal of Engineering and Applied Sciences, Vol. 10, No. 10

[8] Irawansyah, H. Kamal, S, 2015, The study concluded that the value of the thermal conductivity of nanofluids increase along with a rise in temperature and volume fraction, Science And Engineering National vol. 1, pp. 42-47

IOP Conf. Series: Materials Science and Engineering **674** (2019) 012064 doi:10.1088/1757-899X/674/1/012064

Influence of tool speed on the friction stir welding joint of aluminium and steel with single weld line

B Sugito[1,*], A Hariyanto[1], A D Anggono[1], Subroto[1] and A S Darmawan[1]

[1]Department of Mechanical Engineering, Universitas Muhammadiyah Surakarta, Jl. A. Yani, Tromol Pos I Kartasura. 57162, Indonesia

[*] Email: bs110@ums.ac.id

Abstract. The objective on this research is to investigate the influence of tool rotation to the welding strength of Aluminium and Steel. AA1100 and ST37 were material selected for the specimens. The welding process was conducted by using Friction Stir Welding (FSW) method. The tensile strength and hardness of the specimens will be indicators of welding strength. FSW was carried out by using conventional milling machine with tool rotation of 800 rpm and 1250 rpm. Feed rate 10 mm/minute for all process. The tool tilt angle was 1° and 1.8 mm of depth plunge. By using 800 rpm, the average of Tensile Stress was delivered of 82.22 MPa and Strain of 1.09%. While for the use of 1250 rpm, Tensile Stress given 96.23 MPa and Strain 1.14%. The hardnes of HVN on the weld zone was 42.2 and 56.2 HVN for rotation of 800 and 1250 rpm respectively. Based on the Tensile Stress and the Hardness of the weld zone, the FSW process was influenced by tool rotation. The higher rotation given higher Tensile strength.

1. Introduction

In the manufacturing industry is often found the manufacture of products and components that required joining two or more of metals. Most of that can be found in the fields of automotive, shipping industries, aircraft, and construction. Welding is one of the most commonly used connecting methods for joining metal materials. Currently, the manufacturing industry develops welding methods to improve product quality and to decrease the production costs [1].

The metal welding process is classified into two groups, i.e. Liquid State Welding (LSW), and Solid State Welding (SSW). LSW is a metal welding process that is conducted by melting the material part to be connected. While SSW is a welding process performed on the solid phase of the material to be connected. One of the SSW methods is Friction Stir Welding (FSW). FSW is a process of connecting metal by using heat energy generated due to friction between tool and workpiece. This connection occurs due to stirring two-sided pieces of metal that begin to soften due to friction, but still in the solid state. This method was developed by Wayne Thomas in 1991 at TWI (The Welding Institute) the United States [2].

Many researchers have studied the effect of machine parameters on the mechanical properties of AC4CH material by using the friction stir welding method [3]–[5]. Variations of parameters used were round tool, tool shape, and slope tool [6]. The results show that the friction stir welding method can produce a good joining. Tensile strength could decrease by employing the constant feeding and increase tool rotation. The cylindrical pin was given lower tensile strength than the tapered and bullnose tool. The use of tool tilt angle more than one degree can decrease the tensile strength as well [7].

The FSW method can be applied to dissimiliar material between nonferro and ferro. For the example, joining between Aluminium alloy and Steel. The most important problen during FSW is the softening material in the weld nugget. Therefore, it cause to decrease the tensile strength of the specimen due to the recrystallization process in the nugget zone during the welding process.

2. Research Method

The materials used in this research are Aluminium alloy of AA1100 and ST37 Steel. The materials were a plate in 2 mm thickness. An FSW tool is used Bohler K100 steel with the diameter of shoulder 18 mm, pin diameter 4 mm and 1.7 mm of pin high as shown in Figure 1. The tool rotates counterclockwise direction and the tilt angle of a tool is 1°.

ST37 steel was installed in the advancing side during the welding process. The FSW process was performed by using a conventional milling machine with an adjustable head tool. The brand was Richon, X8140A type, and 3-380v 50Hz power. Table 1 shows the FSW conditions of each specimen.

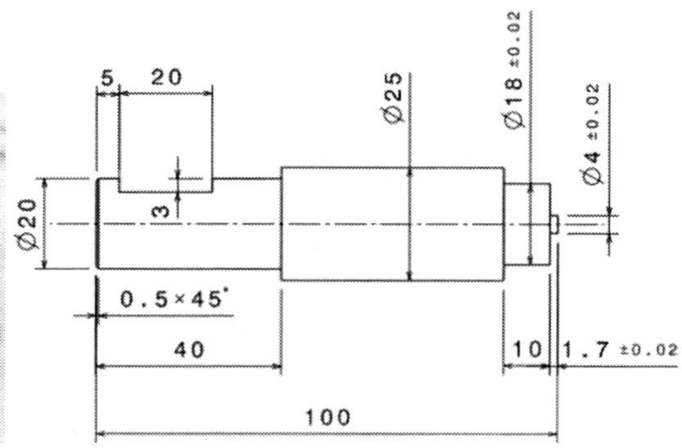

Figure 1. Dimensions of tool in FSW process.

Table 1. Parameters of stir welding.

Materials	Tool speed (rpm)	Welding speed (mm/mnt)	Tilt angle (deg)	Depth plunge (mm)
AA1100 x ST37	800	10	1	1.8
	1250	10	1	1.8

3. Results and Discussion

Figure 2 shows the appearance of the weld surface result of AA1100 and ST37. Figure 2(a) was the result of FSW by employing 800rpm of tool rotation. While the figure 2(b) is for the 1250 rpm. All specimens are then analyzed based on the welding surface visualization. The analysis was conducted to observe the weld surface contour, exit hole, and a defect that may occur during the welding process.

In the specimen, as shown in figure 2(a) and 2(b), it has shown a rough weld surface. Weld flash appeared as well in the weld line. It could be caused by reconstructed soft material (AA1100) out the surface around the tool shoulder [7]. The value of tool tilt angle more than one degree could influence the appeared of weld flash [5], [8].

The exit hole always appears in the friction stir welding as shown in figure 2(b) and 2(c). The hole was created by the pin tool that used in the FSW. That is one of the disadvantages of FSW. The hole can be eliminated by exiting tool after the outside of the material. In the exit hole zone, there is a line between material AA1100 and ST37. That indicates that the stirring was not conducted totally. The welding contour of all specimens was almost same between 800 rpm and 1250 rpm. The gap in the weld nugget was seen closed although the weld surface was not smooth.

IOP Conf. Series: Materials Science and Engineering **674** (2019) 012064 doi:10.1088/1757-899X/674/1/012064

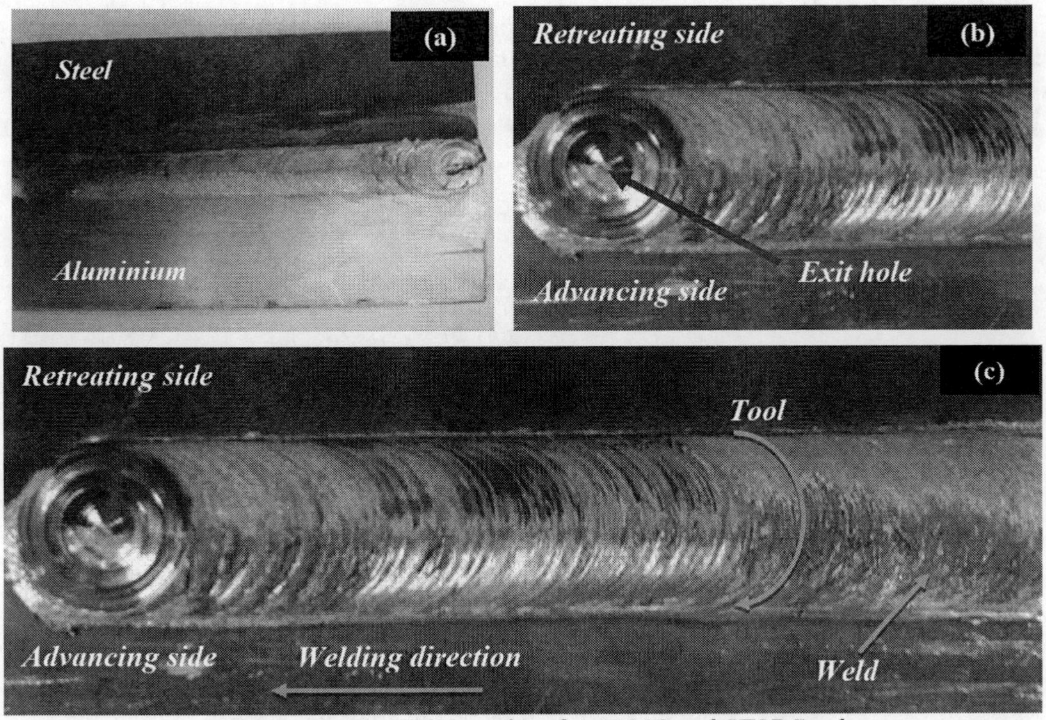

Figure 2. The FSW results of AA1100 and ST37 Steel.

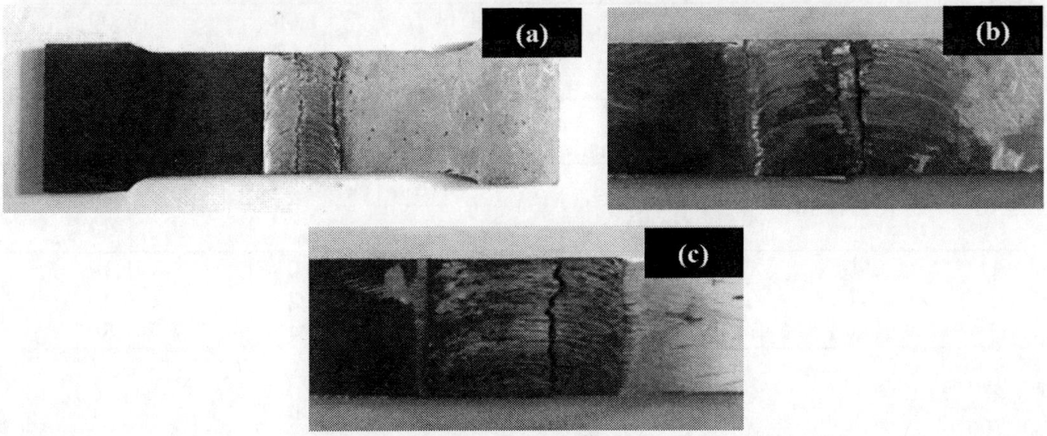

Figure 3. Tensile test specimen and results.

The FSW result is then cut for tensile and hardness test specimen. Figure 3(a) shows the specimen for the tensile test based on JIS Z 2201 standard. The broken zone was occured in the weld nugget for all specimens both 800 and 1250 rpm, as seen in figure 3(b) and 3(c). The broken weld nugget indicates that the welding has lower strength then the base AA1100. It also can be seen in the Table 2 to Table 5 about the results of tensile strength experiment. The experiments of tensile were conducted twice for every specimen. Then, the final value is the mean of two tensile stress. The values of tensile stress were below the tensile stress of base metal AA1100. It can be seen in the Figure 4. The maximum tensile stresses were 82.22 and 96.23 MPa for the use of tool rotation 800 and 1250 rpm respectively. The higher tool rotation can increase the tensile strength of FSW process. In this case, it can increase around 17% of tensile strength.

Table 4 and 5 shows the result of tensile strain for each specimen. All tensiles strains of the specimens were below the base material, AA110 and ST37. According to that results, it can be said than the specimens of FSW were brittle if compared with the base AA1100. The use of tool speed 800 and 1250 rpm is not an influence on the tensile strain significantly.

Table 2. Tensile stress of FSW in 800 rpm

No	Material	A_0 (mm²)	Stress (MPa)		Mean σ (MPa)
			σ_1	σ_2	
1.	Al-Fe	12	78,74	85,71	82,22
2.	ST37	12	270,49	365,53	318,01
3.	AA1100	12	163,48	154,31	158,89

Table 3. Tensile stress of FSW in 1250 rpm

No	Material	A_0 (mm²)	Stress (MPa)		Mean σ (MPa)
			σ1	σ2	
1.	Al-Fe	12	84,26	108,02	96,23
2.	ST37	12	270,49	346,24	303,36
3.	AA1100	12	136,96	149,13	143,04

Table 4. Tensile strain of FSW in 800 rpm

No	Material	L_0 (mm)	Strain (%)		Mean ε (%)
			ε1	ε2	
1.	Al-Fe	32	1,11	1,08	1,09
2.	ST37	32	5,72	9,53	7,62
3.	AA1100	32	4,19	2,13	3,16

Table 5. Tensile strain of FSW in 1250 rpm

No	Material	L_0 (mm)	Strain (%)		Mean ε (%)
			ε_1	ε_2	
1.	Al-Fe	32	1,19	1,10	1,14
2.	ST37	32	5,72	13,2	9,46
3.	AA1100	32	2,37	1,24	1,80

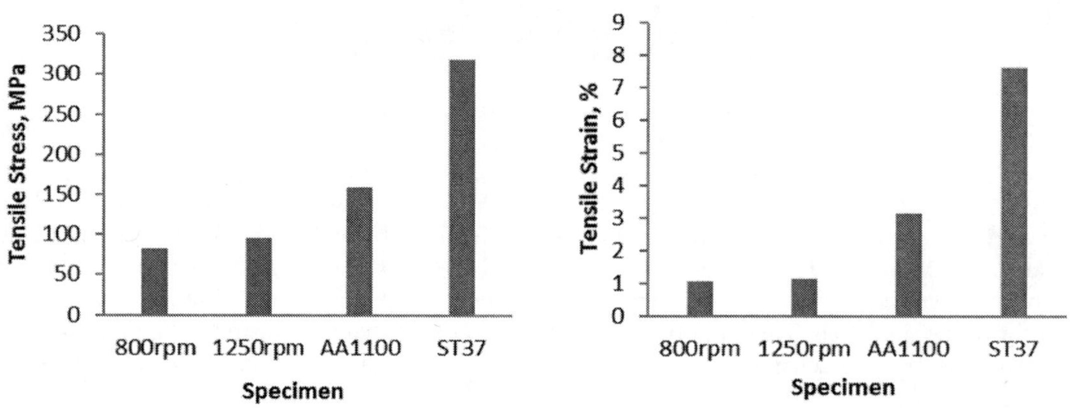

Figure 4. Tensile stress and strain FSW of AA1100-ST37.

Figure 5 shows the result of HVN hardness for the specimen of 800 and 1250 rpm. The use of higher tool rotation speed can increase the hardness of weld zone. The hardness of weld nugget was 42.2 HVN for the FSW with 800 rpm tool rotation. While the 1250 rpm can deliver 167.1 HVN in the weld line zone. The hardness can increase around 300%. It could be caused by the debris of ST37 distributed equally in stiring zone. But, it should be investigated by using scanning electron microscopy (SEM).

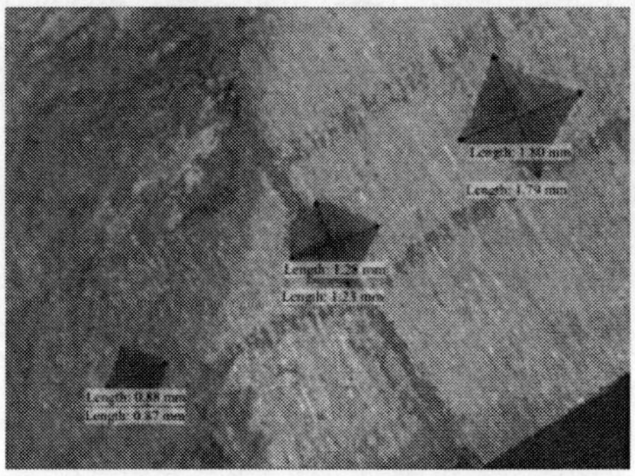

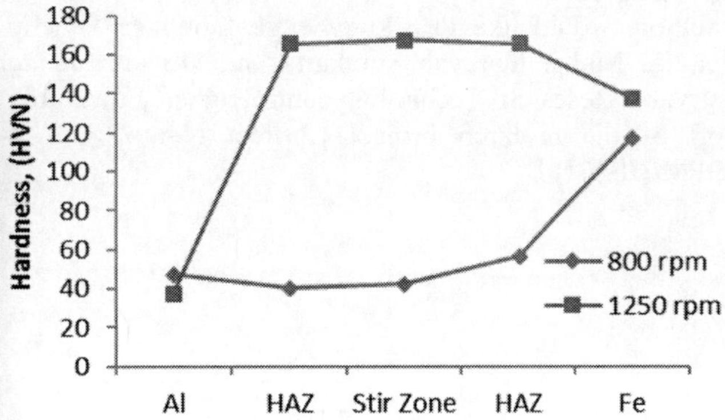

Figure 5. Hardness test of FSW of AA1100-ST37.

4. Conclusion

Welding of Aluminium alloy AA1100 and steel ST37 by using friction stir method was conducted successfully. Joining performance of FSW was influenced by tool rotation. The results were obtained as follows. From the tensile test, the higher tool rotation gives higher tensile stress and strain compared with the lower tool rotation. The hardness of stir zone or weld nugget was shown harder for the use of high rotation tool. Overall preperties, the use of 1250 rpm tool rotation was delivered better performace than the 800 rpm during FSW between AA100 and ST37.

References

[1] X. He, F. Gu, and A. Ball, “A review of numerical analysis of friction stir welding,” *Prog. Mater. Sci.*, vol. 65, pp. 1–66, 2014.

[2] W. M. Thomas, E. D. Nicholas, J. C. Need ham, M. G. Murch, P. Templesmith, and C. J. Dawes, “Friction stir butt welding,” *Int. Pat. Appl. no. PCT/GB92/02203 GB Pat. Appl. no. 9125978.8*, no. 6 December 1991, 1991.

[3] W. Y. Li, T. Fu, L. Htsch, J. Hilgert, F. F. Wang, J. F. dos Santos, and N. Huber, “Effects of tool rotational and welding speed on microstructure and mechanical properties of bobbin-tool friction-stir welded Mg AZ31,” *Mater. Des.*, vol. 64, pp. 714–720, 2014.

[4] X. Liu, S. Lan, and J. Ni, “Analysis of process parameters effects on friction stir welding of dissimilar aluminum alloy to advanced high strength steel,” *Mater. Des.*, vol. 59, pp. 50–62, 2014.

[5] T. Tanaka, T. Hirata, N. Shinomiya, and N. Shirakawa, “Analysis of material flow in the sheet forming of friction-stir welds on alloys of mild steel and aluminum,” *J. Mater. Process. Technol.*, vol. 226, pp. 115–124, 2015.

[6] A. D. Anggono, T. Widodo, B. Riyadi, D. Triyoko, and A. Hariyanto, “Influence of Tool Rotation and Welding Speed on The Friction Stir Welding of AA 1100 and AA 6061-T6,” vol. 20054, 2018.

[7] A. D. Anggono, B. Sugito, A. Hariyanto, Subroto, and Sarjito, “Mechanical Behaviour Investigation Of Aluminium Alloy Tailor Welded Blank Developed By Using Friction Stir

Welding Technique," pp. 2–11, 2017.

[8] B. Li, Z. Zhang, Y. Shen, W. Hu, and L. Luo, "Dissimilar friction stir welding of Ti-6Al-4V alloy and aluminum alloy employing a modified butt joint configuration: Influences of process variables on the weld interfaces and tensile properties," *Mater. Des.*, vol. 53, pp. 838–848, 2014.

Acknowledgement

The authors would like to acknowledge Lembaga Penelitian dan Pengabdian Masyarakat (LPPM) Universitas Muhammadiyah Surakarta and Direktur Jenderal Penguatan Riset dan Pengembangan, Ministry of Research Technology and Higher Education, Republic of Indonesia for the funding support of the research project [Project No: 72.16/A.3-III/LPPM/III/2018 and No: 72.99/A.3-III/LPPM/III/2018].

Strength of polymeric foam composite reinforced oil palm empty fruit bunch fiber subjected to impact load

M Yani[1,*], M A Siregar[2], B Suroso[3] and Arnita[4]

[1,2,3] Department of Mechanical Engineering University of Muhammadiyah Sumatera Utara, Jl. Kapten Mukhtar Basri No. 3, Medan, Indonesia
[4] Department of Computer Science State University of Medan, Jl William Iskandar Pasar V, Medan, Indonesia

[*] Email: m.yani@umsu.ac.id

Abstract. The subjects of this study were polymeric foam (PF) composites reinforced oil palm empty fruit bunches (OPEFB) fiber to dynamic load. OPEFB is waste of Palm Oil Mill which large number. This waste is seen has good potential to be harnessed and developed to become the engineering materials. This composite is processed with mixing of polyurethane (PU) and polyester resin. The paper aims to find the density of PF composite reinforced OPEFB fiber and to investigate the mechanical properties due to dynamic load. The specimens were prepared following the ASTM D1621-00 impact test standard. The method of making the test specimen is done by casting. The ratio of PU and mixing of polyester resin and OPEFB is 45 %: 65%. The results are gotten density 630 kg/m^3, at a level test of 0.5 m the magnitude of the tension that occurs is 226.68 kPa, the average energy absorbed is 75.20 Joule and at the level test of 1m, the magnitude of the tension is 261.43 kPa, the absorbed energy is 184, 68 Joule.

Keywords: OPEFB, impact load, composite, polymeric foam

1. Introduction
OPEFB is a waste of Crude Palm Oil (CPO) processing at palm oil mills with a considerable amount, reaching 1.9 million tons of dry weight per year or equivalent to about 4 million tons of wet weight per year in Indonesia. Conventionally, the OPEFB is mainly used as the burning fuel in boilers to supply electricity for the plantation mills. As one of the agricultural by-products with abundant quantity, the OPEFB also causes environmental problems due to its disposal issues. The potential of fiber originated from the OPEFB to be the alternative, natural composite material. However, this waste is considered to have good potential to be utilized and developed into alternative engineering materials.

The previous investigation about OPEFB has resulted panel board [1], alternative natural acoustic material [2], fire resistance panel [3], fertilizers, mulching material, and reinforcement materials in polymer composites [4][5]. Early development of research has been done on oil palm fiber on its potential as thermal insulating material in construction of buildings [6].

The aim of this study is to investigate obtain strength of polymeric foam composite reinforced oil palm empty fruit bunch fiber subjected to impact load. This pf composite could absorb impact energy. The impact method that applied is free fall. PF Composite reinforced OPEFB can be alternative materials to replace a synthetics polymer foam.

2. Literature Review

2.1. Composite Materials
Composite materials are defined as a mixture of two or more materials that produce new material with properties that are still dominated by the material properties of its formers [7].

Generally, composite materials are formed in two types of phases, namely the phase matrix and the reinforcing phase. The matrix phase is a material with a continuous phase that is always not rigid and weak while the reinforcement phase is always more rigid and strong, but more fragile. The density of composite materials is obtained from ratio of mass and volume. The combination of these two phases produces materials that can distribute the load received along the reinforcement so that the material becomes more resistant to the effect of the load.

Mechanical properties of the composite are influenced by fiber length, fiber orientation, and interfacial adhesion on polymeric alloys reinforced with short glass fiber [8]-[12].

2.2. Impact load on structure composite
Impact load [13]-[15] of the free-fall impact method is a test in which the batter moves down free of the stay condition and then moves and accelerates as the object falls. If the object falls to earth from a certain height is relatively small compared to the radius of the earth, then the object has increased the speed down at the same value every second. This means that the acceleration of the object decreases at the same value if an object is fired at a reduced speed at the same price every second and a constant upward slow down.

To determine the speed of falling objects every second will be obtained by the approach, as shown in Table 1.

Table 1. Time and velocity of falling objects

Time t (s)	0	1	2	3	4	5
Velocity v (m/s)	0	9.8	19.6	29.4	39.2	49

The v-t graph corresponding to Table 1 shown in Figure 1 is a straight line so that the acceleration is uniform. If the air resistance is ignored the free-falling object motion can be calculated by constant acceleration to form a straight line, when the acceleration of a free-fall object equals the acceleration of gravity (g): 1) For downward movement a=+g
2. For upward motion a = - g; and 2) The acceleration of gravity (g), can be viewed as a vector in a straight downward direction toward the center of the earth.

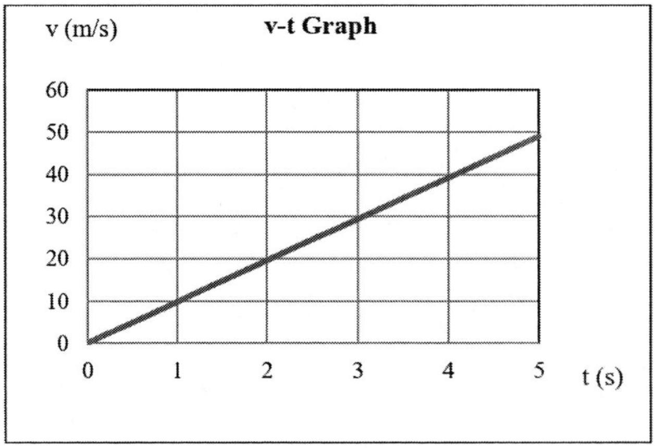

Figure 1. Velocity-time graph

The definition of displacement is a change in position; this is the vector quantity including distance and direction. Speed is the rate of change of position over time. It is also a vector quantity, including distance, direction and time. Whereas the uniform acceleration of the particles undergoing the same velocity change in the same time interval respectively regardless of the difference in time interval, as shown in equation 1.

$$v = \sqrt{2gH} \quad (1)$$

Then the equation 5 is the acceleration of the free-falling object depending on the distance or height of the falling object from the center of the earth, when a solid object falls with medium density, it can be assumed that the object undergoes a uniform acceleration of gravity, for a general sense the scientists take the acceleration rate of gravity g = 9.81 m /s^2.

2.3. *Energy* mechanics

The mechanical energy in an object is fixed provided there is no external force acting on the object. The combined mechanical energy of potential energy and kinetic energy. Potential energy is the energy that an object possesses because of its level.

$$Ep = m.g.h \quad (2)$$

Kinetic energy is the energy that objects possess because of the motion that works on it.

$$Ek = \tfrac{1}{2} mv^2 \quad (3)$$

3. Materials and Method

3.1. *Materials*

3.1.1. *Polymer Composite Material*

The technique of making polymeric composite materials generally does not involve the use of high temperature and impurities. This is due to this material easily being soft or melting [2]. The process of mixing the reinforce into the matrix is done when the matrix is in a liquid state.

The direct pouring method is carried out by attaching or touching the constituent materials to open mould and gradually flattened by using a grader or by outer impacting. This method is suitable for continuous and random fiber types as reinforce.

3.1.2. *Unsaturated Polyester Resins*

The unsaturated polyester resin is a type of thermosetting polymer having a long carbon chain structure. The mechanical data of the matrix material is shown in Table 2.

Table 2. Mechanical properties of unsaturated polyester resin.

No	Mechanical Properties	Unit	Amount
1	Density (ρ)	$Mg.m^{-3}$	1,2 s/d 1,5
2	Young Modulus (E)	GPa	2,0s/d 4,5
3	Tensile strength (σ_T)	MPa	40 s/d 90

Sources: YUKALAC, PT. Justus Indonesia Raya

3.1.3. *OPEFB* Fiber

OPEFB fiber is a natural fiber made from empty bunches of palm oil that is a waste in processing at a palm oil mill. The results of a research conducted by a commercial institution [10] on the chemical composition known that the material content of fiber in OPEFB is the maximum content as shown in Table 3.

IOP Conf. Series: Materials Science and Engineering **674** (2019) 012065 doi:10.1088/1757-899X/674/1/012065

Table 3. Parameter of OPEFB per kg.

No.	Materials Contains	Composition (%)
1.	Vapour	5.40
2.	Protein	3.00
3.	Fiber	35.00
4.	Oil	3.00
5.	Water solution	16.20
6.	Alkali solution 1%	29.30
7.	Dust	5.00
8.	K	1.71
9.	Ca	0.14
10.	Mg	0.12
11.	P	0.06
12.	Mn, Zn, Cu, Fe	1.07
	TOTAL	100.00

Figure 2 and 3 in a row is an OPEFB that has been minced into smaller parts and OPEFB fiber that have been mashed

Figure 2. Strips of OPEFB

Figure 3. OPEFB fiber

Mechanical strength of OPEFB fiber with an average diameter of 0.4 mm has an average elasticity modulus value of 11.88 GPa with average maximum impact stress of 156.3 MPa [11]

3.1.4. Blowing Agent

A blowing agent is a material used to produce a hollow structure on the composite formed. The type of blowing agent used in this study is polyurethane. The material formed from a mixture polyurethane and polymer is called polymeric foam material (PF).The shape of the polymeric foam structure formed is illustrated in Figure 4.

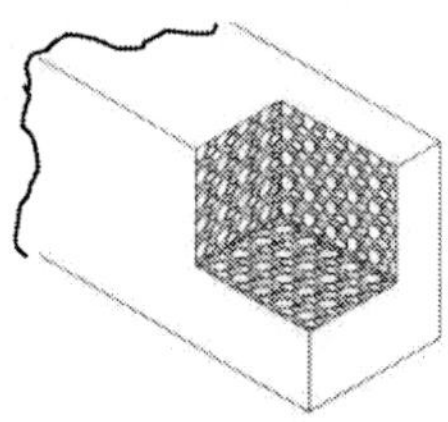

Figure 4. Illustration of polymeric foam material.

The catalyst is a chemical material used to speed up the polymerization reaction of the composite structure at room temperature conditions and atmospheric impurities. Administration of the catalyst may serve to regulate the formation time of BA bubbles, so as not to over-expand, or to harden too quickly, which may result in inhibition of bubble formation.

IOP Conf. Series: Materials Science and Engineering **674** (2019) 012065 doi:10.1088/1757-899X/674/1/012065

3.2. Manufacturing Processes

The manufacturing of dynamic impact test specimens is adjusted to ASTM 1621-00 standard for composite materials as shown in Figure 5. The process of manufacture is as follows:

1) Apply a separator layer on the inside of the mould with mould release wax to easily remove the product from the mould.
2) Preparation of necessary materials is OPEFB fiber, BQTN 157-EX type resin which is unsaturated polyester resin then weighed following the weight of the mixture set.
3) Combine the unsaturated polyester resin and OPEFB fiber and stir thoroughly for ± 2 minutes with an indication that all the fiber are immersed in the resin and coded C1.
4) Mix the catalyst into a mixture of fiber and resins and mix thoroughly for ± 2 minutes with the indicative color of this mixture is dark brown and give C2 code. Do not let C2 be more than 15 minutes, because it will harden.
5) Combine polyurethane-forming material with polyol and isocyanate compositions, and mix thoroughly for ± 0.5 minutes with the indication of polyurethane being formed is cream-colored foam and give C3 code.
6) Insert the C3 mixture into the resin mixture and OPEFB fiber, and stir it evenly over ± 0.5, then mix it with the catalyst and stir until the mixture is evenly distributed for ± 0.5 with indicative of this mixture color light brown and give C4 code.
7) Pour the C4 mixture into the mould as shown in Figure 5 and leave it for 24 hours until it is completely hardened and dry. Then the specimen removes from the mould. The shape of the specimen is shown in Figure 6.

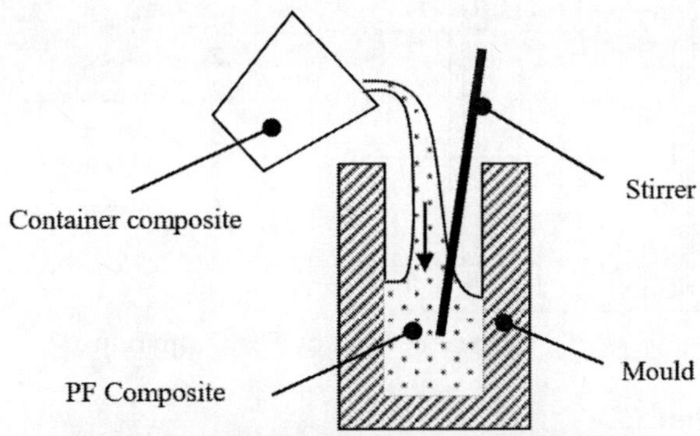

Figure 5. Manufacturing of composite

Figure 6. The impact test specimen of PF composite

The results of specimens of impact test according to ASTM 1621-00 standard with ratio polyurethane and composite reinforced OPEFB fiber 45%;65% are shown in Figure 7 and density ρ = 630 kg/m^3

3.3. Impact Test

Impact test of the specimen with impact load with the free-fall method and samples were taken with 3 samples of each height variation 0.5 m and 1 m. The impactor mass 6.8 kg is above specimen. Set up impact tests shown in Figures 7 and 8.

Figure 7. Set up Impact Test

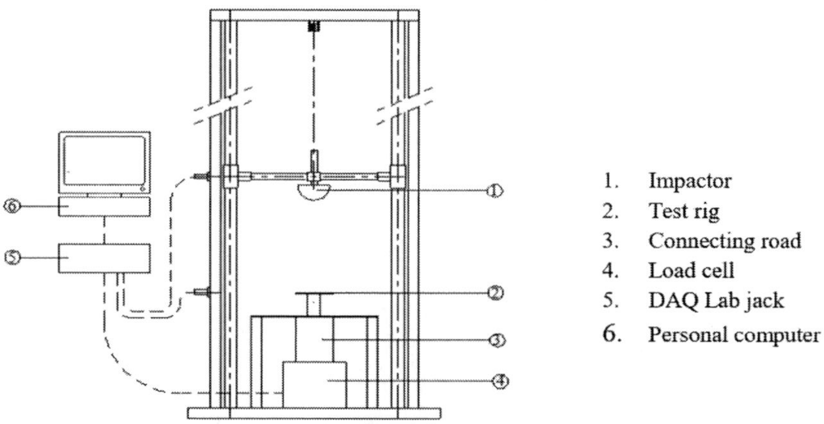

Figure 8. Impact Test Apparatus

4. Results and Discussion

The result of the impact test shown in Figure 9. Variant height of impact load is from 0.5, 1 and 1.5m. The impact load increase if the impactor is higher. It becomes potential energy bigger.

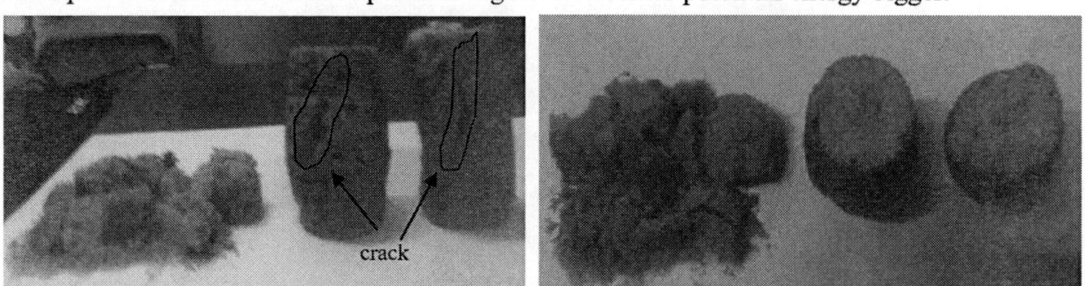

(a) (b)

Figure 9. Fracture of the specimen after impact load, variant height 0.5, 1 and 1.5m (from right to left)

The force-time graph of the free-fall method at 0.5 meter height is shown in Figure 10.

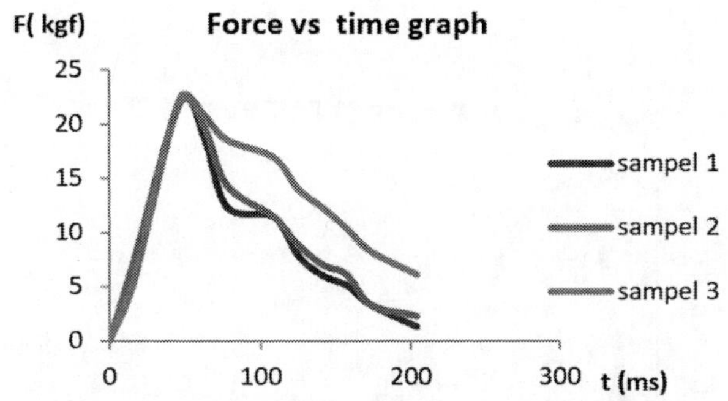

Figure 10. Force-time graph at 0.5 m

In the composite graph shown in figure 10 where the impact test results at 1 meter height are all large forces, dynamic timing and muted wave propagation vary. In sample 1 the force and dynamic time were 25.22 kgf and 31 ms, in sample 2 were 25. 83 kgf and 47 ms and sample 3 were 25.83 kgf and 47 ms so that the average force and dynamic time is 25.63 kgf and 41.67 ms.

In the loading area, the surface area is 961.62 mm^2. A comparison between the force with the cross-sectional area will be generated stress. Whereas energy is obtained from the product of dynamic force with height, the result of this calculation is tabulated in table 4.

Table 4. Results of testing the sample with dynamic free fall at 0.5 m

No.	A (mm^2)	F (kgf or N)	σ (kPa)	E (Joule)	I (Ns)
1	961.62	22.23(216.11)	224.74	108.06	10.16
2	961.62	22.02(218.07)	226.78	109.04	9.05
3	961.62	22.40(219.14)	228.51	109.87	7.03
Av.	961.62	22.22(217.97)	226.68	108.55	9.60

From the result of the dynamic impact test of the free-fall method to all samples, it is found that the force is 22.22 kgf (217.97 N). Thus to calculate the magnitude of the stress that occurs in the sample at the loading area is based on the stress equation is the force divided by the area. The area is obtained on the cross-sectional surface of the sample that is 961.62 mm^2. So the magnitude of the stress that occurs is 226.68 kPa. The stress causes plastic deformation in the sample.

The total impact energy that occurs on an experimental basis is the measured force multiplied by the height. The mean experimental dynamic impact energy is 108.55 Joules. The theoretical dynamic impact energy is the multiplication of mass with the acceleration of gravity and the height of the test. The mean of theoretical dynamic impact is 33.35 Joule. Moreover, the energy absorbed by the sample is the difference in total dynamic impact energy based on experimental and theoretical energies. The average energy absorbed is 75.20 Joules. The data of the three samples are tabulated in Table 5.

Table 5. Results of calculation of the dynamic impact energy of the sample at the height of 0.5 meters

No.	h (m)	M test rig (kg)	Transfer energy (Joule)	Potential energy (Joule)	Absorbed energy (Joule)
1	0.5	6.8	33.35	108.06	74.71
2	0.5	6.8	33.35	109.04	75.69
3	0.5	6.8	33.35	109.87	76.52
Av.	0.5	6.8	33.35	108.55	75.20

IOP Conf. Series: Materials Science and Engineering **674** (2019) 012065 doi:10.1088/1757-899X/674/1/012065

The combination of the three force-time graphs of the free-fall method at the height of the 1 meter test is shown in Figure 11.

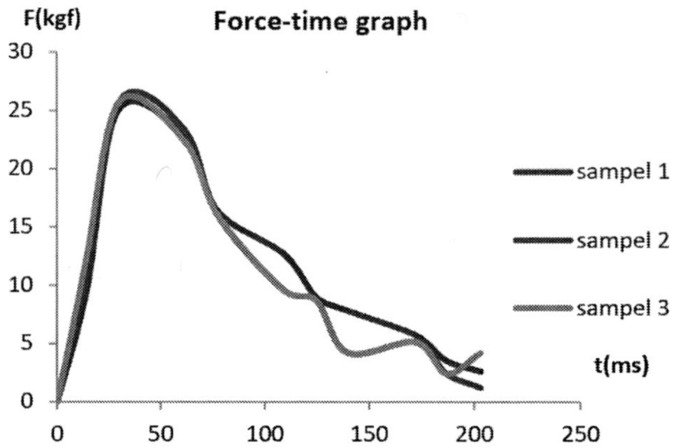

Figure 11. Force-time graph at 1 m

In the composite graph shown in Figure 11 where the dynamic impact test results at 1 meter height are all large forces, dynamic timing and muted wave propagation vary. In sample 1 the force and dynamic time were 25.22 kgf and 31 ms, in sample 2 were 25. 83 kgf and 47 ms and sample 3 were 25.83 kgf and 47 ms. So that the average force and dynamic time is 25.63 kgf and 41.67 ms.

From the result of the impact test, the free-fall method to the sample obtained data that the force is 25.63 kgf (251.39 N). So that the average tension is 261,43 kPa. The impulse that occurs due to the dynamic impact is the multiplication between force and time. At the height of the 1 meter sample test the mean impulse was 9.79 N.s.

Table 6. Results of testing the sample with dynamic free fall at 1 m

No.	A (mm^2)	F (kgf or N)	σ (kPa)	E (Joule)	I (Ns)
1	961.62	25.22(247.41)	257.28	247.41	7.67
2	961.62	25.83(253.39)	263.61	253.39	11.91
3	961.62	25.83(253.39)	263.61	253.39	11.91
Av	961.62	25.63(251.39)	261.43	251.39	9.79

From the result of the dynamic impact test of the free-fall method to all samples, it is found that the force is 25.63 kgf (261.43 N). Thus to calculate the magnitude of stress that occurs in the sample in the dynamic area based on the stress equation is the force divided by the area. The area is obtained on the cross-sectional surface of the sample that is 961.62 mm^2. So the magnitude of the stress that occurs is 261.43 kPa. The resulting stress causes plastic deformation in the sample.

The total dynamic impact energy that occurs on an experimental basis is the measured force multiplied by the height of the test. The average of experimental dynamic impact impediment is 251.39 Joule. The impact energy of theoretical dynamics is the multiplication of mass with the acceleration of gravity and altitude. The mean of theoretical dynamic energy is 66.71 Joule. Also, the energy absorbed by the sample is the difference in total dynamic impact energy based on experimental and theoretical energies. The average energy absorbed is 184.68 Joule. The data of the three samples are tabulated in Table 7.

IOP Conf. Series: Materials Science and Engineering **674** (2019) 012065 doi:10.1088/1757-899X/674/1/012065

Table 7. Results of energy calculations of samples at the height of 1 meter

No	h (m)	Mass of a test rig (kg)	Transfer Energy (Joule)	Potential Energy (Joule)	Absorbed Energy (Joule)
1	1	6.8	66.71	247.41	180.70
2	1	6.8	66.71	253.39	186.68
3	1	6.8	66.71	253.39	186.68
Av	1	6.8	66.71	251.39	184.68

5. Conclusion

Preparation of dynamic impact test specimen based on standard ASTM D1621-00, made by casting method, and has a density of ρ_{av} 630 kg/m^3. At the height of the dynamic load test of 0.5 m, the magnitude of the stress that occurs is 226.68 kPa, the average energy absorbed is 75.20 Joules, and at the height of the test 1m, the magnitude of the stress that occurs is 261.43 kPa, absorbed is 184.68 Joule.

Acknowledgment

The author would like to thank the University of Muhammadiyah Sumatera Utara for financial support, Faculty of Engineering, for providing administration.

References

[1] Ridzuan Ramli, Stephen Shaler, And Mohd Ariff Jamaludin, Properties Of Medium Density Fiberboard From Oil Palm Empty Fruit Bunch Fiber, Journal of Oil Palm Research Vol. 14 No. 2, December 2002, p. 34-40

[2] Khai Hee Or, Azma Putra, Mohd Zulkefli Selamat, Oil palm empty fruit bunch fiber as sustainable acoustic absorber, Elsevier, Applied Acoustics 119 (2017) 9–16

[3] Roshazita Che Amata, Norlia Mohamad Ibrahima, Nur Liza Rahima, Nur Saidatul Balqish Ahmad Tajudina, Khairel Rafezi Ahmad, Fire Resistance of Biomass Ash Panels Used for Internal Partitions in Buildings, Procedia Engineering 53 (2013) 52 – 57

[4] Mahjoub R, Bin Mohamad Yatim J, Mohd Sam AR. A review of structural performance of oil palm empty fruit bunch fiber in polymer composites. Adv. Mater. Sci. Eng. 2013;2013:1–9.

[5] Shinoj S, Visvanathan R, Panigrahi S, Kochubabu M. Oil palm fiber (OPF) and its composites: a review. Ind. Crops Prod. 2011;33(1):7–22.

[6] Asdrubali F, D'Alessandro F, Schiavoni S. A review of unconventional sustainable building insulation materials. Sustainable Matter Technology 2015;4:1–17.

[7] Thermoplastics. In *Handbook of Industrial Materials.* Elsevier Science Publishers Ltd. Oxford, 1992. pp. 369492. 641

[8] Gupta, V.B.. Mittal, R.K. and Scharma. P.K.. Some studies on glass fiber-reinforced polypropylene. Part II: mechanical properties and their dependence on fiber length, interfacial adhesion. and fiber dispersion. Polymer ***Compos.,*** 1989, **10,** 16-27.

[9] Joshi. M., Main, S.N. and Misra, A.. Influence of fiber length, fiber orientation, and interfacial adhesion on poly butylene terephthalate) / polyethylene alloys reinforced with short glass fibers. ***Polymer Compos,. 1994,*** **15,** ***349-358.***

[10] Pande, S. J. and Sharma, D. K., Strength and stiffness of short glass fiber/glass particulate hybrid composites. ***Fihre Sci. Tech&. . 1984,20.235-243.***

[11] Hargarter, N., Friedrich, K. and Catsman. P.. Mechanical properties of glass fiber/talc/poly butylene-terephthalate composites as processed by the radlite technique. ***Compos. Scinc. Techno/..*** 1993. 46. 229-244.

[12] Yilmazer. U.. Tensile, flexural and impact properties of a thermoplastic matrix reinforced by glass fiber and glass bead hybrids ***Compos. Sci. Technol.. 1992, 44,*** 1 19- 125.

IOP Conf. Series: Materials Science and Engineering **674** (2019) 012065 doi:10.1088/1757-899X/674/1/012065

[13] Hine, P.J.. Duckett. R.A. and Ward, 1.M.. The fracture behavior 01 short glass fiber-reinforced polyoxymethylene Composites. 1993. 24.643-649.
[14] Akay. M., O'Regan, D.F. and Bailey, RX, WOF and impact behavior of glass-fiber-reinforced polyamide 6.6 injection molding. *Compos. Sci.* Technol., 1995, 55, 109-I 18.
[15] Ramsteiner. F. and Theysohn. R., Tensile and impact strengths of unidirectional. short fiber-reinforced thermoplastics Composites. 1979.10, 111-119.

AUTHOR INDEX

Institute of Physics
Dirac House, Temple Back
Bristol BS1 6BE UK

ISSN: 1757-8981
ISBN 978-1-7138-0926-5

CURRAN ASSOCIATES INC.
proceedings.com